AF679097

METROS IN INDIA

VOLUME 1

FROM VISION TO REALITY: THE MAKING OF THE KOLKATA METRO

DR. BC ROY

notionpress.com

INDIA • SINGAPORE • MALAYSIA

Copyright © Dr. BC Roy 2026
All Rights Reserved.

ISBN 979-8-90207-357-4

This book has been published with all efforts taken to make the material error-free after the consent of the author. However, the author and the publisher do not assume and hereby disclaim any liability to any party for any loss, damage, or disruption caused by errors or omissions, whether such errors or omissions result from negligence, accident, or any other cause.

While every effort has been made to avoid any mistake or omission, this publication is being sold on the condition and understanding that neither the author nor the publishers or printers would be liable in any manner to any person by reason of any mistake or omission in this publication or for any action taken or omitted to be taken or advice rendered or accepted on the basis of this work. For any defect in printing or binding the publishers will be liable only to replace the defective copy by another copy of this work then available.

Contents

Part 3

List of Figures

Part 1

Part 2

Part 3

List of Tables

Part 2

Part 3

Foreword

Prof SS Chakraborty, FNAE (I), FICE (UK)-Gold Medalist, FIStructE (UK)
Co-Founder & Chairman Emeritus - CES, & Chairman-Burdwan Health Care
Metros in India

It has been my pleasure to know Dr. BC Roy since 1978, when his association with CES began. Over time, what started as a professional acquaintance was transformed into a collaboration that has continued for more than forty-five years. During this period, his steady growth as a versatile engineer has been observed, with contributions made to high-rise buildings, bridges, environmentally sustainable facilities, sports complexes, and more importantly, major transit projects in India and abroad. His commitment to hard work, and constant pursuit of deeper knowledge have consistently been demonstrated.

The foundation of Dr. Roy's expertise in metro systems was laid with his involvement in India's first metro project in Kolkata. Germany was the source of his first metro lesson. For the young engineers, he was considering consolidating all of his Metro-related experience in one location and authored a book on "Indian Metros." This early experience was further enriched through exposure to international practices and decades of professional engagement across the country. A distinctive ability to integrate innovative design with practical execution was displayed in projects of national importance, including the development of metro systems in several Indian cities. It is from this wealth of experience of "Metros in India: Evolution of Indian Metros, Kolkata and Beyond" has been authored, with the intention of sharing valuable knowledge with younger generations of engineers and practitioners.

In this volume, a comprehensive account of India's metro journey has been provided by Indian engineers—from the pioneering work in Kolkata to the extensive networks now seen in Delhi, Mumbai, Bengaluru, Chennai, Hyderabad, Pune, and beyond. The technical dimensions of metro development, such as tunnelling, rolling stock, and signalling systems, have been brought out with clarity, while attention has been given to the social, environmental, and policy challenges accompanying such transformative undertakings. Particular emphasis has been placed on the Kolkata Metro, which served as a crucible of innovation and a source of vital lessons for subsequent projects across the nation.

The expansion of metro networks has been presented not merely as an infrastructure initiative, but as a redefinition of urban mobility in response to rapid population growth and the demands of sustainable development. The broader impact of metro projects on the quality of urban life has been underscored, bearing in mind that these systems are as much about people as they are about engineering.

A distinctive presentation style has been adopted, where practical examples, project experiences, and visual illustrations have been integrated to enhance accessibility. This approach has ensured that the book is not only a reference for policymakers and professionals but also a valuable guide for students and young engineers.

This work will be regarded as a standard reference in the field of urban transport. Inspiration will be drawn by future engineers and planners, who will be encouraged to view metro development as a mission and an absolute essential, aimed at improving urban mobility in Indian cities.

Prof SS Chakraborty FNAE

Preface

With rapid global urbanisation, over half the world's population now lives in cities. In India, nearly 1.46 billion people live in the country, of which 37% (542 million) are urban residents. This scale of growth demands integrated planning and robust Mass Rapid Transit Systems (MRTS) to enhance mobility, living standards, and economic productivity. Metro rail, operating on independent rights-of-way, is central to sustainable urban mobility by offering fast, safe, reliable, and low-emission connectivity. India's metro journey began in Kolkata (construction in 1970, operations in 1984) and has since expanded to 17 cities, highlighting technical, institutional, and financial challenges as well as indigenous innovations and global collaborations. Drawing on the author's direct involvement in projects with CES, AECOM, and Chatterjee & Polk, it covers planning, design, execution, and operations, supplemented by DPR excerpts, tender samples, case studies, and design details. While financial returns remain challenging, metros are indispensable in addressing congestion and supporting economic growth. This book draws on the author's experience and collaborations across Kolkata, Delhi, Mumbai, Bengaluru, and other cities. It covers the planning, design, construction, and operation of metro projects, with case studies ranging from Kolkata's pioneering North-South and East-West Metro lines - India's first underground line and first river-crossing metro—along with metro developments in Delhi, Mumbai, Bengaluru, Chennai, Jaipur, Hyderabad, and Pune in the PPP models. Insights from DPR preparation, tendering practices, depot & station design, tunnelling technologies, and green building concepts are included.

Structured in two volumes, the book combines technical depth with practical case studies to serve as a comprehensive resource for students, engineers, and practitioners engaged in metro development within India's evolving urban landscape.

Author's Note

A

The rapid pace of urbanisation in India, like in the rest of the world, has transformed cities into densely populated spaces with limited room for expansion. Narrow roads and heavy congestion posed enormous challenges for planners, particularly when designing for future needs. The demand for an efficient urban transport solution capable of carrying large numbers of passengers across cities led to the adoption of Rail based Mass Rapid Transit Systems (MRTS). The metro emerged as the most appropriate choice, offering high capacity, safety, comfort, and short headways of 90–150 seconds.

The foundation of India's metro development was laid in 1969 with the establishment of the Metropolitan Transport Project (MTP) under the Ministry of Railways. With the support of Soviet specialists and East German engineers, MTP prepared the first master plan for Kolkata. The North–South corridor of the Kolkata Metro became India's first metro rail project, designed, engineered, and constructed by Indian engineers. Funded through the Ministry of Railways under budgetary provisions, this metro marked the beginning of India's journey into modern urban mobility. The Kolkata Metro relied on broad-gauge technology, familiar to Indian Railways. This avoided the need to design or import new bogies, saving time and resources on validation and durability. India already had a strong tradition of coach manufacturing for broad, meter, and narrow gauges, with ICF Perambur serving as the principal manufacturing hub. Except for advanced features such as automatic door closure, most requirements were indigenously met. The signalling system was based on interlocking technology developed by Indian Railways, and today the North–South Metro is being upgraded to modern signalling standards.

However, a few experts considered the Kolkata Metro obsolete compared to global standards, but history has proven otherwise. Karnataka Government prepared a Project Report for Bangalore Metro, with an idea to use Indian System in bringing down the cost, in early nineties. In 1994 BMRTL prepared feasibility report for Bangalore Metro. Delhi Metro Rail Corporation (DMRC) was established in 1995, and the report of first phase was approved by Union Cabinet in 1996 and began operations in 2002. Delhi Metro initially used broad gauge with imported coaches before transitioning to standard gauge. International firms such as Hyundai Rotem, Mitsubishi, Bombardier, and CAF supplied the early stock, after which Bharat Earth Movers Limited (BEML), in collaboration with Hyundai Rotem, began indigenous manufacturing. Today, BEML supplies metro coaches for ongoing and planned projects, with other Indian firms, such as Tikamgarh Rail Systems, also entering the coach manufacturing sector. Delhi now operates the country's largest metro network, spanning over 350 km with 256 stations, and has ambitious expansion plans under Phase V. Broad-gauge lines primarily use indigenously manufactured coaches, while standard-gauge corridors employ a mix of imported and locally built stock. Except for Kolkata's East–West Metro, which runs on standard gauge, most new corridors under construction or commissioning continue to adopt broad gauge with Indian-built coaches.

A significant hurdle in the early stages of metro development was the absence of a dedicated legal framework for urban rail transport. In Kolkata, the outdated Kolkata Tramways Act was modified to accommodate metro operations, though its provisions were limited to a single municipal jurisdiction. Similar challenges were faced in Mumbai, governed by the Bombay Tramway Act. Delhi Metro necessitated the creation of an entirely new legal framework to allow operations across and beyond the city limits. This was the groundwork for future expansion across the country. Metros in India have expanded rapidly. Today, seventeen Indian cities operate metro systems, with nine more in different stages of construction and planning. Kolkata's North–South Metro, largely constructed underground through the cut-and-cover method and partly with Tunnel Boring Machines (TBMs), set the standard for engineering practices in subsequent projects. Russian engineers initially recommended sheet piles for deep excavation, but when locally available piles proved inadequate, Indian teams introduced diaphragm wall construction with German technical support, a Milestone innovation at that time. Later, advanced tunnelling techniques such as precast concrete segments and TBMs were introduced, particularly in the East–West Metro, which achieved the distinction of making India's first under-river tunnel beneath the Hooghly.

The East–West Metro also pioneered modern construction practices with extensive use of instrumentation, top–down station construction, and experiences of international consultants. However, frequent alignment changes and delays demonstrated the challenges of integrating metro systems within complex urban environments. Parallel to these developments, other cities started planning and implementing their metro projects. Mumbai, Bangalore, Chennai, Hyderabad, Jaipur, and Pune followed, with Hyderabad and Pune's Line 3 adopting the Public–Private Partnership (PPP) model. Hyderabad Metro, executed by the L&T-LED consortium under HMRL, became India's largest PPP-based metro project, whereas Rapid Light Metro system in Gurgaon was the first fully owned by IL&FS in India. Similarly, Pune Metro's Line 3 was entrusted to a Tata Realty–led consortium under Maha-Metro.

In Mumbai, the Metro Planning Group was established in the mid-1990s to prepare feasibility studies. Alternative systems, such as the Monorail and the Konkan Railway's Sky Bus Metro, were also evaluated as feeder services. These experiences contributed to the broadening of India's urban transport strategies beyond conventional metro systems.

The development of metro systems in India has always been closely related to innovations in design, construction, and project management. From diaphragm walls and precast tunnel segments to advanced TBM tunnelling and modern depot design, each project has introduced new practices that were later adopted by various cities. Building condition surveys, an integral part of the East–West Metro, became SOP for assessing urban infrastructure before construction. Condition of neighbouring structures including Heritage Structures along a Metro Corridor and remedial measures are included in section for Mumbai Metro Line 3 in Volume 2.

Over the years, the metro has transformed from a pilot project in Kolkata to a national urban ecosystem. Supported by domestic manufacturing, evolving legal frameworks, and continuous learning from both global and local experiences, the development of metros in India reflects the resilience and ingenuity of its engineers and planners. Indian Engineers carried out lot of Research work on Hanging Metro, under the leadership of Mr Rajaram, former MD, KRCL at KRCL which was not successful in first attempt, though a few cases of such metros are in operation outside India.

B

The Presentation of the book is a bit different from the general book format. The book is presented using a combination of texts, bullet points, and figures. The figures include design sketches, execution plans and many more photos, as the Author's experience is that this approach will provide a better understanding.

While reading, one may experience repetition, which is done to facilitate comprehension. For comprehensive details on the Maintenance Depot or Systems, one may refer to the details discussed for other metros included in Parts 2, 3 and 4. The systems for the Kolkata North-South Metro, the first metro, and the East-West Metro differ substantially. While metros in other Cities primarily follow the Delhi Metro with variations. Author believes, all most all aspects of an Urban Metro planning, design and construction aspects have been included, may not be in one place but distributed among the Metros included. It is particularly for Depot and System designs. The author included most of the references for both texts and photographs. However, a few references might have been inadvertently missed and may be excused.

Safety for construction employees, important issue, in Metro Construction, has been included, although the author's name could not be located.

Metros in India

Volume 1 of 2

PART 1

CHAPTER 1

General

1.1 Introduction

The global population is rapidly on the rise. A century ago, the urban-to-rural population ratio was 1:7 globally. Today, more than half the world's population lives in urban areas. India's population increased from 431 million in 1961 to 1285 million in 2015, and the country's urban population rose from 82 million to 410 million during the same period. With a significant concentration in urban areas and exponentially higher growth, 60% of India's GDP comes from cities and towns.

Four metropolitan cities—Delhi, Mumbai, Kolkata, and Chennai—act as regional Capitals, but several others have also grown in population and size. A corresponding pressure accompanies this mammoth increase, placing demands on resources and the environment.

An integrated urban area agglomeration becomes imperative to meet the increasing demands of a growing population and needs a comprehensive and improved city-based policy that incorporates the necessary infrastructure for adequate commercial and office space, industry, and the neighbourhoods, including public spaces, landscaping, arboriculture, educational institutions, sports and entertainment centers, etc. - all tied together through robust multimodal transport connectivity.

Fig. 1.1: Locations of Metros in India (Ref. urbanrail.net)

While planning decisions must include horizon years and flexibility, the planner must also be sensitive to and include requirements for preserving historical buildings, monuments, and heritage structures within these urban agglomerates. Indian heritage has a huge potential to contribute to livability and sustainable development plans.

In India, as in the rest of the world, rapid urbanisation means cities expand into and consume nearby areas. Heavily congested areas with narrow roads are challenging to plan for, especially for the future. Planning for utilities and transportation structures in such areas warrants much work. Real-time planning is relatively more straightforward and guarantees more prosperity and enhanced security. Planners and Engineers felt the necessity of a capacity transport system to carry passengers in masses from one location to another as an intracity facility, which is why the Mass Rapid Transit System (MRTS). Metro systems, at present, are high capacity, with less headway in the 90-150 seconds range, safe (closed-door operation) and more comfortable.

'The International Association of Public Transport defines the Metro System as urban transport that carries urban passengers and operates an independent right-of-way, separated from road and passenger traffic. A Mass Rapid Transit System is necessary. The first metro in the world, **London,** was proposed in 1830, construction started in 1884, and it started operations in 1863. Since then, there has been all-around development. After London began operation, the metro was extended in 1890 & 1896 (Glasgow), followed by the implementation of the Hungary Metro in 1896, the Chicago Metro (USA) in 1897, Paris (France) in 1900, Berlin (Germany) in 1902, Moscow Metro (Russia) in 1935, Kolkata Metro (India) 1984 and at all major cities in the world.

Fig. 1.2: View at Platform Level of a Kolkata Metro Station

Metro rail projects are one of the most important components of urban transportation infrastructure in India. It provides a fast, reliable, safe, and environmentally friendly mode of transport for large commuters, connecting different parts of a congested urban agglomerate within the shortest possible time. Metro rails contribute to economic growth, social development, and integrated urban planning of a region.

1.2 Need for Metro

The Public Transport System is an efficient user of space and has a reduced level of air and noise pollution. As the city population grows, the share of public transport, whether road or rail-based, should increase. Experience has shown that, in cities like Pune, where roads do not have adequate width and which cater to mixed traffic conditions comprising slow and fast-moving vehicles, road transport can optimally carry 8,000 persons per hour per direction (phpdt). When traffic density increases beyond this level, the average speed of vehicles comes down, journey time increases, air pollution increases, and commuters are put to an increased level of inconvenience. Thus, when on a corridor, traffic density during peak hours crosses this figure, provision of rail-based mass transport, i.e. Metro system, should be considered.

Advantages of the Metro System require 1/5th of the energy per passenger km compared to a road-based system. It does not cause air pollution in the city, but it reduces the noise level. The metro system does not occupy space, in the case of the underground system and only about 2 metres of the road, if elevated. A metro system is a high-capacity public transport system that operates on an exclusive right-of-way and is generally a rail transit. It can be at grade, elevated, underground, or a combination. In 1863, the first metro started operating on a partially underground corridor. Planning and implementing a metro system involves studies in different stages, like Feasibility to include Traffic, Route planning, Engineering, & Design of various disciplines, estimating passenger traffic and coaches to carry passengers, Civil & Structural, Power to run, Signalling and Communication, Stabling and Maintenance, Cost and Financial Models.

Route/Alignment studies include analysis and planning to provide an approach for selecting routes resulting in a system network in an urban environment and integrating the existing road network system. Study results should be accurate enough to ensure efficiency, connectivity, and the proper location of Stations. The MRTS system is environmentally friendly and should be integrated with the existing road network, including its development plans.

Fig. 1.3: Need for Metro in India (Ref. www.oneindia.com)

The metro system and network should be accessible, connected, safe, adequately secured, and affordable to many. The appropriate design of the Metro reduces Traffic Congestion and provides efficient Transportation and Environmental Benefits. The system offers high-capacity corridors and passenger carriers, low energy consumption, increased speed, and reduced journey time, covering less ground space.

The Metro System, commonly known as MRTS, has provisions for different Capacities to satisfy the city's geography, population density, Budget constraints, and Technological advancements. There are three significant alternatives: i) HCRT-Heavy Capacity Rapid Transit has a large and powerful metro that operates on a dedicated corridor. ii) MCRT- Medium Capacity Rapid Transit for medium population urban areas, between heavy and light capacity transit systems. iii) LCRT- Light Capacity Rapid Transit.

In comparatively less populated cities, miniature, lighter trains could be used on a dedicated corridor. However, they need Rapid Transit Systems, and capacity is required more than for BRTS (Bus Rapid Transit System).

1.3 Ridership

Metro Systems are the backbone of urban transportation. As a Low-Carbon Transport System, they are an eco-sustainable solution for modern sustainable solutions. During the preparation of the Detailed Project Report (DPR), important parameters for the metro system included in Chapter 3, a comprehensive traffic study of existing traffic, zones, and corridors and its projections for the future, horizon year, is one of the prime factors in working out corridor-wise, appropriate transit system. The estimated traffic for the horizon year is worked out considering the actual average growth in the previous 10/15 years, as available. While conducting detailed traffic and its projection, the city is divided into different zones and possible routes based on past studies.

To brief, for Delhi Metro, the original feasibility study for developing the metro system estimated 31 lakhs of commuter trips for Phase-1 by 2005 and 126 lakhs by 2021. The figures were later reduced to 22 lakhs by 2005. For Phase-2, in 2005, it was estimated as 31 lakhs by 2021 and revised during the projection of Phase-3, reducing projected traffic for Phase-1 by seven lakhs, 18 lakhs for Phase-2 and 23 lakhs for Phase-3. The latest projection in 2018 was reduced to 41 lakhs for Phases 1, 2, & 3 against 48 lakhs as estimated. Later, the definition of Ridership was changed; trips made on separate routes counted as separate trips, while earlier, it was estimated as total trips between origin and destination. The same has increased ridership by 50%. The actual ridership in 2019-20 was 57 lakhs per day. The total length, without Phase -4, is 346.5 Km. The daily ridership of the Delhi Metro is around 66 lakhs in total, and on February 13, 2024, it was 72 lakhs.

Fig. 1.4: Typical Ridership – Metros in India (Ref. www.oneindia.com)

The Metro DPR estimated the daily traffic for the Mumbai Metro Phase 1 to be 43 lakhs in 2011 and 67 lakhs in 2021. The Mumbai Metro opened in 2014, and the actual average daily ridership was 34 lakhs in 2018 and 45 lakhs in 2019.

Kolkata metro saw record growth in May, June & July 2024, and ridership was 530 lakhs, more than estimated; however, the estimate did not include any period and 460 lakhs in the blue line. On September 30th, the total network carried 170 lakhs of passengers for partially operated corridors except the completed Blue Line of 32.33 Km. Partially completed corridors Green Line, Line-2 carried in three months 36 lakhs, purple line- Line-3 recorded to carry 48000 passengers and 89,000 passengers in orange line—the current trends in Kolkata Metro commuter preferences and operational success. **(Ref. Wikipedia and Travel and Tour Guide)**

1.4 Passenger Safety During

In general, Indian Metros are considered for travel, including night travel. The metro system operates with security measures, including surveillance and personnel, to ensure passenger safety. However, personnel meant for safety need to be stationed in well-lit areas, and passengers need to remain vigilant. during peak hours. When the metros are crowded, sometimes more than the capacity of the coaches, passengers should avoid metro travel if possible. The well-maintained infrastructure and adherence to safety standards contribute to safe commuting.

Commuters need to follow safety rules. They should be aware of the platform edge, queue up for boarding and alighting, avoid no-entry zones, use handrails, be cautious about gaps, know the emergency exit, and follow digital displays and announcements. Use designated areas for boarding and alighting, beware of pickpockets, and remain vigilant about one's belongings. One should report suspicious objects and activities.

Passengers should follow signage at stations, understand colour codes, plan for interchanges, use Mobile Apps and online resources, and remain alert on escalators and/or stairs; holding the railing is necessary. They need to follow Staff Instructions and be respectful and considerate of fellow passengers.

1.5 Metro System in India

Urban mobility includes the present-day scenarios of three major Indian metropolises: Delhi, Mumbai, and Kolkata, the most populous. This book touches upon a few issues that one another megacity faces—Chennai. The role of the Metro system in various cities worldwide, including in India, with increasing traffic demand and vehicles in different modes of traffic creating traffic congestion, accidents, and environmental pollution, is to improve the city's mobility and environment.

The book is primarily based on the author's experiences and recent books and literature, which have been referenced with available databases. It aims to provide knowledge and assistance to young engineers, postgraduate students, and professionals working in similar domains in developing the concept of urban masterplans in a brownfield environment.

Fig. 1.5: Views of Indian Metros (Ref. www.rediff.com) **Fig. 1.6: Internal View of Indian Metro**

This book comprehensively examines the transport system to urban growth, remarkably rapid globalisation, growth in the Indian Metropolis, embodied energy, green transportation, and multimodal integration. India is an ancient civilisation with centuries of architectural heritage. Its transport infrastructure requires strategies and efforts by architects and engineers to preserve this heritage and study the effect of infrastructure development on it. Special care also needs to be taken about religious structures.

The study examined existing traffic scenarios, examining different modes of transport on land and water, and sought to answer how to develop an affordable integrated transport network under the present constraints. The data that has been used is presently available, but this may change during the commencement of a new project. It is estimated that Delhi's population will grow to become the largest urban area in the world by 2030, overtaking Tokyo, which has a population of more than 37 million. Delhi and Mumbai will closely follow with 27 million, and Kolkata will have nearly 22 million by 2025.

Fig. 1.7: Views of the Indian Metro during Construction

City planning is becoming increasingly difficult due to the scarcity of open spaces and old utility systems. Ageing or closed industries also impact the ease of planning. Reducing rehabilitation and resettlement of people affected by transport projects and reducing road casualties must also be factored in.

Much is needed to improve pedestrian facilities, especially in cities with inadequate mass transit systems. 'The International Association of Public Transport' defines the Metro System as Urban Transport that carries urban passengers and operates an independent right-of-way, separated from road and passenger traffic. The rail-based transportation system, its financial aspects, and economic impacts are further examined. Although a few MRTS projects have been attempted with private financing in the rail sector, these were only somewhat successful, with two out of three projects surrendered to the government by the Concessionaires. PSUs now run them. However, lately a few MRTS projects in PPP mode are in implementation stage and in operation. One of them is Hyderabad Metro which is in operation.

The pros and cons of all types of urban rail-based systems, such as Mainlines (not included), Suburban Railways, rail-based urban transit systems, light transit systems, tramways, monorails, and the metro rail system, have been examined. As such, though the Metro system is generally not financially viable (from an FIRR perspective), local governments are introducing the system to solve mobility issues in metropolitan areas. The funding of metro rail systems is usually through external loans, while the regional and central governments jointly fund a few.

Delhi has the most extensive metro network of all cities in India, with over 400 km of metro rail. Passenger coaches for the primary and suburban lines, one of the most extensive rail networks in the world, are manufactured in India and operate on broad-gauge track. India's first metro network in Kolkata, built in 1984, had rolling stock that was manufactured in India and operated on broad-gauge tracks. Similarly, initial corridors in Delhi also operate on the broad gauge (Indian gauge). Currently, coaches for most of the new corridors are being imported and operated on standard-gauge tracks. Other than the East-West Metro, which runs on standard gauge in Kolkata, all remaining corridors currently in various stages of construction/commissioning will operate on the broad gauge with indigenous coaches. However, with the increase in rolling stock manufacturing in India, the trend is changing, and new metros are designed using indigenous coaches.

India's first metro system was planned, designed, and implemented entirely by Indian engineers. The Union Ministry of Railways funded the project and developed the interlocking signaling system to improve mainline operation and use in metros. After the first metro, most of the subsequent metros were designed and implemented by teams comprised of both Indian and Foreign Experts. The first technology to use precast concrete tunnel segments was developed and used successfully in Germany across the river Main. It was used in India for a small section of the N-S Metro at Chitpur railway yard for the first time. However, during the erection of the precast segments, the team faced problems. With the help of Russian experts, the remaining small portion of the tunnel corridor was implemented using the cast iron segments method.

Fig. 1.8: India's First Metro Built with Private Sector Participation

Fig. 1.9: Depot of First Indian Metro

The East-West Metro in Kolkata is nearing completion. It is the first metro in India to make an underground crossing of a major river, the Hooghly. The metro corridor is partly elevated and partly underground. The underground section uses Earth Pressure Balancing Tunnel Boring Machines (TBM), while the underground stations are built using the cut-and-cover method.

In summary, the rail-based mass transit system is the best option for solving the transport issues in most urban areas, though the system is not financially viable. The metro system is an environmentally friendly, faster, comfortable, affordable, and safer public transport system. As a public transport system, the metro railway changes the general public's mindset to shift their travel from personal transport to MRTS, providing an easy, faster, and time-bound system and decongesting the thickly built urban areas.

Light Rail systems are mostly fenced and can be built with a complete right-of-way if preferred. Different types of Urban Rail Systems in India are:

- **Rapid transit:** Rapid transit, or the metro in India, is an urban high-capacity rail system commonly operated in metropolitan cities. These systems are segregated from Indian Railways and have their own right-of-way.
- **Suburban Railway:** Suburban rail, popularly known as the local train, is an urban rail transit system that connects the suburbs to the city's centre. These systems are linked to and operated by Indian Railways.
- **Medium-capacity rail:** It is a rapid transit (metro) system that has a capacity higher than light rail but lower than a rapid transit system to serve a medium demand.
- **Light rail:** Light rail is used in cities have low demand. It is a combination of rapid transit and tram systems. It has a higher capacity and speed compared to tram services and has dedicated tracks that are mostly fenced.
- **Monorail:** This system has trains running on a single rail/beam. It has found its application in medium-capacity transport, but due to low efficiency and high costs, it is yet to be popular in India.

- **Regional Rapid Transit System:** This system is operated either between two similarly sized cities, close to each other or between a larger city and smaller cities nearby.
- **Tram:** The systems are one of the oldest modes of urban transport in India. They are low-capacity, slow-moving trains that run on tracks that are embedded in the urban streets.

The following chapter includes the evolution of Indian Metros, followed by a chapter on how to develop a metro system, and important parameters need to be studied to design a metro transport system. Chapter four is the prelude to Chapter five, from which the selective metro corridors in operation in India are detailed.

CHAPTER 2

Evolution of Rapid Transit System in India

2.1 Evolution

Urban Transit is one of the primary systems for transporting mass within thickly populated intracity areas and from one area to another in neighbouring thickly populated cities. Urban transit in India is vital for intracity transportation in highly populated towns and cities. According to reports published in 2021, 2.63 billion people travel annually in most of the Indian metro systems. The Ministry of Urban Development is a nodal agency for coordination, appraisal, and approval. However, the Kolkata Metro network is under the Ministry of Railways. This book covers Rail, the transport system and heavy metro, providing higher passenger carrying capacity, faster, smoother, safer, non-polluting, and energy efficient. The author's book on **urban mobility** includes comprehensive multimodal transport systems, including different modes of traffic.

2.2 Briefs of Indian Metros

2.2.1 Kolkata Metro

In September 1919, during a session of the Imperial Legislative Council at Shimla, India, a committee was set up by W. E. Crum that recommended a metro line for Kolkata, the first time in India to transport passengers by Metro. In 1921, Harley D Hay proposed the East-West Metro crossing the river, as in the Crum proposal, connecting Howrah and Kolkata. In 1923, it was decided not to take up the proposal due to a shortage of funds. In 1947, the planners tried to explore introducing a rail-based Rapid Transit System (RTS), which gave birth to the Circular Railway.

In 1949, the then Chief Minister of West Bengal, Dr Bidhan Chandra Roy, reconceived the proposal to connect West with East of Calcutta by a Metro aligning beneath the Hooghly River to solve traffic problems besides the augmentation of Public Transport System as Kolkata (then Calcutta) had a meagre road area available for road transport. A French group of experts conducted a study and prepared a report proposing two underground routes: Howrah Maidan to Sealdah and Paikpara to Kalighat. However, no proposal was made for the difficulties in implementing underground corridors. The government of West Bengal mooted the following proposal for a metro system in 1949-50, and French experts conducted a survey. However, the proposal could not be brought into effect. In 1965, the Planning Commission set up a team to assess the transport facilities in Calcutta (Kolkata), Delhi, Bombay (Mumbai) and Madras (Chennai).

In 1969, the Metropolitan Transport Project (MTP) was formed to work out alternative solutions. The MTP, with the help of Soviet specialists and East German engineers, prepared a master plan that included five corridors for the city of Kolkata. Out of five corridors, three were mooted for implementation. Three corridors were i) Dum Dum – Tollygunge line (today named as North-South Metro Corridor), ii) Bidhan Nagar- Ramrajatala (known as East-West Corridor), and iii) Dakshineswar _ Thakurpukur corridor, which is now divided as Noapara- Dakhsineswar, part of North-South Metro and a corridor from Joka to Daljhousie (BBD Bag) (Now terminating at Esplanade). The highest priority was given to the North-South metro, and the Government approved the report and the alignment. In June 1972, then Prime Minister Indira Gandhi laid the foundation.

The ambitious metro system began with the construction of the N-S Metro in 1972. On 24 October 1984, India's first metro system became operational in Kolkata.

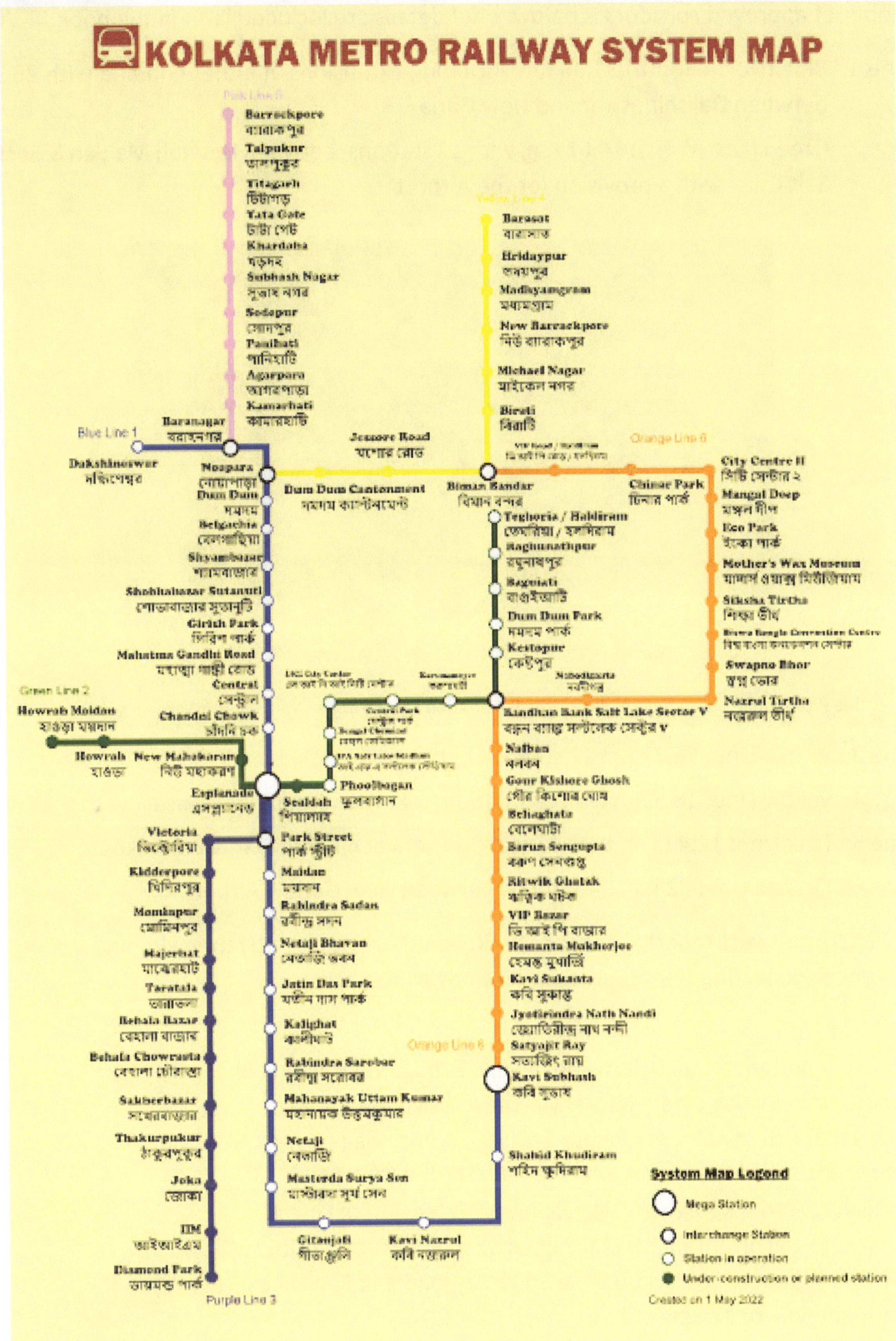

Fig. 2.1: Kolkata Metro Network Plan

The number of approved corridors is below; brief details are included later in the book.

Line 1: Blue (North-South):31.36 Km, including extensions at either end and with 26 stations between Dakshineswar and New Garia

Line 2: Green (East-West): 14.67 Km, with 12 stations, between Howrah Maidan & Sector V, at Salt Lake, with a provision for the Airport

Fig. 2.2: Hon'ble PM of India Indira Gandhi, laying the Foundation Stone For the North-South Metro Kolkata

Line 3: Purple Line: 16.72 Km 12 stations between Joka-Esplanade

Line 4: yellow Line: 18.5 Km with ten stations between Noapara- Barasat

Line 5: Pink Line: 12.42 km, with 11 stations between Baranagar & Barrackpur

Line 6: Orange Line: 32 km, 24 stations between New Garia – Airport.

In 2012, RITES prepared a report to enhance the Kolkata Metro system by surveying and recommending 14 new routes connecting the suburbs, and the key routes are,

1. Majerhat to Ruby via Kalighat and Ballygunge
2. Basirhat to Mahanayak Uttam Kumar, Kavi Subhas via Haroa & Bhangar
3. Joka to Mahanayak Uttam Kumar via Thakurpukur
4. Kavi Subhas to Canning via Rajpur-Baruipur-Uttar Bhag
5. Joka to Diamond Harbour along Diamond Harbour Route
6. Barasat to Barrackpore via State Highway (Extension of Line 4)
7. Barackpur to Kalyani via Kalyani Expressway (Extension of Line 4 & Line 5)
8. Madhyamgram to Barrackpore via Sodepur Road and Kalyani Expressway
9. Karunamoyee to Kolkata Station
10. Howrah Maidan-Shalimar-Santragachi via Kona Expressway
11. Santragachi to Dhulagarh (Extension of Line 2)

12. Howrah Maidan to Dankuni via Benaras Road
13. Howrah Maidan to Srirampur to Chandan Nagar via NH 2
14. Howrah Maidan to Belur

Together, these lines represent a significant leap forward in Kolkata's urban transit landscape, with each corridor tailored to meet the city's varied demographic and logistical demands. The integration of elevated, at-grade, and underground sections showcases the adaptability of the engineering teams in overcoming geographical, infrastructural, and regulatory challenges. Notably, the projects have required close collaboration between multiple government agencies and innovative solutions, such as the re-routing of lines due to urban constraints and the use of advanced tunnelling techniques under rivers and densely built-up areas.

Fig. 2.3: Typical Entry/Exit of Kolkata Metro Line 2.

As the metro network expands, it not only alleviates surface congestion but also acts as a catalyst for economic growth and urban renewal in the city's peripheral regions. The strategic placement of depots, interchange points, and the focus on seamless connectivity to major railway hubs and the airport reflects a comprehensive approach to building a truly multimodal transport system for Kolkata's future.

Kolkata Metro has six depots in operation: Noapara for the blue, yellow, and pink lines; Tollygunge for the Blue line; New Garia for the Blue and Orange lines; New Town for Orange Line, Central Park Depot at Salt Lake City for the Green line; and Joka for the Purple line. In addition, the Airport Depot for the Yellow and Orange lines and the Depot at New Town for the Orange line are under construction.

2.2.2 Delhi Metro

Delhi joined Kolkata next when the concept of mass transit emerged from a traffic study. Between 1969 and 1984, several committees from different government departments were commissioned to examine alignments and technology. In 1984, the Delhi Urban Arts Commission planned a multi-modal transport system for Delhi, including three underground metro corridors. The Government of India, under the then Hon'ble Prime Minister of India, and the government of Delhi set up the Delhi Metro Rail Corporation **(DMRC)** in May 1995.

The construction of the Delhi Metro began on October 1.1998, and the first line was inaugurated on the 24th. December 2002. Delhi metro corridors were planned to be implemented in three phases: I, II & III. Later, Phase IV was introduced in the Master Plan. Phase IV is under implementation. Phase V is the planning stage. As of now, Delhi Metro has the largest Metro Network in the country. The master plan of the Delhi Metro is included in Fig. 2.2.

Delhi Metro currently comprises a network of 389 Km with 285 stations. It stretches to Noida and Ghaziabad in Uttar Pradesh, Gurgaon, Faridabad, Bahadurgarh, and Ballabhgarh in Haryana. Construction works to build the rail mass transit system began in October 1998, following more than 40 years of studies. Phases one and two of the Delhi Metro were fully operational by January 2013. Funding principally came from a Japanese loan and equity infusion by Indian public funds. Delhi Metro was designed to be integrated with other public transport. DMRC signed an agreement with bus operator Delhi Transport Corporation (DTC) to incorporate management and through ticketing. In March 2010, DMRC partnered with Google India (through Google Transit) to provide train schedules and route information to mobile devices with Google Maps. The free service allows commuters to get the latest service information and plan their journeys.

Delhi enjoys a good infrastructure level, but the rapid rise in population and the extent of the urban area are putting tremendous pressure on it. Transportation, reckoned as a linchpin in the developmental program, has not kept up with the city's growth. The clear indication is that the number of vehicles plying in Delhi outnumbers that in Mumbai, Chennai, and Kolkata. Delhi Metro is India's largest and busiest rapid transit system, connecting the country's capital region with satellite cities.

Fig. 2.4: Station of Delhi Metro (Magenta Line)

The Delhi Metro system has a mix of elevated, at-grade, underground corridors and stations and a blend of Broad Gauge and standard gauge, like the Kolkata Metro.

Fig. 2.5: Shows an Elevated Corridor Crossing the Existing Road--Common in Indian Metros

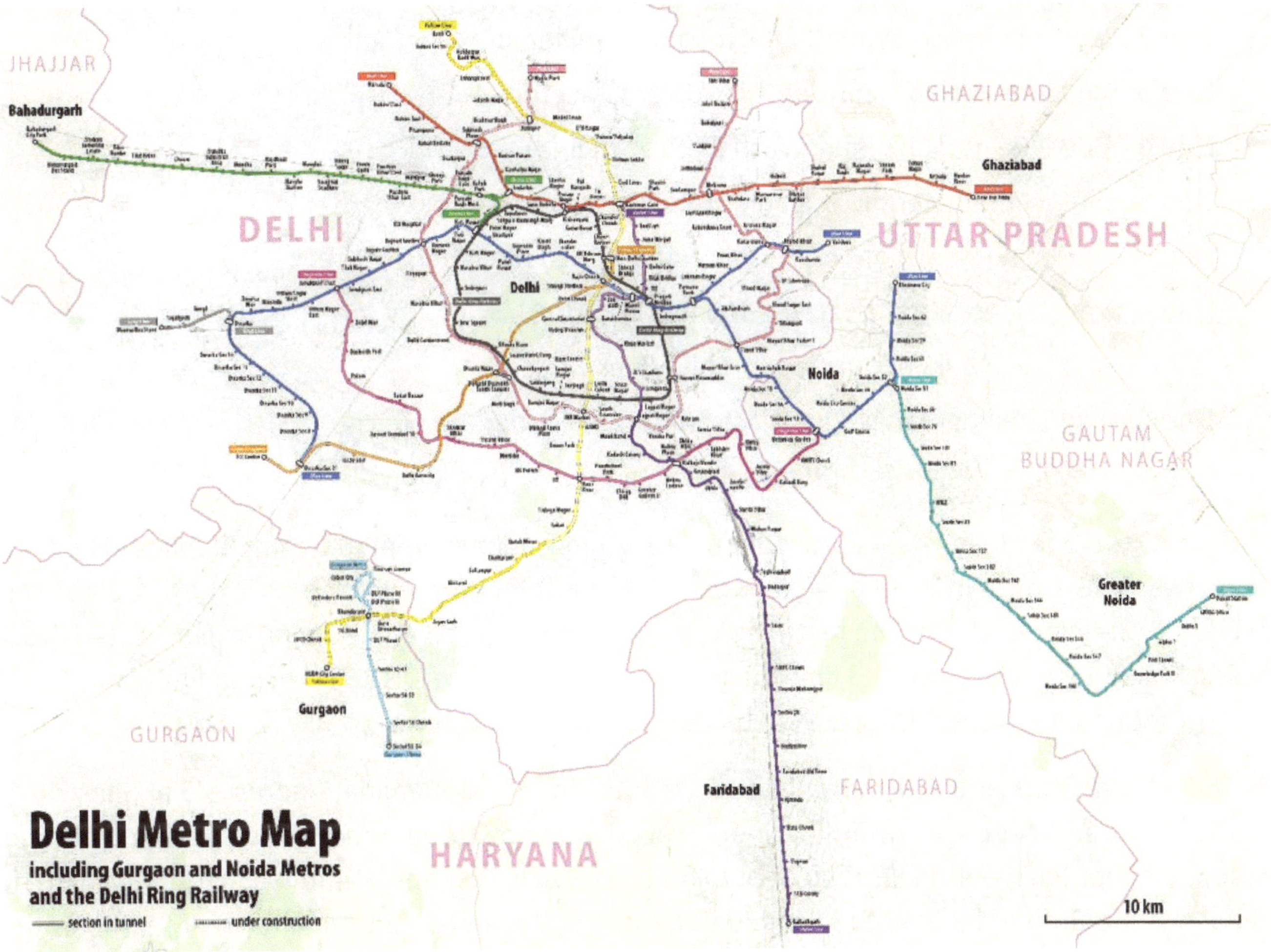

Fig. 2.6: Delhi Metro Master Plan

In September 1996, the Union Cabinet approved three lines included in Phase 1, and in 1997, Japan granted loans. Construction began on October 1, 1998.

The first line, the Red Line, was inaugurated by the then-Hon'ble Prime Minister of India on December 24, 2002. The DMRC hired Hong Kong MTRC as a technical consultant for construction techniques and operations. The Author led a team from **CES** for the elevated corridor and stations for line 1. The same team was involved in two sections of underground routes and stations with the German Consultant to design, detail and schedule items & quantities besides the consultant estimate, the first two underground sections, MC 1A & MC 1B. However, two pre-qualified groups joined during the implementation stage, and a different consultant carried out the detailed design. Phase 1 encompassed initial sections of red, yellow, and blue lines. The then-Hon'ble Prime Minister of India inaugurated the first phase of the underground line and was put into operation in December 2002. The remaining sections opened in October 2003 and March 2004. Phase 2 corridors started operation in June 2008 & August 2011.

Lists of Delhi Metro Corridors are:

Line 1- Red line-Sheed Sthal-Rithala

Line 2- Yellow line- Samaypur Badli-Huda City/Millenium City Centre

Line 3- Blue line- Noida Electronic City-Dwarka Sector 21

Line 4- Blue line- Yamuna Bank- Vashali (Branch line)

Line 5- Green line- Inderlok- Brigadier Hoshiyar Singh

Line 6- Violet line-Kashmere Gate-Raja Nahar Singh (Ballabhgarh)

Line -- Orange Line-Airport Express-Dwarka Sector 21

Line 7- Pink line- Mukundpur Depot- Shiv Vihar

Line 8- Magenta line- Janak Puri West- Botanical Garden

Line 9- Grey line- Najafgarh- Dhansa Bus Stand

Break-up Delhi metro lengths are Redline-34.55 Km, Yellowline-49.02 Km, Blueline-64.62 Km, Greenline-28.78 Km, Violet line- 46.34Km, AExpress-22.91Km, Pinkline-59.24Km, Majentaline-37.44Km & Greyline-5.19Km. The metro system was implemented and is being implemented in four Phases. Phase 1 is 64.75 Km,123.3 Km; Phase 2 is 123.3 Km, Phase 3 is 162 Km, and Phase 4 is 103.93 Km. Currently Delhi Metro Network Plan has been added with Phase V.

The Delhi Metro currently has more than 15 Depots. For line 1, Redline, depots are at Shastri Park (in operation) and Narela (future); for line 2, Yellow line has three depots in operation, which are Sirsapur, Khyber Pass, Sultanpur; two depots for lines 3 & 4, Blueline, Yamuna Park & Najafgarh, for Greenline, line 5, two depots Mundka and Bahadurgar. Redline-34.55 Km and rh are in operation; the third is planned for the future. Depot for line 6, Violet line, two depots are in operation, one at Sarita Vihar other one at Neelam Chowk Ajronda; for the Airport line, the depot is at Dwarka sector

21, for line 7, Pink line depots are at Mukundpur and Vinodnagar, Ghazipur, for Magenta line, line 8 the present depot is at Kalindi Kunj the other one will be constructed in Mukundpur and for the line 9, Grey line depot is at Najafgarh. Ghitorni Depot was designed by **CES.**

For Airport Express, eight rakes with a six-coach formation were supplied by CAF Beasain of Spain. The unique features of the coaches are that they are padded with fabric seats to reduce noise levels and are equipped with an LCD screen for the flight.

2.2.3 Gurugram (Gurgaon) Metro Line

Gurgaon Metro is a light metro system serving Gurgaon, Haryana, India. Rapid Metro connects the city's commercial areas and acts as a feeder link to the Delhi Metro, which has an interchange with its Yellow Line at Sikanderpur Metro Station. The author's team was involved as the Lender's Engineer.

Fig. 2.7: Rapid Metro at Gurgaon

Rapid Metro Gurgaon is a light metro system serving the city of Gurgaon. Rapid Metro connects the commercial areas with an interchange with Delhi Metro's Yellow Line at Sikanderpur metro station. Built by Rapid Metro Gurgaon Limited (RMGL), the system was the world's first fully privately financed modern light metro system by IL & FS. It was not the first privately financed rapid transit system. The first phase of the system opened in November 2013. The second phase began commercial operation in March 2017. In September 2019, IL&FS announced that it did not have the resources to continue running the Rapid Metro. After a short dispute with the Haryana government and a court ruling from the Punjab and Haryana High Court, the Delhi Metro Rail Corporation took over the operation. Rapid Metro has a total length of 12.85 kilometres, serving 11 stations. The system is fully elevated using standard-gauge tracks. The trains are composed of three cars. The power is supplied by 750-volt direct current through the third rail.

In April 2010, Siemens was awarded a turnkey contract to build five three-car metro trains. Siemens sub-contracted CRRC Zhuzhou Locomotive for 5 aluminium-bodied air-conditioned trains. The first three-coach train set was built in China. RMGL ordered an additional seven three-car metro train sets for the second phase of the metro. The final 4 of 7 rakes arrived in February 2016.

Each train has three coaches. The total length of a 3-coach train is 59.94 metres. The coaches are 2.8 metres wide, have roof-mounted air conditioning and have 4 doors on each side of each coach. Each train has a capacity of approximately 800 passengers and was designed to carry 30,000 passengers per hour. The Haryana Mass Rapid Transit Corporation (HMRTC) planned to build an additional metro network spanning 188 km. Gurugram Metro Rail Limited (GMRL), similar to the Delhi Metro Rail Corporation, is responsible for constructing, maintaining, and operating this metro line.

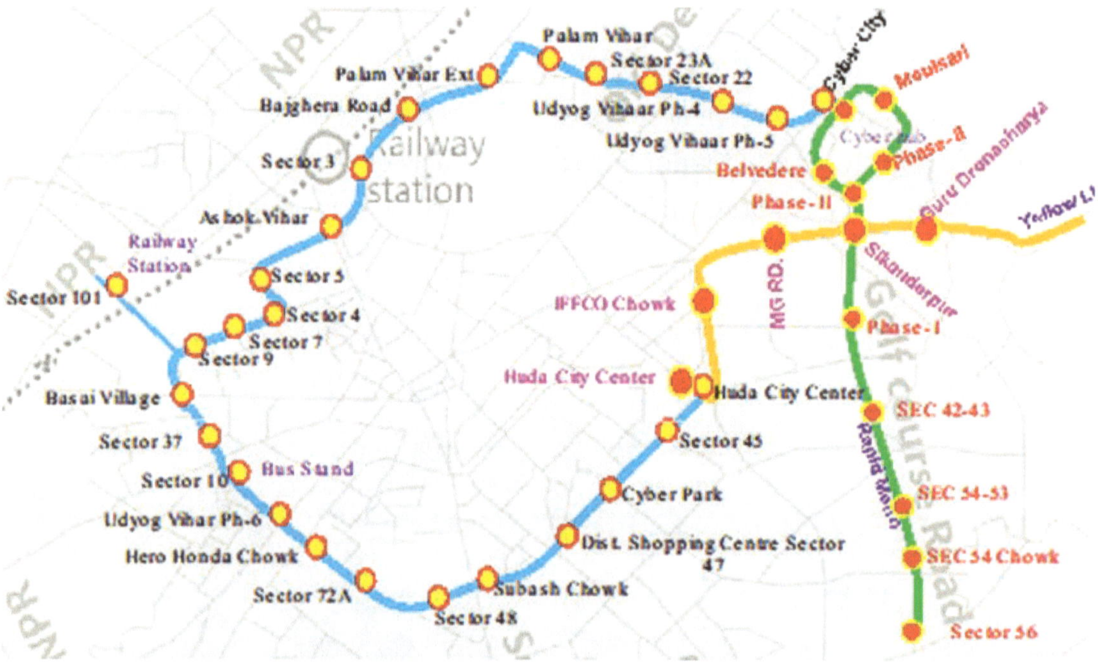

Fig. 2.8: Gurugram Metro Network

Fig. 2.9: City Center Station at Gurgaon

The newly approved Gurgaon Metro will connect Huda City Centre to Gurgaon Cyber City via Old Gurgaon. The Union Cabinet has approved the project. Once the project is complete, it will connect the two sides of the Gurgaon city. The total length of the Metro is around 28.5 km. The length of the main line will be somewhere around 26.65 km, and the remaining 1.85 km stretch will be the spur constructed from Basai village to Dwarka Expressway.

2.2.4 Bangalore Metro

In October 2018, the third-longest metro network started operation in Bangalore (Bengaluru). The Bangalore Metro Rail Corporation Ltd. (BMRCL), a joint venture company between the GoI and the Government of Karnataka, were formed to implement the Metro Network. The DMRC prepared the Detailed Project Report and submitted it in April 2006 and approved by the Union Government Cabinet approved the plan. The Bangalore Metro has been planned to be implemented in two phases, Phase 1 & Phase 2. Phase 1 includes the Purple Line (East-West Corridor) and Green Line (North-South Corridor), covering 42.30Km. Phase 1 of the metro system was designed as an Elevated Corridor of 33.48 Km and 8.82 km Underground. The purple line comprises 13.30 Km as elevated and 4.8 Km as underground, with a total of 17 stations for a total length of 18.10 Km. The green line, with a total length of 24.20 km, includes 20.20 Km as elevated and 4.0 Km as underground. Phase 2 of the Bangalore metro, January 2014, includes 24 GoI-approved extensions 1 and 2 of the purple line, 8.81 km and 15.50 km, and extensions 1 and 2 of the green line, 6.29 km and 3.77 km, respectively. The author was the Project Director and Team Leader for the elevated Nayandahalli (NYN)Station corridors to Sumangala and Rajarajeswarinagar (RRN) to the Chetna Station elevated corridor and stations. Brief details are included later in this Book.

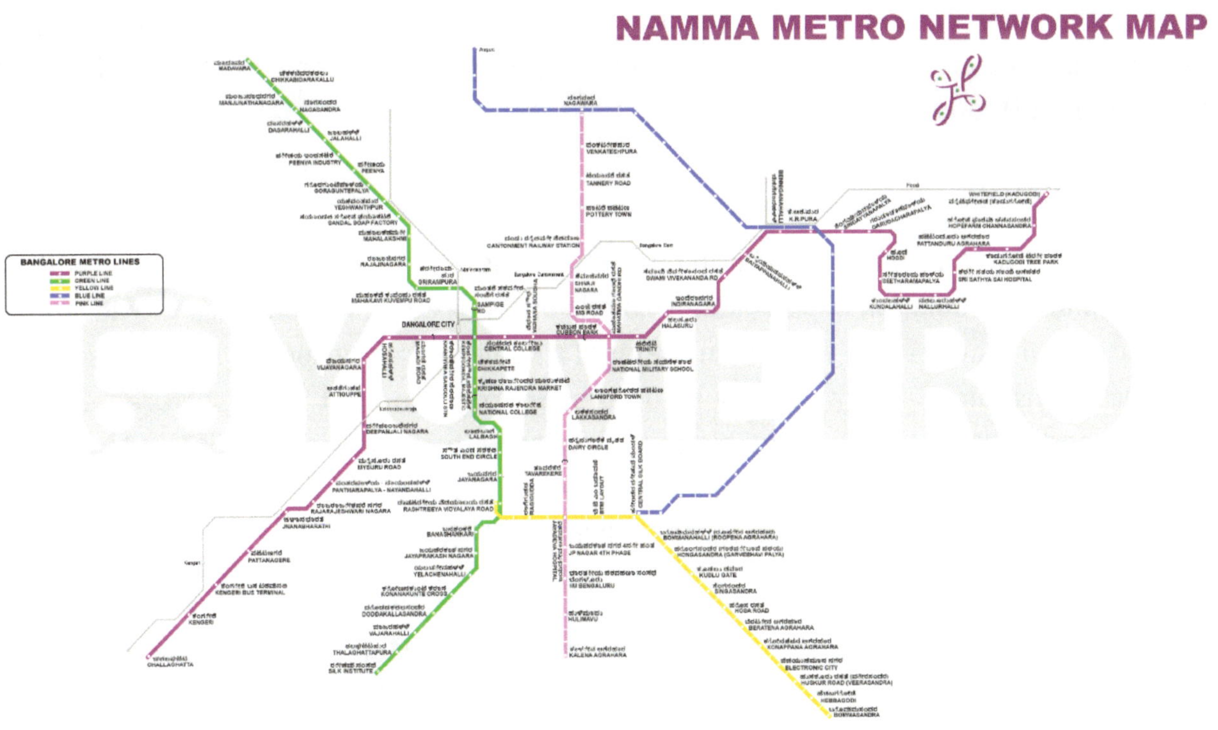

Fig. 2.10: Metro Network Plan- Bengaluru (Ref. yometro.com)

For Bangalore Metro, BMRC procured 50 rakes of three-coach units from BEML- Hyundai Rotem, with 20.8 X 2.88 m X 3.8m high coach dimensions. Each coach has a seating capacity of 50 and 306 standing passengers, calculating height passengers/sq.m. Later, BMRC ordered rakes with six coach trains operated by 750 V DC with a third rail bottom power collection system.

Fig. 2.11: Bangalore Metro-Two Corridors, One Crossing Over the Other

Fig. 2.12: Bangalore Metro Station in Green Line

Bangalore Metro opted for several depots for its operation. Phase 1 of the metro includes two Depots, one at Balyyappanhalli and the other at Penya. For phase 2, BMRC planned a Depot at Silk Institute on the green line and Challaghatta and Kadugodi on the Purple line. For new lines for phase 2, additional depots are being built at Kothanur (Purple line) and Hebbagodi (Yellow line), besides another depot at Doddajala.

2.2.5 Mumbai Metro

Being the capital of Maharashtra, Mumbai is among the largest cities in the world, with a total metropolitan area population of near about 20 million as of 2011. Mumbai has the advantage of a high public modal share of 88% who are in favour of a public mass transport system. Due to the city's geographical constraints and rapid population growth, road and rail infrastructure development, like in Delhi and Kolkata, has been unable to keep pace with growing demand. The Mumbai Metro System is designed and planned to be implemented in fourteen phases, based on contemporary traffic, to reduce traffic congestion, increase mobility, and reduce travel time.

For the Mumbai metro, the Author was involved as Project Director cum Team Leader both for Thane MRTS and Andheri Ghatkopar Metro, the first metro in Mumbai, which started operation in 2000. MSRDC awarded CES the task of preparing the Detailed Project Report (DPR), of Thane MRTS, for which RITES prepared the initial report. In association with PB, CES prepared the project report to enable the MSRDC & Thane Municipal Corporation to take up the project for implementation.

The report included the modified and extended Alignment, Detail Topographic & Geotechnical Survey, Traffic surveys including willingness to pay, Ridership, Detailed architectural, structural, MEP, Environment, Traffic dispersal plan during the implementation & operation of the system, System Engineering Economical & Financial and Legal & Institutional aspects.

Fig. 2.13: Mumbai Metro Station at Gundavali

MMRDA, the planning authority for the Mumbai Metropolitan Region, awarded CES the job of the Andheri-Ghatkopar metro system. MMRDA was responsible for planning MRT corridors for the Mumbai Metropolitan Region. DE-Consult and TCE, an Indo-German Technical Cooperation program, conducted a feasibility study between 1997 & 2000. MMRDA decided to take up this line for implementation and appointed CES to prepare the detailed Project Report and include all Surveys & investigations, detailed design & drawings of Architectural, Structural, MEP, and Environmental aspects. In addition, the consultants prepared the Systems, AFC, Economic & Financial Analysis, Legal and Institutional and all other elements like Tender Documents, to complete the Project Report to facilitate the call of tender in continuous discussions and presentations to all concerned Authorities during 2002-2004 to update the project report.

The Mumbai Metro network includes the following corridors:

Line 1	Versova-Andheri-Ghatkopar
Line 2	Bandra-Kurla-Mankhurd
Line 3	Colaba-Bandra-Charkop
Line 4	Charkop-Dahisar
Line 5	Ghatkopar-Muluhnd

Line 6 BKC-Kanjurmarg

Line 7 Andheri(E)-Dahisar(E)

Line 8 HutatmaChowk-Ghatkopar

Line 9 Sewri-Prabhadevi

Mumbai's metro system includes depots like Aarey, built on 30 Hc of land, and Dongry depot, on 4.4 Hc. of land, Vikhroli, Mogharpada, Kasarvadavali, Mandale and Uttan. Rolling stock supplied by BEML, 63 rakes with six coach combinations. Traction Power is 25 KV AC overhead.

Fig. 2.14: Mumbai Metro Network

2.2.6 Jaipur Metro

The Jaipur Metro is the sixth metro in operation in India. Construction on the most elevated part of the first line, called Phase 1A, comprised a 9.63 km route and nine stations. Construction was started in November 2010 and completed in 2014. The Jaipur Metro began commercial service in June 2015, 2.34 km long and two stations after Kolkata, Delhi, Bangalore, Gurugram, and Mumbai. Phase 1-B, from two stations, began operation in September 2020. Phase 2 has a length of 23.099 km and 20 stations. is planned to be completed by 2021.

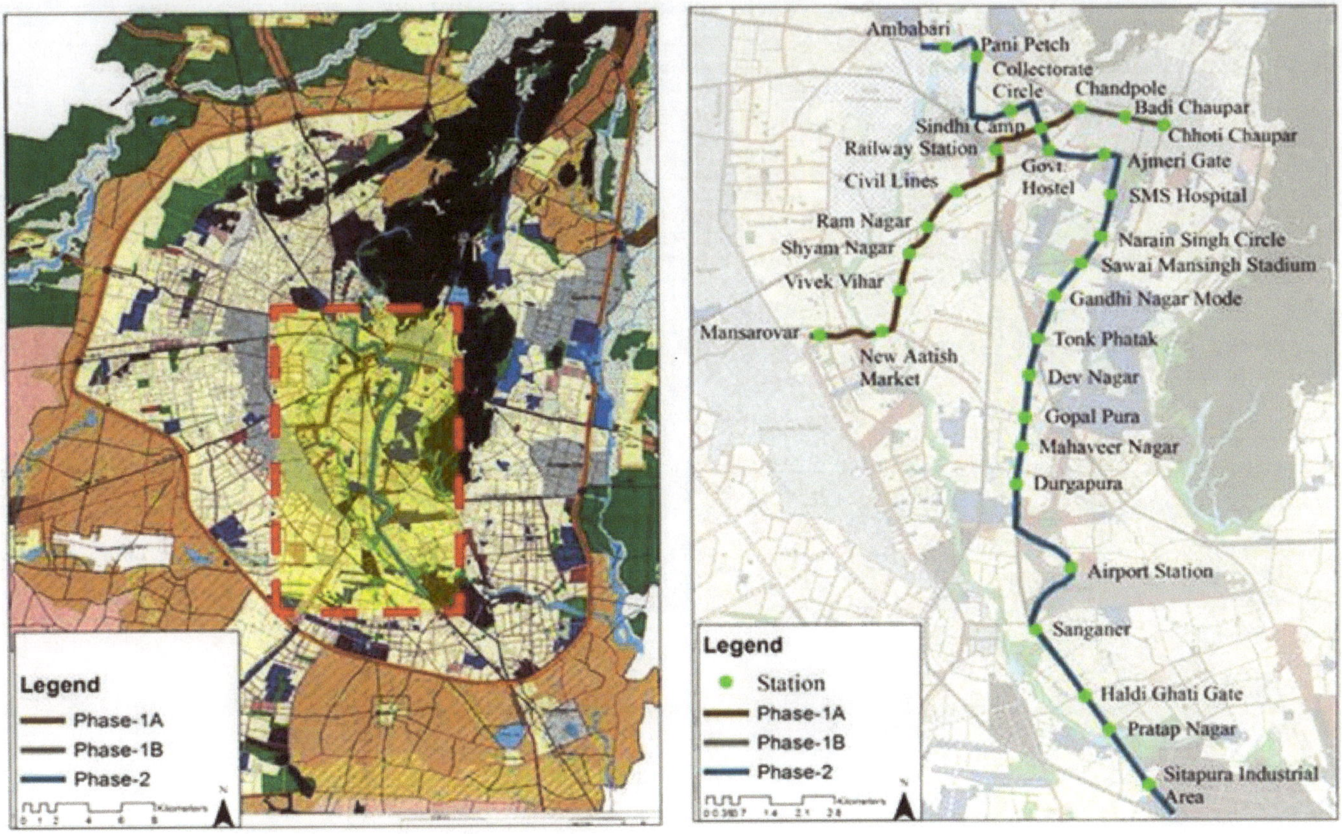

Fig. 2.15: Plan of Jaipur Metro (Ref. www.researchgate.net)

2.2.7 Chennai Metro

On November 1, 1995, the **Chennai MRTS** (elevated Railway) began operating, becoming India's first elevated railway. Between Chennai Beach and Chepak, it extended to Thiruvamayilai in 1997 and Velachery in 2007. The Author was also involved in the Projects and designed two Stations, Mandaveli and Greenways. The elevated station is built on the eastern bank of Buckingham Canal, adjacent to the junction of the Buckingham Canal with the Adyar River. Like most other stations along the Chennai MRTS route, **Greenways** Road features two side platforms, length of platform being 280m. **Mandaveli** is a railway station on the Chennai MRTS. Located across Venkatakrishnan Road in Mandaveli, it exclusively serves the Chennai MRTS.

Fig. 2.16: Chennai Metro Network Plan (Ref. themetrorailguy.com)

Fig. 2.17: Chennai Metro in Operation (Ref. www.kgbuilders.com)

The Chennai Metro a modern metro (Phase-1), based on the Delhi Mass Transit Corporation (DMRC) report, was approved by the Government of Tamil Nadu in 2007. Construction for Chennai Metro Phase 1 (45.1 km) started in April 2009, with the first section of the Blue Line (Line-1) between Koyambedu – Alandur opening in June 2015. Further sections opened one after the other between 2015-2019, with the last section connecting AG-DMS to Washermanpet opening in February 2019.

In December 2015, a 9 km extension of the blue line toward Wimco Nagar was approved as part of the Phase 1 Extension project. Construction on this extension started in 2016, and the line opened for commercial operations.

In 2019, Chennai Metro Rail Ltd. appointed Larsen & Toubro to create a feasibility report for extending Chennai Metro's Blue Line from the Airport to Kilambakkam. This proposed 15.3 km extension was to have 13 elevated stations along GST Road as detailed in the Detailed Project Report (DPR).

The first phase 1 was 45.1 Km with 25 Km of underground Washerman pet. The second corridor was planned to connect St. Thomas Mount with Chennai Central. CMRL was established in December 2007 and appointed to execute.

The first phase of Metro was funded by JICA (59%), GoI (15%), and the remaining by the Govt of Tamilnadu. Phase I is comprised of a green line and a blue line. The Blue Line is planned to extend to the Airport in 2021. In July 2016, GoTN approved phase II of Metro lines with a total length of 106 km.

Fig. 2.18: Tirumalai Station in Chennai

Fig. 2.19: Wimconagar Station in Chennai

Chennai Metro Phase 1 has 40 stations for two lines: 22 underground, 200m wide, and 6.1 m deep, and 18 elevated, composed of three floors. Alstom supplied the Rolling Stock, and the first nine metro rakes, each consisting of four coaches, were imported from Brazil. The remaining coaches are being manufactured in Chennai itself. Traction Power is a 25 KV AC traction catenary system. Metro rakes include one compartment of First class with provision for women passengers. The Chennai metro depot at Koyambedu includes test tracks, and Elevated Depot at Wimco Nagar, built on 8.6 acres of land and a third depot is planned at Madhaban (Phase 2).

2.2.8 Hyderabad Metro

The city's population was increasing rapidly, and the need for MRTS was felt, and in 1988, RITES conducted a feasibility study and proposed implementing 22.0 km of LRTS. Later, DMRC prepared the DPR for Phase 1 in 2003. In 2017, the Commissioner of Railway Safety approved the 12.0 km section between Miyapur and SR Nagar, 10 km between SR Nagar and Mettuguda, and 8 km between Mettuguda and Nagole. The Hon'ble Prime Minister of India inaugurated the first corridor in Hyderabad. The metro system includes three depots. The Miyapur Depot near Calvary Temple and Uppal Depot at Nagole Road are in operation, and the proposed Falaknuma Depot is planned on 17 acres of Land.

The Hyderabad Metro is implemented on a public-private partner basis (PPP). Hyderabad metro is the second MRTS system implemented on the PPP model after the Andheri -Ghatkopar-Versova metro line in Mumbai. Public-private partnerships fund the state government, hold an equity share, and perhaps have the most extensive design, which is 30 km long and has 24 stations elevated metro, built, financed, operated, and transferred (DBFOT). L&T and Metro Rail Hyderabad Ltd jointly implemented the job.

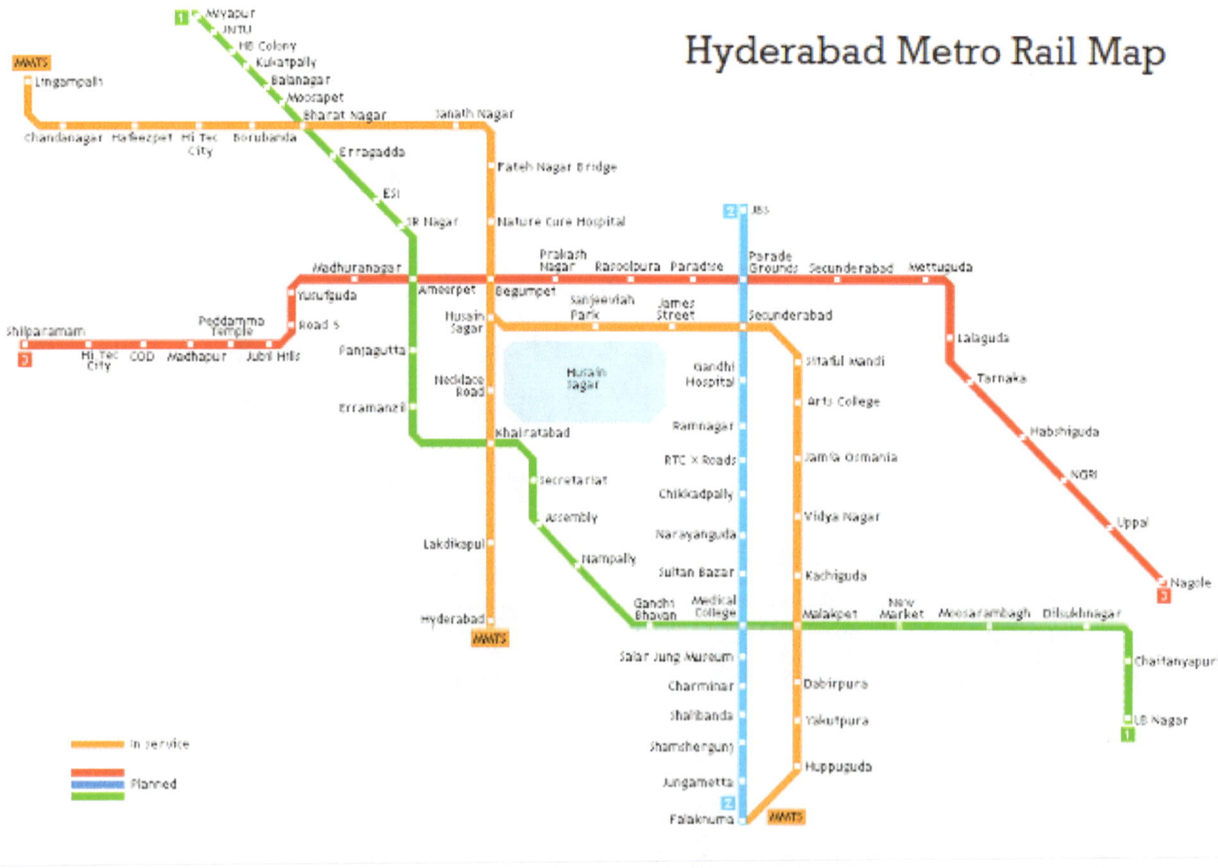

Fig. 2.20: Hyderabad Network Plan (Ref. hyderabadmetrorailworks.blogspot.com)

The metro construction started in June 2012 in two phases, Phase 1 & 2. Phase 1 includes three red, green, and blue lines, totalling 72 km. Phase 1 is again implemented in stages 1 & 2, 3/1 & 3/2, and 4/1 & 4/2. Currently, two lines are in operation: Red, Line 1 (Red), and Line 3 (Part of Line 3). Phase 1 was implemented in PPP mode. Phase 2 is being taken up by the Government of Telangana and is planned to be implemented in seven stages. The metro operates on a standard gauge with a ballast-less track with a 3.5-6.5-minute headway. Traction power is 25 KV, 50 Hz AC through an overhead catenary.

Fig. 2.21: Elevated Station for Hyderabad Metro

The Hyderabad Metro project is being implemented in Phase 1 (71.21) and is scheduled for Phase 3 by the Govt. of Telangana. Phase 1 elevated corridor includes Red Line, 29.21 Km with 27 Stations; the Green Line, 15 Km long with 16 elevated stations; and the Blue Line, 27 Km long, with 23 elevated stations in operation. Total Stations in Phase 1 are 66.

2.2.9 Kochi Metro

The foundation of Kochi was laid in September 2012, work started in June 2013, and it began operating in June 2017. Kochi metro is the first metro in India implemented in four years, with a 25.8 km elevated length and 24 stations. Kochi Metro is planned to be implemented in three phases. Phase 2 is scheduled to be implemented by KMRL, i.e. Kerala Metro Rail Limited. Phase 2 is the extension of the existing metro by 11.2 km starting from Jawaharlal Nehru Stadium to Info Park via Kakkanand, including 11 stations. Phase 3 consists of the extension to Kochi International Airport.

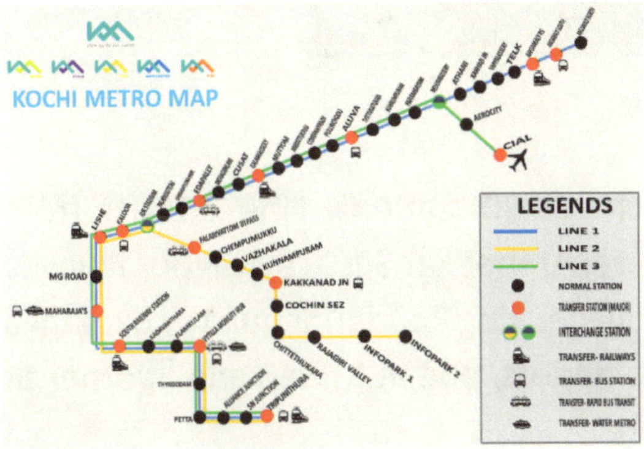

Fig. 2.22: Kochi Metro Route Plan (Ref. www.skyscrapercity.com)

Fig. 2.23: Kochi Metro in Operation (Ref. Metrorailtoday.com)

2.2.10 Nagpur Metro

The Government of Maharashtra approved Nagpur Metro in early 2014, and the Ministry of Urban Development, the Government of India, provisionally approved it in the latter part of the year. DMRC submitted the Project Report to Nagpur Improvement Trust, the nodal agency. The Project Report included two lines: one from Automotive Square to Mihan Metro Depot and the other from Prajapati Nagar to Lokmanya Nagar. **Line 1, Orange line,** North-South Corridor travers section beyond Airport Station, including 20 stations, mostly elevated.

It has a total length of 22.3 km and is elevated except for a 4.6 km at-grade section beyond Airport Station. It includes 20 stations, primarily elevated (15nos).

Fig. 2.24: Nagpur Metro Network (Ref. www.reddit.com)

Line 2, Blue line, is East-West corridor for 19.4 Km that includes 20 stations. Line 2 starts from Prajapati Nagar in the East and runs through Vaishnodevi Square, Ambedkar Square, Nagpur Railway Station, Cotton Market, Institute of Engineers, Dharam Peth College, Ring Road Junction to Lokmanya Nagar. Line 2 is elevated. Nagpur Metro operates through two depots, one at Mihan and another at Lokmanya Nagar. Two lines are partially (24.5 km) in operation.

Fig. 2.25: Nagpur Metro Corridor

2.2.11 Ahmedabad Metro

Ahmedabad Metro, MRTS, is being built for two cities, Ahmedabad and Gandhinagar, in Gujrat. Gujrat Metro Rail Corporation (GMRC), a Special Purpose Vehicle Company, was established in 2018 to implement all metro projects in Gujrat. The Ahmedabad metro is being implemented in two phases.

Fig. 2.26: Ahmedabad Metro line

The Phase 1 corridor is partly in operation and partly under construction. The total length of the corridor is 40.0 Km, of which the elevated section is 33.5 Km and the Underground is 6.5 Km. Phase 1 includes two lines. Line 1, the East-West corridor, is 21.16 km long with 17 stations.

Fig. 2.27: Ahmedabad Metro being inaugurated by Hon'ble PM of India

There are six stations in operation, and the remaining 11 are at the construction stage. The North-South corridor, line 2, is 18.87 in length, with 15 stations, of which four are in operation. The Ahmedabad metro runs on standard gauge and with Traction & power supply through 750V DC, third rail. A section of the Metro was opened to the Public in March 2019. The Author's team was involved in preparing the Design and Drawing for section with difficult span configuration and for Systra.

2.2.12 Other Indian Metros

At present, Metros are being operated in different cities. Cities are Kolkata, Delhi, Bangalore, Hyderabad, Chennai, Mumbai, Ahmedabad, Nagpur, Pune, Noida, Kochi, Lucknow, Gurgaon, Jaipur, Navi Mumbai, Kanpur, and Agra. In addition, many other cities are in the process of implementing the Metro system besides the Metros currently in operation. In other cities in India, metro lines under different stages of implementation are Indore, Bhopal, Varanasi, Kozhikode, Vijayawada, Meerut, Visakhapatnam, Coimbatore, Patna, Navi Mumbai. In 2006, the National Urban Transport Policy was proposed for implementation in all cities with a population of 2.0 million. Later, the Government of India implemented metros in all cities with more than 1.0 million population. It is interesting to note that the first metro, N-S Metro, was designed and implemented by Indian Engineers, including experts of all disciplines; besides the recently opened metro between Joka and Khidirpur, an elevated section (remaining between Khidirpur to Esplanade under implementation is Underground) and New Garia –link with the Airport station underground. Expatriates are involved in most of the other metros in the country.

CHAPTER 3A

Development of Urban Mobility Plan

3A.1 Introduction

The planning and design of Metro Rail Projects involve conducting feasibility studies (Feasibility Study Reports), selecting routes and alignment, preparing a Detailed Project Report (DPR), Designing Stations and depots, and Choosing Rolling Stock.

Studying and finalisation of all other important parameters like Power, Traction, Signaling & Communications, Automatic Fare Collection (AFC) others to complete a system to operate.

The metro system for an urban area should be part of a comprehensive transit plan, i.e. the Mobility plan of the area. Development of an Urban Mobility Plan to include

a. Plan to include past trends, existing conditions, and future demand, Socio-economic conditions, land-use pattern (existing & future), traffic and transport demand analysis (existing & future), and available technology
b. Sustainable Transport Scenarios
c. Development of an integrated urban mobility plan
d. Regulatory and Institutional scenarios
e. Development of an integrated fare policy
f. Implementation Plan.

Note: *For* urban mobility, refer to "Urban Mobility" by Dr Bidhan Chandra Roy, PRK Murthy, *and* Vikram Pawar.

CHAPTER 3B

Metro Systems and Engineering

3B.1 Introduction

Developing and operating existing metro systems in India provides valuable insights into improving the planning and investments of upcoming systems. The planning and performance of the existing and forthcoming Mass Rapid Transit System (MRTS) fare well operationally compared to the best international practices. Significant scope exists for improvement in both technical and financial aspects. Concerned, authorities are aware that metros will continue to face economic and financial crises.

Planning a metro system involves various aspects during feasibility stages, route planning, design, construction, and operations. It includes population density, traffic patterns, connectivity, and environmental impact. The aim is to provide an efficient and sustainable transport system. The development of a metro system includes the design and construction of an elevated and underground transporet network in an urban area, and the system is to include route alignment, track, stations, trains, power, signaling, and other infrastructure components, like fare collection, depot, administration office, safety & securities, legal, etc. and operations. The system has the critical role of reducing traffic congestion, improving the environment, and increasing mobility and connectivity. The metro system includes Trains, Tracks, and Traction Power through a third rail mounted at the track or a catenary suspended from a pole fixed along the track at a suitable spacing. The third rail is used to reduce tunnel size and prevent overhead wires. Using overhead catenary-shaped wires allows higher power supply voltage, whereas the third rail provides low voltage, 750 V, DC.

Underground tunnels move traffic below road level, avoiding delays caused by traffic congestion and leaving road space for vehicle and pedestrian traffic. In areas of highland prices, densely built-up areas using existing/upgraded land use plans make underground and elevated sections economical for the transportation system wherever possible.

3B.2 Implementation of a Metro System

Implementing a Metro System starts with a concept plan, feasibility study, detailed engineering, construction, installation of tracks, trains, traction power, signalling and communications, depot, administration building and other infrastructure, and trial runs, including checking the total system and its safety.

To prepare a complete, workable Project Report, including components of metro projects, are

1. Introduction
2. Proposed Alignment
3. Survey & Investigations such as Topographic, Traffic, O-D & Willingness to pay, Ridership Analysis and others
4. Geotechnical Survey, Seismic & Environmental survey
5. Finalisation of alignment & Formation level considering At-grade (if land is available), Elevated, Underground and/or a combination of all (like First Indian Metro, North-South Metro Kolkata)
6. Station Buildings, elevated, underground and at-grade, including Civil, E&M, ECS, TVS, Lift and Escalators
7. Shifting of existing Utilities
8. Permanent way
9. Traction and Power Supply
10. Signalling and Telecommunication
11. Rolling Stock
12. Automatic Fare collection system
13. Platform Screen Door
14. Multi-modal integration
15. Land, Resettlement & Rehabilitation
16. Cost
17. Economic & Financial
18. Legal
19. Institutional
20. Inspection, Checking and Approval for Operations
21. Operational & Maintenance and Performance

3B.3 Performance Analysis

A metro's performance is measured by its operational indicators, which are i) Daily Ridership, ii) Ridership per km of the network, and iii) the comparison between the estimated ridership and the ridership achieved during operation.

3B.4 Economic and Financial Viability

The financial analysis of most of the metros is carried out separately from the perspective of costs and revenues, and financial performance is benchmarked against national & international peers. Most metro systems are not financially viable and are unattractive to private entrepreneurs. Very few entrepreneurs took the challenge, but many of them surrendered. However, economic viability, in most cases, is high and attractive. Financial benefits, including economic growth, increase in land value, reduction of travel stress, lifestyle aspects, competitive edge efficient use of land resources, and many other intangible benefits, are not covered in brief but are exceptionally important for the city.

3B.5 Geotechnical Investigations

Like any other project, the assessment of the geological conditions of a metro is crucial for both elevated and underground. This includes soils of various types, such as silty, sandy, soft/medium/ complex clayey layers, and smooth, medium, and hard rocks. The investigation is to classify soil, check the properties of soil, note the water level beneath the ground, evaluate the physical qualities to find the appropriate location and its suitability for foundations, compute settlement and investigate structural safety. Under the geotechnical survey, the team must conduct standard penetration tests, collect soil samples, record the water level, and do other things as needed for the project. At the end of the sample collections, the team analysed and performed necessary tests to prepare a report recommending the location and type of foundations, capacity, design criteria, approach, and estimated settlement.

3B.6 Alignment and Formation

3B.6.1 An elevated (Viaduct) railway is a rapid transit system that runs over elevated tracks above surface-level street roads using steel, concrete, or steel-concrete gauge. Elevated structures or viaducts and metro systems supporting the tracks and allowing the trains to operate are generally cheaper than other modes of construction. Viaducts are typically constructed using various methods, including precast and/or prefabricated sections like beams, girders and precast segmental box/U-Girder construction.

The corridor consists of short to medium span lengths, notable spans as per site, like clearing roads, rail, utilities, and others below the elevated corridors. The superstructure consists of different geometrical forms, such as Girders, in situ or precast or prefabricated, box sections, either one or multiple, depending on the structural design, either segmental or span-by-span, supporting the tracks for the Rail to run. The substructure comprises a Pierhead/cap and pier (single & multiple) that transmit the loads to the foundation, type & details as per design. This method is popular because it is cheaper and faster, minimises existing traffic hindrances, and has fewer environmental hazards.

3B.6.2 Underground corridors can be of two significant types: i) using the cut-and-cover method and ii) using the Tunnel Boring Machine (TBM)

i. **Cut-and-cover** corridors, including ramps, are constructed by digging up city streets and later rebuilding them over the newly built underground corridors. The cut-and-cover method is comparatively cheaper, but as this type of construction is at the street/ground level, it needs elaborate arrangements for existing traffic and utility diversion. Also, the disposal of excavated soil can be done in restricted & predetermined hours. It is possible to only to a certain depth.

ii. **A Tunnel Boring Machine** (TBM) excavates tunnels with a circular cross-section through various soil, sand, and rock strata, including hard rock. The tunnel's lining can be precast concrete or iron segments (as used in North-South Metro, Kolkata). TBMs have the advantage of limiting disturbance to the surrounding ground and producing a smooth tunnel wall, significantly reducing the cost of the lining. TBMs are efficient and result in shorter completion times, making them most suitable for heavily urbanised areas. Cross Passage is the connecting passage between two parallel tunnels provided for emergency and maintenance. In case of TBM used for underground alignment, horizontal boring is required for each line (Up & Down) with a gap of at least one diameter of TBM Bore and for cross passages required at regular interval, method of NATM must be used for construction of Cross/Connecting passages.

iii. **Stations** for both construction approaches must be made from the surface level. For elevated stations, the space required for construction and movement of construction equipment must be worked out. Space needed for underground stations with bottom-up is like elevated stations. However, for stations constructed with top-down, the construction approach and space at the surface level will be the same, but it will be required until the top slab is cast, and backfilling is done. After reviving the surface level to its original condition, the space at the surface level can be considerably reduced to continue with the remaining construction. During the construction of stations, Electrical & Mechanical systems, Fire alarms, Detection and Suppression systems, VAC systems, and all other facilities of a station are to be installed. Finishing works can be taken after the construction, and during this stage, space covered at the surface to facilitate the remaining works, including signage, can be reduced substantially. Construction of Underground Stations can be done using Cut-and-Cover approach or NATM.

iv. **Permanent way/Track works** intended to carry metro trains can be taken up after the plinth is completed on the finished deck/track level. A plinth to support tracks can be used. A track can be ballasted or ballast-less. Ballast-less tracks are laid on the Mainline, while the ballasted tracks are adopted at the Depot and stable yard. The ballastless track is the railway track, whose bed is composed of concrete and bituminous mixture and is made of steel rail, fasteners and concrete slab. Ballasted track is a traditional track commonly composed of Steel Rails, Sleepers, Fasteners and Ballast beds.

v. Electric power is meant to operate the metro system. Three **traction** systems are used in India: Traction and Power Supply. The first is a 25 AC Overhead Catenary System (OCS), the second is a 1500 Volts DC overhead catenary, and the third is a 750 V DC **Third rail** System. The third rail, which has 750 V DC, is extensively used worldwide. In Kolkata, India, all lines of the metro system use a third rail. The third rail system has the advantage of being free from visual hazards. The **25 KV traction** requires catenary masts both for at-grade and elevated systems, which are visual obstructions and change the skyline of the urban area. The system also needs heavy transformers and converters. Use of OHE in Indian metros are also common.

vi. **Flexible OCS (FOCS)** traction is used for the Elevated and grade section, while **Rigid OCS (ROCS)** is used for the Underground section using TBM.750 V DC Third Rail Traction is used for the elevated, at-grade and underground sections

vii. The **traction and power supply** system consists of traction, OCS, a third rail, an Internal power distribution system, Receiving Substations (RSS), high-voltage cabling between the source and RSS, and a Supervisory Control and Data Acquisition System (SCADA). An Auxiliary Substation is installed in both reaction systems. Electric traction is used for high acceleration and pollution-free services in urban areas.

Fig. 3.1: View showing the Third Rail along the Track

Fig. 3.2: Battery Room for Emergency Lighting

viii. **Rolling Stock** in the metro system refers to vehicles (trains) that move on tracks and include both powered & unpowered railroad cars and coaches. Metro coaches for the first Indian metro was manufactured indigenously, and the rolling stock for the first metro across a major river was also made in India. However, many current metro systems in operation and implementation use imported coaches, though India can manufacture modern coaches nationwide. In Kolkata, most of the Metro corridors use Indigenous coaches.

ix. The **Signaling** system ensures safety in train operation and an efficient train control system for movements. The system stipulates that Continuous Automatic Train Control **(CATC)** works on communication-based train control systems **(CBTC)** to move trains between two sections based on moving blocks. CATC has subsystems i) Automatic Protection System (ATP), ii) Automatic train operation system (ATO), and iii) Automatic Train Supervision (ATS). Continuous Wireless communication systems are used between the wayside and train-borne CBTC systems. The CATC system is used for continuously controlling and monitoring movements through an automatic system for controlling and monitoring train movement by Automatic Train Protection System. ATP includes i) Cab Signalling, ii) Track-related speed profile generation, iii) Continuous monitoring of braking curve, iv) Monitoring of maximum travel speed. Detection of over-speed vi) Maintaining safe distance vii) Monitoring of stopping point viii) Monitoring of travel direction. ATO operates the trains automatically from station to station within the safe envelope of ATP. The system opens the doors automatically while the driver closes the doors to depart. For driverless operation doors get opened and closed automatically. ATS provides the functions of i) Automatic route setting, ii) Automatic train regulation, iii) Continuous tracking of train location, iv) Display panels, v) Adjustment of train dwell time. vi) Passenger information display system, vi) Computation of train schedules.

x. **Telecommunication** systems are the communication backbone for the Signaling systems and other systems like SCADA, AFC, etc. Telecommunication services are needed to meet the metro network's operational and administrative needs. The components of the Telecommunication system include

   1. Fabric Optic System
   2. Telephone Exchange
   3. Mobile Radio Communication
   4. Passenger Announcement System
   5. Passenger Information Display system
   6. Centralised Clock system
   7. Close Circuit Television System
   8. Network monitoring and management

xi. **Automatic Fare Collection (AFC) System** is used nationwide. Metro systems in Hyderabad and Kochi used the AFC system on PPP (public-private partnership) basis. The automated Fare collection system is an intelligent payment method that uses pre-paid cards through an electronic system or gates to allow commuters to access and depart from different transportation options in the city besides the metro network.

xii. **Platform Screen Door:** During the country's metro's initial stages of operation, there was no provision for a Platform Screen Door. Later, a platform screen door was introduced to prevent people or objects from falling or trespassing on tracks. S creen doors are essential both for elevated and underground metro. For elevated metro tracks, including station platforms, screen doors up to required height is used for passenger safety, for un air-conditioned stations. For air-conditioned stations screen doors are extended up to roof or false ceiling.

3B.7 Depot cum Workshop

Apart from Station Buildings, Substations for the operation and maintenance of metro services, Depot cum Workshop, which includes stabling and maintenance of the coaches is an essential part of the system. Preventive and corrective maintenance of metro coaches is done in inside the depot having required infrastructure including stabling, cleaning and maintenance of Metro coaches and rakes.

Fig. 3.3: 3D Top View of a Constructed Depot

In addition to Civil and E&M, Tracks, including support structures, Traction and Power supply, Signalling, and Communications maintenance facilities are istalled in the depot.

Fig. 3.4: Hyderabad Metro Depot

Various optional structures in a depot include

1. Workshop
2. ETU Maintenance Building
3. Stabling Yard
4. DCOS store and property development
5. Administrative Building and Operation & Control Centre (OCC)
6. Chemical Store

7. Underground and Overhead Water Storage
8. Canteen
9. Driver's Booking Office
10. Cable Store and External Services
11. TSS and ACS
12. Effluent Treatment Plant
13. Pumproom
14. Waste Storage
15. Guard House

i. **Environment**

ii. **Estimated Cost**

iii. **Other services required for MRTS**

- Shifting of Utilities
- Multimodal Integration
- Last mile Connectivity
- Staff Quarters
- Security
- Project cost (Executed)

iv. **Inspection, Checking and Approval by Approving Authority (CRS)**

The Commissioner of Railway Safety (CRS) will provide safety certification and technical clearances for Indian metros. In the process, they may prepare a Report to include: i) Introduction, ii) Procedure, iii) Scrutiny of Schedule of Dimensions (SOD), Scrutiny of Documents, a) Rolling Stock, Traction & Power Supply, Structures such as Metro corridor at – Grade, Elevated, Underground, Stations including Ancillary Buildings, Signalling & Communications, Test Reports and Certification.

v. **Maintenance Manual**

The Maintenance Manual should include design parameters, drawings, specifications, and the maintenance aspects of all elements, big or small.

In Summary, the Structure of a Detailed Project Report (DPR) consists of the following.

1. Introduction
2. Topographical and Geotechnical Report
3. Traffic Demand Forecast
4. Alignment and Geometric Design
5. Design Basis
6. System Design

7. Civil Engineering
8. Environmental Assessment
9. Station Planning
10. Train Operation Plan
11. Rolling Stock
12. Power Supply
13. Ventilation and Air-Conditioning
14. Signaling and Communication
15. Cost Estimate
16. Financing Option and Fare Structures
17. Implementation Plan
18. Executive Summary.

CHAPTER 4

Kolkata Metro and Status as of Today

4.1 General

MTP prepared a master plan in 1971 that envisaged the construction of five rapid transit lines for Kolkata: three north-south and two east-west, totalling a route length of 97.5 km. **Fig. 4.2.** shows the Master Plan of the Kolkata metro with six lines: blue, Green, Purple, Orange, Yellow, and Pink. **Fig. 4.1** shows a typical view of line 1 during operation.

The primary characteristic interchange points in the city were three, with north-south lines. In 1971, a team of experts from the USSR made several substantial recommendations regarding the future rapid transit system in light of possible growth of the city areas, the city pattern, land use, and road and railway systems—the most important being that a very high priority had been given to the first line, which is the busy North-South axis between Dum Dum and Tollygunge over a length of originally, 16.55 km.

Fig. 4.1: N-S Metro Station During Operation in the Nineties

The Kolkata Metro network master plan is included in Fig. 4.1.

Fig. 4.2: Kolkata Metro Network with six routes

Kolkata Metro operations started in the early nineties as a part of the Integrated Public Transit System, conceptualised in the early seventies to cater for the growing need for transportation facilities in the city. This section briefs excavation using a rigid diaphragm wall and the construction method adopted for predominantly soft alluvial soil of Normal Calcutta Deposits below one of the most thickly built-up cities in the World.

As a relief measure to the growing demand for transportation facilities, a Master Plan was developed by National and International Committees through several studies, proposing a total length of 97.50 km for the Rapid Transit System in the City. Out of which 16.693 km was implemented in the first phase. The alignment for the first phase, included in Fig4.3, was divided into several sections for implementation and consisted of 1.288 km elevated, 0.270 km at grade and 15.135 km underground portion.

Most of the underground portions, except for sections where bore tunnelling had been planned, were initially proposed with excavation using strutted steel sheet piling. However, sheet piles of capacity available in India at the implementation stage were unsuitable for braced cuts with deep excavation. They could only be adopted in Section 2 between Dum Dum and Belgachia Station. As such, higher capacity sheet piles were proposed for deeper excavation, warranting import. With indigenous sheet piles, excavation was taken up in Section 2 of Metro Corridor, as the excavation depth was less for partly elevated and partly underground sections. Adopting higher capacity imported sheet piles was found not to be cost-effective. Accordingly, M/s Chatterjee Polk Pvt. Ltd., in association with Phillipp Hotzmann AG, Germany, submitted proposals for design and construction with two alternatives using indigenous rigid diaphragm walls. Alternatives were proposed with semi-integrated and integrated structural systems. After the initial assessment, the semi-integrated structural system with a diaphragm wall was adopted for Section 11, between Park Street and Maidan stations. However, the delay in importing specialised equipment for diaphragm walling led to the development of an alternative with open excavations for the General Tank Area.

Besides the design approach and construction techniques, monitoring subsoil behaviour during construction in thickly built-up areas was paramount. A program was devised to study subsoil behaviour in the Maidan area. Panels 18, 19, and 20 were selected between chainage km. 10/098.60 to km. 10/086.10 indicated as Test Section in Chapter 5. As the decision on instrumentation was taken after the execution of the diaphragm wall, simple instruments could only be installed to measure strut load, ground settlement, and diaphragm wall movements. The instrumentation details were planned by Dr. BC Roy (the Author) and carried out jointly by the author and Dr N. Som, Consultant to Metro Railway, Calcutta and Professor, Jadavpur University. Subsequently, Prof. N. Som and his team from Jadavpur University continued with the investigations in the most thickly populated section between MG Road Station and Girish Park Station, identified as Test Section 2 in Chapter 5.

Currently, all the projects are under various stages of operation and implementation. Kolkata Metro, being the first Metro in the country, was more of a learning and gathering of experience for Indian Engineers (planning, design & implementation by Indians). Construction of the first 17 km underground (partly elevated, partly underground & partly at grade) railway took a long time to operationalise the complete line. The government of India financed the project, which the Ministry of Railways operated. The reasons for the delay in completing the project are mentioned below:

- Delay in land acquisition.
- Long processing time for importing construction machinery.
- Ineffective and time-consuming technical transfer of construction skills.
- Existing old buildings (Vulnerable to vibration) along the work site.
- A complicated underground sewerage network and water mains exist beneath the work site.
- Limited working space availability due to road congestion and limited working hours.
- For long-distance transportation of construction materials
- Interruption of construction works due to rain and long rainy seasons.

(Observations by JBIC)

Besides a **Lack of appropriate & consistent Fund flow.** The project was funded by MOR & GOI.

Since the formation of the MTP R in 1969, Kolkata Metro has been under the Indian Railways. It is the only metro in India controlled by the Indian Railways. Kolkata and Delhi Metro use both Standard Gauge and Broad (Indian) Gauge. In addition, the Rolling Stock for all the routes in Kolkata was manufactured in India. Kolkata Metro opted for meter-gauge coaches on broad gauge for line 1, and rakes were custom-built with a special assembly.

MTP prepared a master plan in 1971 that envisaged the construction of six rapid transit lines for Kolkata: 3 north-south and two east-west, totaling a route length of 97.5 km. Fig. 4.1 The network plan of the Kolkata metro has six lines: blue, green, purple, orange, yellow, and pink.

The primary characteristic interchange points in the city were three, with north-south lines. In 1971, a team of experts from the USSR made several substantial recommendations regarding the future rapid transit system, considering possible growth of the city areas, the city pattern, land use, and road and railway systems—the most important being that a very high priority had been given to the first line, which is the busy North-South.

Briefs on different corridors included in the Kolkata Metro are as follows.

4.2 Line 1 North-South, including extensions at either end – blue line

Line 1 has a total length of 25 km with 23 stations, of which 15 are underground. It combines elevated, at-grade, and underground sections and uses 1,676 mm broad-gauge rolling stock. This underground railway is the first in India, with operations commencing in October 1984. Being the country's first and fully indigenous effort, the construction of the Kolkata Metro was more of a trial-and-error affair, and as a result, it took nearly 23 years to construct a 17 km underground railway. In addition to the reason for the delay as mentioned, there are other reasons too, as given below,

The extensions of Line 1 in the south direction from Tollygunge to New Garia (5.85km), now in operation, and in the north from Dumdum to Dakshineswar (5.2 km) are complete. Construction of this stretch started in February 2010 and will connect the metro line from Baranagar to Barrackpur (12.5 km). Noapara Station is the largest station on line 1 of the Kolkata Metro. It has a provision to connect with line 4. Line 4 starts from Noapara and continues until Barasat via Kolkata Airport. The station includes four platforms, 182 meters long. It serves the local areas of Noapara, Tobin Road, Banhoogly, and Barasat. Unlike other metro stations, its two-story building has six escalators, four lifts, and 31 CCTV cameras. Fig 4.5 shows a view of the station. The station abuts the Noapara Depot, spread over 147 acres of land, and is one of the biggest Depots in the country. The depot can hold 40 rakes at a time. Newly received rakes for East-West are also tested here. The depot is connected to the Chord line to have direct access from the existing Railway system. **Fig. 4.4** shows a view of the Noapara Depot and final alignment of line 1 is included in **Fig. 4.3.**

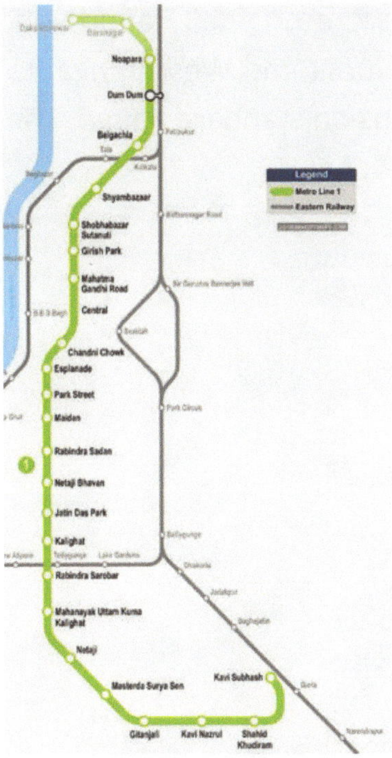

Fig. 4.3: Line 1 is Under Operation

Fig. 4.4: Noapara Depot

Fig. 4.5: First elevated Metro Station at Dumdum (Line 1)

4.3 Line 2 East-West - green line

East-West Metro, 16.55 Km, an Rs. 48,745.8 million project to connect Kolkata with Howrah by an underwater metro (first time in India) line, is under operation, and the details are given later in this volume. It was necessary to have Metro connecting Howrah, BBD Bag, Esplanade, Sealdah, Salt Lake, and the Airport.

Kolkata East-West Metro line forms the second of the three lines that were originally approved. The project is part of a comprehensive initiative by the governments of India and West Bengal to modernise and expand the transport infrastructure in Kolkata. The line runs on standard gauge. The alignment is included in **Fig. 4.6**, and proposed extension included in **Fig. 4.7**.

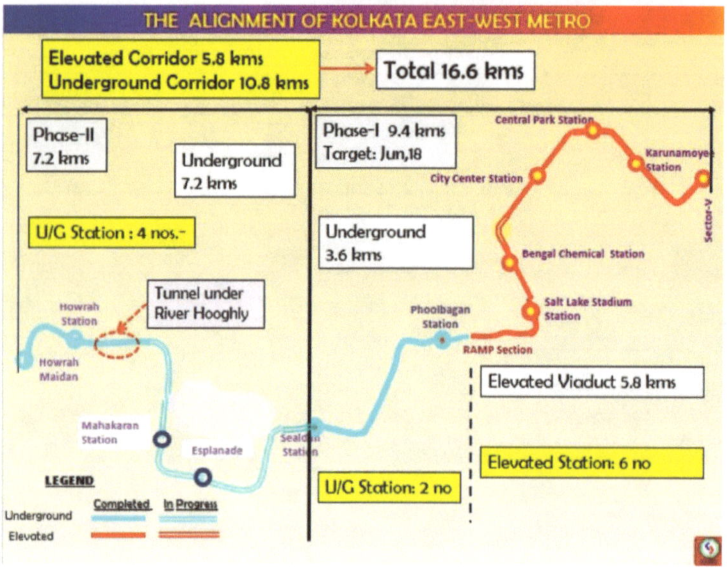

Fig. 4.6: East-West Metro Alignment

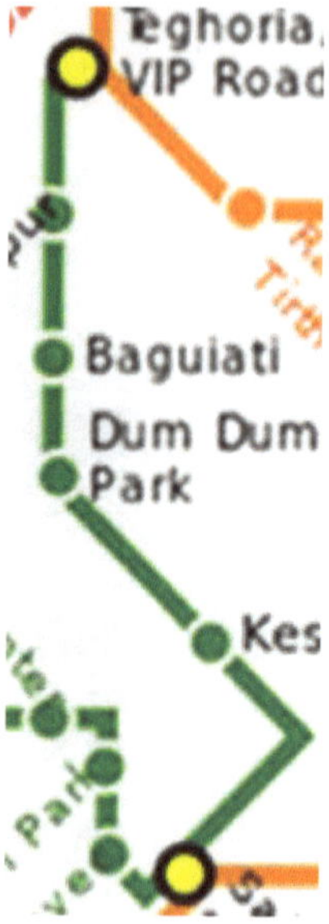

Fig. 4.7: E-W Metro Extension

Technological innovations and challenges are the hallmarks of river crossings in and around Kolkata. The dual tunnel, with an outer diameter of 5.6 meters (inner - 5.2 m), was constructed at a depth of about 32 m from the river surface and 13 m (top tunnel tube) below the riverbed, connecting both sides of the river. The constructed section is included in **Fig. 4.6** and the planned extension in **Fig. 4.7. Figure 4.8** includes a photograph of the Hon'ble PMof India during his travel by East-West Metro.

The first metro across a major river in India.

Fig. 4.8: Hon'ble PM- Riding Kolkata Metro

4.4 Line 3 Joka-Esplanade - purple Line

Metro Line 3, or the Purple Line, 15.08 km long, starts from Joka and terminates at Esplanade. The esplanade is the transfer station and connects three lines, Lines 1, 2 &3. The stretch between Joka and Taratala, a 6.5 Km section, was put into operation in December 2022 and extended up to Majerhat by March 2024 and put to operation.

The section between Joka and Majerhat is elevated, and the rest between Majerhat and Esplanade is underground. At present, Twin Tunnels is implementing the underground section. In the Master plan, the corridor was planned to terminate at Dalhousie, but later, it was decided to terminate at Esplanade. The Hon'ble Prime Minister of India inaugurated the metro corridor. The corridor is planned to extend to IIM, Joka and Dimond Park. A separate Depot is constructed at Joka.

In summary, the Joka-Majerhat section is under operation, the Majerhat-Esplanade Section is under construction, and the Joka-Diamond Park section has been approved. Line 3's total length is 18.47 km. A separate Depot is being constructed at Joka. **Figure 4.9**: includes Line 3, Elevated Portion.

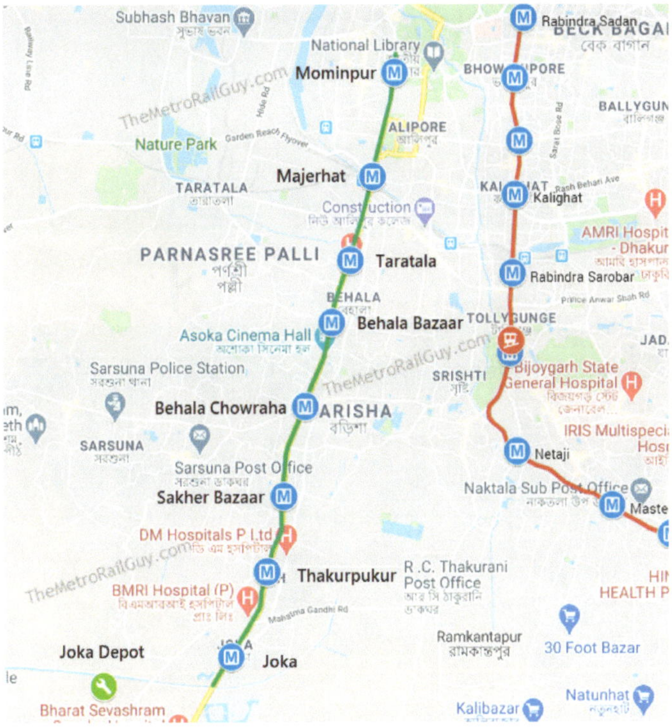

Fig. 4.9: Line 3, Section, under operation (Underground Section to Esplanade not included)

Fig. 4.10: Line 5; Baranagar-Barrackpore

4.5 Line 4 Noapara- Barasat - yellow line

The Kolkata Metro Line 4 (Yellow Line), 18.5 km long, starts from Noapara Station and terminates at Barasat. Line 4 is primarily underground and partly elevated at Jessore Road. This Line integrates with circular rail at Dumdum and proceeds to the International Airport, which is 6.249 km long and extends to Barasat. The project is being executed by Metro Railway, i.e., the Indian Railways. The project from Noapara to Barasat via Netaji Subhash Chandra Bose International Airport has also been sanctioned, and construction work has been completed. It connects Jessore Road and Jaihind station at grade with the station at the Airport and continues underground until Barasat for 16.9 km.

As the Airport Authority of India objected, the corridor will have a grade station and underground stations at Jessore Road and Airport, respectively, and will be extended to New Barrackpore. RVNL is executing the corridor in three phases: i) Noapara to Airport ii) Airport to New Barrackpore and iii) New Barrackpore to Barasat. Phase i) is under construction and in an advanced stage. Phase ii) The section was planned originally to operate on an elevated track, but due to objection from the Airport Authority, the section was changed to underground. Phase iii) was also initially planned to be elevated. But the problem of land accusation in this built-up area changed to underground and will be taken up later.

Noapara-Barasat alignment is shown in **Figure-4.11.**

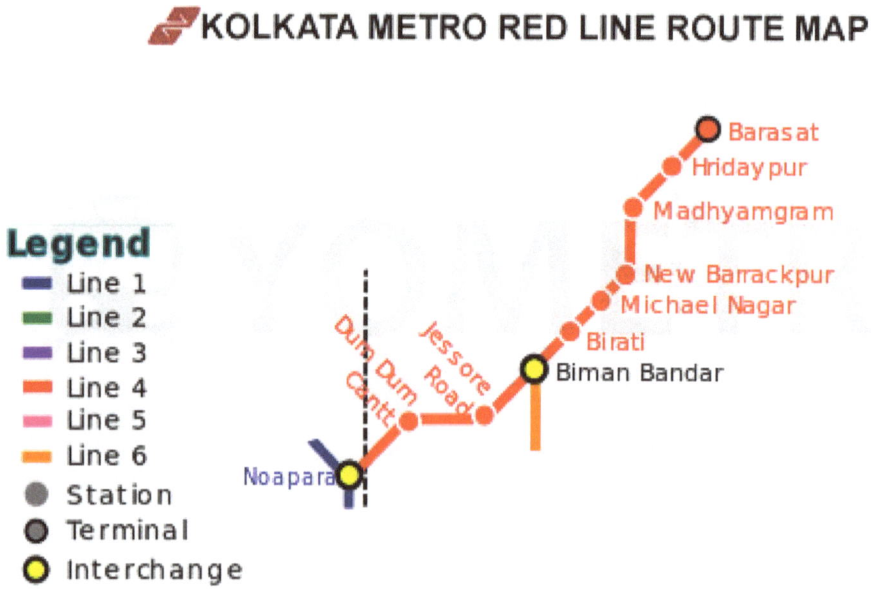

Fig. 4.11: Line 4. Noapara-Barasat

4.6 Line 5 Baranagar-Barrackpur - pink Line

Line 5 is an extension from Baranagar to Barrackpore **(Fig. 4.10),**12.40 km long, being executed by RVNL. The corridor was planned through Barrackpore Trunk Road (BT), and geotechnical investigations started. The project was approved during 2010-11. However, due to the existing large diameter main Water line along this road, rerouting has been planned through Kalyani Expressway. Work will be resumed after the utility pipes are relocated. The route is proposed to be extended to Kalyani, a pre-planned large Residential area & to locate several Industrial sectors. Line 5, or Pink Line, is an elevated 12.45 km corridor starting from Barasat. at the Airport. There will be 11 metro stations in this corridor.

4.7 Line 6 New Garia – Airport – orange line

Kolkata Metro Line 6 or Orange Line, 32 Km long, starts from New Garia and terminates at Kolkata Airport via Salt Lake City and Newtown. This corridor has planned with three Depot. One is at New Garia, the second one is at Newtown and is under construction, and the third one is at the Airport and is constructed. This metro corridor connects Line 1 at New Garia. The Orange Line is primarily elevated and planned for 23 elevated stations. One station on the grade and the airport station are underground. Airport Station is an interchange with the Noapara station link to Barasat. This station will have a yard and the largest underground facility with a 550m X 41.6m stabling yard, including a reversal facility in the country. The stretch between New Garia and Ruby Hospital (Hemanta Mukhopadhyay) and extended to Beleghata is in operation and inaugurated by the Hon'ble Prime Minister of India in March 2024. The stretch between Rubi and Beliaghata is ready and to started operation in 2025, Inaugurated by the Hon'ble PM. The essential structural issue in this corridor is a 76m span over Tagore Park culvert. This viaduct span was designed as a single-span open-web girder steel bridge. Line 6 will have an interchange at New Garia (Kavi Subhash) and Suburban Rail (Sealdah South Section). Work on the 32-km metro connector between New Garia and the Airport via E M Bypass, Salt Lake and Rajarhat are in an advanced stage of construction. It will help reduce travel time between the southern fringes of Kolkata and the Netaji Subhash Chandra Bose International Airport. The 32-km link between Kavi Subhas (New Garia) and the airport will have 24 stations, with the terminal Airport station underground, included in **Fig. 4.12.**

Fig. 4.12: Line 6-Kavi Subhas-Airport

The total anticipated length (as per the Master Plan included in Fig. 4.13) of the metro network in Kolkata will be about 130 km, excluding the approved extensions and 14 future corridors.

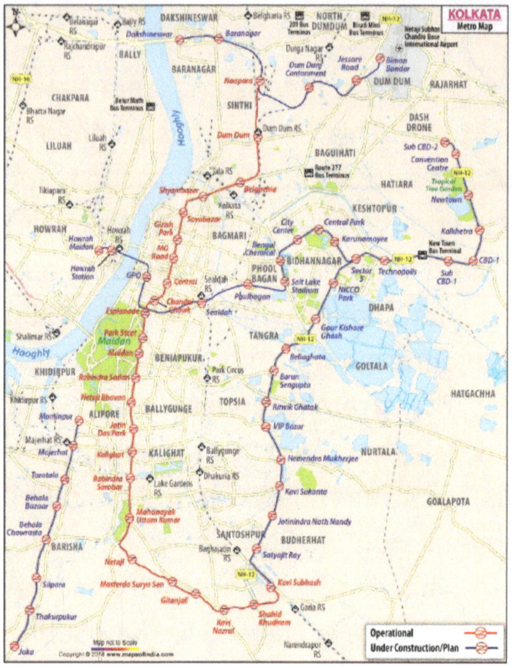

Fig. 4.13: Master Plan of Kolkata Metro without future corridors

PART 2

CHAPTER 5

North-South Metro

5.1 Introduction

Indian underground metros are evolving to adopt various designs and construction techniques based on site conditions, project types, materials, construction equipment, and engineering solutions. Complexity and land availability are crucial for the design and construction methodology.

A rapidly developing megacity like Kolkata has an extensive urban conglomeration surrounding its core, housing millions of people. Working on a transportation solution for a city that is more than 300 years old is a daunting task. Several metro corridors are planned in the city. This North-South Metro, Line 1, designated as the Blue Line, is the first metro in India. East-West Metro is the second metro corridor in a Master plan of six approved metro corridors.

Kolkata developed along the Hooghly River. After 1947, transport facilities needed to be upgraded faster to meet the rising demands. Two major rail terminals, Howrah on the west of the River Hooghly and Sealdah on the east **(Fig. 5.1),** were designed to serve millions of suburban passengers daily.

Kolkata, a major metropolitan city of India, has a population of more than 20 million living in the Kolkata Metropolitan Area (KMA in 2024), which covers an area of more than 1876 sq. km and has a population density of 24000 persons per sq. km, which is one of the highest in the world. The rapid population growth in and around Kolkata has made it increasingly difficult to maintain the minimum acceptable level of urban services, and most of them show visible signs of deterioration. Transportation services are one of the worst affected.

Fig. 5.1: Howrah and Sealdah Terminals

Subsequent efforts to solve the problem by augmenting public transport vehicles barely touched the fringes of the problem, as the roads accounted for only 6.2% of the surface area. This needed to be more grossly sufficient compared to 30% in Delhi and up to 25% in other cities. The average weekly passenger volume increased by 100% from 1960 to 1970, and in the early twenty-first century, rose by more than 190% in bus travel, while tram travel remained stationary due to capacity saturation. Meanwhile, the number of suburban rail passengers for Sealdah and Howrah stations increased by 323% between 1960 and 1970 and continued to rise rapidly. Kolkata tops India with 2448 cars per Km.

This automatically means moving people and goods across an expanding landscape. And, as it expands, the things to be crossed loom ever more significant. Metro lines are no exception. In a congested city like Kolkata, the crossings may be short but are complex. Kolkata is a linear city that developed in the north-south direction. But since the sixties, Kolkata has been spreading in all directions. Consequently, the East-West travel patterns are becoming more intense. At the same time, rail-based systems are under different planning and implementation stages, besides other developmental initiatives, particularly in the road sector, on both banks of the River Hooghly. As cross-river traffic increases continuously, water transport is taken care of.

In 1969, the Metropolitan Transport Project (MTP) was established under the Ministry of Railways (MOR) to explore alternative transport solutions. With the assistance of Soviet specialists and East German engineers, the MTP prepared a master plan for Kolkata. The responsibility for operating the Kolkata Metro network was subsequently handed over to Metro Railway under the MOR, which also funded the construction of the North-South Metro from its budgetary provisions.

For India's first metro rail, Indian engineers worked on ridership projections, engineering, construction, and project management. They opted for the broad-gauge railway, a familiar and well-understood technology, relying entirely on domestic resources rather than imports. This decision eliminated the need to design new bogies or import them, saving rigorous validation protocols and extended durability testing. India has a long history of manufacturing coaches for broad, meter, and narrow gauges, with coaches from ICF, Perambur, largely meeting the requirements, aside from a few areas, such as automatic door closure. The signaling system utilised the interlocking system, already established by Indian Railways.

The book includes insights into metro projects led by the author during his tenures at PH (Germany), Chatterjee & Polk, CES, and AECOM. Its content differs from a typical book format, encompassing nearly all disciplines required for metro project implementation. It includes a brief on the Detailed Project Report (DPR) for the Thane MRTS and the Andheri Ghatkopar project. For the North-South Metro, excerpts from tender documents and sample drawing calculations are provided, allowing readers to assess improvements in tendering procedures and design approaches compared to the initial metro projects.

During the preliminary design for the North-South Metro, Russian engineers recommended using sheet piles for deep excavation in cut-and-cover construction. The sheet piles available in India at the time were too short, and the available sections modulus was grossly inadequate for metro corridor excavations, necessitating the import of sheet piles more than 16m deep. For the first sections of the North-South Metro's Maidan Section, Chatterjee Polk Pvt. Ltd. (CPPL), led by the author and supported by Philipp Holzmann, Germany, submitted a design-and-build tender proposing concrete diaphragm walls instead of sheet piles. This tender with a design-build scheme was accepted by MTP (R), which had a strong implementation team that included SS Mukherjee, HD Bhowmik, E. Sreedharan, Mr. Phadke, AK Sen, Satyen Bhowmik, Satya Pal, Tathagata Roy, and KJ Singh. This team was actively involved in planning, design, construction, and proof-checking CPPL's design. The CPPL team also included Rodio Hazrat & Dalmia Construction (initial stage), KN Dadina, PFCC, and a specialist steel fabrication sub-contractor. Philipp Holzmann provided advisory support, with Prof. Simons visiting Kolkata and Mr. Martinich assisting during the initial DW construction, along with an imported clamshell bucket for DW trenching.

Another significant challenge was the absence of a legal framework for operating rail-based urban transport in India. For the Kolkata Metro, the existing Kolkata Tramway Act was modified and approved for use, although it did not cover operations beyond a single municipal corporation or municipality. Similarly, Mumbai had the Bombay Tramway Act. Delhi's government had to develop a new legal framework to facilitate operations within and beyond the city limits.

Kolkata East-West Metro project (Details included in Chapter 6) is an essential urban engineering project in the region that involves numerous challenges, including crossing the Ganges (Hooghly). Twin tunnels of East-West (E-W) Metro pass below the river, 520m wide and approximately 14m below the riverbed. Tunnels are 5.55m in diameter and are made of precast concrete segments. For the first time, precast concrete segments were introduced in the mid-seventies under the River Main in Germany, and the Author happened to be there with Philipp Holzmann, Germany did the design and execution of the Earth Pressure Balancing-Tunnel Boring Machine (EPB-TBM).

Kolkata's East-West Metro, which runs on standard gauge, all other corridors currently under construction or commissioning will operate on broad gauge with indigenously manufactured coaches. BEML has started manufacturing Metro Coaches and can supply Metro Coaches for metros under construction and in the planning stages. Titagarh Wagon Kolkata has started manufacturing Metro Coaches.

Assessing the condition of buildings/structures abutting and nearby areas is important for implementing Metros in urban areas. Lack of proper condition survey of abutting structures was one of the reasons in delaying of North-South Metro construction. For the East-West Metro, essential aspects of building condition surveys, including types and extent of damage, are discussed, outlining critical outputs for engineers to take appropriate measures during design and construction, and included in chapter 6.

5.2 Master Plan

Kolkata lies about 200 Km north of the Bay of Bengal. The city has grown along the banks of the Hooghly River in a north-south direction, with the neighbouring city of Howrah on the other riverbank. Kolkata is one of the major industrial and commercial cities in India. The tremendous population growth in the mid-nineties has put enormous pressure on the systems, including the transport sector. This intensifies the demand for transport systems, among other infrastructures. The government started working on how to meet the demand.

A master plan was prepared to include corridors of rapid transit lines, three north-south and two east-west, for a total route length, approximately of 137km by 2027. The comprehensive planning effort covered the future rapid transit system, considering possible growth in the city area, the city pattern, land use, and roads and railways. The Kolkata Metro was planned to serve the city and the Kolkata Metropolitan Region. It was initially planned in the 1920s, but construction started in the 1970s. The first stretch, the underground section from Bhawanipur to Esplanade, opened in 1984.

And this is the first metro corridor in India. Kolkata Metro Rail Corporation (KMRC) owns and operates under the Ministry of Railways (MOR). The MOR set up MTP and MTP(R). In turn, the Soviet Specialists, Lenmetroproekt, and German Engineers prepared a master plan. The master plan had five corridors, two of which were selected for construction in the initial phase. In 2011-2012, the Ministry of Railway announced the construction of five more corridors besides the North-South corridor. The master plan was upgraded as demand grew. The highest priority was given to the North-South Metro.

Line 1, North-South Metro, is 33.13 km long, including its extension of 16.58 km on both sides and has 26 stations. This corridor was designed as a combination of underground, elevated and at grade, with 15 underground stations. It was India's first underground metro railway, and operations commenced in October 1984. Being the country's first and fully Indigenous, the construction of the Kolkata Metro was more of a trial-and-error affair, and it took a long time to complete the 16.5 km, originally planned, underground N-S metro between Dumdum and Tollygunge.

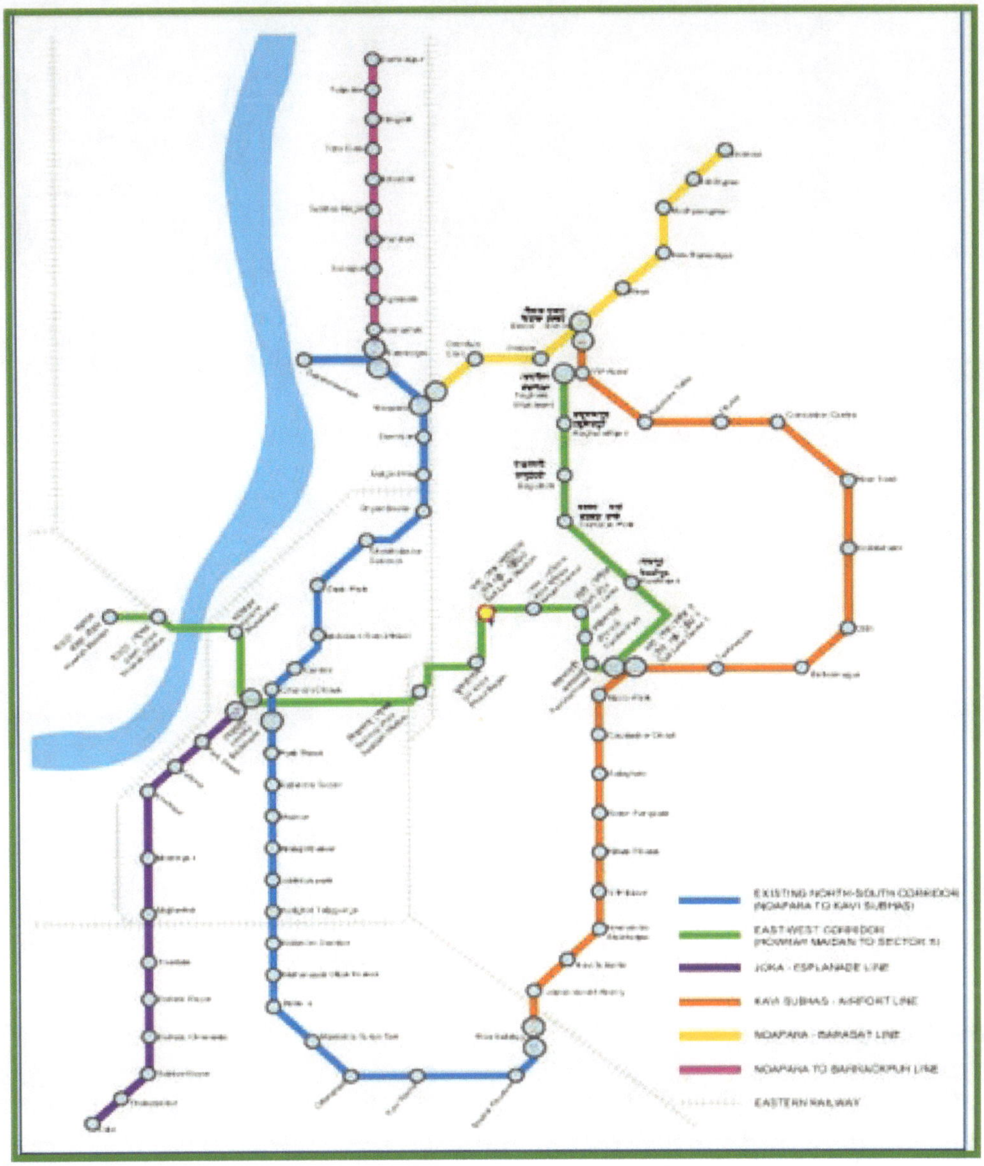

Fig. 5.2: Network Plan of Kolkata Metro

The master plan of the metro network was made in 1971 and included Lines 1 and 2. Line 2, designated East-West Metro, connects Salt Lake's office district with Kolkata and Howrah's twin cities, crossing the Hooghly River. The third Line between Joka and BBD Bag was later changed to Esplanade instead of BBD Bag. By 2011-2012, the Railway Ministry had announced plans to construct five lines, including the extension of line 1, which was originally planned between Dum Dum and Tollygunge. Line 1 was planned to extend from Dum Dum to Dakshineshwar in the north and Tollygunge to New Garia in the south. The Kolkata Metro network plan is included in **Fig. 5.2**, and the North-South Metro is between Dakshineswar - Dum Dum Tollygaunge, and New Garia.

The Kolkata Metro currently operates with four lines: Blue Line (Line 1), Green Line (Line 2), Purple Line (Purple Line (Line 3) and Orange Line (Line 4). These lines have a total of 50 operational stations with 29 stations under construction.

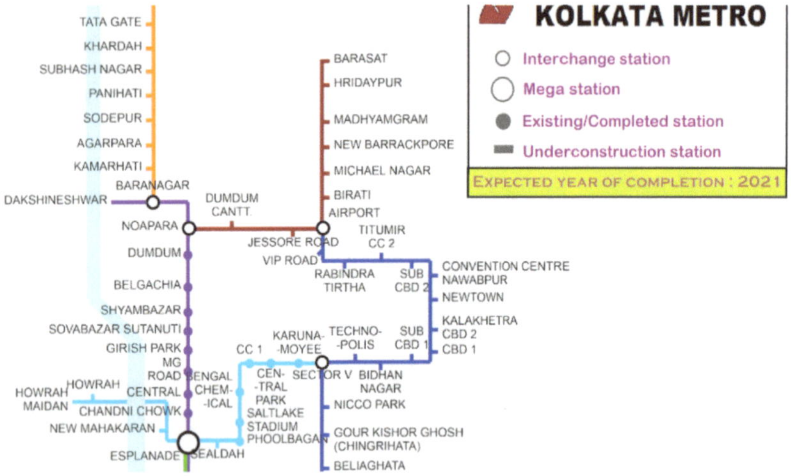

Fig. 5.3: Interchange Stations

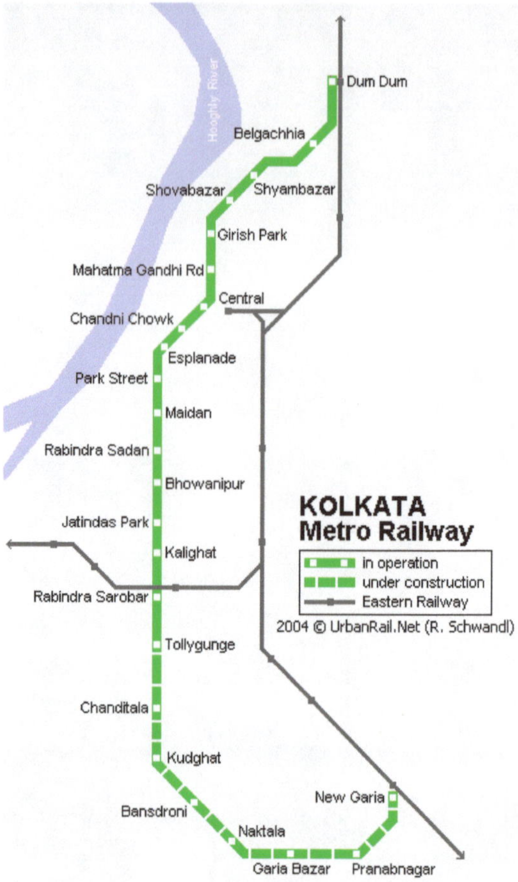

Fig. 5.4: Kolkata North-South Metro Route-Original

5.3 Brief on North-South Metro

The **North-South Metro was the first metro system developed and implemented in India.** Construction began in 1973, and the first section between Esplanade - Bhowanipur (3.4 km), the first Metro in India, was opened for public service in 1984. The line extended to the planned extent of 16.45 km between the two terminal stations, Dum Dum and Tollygunge in 1995. It extended beyond Tollygunge (Uttam Kumar Station) to Garia Bazaar (Kavi Nazrul Station) on 29th September 2009 and further to New Garia (Kavi Subhash Station) on 7th October 2010, covering an additional 8.5 km. **Figure 5.3** shows the interchange locations of N-S Metro, and the alignments included in **Fig. 5.4** were planned to penetrate the principal residential, industrial, and business centers and run along the existing roads to reduce land acquisition. The stations were strategically placed to capture/provide service to commuters and were consistent with station spacing. The line was primarily underground, and the diaphragm wall technique was used extensively in different combinations across various segments, like semi-integrated, integrated, etc. Indian engineers innovated to achieve economy and enhanced performance, incorporating local materials to reduce foreign exchange outgo.

The estimated original project (Dum Dum – Tollygunge) cost in 1971-72 was Rs 140 crores, which escalated to Rs 1707 crores in 1997-98. The foreign exchange from OECF funding involved 4800 million yen (Rs. 100/—to 2400 yen) in Japanese currency, which includes Consultancy and Personnel services besides equipment and materials. In addition, 55.5 million in French Francs, 2953 million Australian Dollars, and 7406 million Danish Kroners were involved. Dollars were utilised in importing airlocks and towards personnel service. **Fig. 5.5** shows an internal view of the Coach manufactured in Perambur, Chennai. **Figures 5.6 & 5.7** include views during operation in the early eighties, and **Figure 5.8** shows the Rake at Platform.

Fig. 5.5: Internal View of Line 1 Coach

Fig. 5.6: N-S Metro Station During Operation in PlatformIn the Eighties

Fig. 5.7: N-S Metro – Concourse and Stair to

Fig. 5.8: Metro Rake at Station Platform

5.4 Subsoil Along Metro Corridor

The Kolkata (Calcutta) Metropolitan Area (KMA) is in the low-lying areas of the deltaic region and spreads over both sides of the Hooghly River. Hooghly runs from north to south. Four iconic bridges connect the two banks. The terrain of both banks is flat with an average height of 6 m. The highest recorded temperature is 43.9 °C, and the lowest is 6.1 °C. Monsoon season is generally from May to October, and the highest humidity is 100%. The city has an ancient, combined sewer system that is inadequate to tackle heavy rain. India lies on the earthquake belt; however, Kolkata is in Zone 3 with a horizontal coefficient that varies. In general, the subsoil in the Kolkata area consists of weak cohesive soil strata in the upper layers, except for a thin surface crust, which appears firm due to desiccation. The upper strata consist of silty clay/clayey silt, particularly soft in the top 12-15 m, with decayed vegetation. The consistency improves in the deeper strata, and alternate layers of firm to stiff silty clay with laminations are found below 15 m depth. Thereafter after dense sand is found below 25 m depth. Below 32 m depth, the subsoil consists of stiff to hard mottled brown silty clay/ clayey silt, followed by dense sand. This is known as the Normal Calcutta deposit.

Cementation Co. and K N Dadina Ltd. conducted geological investigations along the alignment, including in-situ and laboratory tests. The investigations were carried out generally up to a depth of 30m and at 100m intervals. The borehole records reveal that the subsoil is soft in the upper layer except for the top crust, which is moderately stiff to firm, and consistency increases with depth, becoming stiff to hard clay. Kolkata soil is generally divided into two categories, 'Normal Kolkata Deposit' and 'River Channel Deposit'. The Kolkata deposit consists of clay and silt down to great depth, with laminations, lenses and layers of sand and silt from approximately -15 to 25m of depth. The design and details of the N-S metro were based on section-wise geological data. Typical soil data is shown in **Fig. 5.9** for a part of the Maidan between 0.04 & 0.06, and the vertical coefficient is 0.5 of the horizontal coefficients.

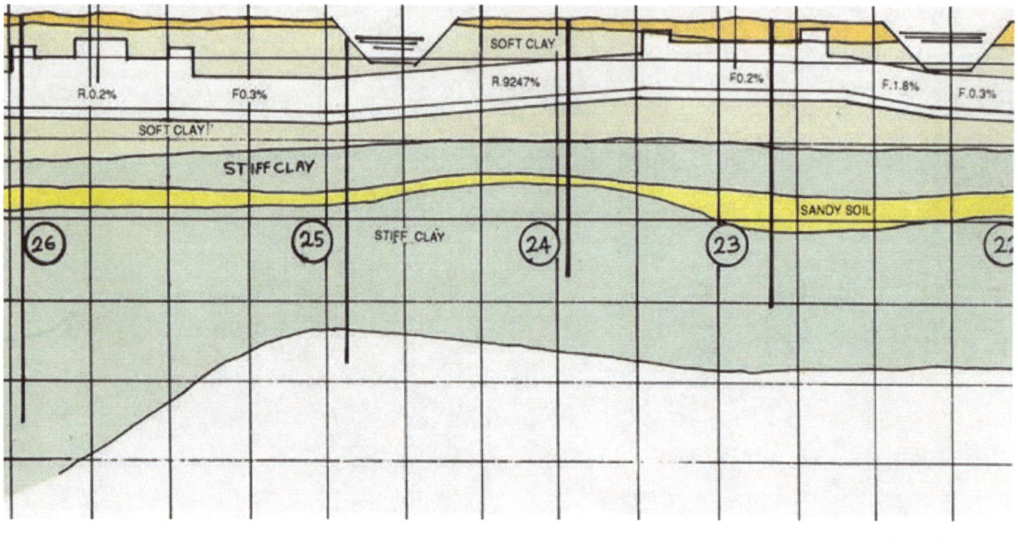

Fig. 5.9: Soil Profile at Maidan Area

The subsoil investigation at the station locations revealed six strata, as given below. In addition, where the subsoil profile indicates abrupt changes, further investigations were carried out at closer intervals.

- Stratum 1: Light brown or browning grey silty clay, clayey silt and sandy silt, including occasional silt fine sand (1.5 m)
- Stratum 2: Dark or grey silty clay with semi-decomposed timber pieces (1.75m).
- Stratum 3: Bluish grey silty clay and mottled brown or grey silty clay with kankar (10 m).
- Stratum 4: Brown or yellowish-brown sandy silt, silty fine sand with lenses or pockets of brown and grey silty clay (10m).

- Stratum 5: Mottled brown or grey clay, grey and brown silty clay, often with laminar character, all with rusty brown pockets of brown silt (12 m).
- Stratum 6: Brown or light brown silty fine to medium sand (15 m).

In addition to the above, the actual physical conditions of the required strata subjected to long exposures in the open excavation were determined, as were the stratum's permeability characteristics concerning the dewatering requirement during construction. The subsoil water level was approximately 1-1.5 m below ground level, and PH values varied between 6.7 m to 8.25. In most cases, water was alkaline, and Sulphur trioxide ranged between 5 and 125 ppm, averaging 30 ppm.

There were variations noted in a few places between the strata given in the Bore Hole data and the strata noted during actual excavation; a sample of such variation is included in **Figure 5.10.** However, the design was based on strata encountered during excavation and taking data from the report of similar strata.

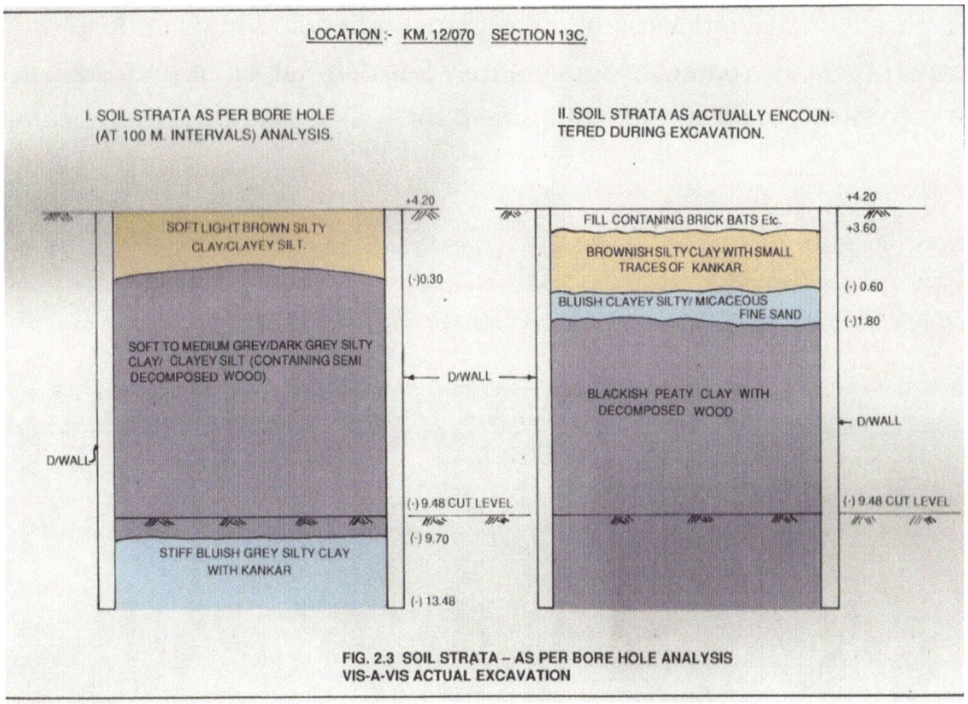

FIG. 2.3 SOIL STRATA – AS PER BORE HOLE ANALYSIS VIS-A-VIS ACTUAL EXCAVATION

Fig. 5.10: Soil Strata as per Bore Hole vis-vis Actual Excavation.

5.5 Section-Wise Alignment

For the system to be effective and efficient, the rapid transit system must be aligned so that it interpenetrates the principal residential, industrial, and business centers. The stations should be located adjacent to groups of buildings with the highest number of potential transit riders, consistent with station spacing. In addition, alignment must be chosen along the existing roads, requiring minimum land acquisitions. The alignment of routes and location of stations was critical in the CBD, where a close grouping of multi-storied buildings caused the heavy concentration of people and public facilities.

The extended section is from Tollygunge Depot to New Garia. The Shyambazar to Dumdum section also extended to Dakshineswar. The alternative alignments, studied for the N-S metro between Dum-Dum and Tollygunge were 1) the corridor along the Eastern Railway line, Belgachia Road, and R G Kar Road, 2) the corridor along Dum Dum Road and Barrackpore Trunk Road, and 3) the corridor along Dum Dum Road, Northern Avenue, Belgachia Road, and R G Kar Road. Alternatives 2) and 3) were discarded due to the difficulties of providing interchange at Dum Dum Junction station, while alternative one was adopted. Different sections of Alternative One, Noapara Car Shed-Dumdum-Dumdum-Shyambazar-Esplanade-Tollygunge, are briefed below.

i. Noapara Car Shed - Dumdum - Shyambazar Section

North-South Metro started at Dumdum, abutting the existing station of the Suburban Railway System, and was designed as an elevated station. Two lines connect the corridor to the Car Shed Complex at Noapara **(Fig. 5.11A).** The car shed starts at 1.240 km from the northern terminal station at Dumdum, later changed to an intermediate and a transfer station to suburban lines and the main line. The two lines cross the Eastern Railway embankment by constructing a twin tunnel underpass. The area of the Car Shed is between two suburban corridors, namely Sealdah-Ranaghat and Sealdah-Bongaon. The underpasses were constructed using the blade shield tunnelling method without compressed air.

Fig. 5.11A: Tracks Leading to Car Shed

Blade shield tunnelling" is a type of shield tunnelling where a rotating **cutting wheel** equipped with blades is used to excavate the tunnel face, and the machine's shield provides a protective shell around the excavation area. The process involves the blades breaking up the ground, a screw conveyor or muck ring transporting the excavated material away, and the tunnel lining, often concrete segments, being installed behind the shield to support the tunnel walls.

Fig. 5.11B: Temporary Car Shed at Dumdum with a stretch of Ballastless Track

Later, the metro corridor was extended in the direction of Dankuni to Dakshineshwar, the terminal station **(Fig. 5.12)** in the north, of Line 1. In addition, a new corridor emerged from Noapara, connecting the Airport through the Dumdum Cantonment and later continuing to Barasat (Yellow Line).

Fig. 5.12: Terminal Station at Dakshineswar after Extension

This section of the metro corridor started as an elevated section, from Dumdum a temporary terminal station, transfer station for two suburban rail corridors and the main line, with minimum hindrances to operation of the existing suburban corridors. On the south, the corridor runs elevated to a limited length **(Fig. 5.13).** It continues as an underground section, connecting through the ramp, partly elevated and partly underground, to connect to Belgachia station and continue underground to Shyambazar. The corridor, across the railway yard located in Chitpur; was constructed as a tunnel section using the Earth Pressure Balancing Tunnel Boring

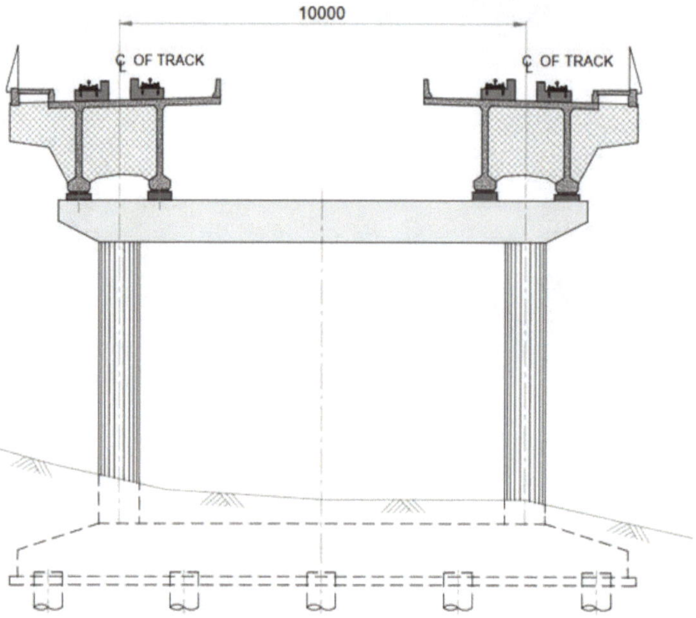

Fig. 5.13: Elevated Corridor Between Dumdum & Belgachaia

Machine (EPB-TBM). This section was constructed using precast RCC lining segments and with cast iron segments for a small part. Section of underpass constructed across the embankment, included in **Figs 5.14 A & B.** A temporary stabling was established **(Fig. 5.15)** ahead of Dumdum metro station for the operation during initial stage. The details are included later in the section.

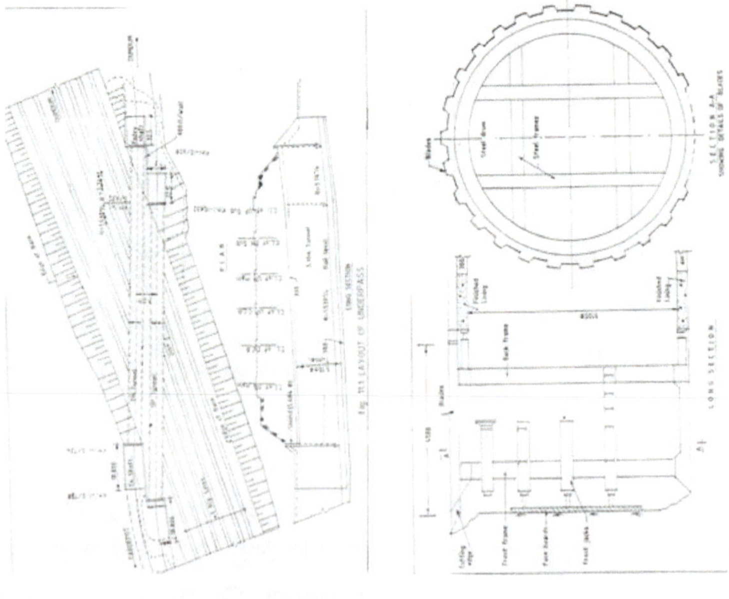

Fig. 5.14 A & B: Underpass Across Embankment- Plan and Section

Fig. 5.15: Metro Rake Towards Temporary Depot

ii. Shyambazar – Esplanade

Three alternative corridors, for this section were studied, which are 1) along Bidhan Sarani – College Street – Nirmal Chandra Street – Lenin Sarani; 2) along Bhupen Bose Avenue – Jatindra Mohan Avenue – Chittaranjan Avenue; and 3) along Bhupen Bose Avenue – Jatindra Mohan Avenue – B K Pal Avenue – Rabindra Sarani – Bentick Street. Alternatives 1) and 3) were discarded due to various factors like carriageway width, rehabilitation, land acquisitions, large-scale demolitions, unavailability of satisfactory soil for shallow tunnelling and cost considerations. In contrast, alternative two was adopted due to the availability of broader carriageways.

This section of the corridor is underground (UG), and both the stations and corridors were constructed using the cut-and-cover method, using Diaphragm Walls (DW) on both sides of the corridor, as an earth-retaining structure and the station boxes, connecting the corridor at both ends. Construction was done with in-situ reinforced concrete. Both the DW and RCC box walls were designed for total earth pressure. In this section, many old multi-storied structures had minor to significant effects on the Structural design and finishes. A Top view for surface level near the underground Shyambazar station is included in **Fig. 5.16.**

Fig. 5.16: Underground Metro Site below Shyambazar Crossing.

iii. Esplanade and Acharya Jagadish Bose Road

Two alternatives were studied in this section: 1) along Jawaharlal Nehru Road and 2) along Red Road, Khidirpur Road and Casuarina Avenue. Alternative 2 was discarded, and alternative 1 was favoured as existing roadway utilisation could be more than 90%. Travel speed on the road was as low as 15 to 20 kmph, so capturing ridership on the metro would be easier. The construction of UG sections except in Maidan Area and at grade near Tollygunge Station was done using Cut-and-Cover, like that in section (ii).

iv. Acharya Jagadish Bose Road and Hazra Road

Three alternatives were considered along the existing roads, which are 1) Sarat Bose Road, 2) A approach Ashutosh Mukherjee Road – Shyama Prasad Mukherjee Road, and 3) Harish Mukherjee Road. Alternatives 1 and 3 were discarded as the system would run significantly away from the city's central and more heavily populated parts. Alternative 2 was favoured since it was more central. The roadways were utilised for building UG corridors, between 70% and 90%, and travel speeds were between 20 mph and 24 mph, making capturing ridership on the metro easier.

v. Hazra Road and Tollygunge tram depot

The favoured alignment was along Shyama Prasad Mukherjee Road—Deshapran Sasmal Road as the existing roadwaycould be utilised between 70% and 90%, and travel Speeds were between 12 mph and 15 mph. Ridership on this section was expected to be more. This section is partly UG and partly at Grade. Tollugunge Station is at grade

vi. Baranagar to Dakshineswar Metro Extension.

For the section Rail Vikas Nigam Limited (RVNL) awarded the detail design to M/s **Ayesa** Ingenieria Arquitectura S.A.U. The same job had previously been awarded to M/s Mott MacDonald, and appointed M/s CCCL & amp; SAM JV, as contractor for 1.574 km, and 2 elevated stations.

The 4.1 km extension is linked with northern suburbs of Kolkata. Detailed Design involved was for a viaduct of 2.343 km. with PSC girders, Embankment for 1.626 Km at-grade, and Stations for the extension from Noapara (excluding) to Baranagar – Dakshineswar.

As such the section was awarded for detail design and drawings for Stations, Viaduct. train stoppage buffer end/reversal/operation facility, connection to Barrackpore line, connection to Noapara depot, tracks on at-grade/surface/embankment formation. Railway and Highway Crossings, Expressway diversions, slip road including Dunlop Bridge and allied facilities for passenger movement to & amp, from stations. Existing metro network, significantly improving public transport for thousands of commuters.

vii. Important Features of the Section

- **CMRS Inspection:** February 05, 2021.
- **Key Stations:** Baranagar and Dakshineswar (elevated).

- **Alignment:** Primarily elevated, constructed on viaduct.
- **Original Contract:** INR 1,59,48,800.
- **Revised Contract:** INR 2,09,66,054.
- **Original Project Duration:** 32 months (up to 26/04/2017).
- **Construction & amp:** Delivery
- **Employer:** Rail Vikas Nigam Limited (RVNL).
- **Construction Contractor:** M/s CCCL – SAM India (JV).
- **Detailed Designer (DDC):** M/s Ayesa Ingenieria y Arquitectura S.A.U.
- **Challenges:** Land acquisition delays, and complexities of alignment near Belgharia

viii. Key Benefits & Impact

- **Enhanced Connectivity:** Provided seamless metro access to Baranagar, Sinthee, Dakshineswar, and adjoining areas.
- **Decongestion:** Reduced road traffic pressure on Barrackpore Trunk (B.T.) Road.
- **Improved Accessibility:** Significantly improved access to Dakshineswar Kali Temple, boosting religious tourism.
- **Future Integration**: Baranagar station designed as an interchange with the upcoming Pink Line extension to Barrackpore, strengthening long-term connectivity.

5.6 Planning Approach

The MRTS finally adopted for Kolkata is a double-track system with steel wheels on steel rails (with 8-car trains), mostly underground. **Fig. 5.17** shows the train at underground station, at Shyambazar. Station locations have been chosen at recognised and important road junctions or near essential places of public interest, and commercial & population density in the area, following the existing tram and bus stops.

The estimated average weekday passenger volume at the maximum load point on the chosen alignment was 0.826, 1.002, and 1.116 million, respectively, for 1978, 1983 and 1990. For the maximum peak hour, sectional volume on an average weekday was 51,000, 60,000 and 67,000, respectively. It means, 8-car trains run every 2 minutes. The headway to be accommodated on a 170-metre-long station platform was designed to cope with the traffic loads. Manual driving of trains was adopted, with the provision of automatic speed and stop controls in the future. Tunnel and station design parameters were accordingly fixed. While the coaches' dimensions were determined to the running tunnel's dimensions, station's size depended upon the projected passenger volume to be handled. In addition, average speed, station-stop time, comfort considerations, etc., were studied for the final system configuration.

Fig. 5.17: UG Shyambazar Station

5.7 Traffic Study and Planning

Traffic and characteristics are the most critical components for planning and designing RTS. A comprehensive traffic study of traffic & travel mode was carried out. From the collected data, the traffic volume and pattern that the RTS may attract. The redistribution over other modes, which would occur with the introduction of RTS. In the mid-sixties, CMPO (Calcutta Metropolitan Planning Organisation) conducted an exhaustive traffic study and prepared a comprehensive report, which included the city's mobility.

According to the Origin and Destination (OD) Survey conducted by the CMPO in 1964, there was a concentration of transit passenger trips in the east-west corridor between Howrah and Sealdah, the principal rail terminals and the north-south corridor along the Chittaranjan Avenue-Jawaharlal Nehru Road-Asutosh Mukherjee Road alignments. The projections revealed that the movements in these corridors would remain dominant despite the dispersal of transit movements on account of growth in fringe areas of the Kolkata-Howrah complex, including the Salt Lake reclamation area, Behala and others.

The surveys conducted by CMPO showed 127 sq.km. The area of Kolkata and Howrah contained a population of 3.4 million and increased to 3.7 million in 1971. Results of the study are included in **Tables 5.1, 5.2, 5.3 & 5.4.**

Table 5.1: Year-wise Population Surveyed and Projected

Year	Population in study area (millions)	Average week day volume of transit trips (millions)
1971	6.4	4.6
1976	7.4	5.0
1983	9.3	6.1
1990	11.6	7.5

Table 5.2: Year-wise Trips Assigned

Year	Study area population (millions)	Average week day volume of transit trips (millions)	Trips assigned to RTS north-south (millions)
1976	7.4	5.0	1.4
1983	9.3	6.1	1.8
1990	11.6	7.5	1.94

Table 5.3: Year-wise Transit Trips & Assigned Trips in N-S Corridor

Year	Study area population (millions)	Average week day volume of transit trips (millions)	Trips assigned to RTS north-south (millions)
1976	7.4	5.0	1.4
1983	9.3	6.1	1.8
1990	11.6	7.5	1.94

Table 5.4: Estimated Station-wise Passengers

	Stations	Predicted station entry in one day No. of passengers in 1990
1.	Dum Dum	1,23,000
2.	Belgachhia	39,000
3.	Shyambazar	1,43,000
4.	Shovabazar	36,000
5.	Girish Park	40,000
6.	Mahatma Gandhi Road	1,02,000
7.	Central	3,74,000
8.	Chandni Chowk	35,000
9.	Esplanade	1,23,000
10.	Park Street	51,000
11.	Maidan	30,000
12.	Rabindra Sadan	44,000
13.	Bhowanipur	62,000
14.	Jatin Das Park	83,000
15.	Kalighat	1,71,000
16.	Rabindra Sarobar	73,000
17.	Tollygunge	1,96,000

Fig. 5.18: Kolkata Traffic During the Seventies

Kolkata has two principal traffic corridors—one along the North-South and the other along the East-West. The passenger flow is tidal, directed towards the Central Business District (CBD) from both the northern and southern reaches of the metropolis in the morning and reversed in the evening. Traffic volume per 1966 projection was 85,000 at CBD and outbound passengers—77,000 per hour. The maximum recorded transit passenger flow was 3,14,172 at Howrah. A daily total of 14,55,562 transit passengers was recorded, which included 7,70,891 by trams and 6,84,671 by buses. Kolkata traffic and congestion during the seventies are shown in **Figs 5.18 & 5.19.**

As per the Origin-Destination (OD) survey by CMPO, there was a concentration of Passenger trips on the corridor connecting two major Rail Terminals, Sealdah and Howrah, in the east-west direction and along the Chittaranjan Avenue—Jawaharlal Nehru Road—Asutosh Mukherjee Road corridor, in the north-south direction. **Figure 5.20** shows the traffic desire mode for the KMA and indicates the two principal directions of movement: North-South and East-West. The number of passengers increased by 100% between 1960 & 1970. Bus passengers increased by 170%, and most passengers travel by Public Transport. The trend of using public transport has continued to this day and will continue. With road area available in Kolkata at 6.2%, and with the population growth rate, Kolkata needs a high-capacity mass transit system that can transport large volumes of passengers from one section to another.

Therefore, a high-capacity transit system was required as early as the sixties. Accordingly, one of the two busiest corridors, the North-South line of Kolkata Metro between Dum Dum and Tollygunge, was taken up and designed and constructed for a high traffic volume, according to traffic figures obtained from traffic surveys. It was designed for an initial traffic of 1.2 million passengers per day at the start of the system and a final figure of 1.7 million when the network of 3 lines (N-S) was planned to be completed for the Metro.

Fig. 5.19: More Views of Kolkata Traffic Scenario

The increase in traffic between 1960 and 1970 was very high. The daily traffic of passenger trips increased from 1960 to 1970 by 100%. The rise in Bus passengers was 193%, and in suburban traffic, including both Sealdah and Howrah, it was 323%.

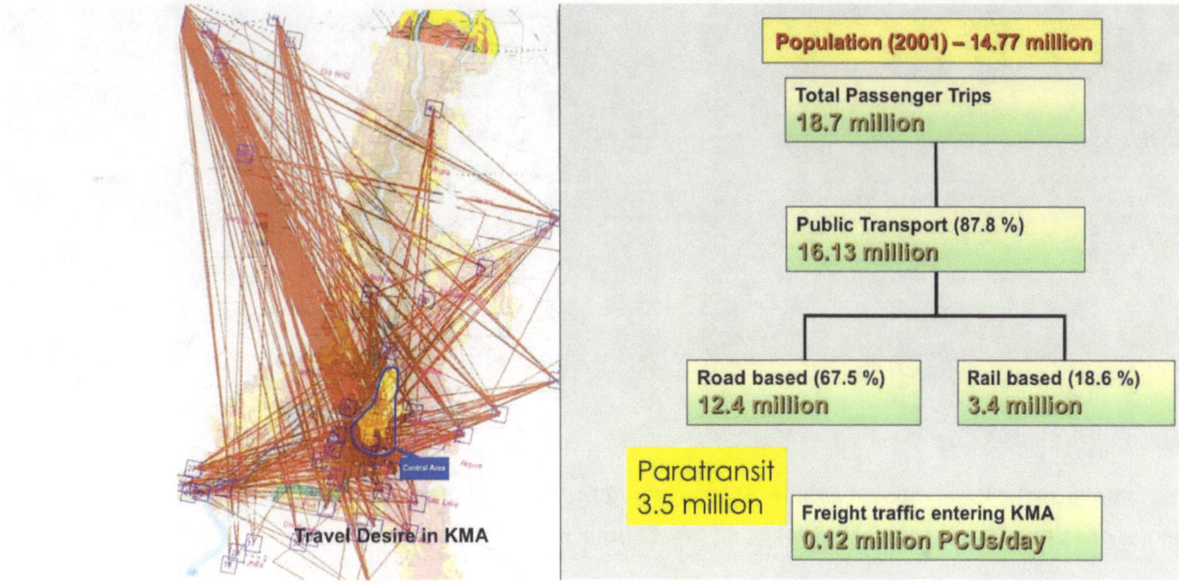

Fig. 5.20: Travel Desire

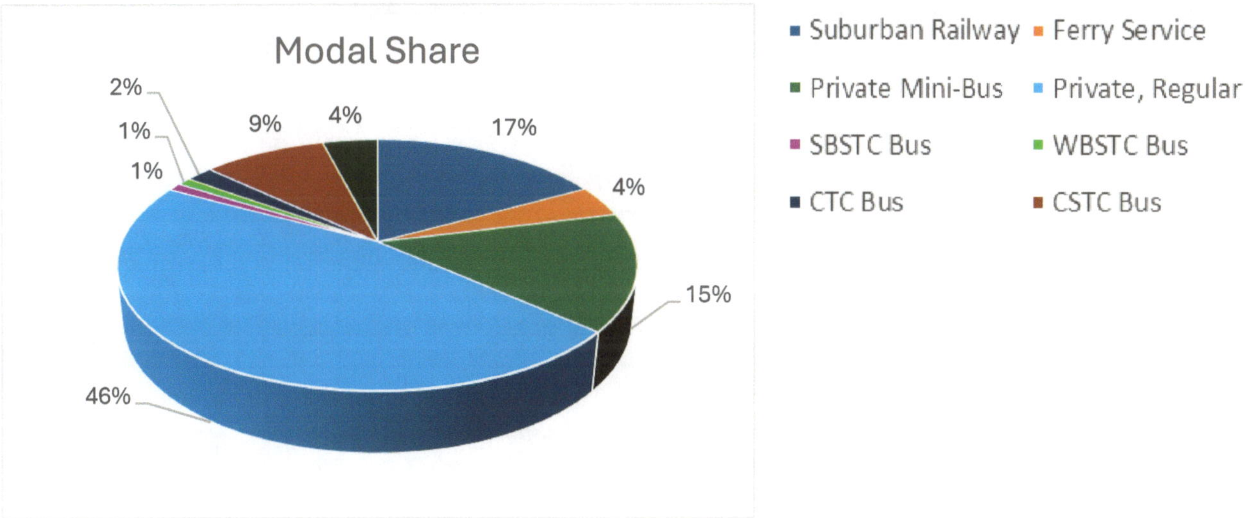

Fig. 5.21: Travel Mode (1970)

Modal Share, included in **Fig. 5.21**, shows the share distribution between different modes of traffic during the 1970s.

The estimated average weekday transit trips in 1971 were 4.6 million, projected to exceed 5.0 million, 6.7 million and 8.3 million by 1978, 1983 and 1990, respectively. The average weekday transit trips assigned to Dum Dum – Tollygunge Rapid Transit System were about 1.3, 1.5 and 1.7 million for 1978, 1983 and 1990, respectively. The projected trip for 2007 is 1.85 million, and the average trip length is 6.8 km. This high traffic volume is needed due to the consideration of station design for providing facilities for handling the high passenger volume during the eighties, **Fig. 5.22. Figure 22A includes Metro**

Fig. 5.22: Traffic conditions during the eighties at a Metro Station (Immediately after the opening of the Blue Line in 1984)

Fig. 22A: Metro during Operation through Tunnel in Chitpur Yard

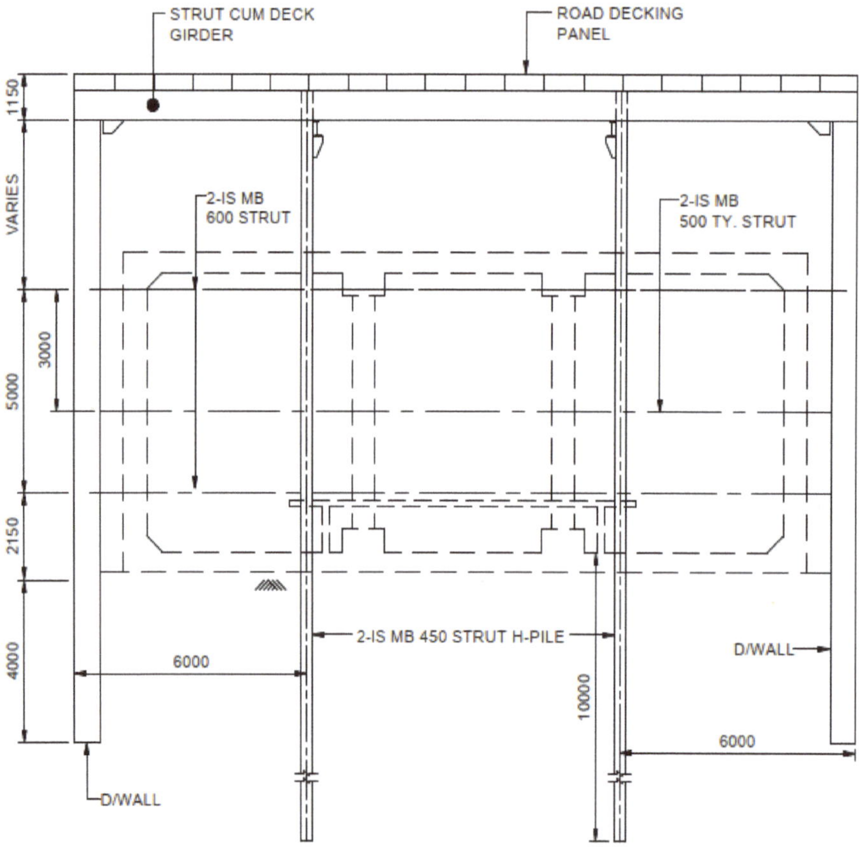

Fig. 5.23: Temporary Deck- Width to Accommodate Tram & Road

5.8 Traffic Diversion

The guidelines adopted at construction sites involve public roads and sufficient carriageway width, preferably a two-lane carriageway, one for each direction, which should be available for the interference-free movement of existing traffic. There should also be separate spaces for moving construction vehicles and equipment. At a few locations where the carriageway width could not be worked out even after including part of a footpath, the carriageways were provided, along with temporary Road Decks **(Fig. 5.23),** were to be built parallel to/over the excavated section. The tram tracks had a part footpath, carriageways, and a temporary road. More on traffic diversion and temporary traffic decking is included in the latter part of the book.

5.9 Utility Diversion

Utility diversion was one of the most critical and complex operations for cut-and-cover construction. It was undertaken before the operation started for Diaphragm walls, Sheet piles, H-piles, and timber lagging take-up before excavation in the case of open excavation. Utilities included Trunk Sewers, the Main Sewer and Drainage system, Water Mains and distribution, Electrical power cables, Telephone cables, and Gas Mains and distribution of pipelines. The general layout of the trunk and main water mains was available with the Calcutta (Kolkata) Corporation. But there were numerous other subsidiary lines, crossings, etc., which were uncharted and encountered during the excavation. The difficulty of relocation of utilities was more pronounced in the matter of water. The first step for handling utilities mapping of existing ones was collected from the Municipal Corporation as available; the missing ones were identified (uncharted), and the existing utilities were made complete, to the extent possible.

In the alignment between Esplanade and Maidan, where the worksites lay in the open area, utilities were mostly out of the way. A few power cables of the Calcutta Tram Company (CTC) were encountered in the Esplanade Tram Terminal area. One of the transformers of Calcutta Electric Supply Corporation (CESC) in this area had to relocate.

While working on diversion plans for utilities, all future schemes were collected and integrated with those for existing utilities for the affected sections on the RTS corridor. The critical issues and difficulties were related to the relocation of the water line distribution, as numerous uncharted subsidiary lines and crossings were encountered during work on the RTS corridor. The diversion plan had to get approval from the concerned authorities before the commencement of site activities. **Figures 5.24, 5.25, 5.26 & 5.27** include photographs of existing utilities and their support

Fig. 5.24: Complex Tangle of Utilities on N-S Corridor

Fig. 5.25: Water Mains Supported on Struts

Fig. 5.26: Hanging of Cross Utilities

Fig. 5.27: Existing Electrical Lines on Timber Frame

Figure 5.28 includes the existing Water Main supported by a hanger made of a structural steel frame hanging from the slab cast below the level of the water main and ground level, **Figure 5.29** shows the enlarged detail of the main support and **Figure 5.30** includes support from DW.

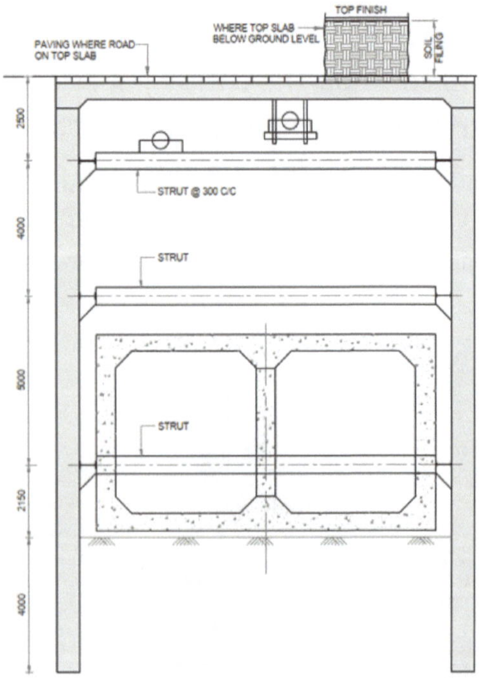

Fig. 5.28: Existing Water Main Accommodated on Frameworks During Excavation

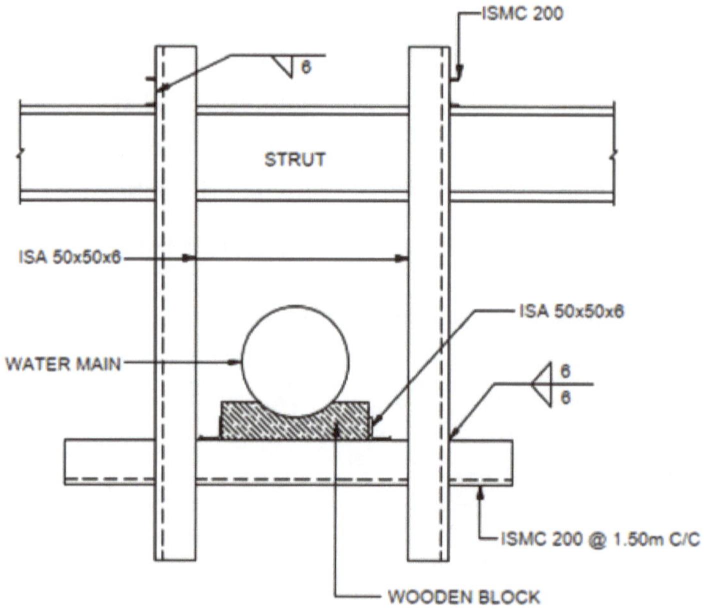

Fig. 5.29: Detail of Hanging Support (Ref.Fig. 5.28)

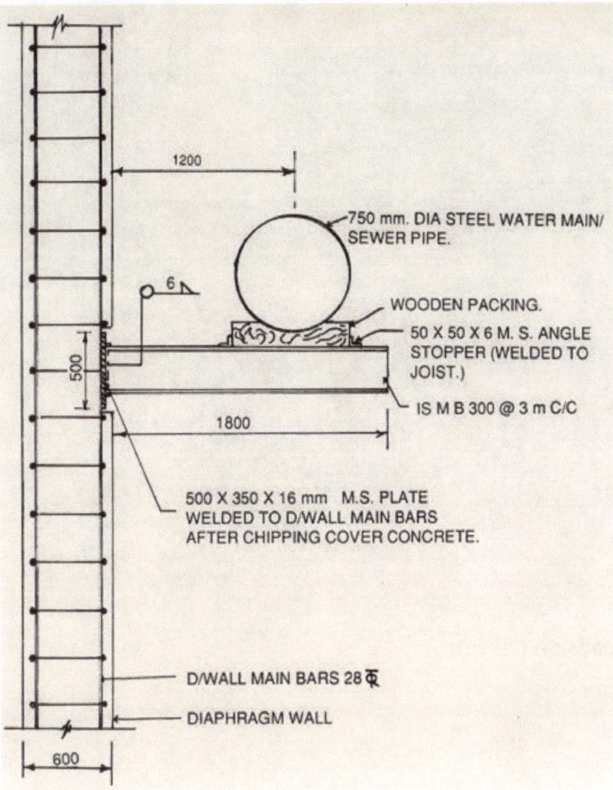

Fig. 5.30: Enlarged Detail of Water Main Supported from DW.

5.10 Land acquisition

Most of the North-South Metro corridors are underground and traverse through the City roads, the Maidan area, and land acquisition was not a critical issue. In addition, 100+ acres of government land were available along the RTS corridor, and the corridor at Dumdum end traverses through Railway land. Private properties, including Tollygunge Club, require about 200 acres. Land acquisition was primarily needed for Substations, Entry-Exit Structures, and Ventilation shafts. However, covered porticos **(Fig. 5.31)** over the footpath had to be demolished in many places under the Metro Railway Act.

Nearly 261.00 acres of land were acquired for the car shed at Dumdum, which includes a part of the Eastern Railway Embankment. Considerable land acquired belongs to the Calcutta Municipal Corporation and the Baranagar Municipality. The alignment, as it approached Belgachia Road, went underground, and the last 200 meters on the approach to Belgachia Road, a heavily built-up area came in the way, and acquisition was resorted to acquire properties covering 7.15 acres.

During the construction of the North-South Metro, Esplanade was an important terminal for Calcutta (Kolkata) Tramways, and there were two turning circles sprawling over the construction area. These had to be remodelled to make room for the construction area for DW. Several shops in the Maidan Market were a barrier to the construction of DWs for about 100 meters. The shifting of these shops got severely bogged down with litigation. Eventually, the case went up to the Supreme Court, and vacation was affected under the court's order.

Fig. 5.31: Demolition of Existing Roadside Balcony

5.11 Salient Features

(Ref. The Calcutta Metro)

General:

Length-Between-Dumdum-and-Tollygunge–16.45-Km

Total Length Excluding Depot Connection – 17.85 Km

Length Including Depot Connection -- 18.062 Km

Length Including Extensions—32 Km

No of Stations – 17, Underground – 15, Elevated – 1 & At Grade --1

Distance Between Stations – Maximum: 2.15 km, Minimum: 0.597 km, & Average: 1.028 Km

Gauge – Broad Gauge (India Gauge):1676 mm (5 ft-6 in.)

Headway—Designed for—90 secs., 12yrs after opening – 116 secs

Platforms – 170 m, Side Platforms: 5.0 & 7.0 m (For Dumdum), 2x5,5 (Park Street)

– Island Platform; (8.26m) + 2 Side Platform (Central), Others: 10.26 & 8.26m.

Horizontal Alignment – Radius of Curvature: Minimum-200 m & Maximum-1000m

Gradient – steepest 1 in 50, Depot Connection – 1in 25, Maximum Track Gradient -2%

– Maximum Gradient to Connect Depot: 4%, Vertical Curve – 2000m

Track Structure: Ballastless, at Depot -- Ballasted

Rolling Stock:

Number--12 yrs after opening –304

Rake Composition—8 coaches, Driving Motor Coach-2, Non-Driving Motor Coach-4 & Trailer Coach--2

Train Capacity—2558 Pax., Sitting- 48, Standing-278 (Motor Coach with Less Capacity

Coach Dimension—20.40 m X 2.74m X 3.70m

Axle Load – 17.0 tons

Speed – Maximum—80 Km/hour, Schedule – 30 Km/hour

Acceleration – 1.1 m/sq. sec, Deceleration – 1.1 m/sq.sec

Station Stop Time—30 secs

Traction System—750 V DC

Braking—Electro Dynamic with Electro Pneumatics, Barking Brakes

Current Collection—Third Rail, -- Return Through—Running Rail

Door Opening—Electro Pneumatic

Power Supply:

Anticipated Maximum Power Demand – 65 MVA

No, of Auxiliary and Mid-Section Substations (11KV/415 AC) –27 Nos.

Ventilation and Air Conditioning – Maximum Permissible Temperature – 30 degrees C

Maximum Relative Humidity – 87 %

Maximum Air Changes – 29 times/hour

Signaling and Telecommunication – Type: Cab Signaling with Standby Line Side Signals

Train Control System -- Automatic Train Control with ATP, ATO, ATS & Stand by Block System

Telecommunication – Automatic Dialling, Train Radio Communication, Public Address System

5.12 Technological Overview

Metro construction is complex, requiring the application of several cutting-edge technologies in civil, electrical, signaling, and telecommunication engineering. Indian engineers, backed by their own experience and supplemented by references to systems abroad, adopted advanced technologies in the fields in India. The primary mode of construction applied was cut and cover using a Diaphragm wall (DW), Sheet piles, H-piles & Timber lagging, and others. Extensive decking was used to keep the traffic flowing smoothly over the cut while the construction work was underway below. Major features were:

a. Reinforced concrete RTS box by construction cut and cover method using DW, Sheet piles, H-Piles & Timber Lagging and using cast-in-situ contiguous bored piles
b. Reinforced concrete Station box by construction cut and cover method using DW, Sheet piles, H-Piles & Timber Lagging and using cast-in-situ contiguous bored piles
c. Cast in situ RCC Box with open excavation in Tank Areas **(Fig. 5.44)**
d. Shield tunnelling using compressed air and airlocks
e. Driven tunnels with cast iron or reinforced concrete lining
f. Elevated length carried over reinforced concrete girders
g. Surface/trough
h. Extensive decking to keep the traffic flowing over the cut during construction in progress underneath
i. Ballast-less track using elastic fastening, rubber pads, epoxy mortar and nylon inserts
j. Safety of passengers
k. Air-conditioning & ventilation system for environmental control of stations and tunnels
l. Third Rail's current collection system for traction
m. Underground substations with dry transformers and SF-6 circuit breakers
n. Tunnel-train VHF-radio communication systems
o. Micro-processor-based train control and supervisory remote-control system for substations
p. Automatic ticket vending and checking system.

5.13 Elevated Section

For the N-S metro, concrete was mainly used for permanent structures, and steel. It was used for temporary structures. When design and construction were undertaken in the late 1970s, RMC (Ready-Mix Concrete) was not yet available. During those days, most projects were implemented using on-site concreting and pouring manually. As such, the concrete used in the N-S Metro was traditionally manufactured on-site. The mix design was based on volume mix. For the metro construction, the volume mix was converted to the equivalent weight mix and produced by an electrically operated Pan Mixer with a 30 cu.m per hour capacity. In the 1970s, however, the use of batching plants for concrete production and precast concrete elements started for big projects, enabling higher strength and improved quality of concrete. In recent years, the use of precast concrete elements/ segments and prefabricated steel structures has become popular in urban structures, particularly elevated metro and highway corridors, to improve the speed of construction and the environment. The modern trend of using precast segments has been discussed in Chapter 6.

The 16.55 km N-S metro corridor was planned in three sections: elevated, underground, and at-grade, along with connecting ramps. Out of 16.55 km, 1.585 km of elevated section designed, 920 m for a double-track elevated running section and 665 m stations, including immediate approach sections at either end. The running section of the elevated corridor supports the entire dead load, the superimposed load, and the load of running trains.

The elevated structure **(Fig. 5.32)** activities comprise followings in sequence.

i. Borrow pits along the corridor were filled up and compacted with fresh earth to a level.
ii. Group of piles for each track, supporting RCC circular piers spaced 4m center to center in the transverse direction, and supporting the deck girder and slab. Piles are cast-in-situ circular piles encased with removable steel tubes, including a steel shoe, with an average depth of 20 m.
iii. Reinforced Pile Cap of 7.51m X 2.5 m wide to support the Piers
iv. Twin RCC circular Piers on pile cap to support the transverse girder (capping beam).
v. Capping beam connecting piers in the transverse direction to support the two longitudinal girders spanning 10 m.
vi. Two longitudinal girders connecting two consecutive pier caps in the longitudinal direction for each track.
vii. Deck slab for each track on two horizontal girders.
viii. Track on each Deck

Fig. 5.32: Part of Elevated Corridor

Fig. 5.33: Elevated Dumdum Station – Platform

Analysis, design, and details for elevated structures were carried out using prevailing practice, provisions of Indian Codes available during the seventies, and international practice commensurate with design parameters in the Tender Documents prepared by MTP (Railways). Essential items from the Tender documents prepared by MTP (First in the Indian UG Metro projects) are enclosed for reference in Annexe 1.

After the borrow pits were filled up, compacted and levelled, piling of the column foundation was taken up. The piles were 530mm in diameter. Piles were driven to an average depth of 20 m. Piles were in situ, and steel casings of appropriate diameter were driven to the required depth. The piling was done using conventional practice, using a tall rig to drive temporary steel tubes with a cutting edge at the bottom end, followed by removing the muck, placing reinforcement cages, and pouring tremie concrete with a 150 mm slump. The steel tube was withdrawn during the concrete process. This method was followed for 920m on the southern approach to Dumdum. For the remaining 400 m towards Dumdum station **(Fig. 5.33),** boring for the full depth was done by Direct mud circulation, due to a shortage of space to erect a tall rig. Bentonite slurry was used to protect the surrounding soil from collapsing. The other elements of the elevated corridor, such as pile caps, piers, transverse and longitudinal girders, and the deck, were cast using steel/timber shuttering, as per prevailing practice, including temporary supports with M20 grade of controlled concrete.

i. Bearing

Reinforced concrete girders were fixed at one end and free at the other end. Since the greasing of the bearing plates would be difficult, specially designed stainless steel bearing plates were used for support (**Fig. 5.34).** The station portion, including its immediate approaches, was designed and built using the conventional method of in-situ construction to carry all Dead Loads (DL), superimposed dead loads (SIDL), Live Loads (Train loads, etc.), and all loads as per the design parameters.

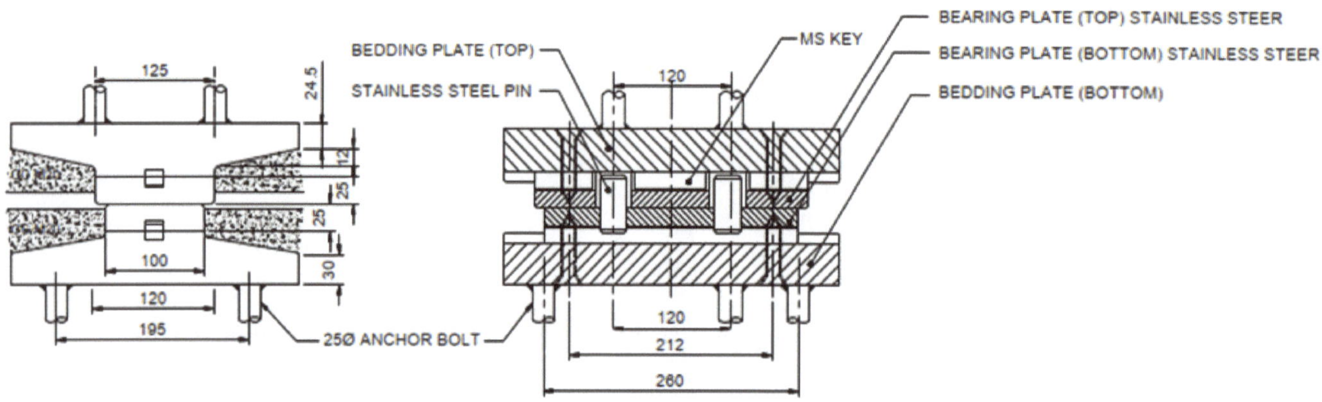

Fig. 5.34: Details of Stainless-Steel Bearing Plates

For the Station portion with immediate approaches covering a total length of approximately 200 m, a continuous RCC deck was used for the total structure. The structure in this section comprises the ground floor, unpaid areas, which hold the public concourse, booking office, other station services, and a Stair Hall to connect to the Platform level. The elevated deck at platform level, with tracks, was the paid area and was covered by a continuous roof. The station structures were designed for two additional floors for future use.

The station structure comprises the entry structure to the Ground floor, including the Concourse, passengers' holding areas, booking offices, and all other operating facilities. Over the concourse, an elevated deck was built for platforms, including facilities required to operate the Metro. The platform level was covered by providing a continuous roof. At the Ground level, two subways, each 7 m wide, with a clear height of 2.6 m and a length of 48 m, were built across the suburban railway on an embankment to provide access to the Metro System. The Depot and car Shed for the N-S metro were planned towards the north, and the elevated approach had to be lowered by 10m to cross the suburban railway embankment to the car shed. The crossing was done by constructing a tunnel using the Jack-pushing method across the barrier, accommodating many tracks for the Suburban and Main Railway.

Fig. 5.35: Final Alignment of Line 1

Fig. 5.36: Extension of Line 1 from Noapara to Dakshineswar

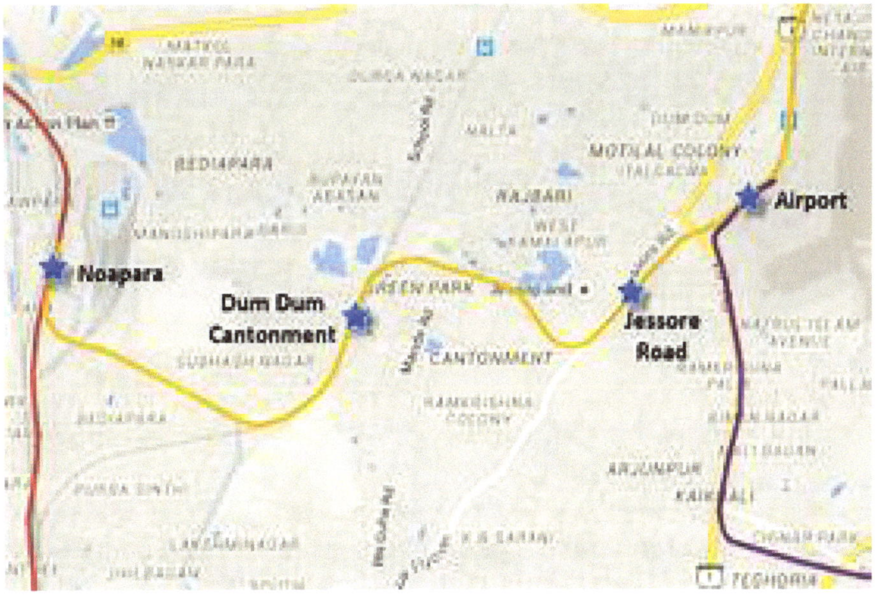

Fig. 5.37: Corridor Between Noapara and Airport

Figure 5.35 includes the final alignment, including extensions on both ends. **Figure 5.36** includes the extension to Dkshineswar through Noapara. And **Figure 5.37** includes the corridor between Noapara and Airport.

5.14 Design Basis Between Tollygunge and New Garia

(Ref. BSES on behalf of RITES)

As described earlier, Line 1 has been extended to both ends of the original Dumdum-Tollygunge corridor: 1) Tollygunge to New Garia and 2) Noapara to Dakshineswar. Between the Dumdum and Noapara sections was developed during the implementation of Noapara Depot. As such, Noapara has a station on Line 1 and a line connecting the Airport, and short section as part of the Barasat corridor. Construction of the Airport – Birati section has been awarded recently, and this section of the Noapara-Barasat Corridor is designed as an underground corridor. The extension between Noapara and Dakhineswar has already been completed, and the operation has also started. Alignment of the extension is included in **Fig. 5. 36.** The extension of the North-South Metro between Tollygunge and New Garia is Elevated and includes 5 Stations. All stations are designed based on RCC column-beam structures. The frames had been modelled as a plane frame structure in STAAD. Loads corresponding to the spacing of columns, and the structure was analysed. The above model was applied for the analysis of the transverse beam & column. The longitudinal beam has been analysed as a continuous beam. The design of the pile, pile cap, column, slab & beam was done according to IS 456 in the Limit State Method. Stations included in this section are Netaji, Surya Sen (Basdroni) **(Fig. 5.38),** Gitanjali, Nazrul, and Khudiram.

Fig. 5.38: Basdroni (Surya Sen) Station.

5.14.1 Basic Design Data and Assumptions

The analysis and design of the station building had been based on IS: 456, IS:800, IS:875, IS:1893, IS: 2911 Part-I to IV, IS:13920, SP:16, etc. The materials used were concrete and steel.

1. M-35 Grade of Concrete used for superstructure, pile cap & retaining wall.
2. M-25 Grade of Concrete used for pile only.
3. All P.C.C. work was done in a 1:4:8 mix.
4. HYSD bars (Grade Fe 415/TMT Bar) conforming to IS: 1786 were provided as reinforcement

Figure 5.39 includes the completed corridor between the Tollygunge to New Garia, already completed and **Figure 5.40** includes the Noapoara Station.

Fig. 5.39: Corridor Between Tollygunge & New Garia

Fig. 5.40: Naktala Station

This design approach pertains to the determination of reactions in piles under different pile caps, the design of piles and the pile cap for Gariabazar metro station. Loadings had been taken as per the "Design Basis Report for Station Building Structures" prepared by RITES. The load in the TSS portion for electrical equipment was as per the details of electrical equipment by RITES, issued by the "Electrical Department Metro Railway, Kolkata". The layout of electrical equipment, based on which the load calculations have been made, is attached to this document. The electrical equipment loads coming between two transverse grids had been taken on the grid by static distribution, considering spans on either side of the grid as simply supported. **Figure 5.41** includes the photograph during external finishing was on.

Fig. 5.41: New Garia Station during Finishing Works

5.14.2 Excavation in Tank Area and Concreting

Open excavation was used to construct RTS (Rapid Transit Box) boxes between Park Street and Maidan Stations for the running corridor across General's Tank. The section is 600+ m long, encompasses a substation area, and is found at a level satisfying the approved drawings. Design and drawings by M/S Chatterjee-Polk Private Limited, appointed as Design-Built Contractor, led by the Author and checked and approved by MTP R. The construction of this section, through the Maidan, abutting the western side of the JLN Road (Jawaharlal Nehru Road), was originally proposed by DW both as a shoring wall and as a semi-integrated system. However, the delay in importing specialised equipment for the diaphragm walling led to innovative designs for open excavations in the general tank area. Using TBM in an isolated section of the tank area was also not feasible, so precast or bored piles/timber piles or concrete/steel liners were studied.

In Tank area excavation was required up to 8 to 9m below tank bed along the north-south direction. Constraints for open excavation were i) depth of slush at tank bed ii) distance between alignment and tram track iii) road abutting tank iv) thickly built-up area with high rise buildings v) Cut of level of Metro Box vi) restriction of type and weight of construction equipment.

Besides being free from physical encumbrance, the tank area had the advantage of a lesser extent of actual excavation. The water-retaining body **(Fig. 5.42)** already had a low bed level with associated removal of top slush for 1.0 to 1.5 m. Top slush removal was also a must, even for diaphragm wall or sheet piles (to excavate liower depth) or H piles and timber lagging, construction to enable movement of heavy equipment. Open excavation was proposed in the tank area based on discussions with MTP authorities. The slope with the desired inclination was not achievable. Limitations were the fixed alignment, fixed level of the RTS box, the abutting tram tracks at the edge of the tank and the overcrowded J N Road with several tall structures. The site had space limitations, providing a stable slope with one vertical to 2.5 horizontal approximately was not feasible. The primary considerations for the safety of sloping excavations were

1. Overall stability of the slope
2. Heave at the founding level
3. Local failure at the toe due to stress concentration
4. Use of appropriate locally available materials
5. Time and Cost

Geometrical configurations posed a significant challenge in ensuring the stability of open excavations. Poor soil characteristics also restricted the gradient of slope that could be adopted.

Fig. 5.42: General Tank – During Dewatering

Fig. 5.43: Driving of Timber Piles

Considering the observed constraints and the possibility of working a stable slope based on stability analysis, stability against total slip was more critical than that of partial collapse. The criteria for the stability analysis of the sloped excavations adopted were

i. Assessment of the maximum permissible stable slope imax for the strength parameters of the soil and the required factor of safety
ii. Assessment of the minimum Factor of Safety (FOS) for a stable slope of the given soil strength

The analysis of slip planes could only achieve a maximum safety factor of 1.15. With safety factor of 1.15, the maximum feasible excavation depth was limited to 5.0 m using two slope profiles of one vertical to 1.5 horizontal slope and a berm of 2m in between. Berm helped to reduce vertical load and flaten the the resulting virtual slope. Combination of two slopes with a berm in between proved to be better than one slope configuration. A space minimum of 1m was required at lean concrete level to facilitate construction of RTS box. A milder slope and/or wider berm could not be worked out due to available distance between RCC box edge and Tram track.

In addition, a vertical interception was necessary to cut the slip circles which were done with a wall, introducing them at berm. Intrcepting wall was designed with timber piles driven contiguously at berm level. Timber piles were driven before introducing the lower-level slope. After ascertaining the locally available 200-250 mm diameter Sal timber logs in 6.0 to 7.0 m lengths. Timber piles required a depth of minimum 12m, each made of two logs. For the required 12.0m to 13.0 m long piles, those made with an intermediate joint using steel collar. Piles were driven using local methods and included in figure **Fig. 5.43.**

Joints made were found to be okay and tested successfully for minimal interference in driving the Sal piles by hammering. The Sal Piles were designed with strutting to support the backfill at the top, avoiding a cantilever length of 3.0 to 4.0 m, so that deflections at the top of the timber piles were acceptable. Timber piles were installed along the longitudinal direction as per approved drawings, with their intermediate joints and collars carefully positioned to intercept potential slip planes and distribute loads efficiently. Once the piles were in place, strutting and backfilling operations proceeded systematically, maintaining continual monitoring to detect any early signs of movement or instability.

Attention to environmental and logistical constraints—particularly the avoidance of the rainy season and optimisation of available workspace—proved crucially. Throughout the construction, safety remained of paramount importance, with constant monitoring of ground conditions and timely adjustment during the construction activities as needed. The integrated use of local Sal timber piles not only helped to maintain the schedule in the absence of mechanical excavation but also delayed initiating soil movement. Late initiation and slow movement helped to complete the construction of the Metro Box in the tank area. **Figure 5.44**. shows stage construction.

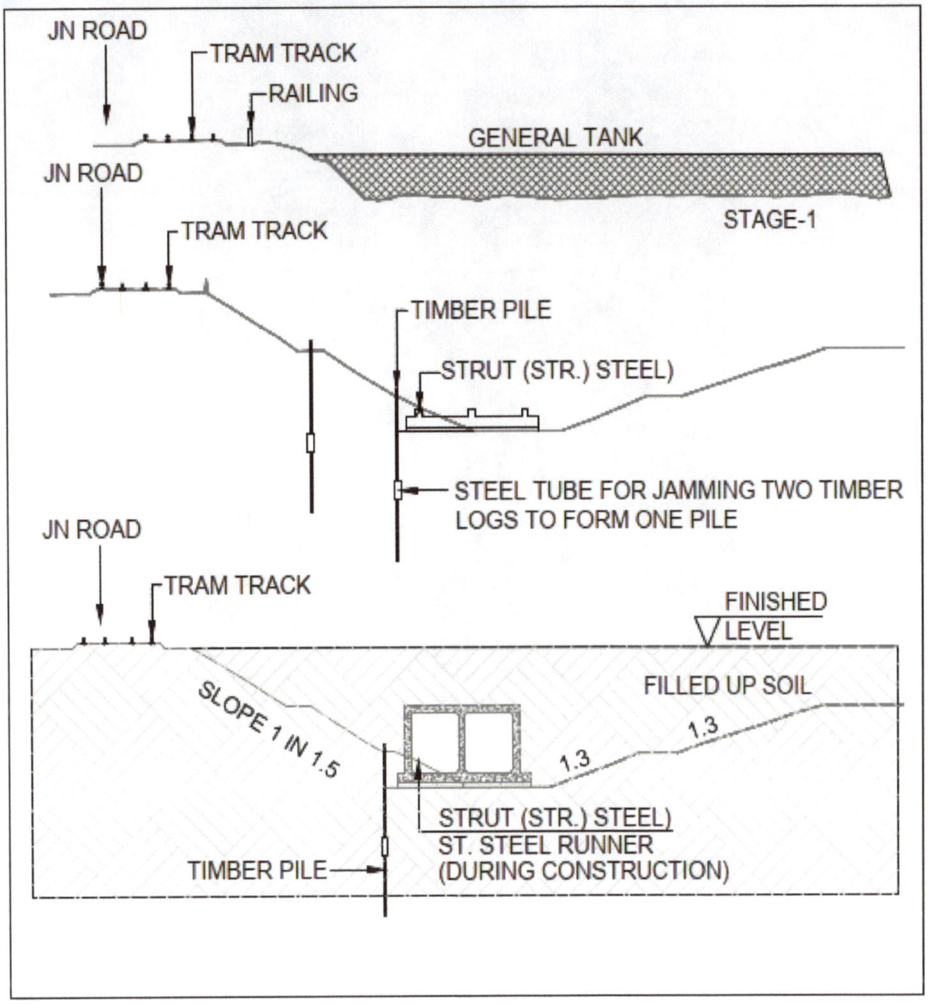

Fig. 5.44: Design Sketch of Construction Stages in General's Tank

The design ensured robust structural support but also exemplified sustainable resource utilisation tailored to site-specific challenges.

In the General Tank Area, these measures were complemented by adaptive planning, enabling rapid progress despite tight timelines. Every phase, from pile driving to strutting and berm installation, contributed to a comprehensive system designed to withstand the unique demands of the project environment.

Figure 5.44 shows the innovative approach adopted in the General Tank Area. However, completing the whole exercise in the minimum possible time was critical. Besides avoiding the rainy season, it would be possible to start only when all job elements before excavation are ready.

Figures 5.43, 5.44 and 5.45 include a steel strut used near Park Street, Bottom Timber Piles during excavation in the tank area, first with a single timber pile and the next with two piles. On-site execution of this slope stabilisation strategy required meticulous sequencing.

Fig. 5.45: Inclined Steel Struts at Park Street Station

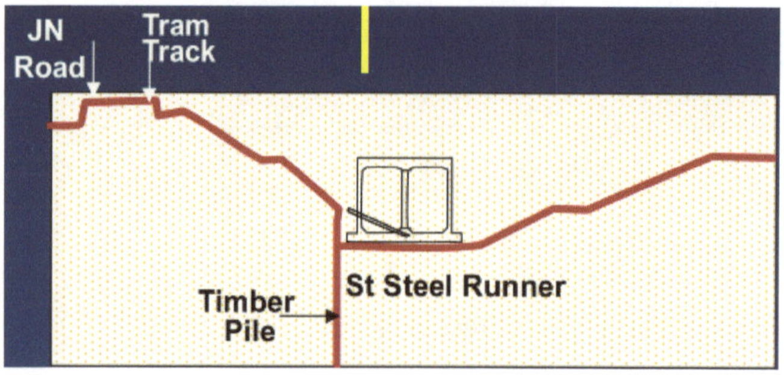

Fig. 5.46: Bottom Timber Piles During Sloped Excavation

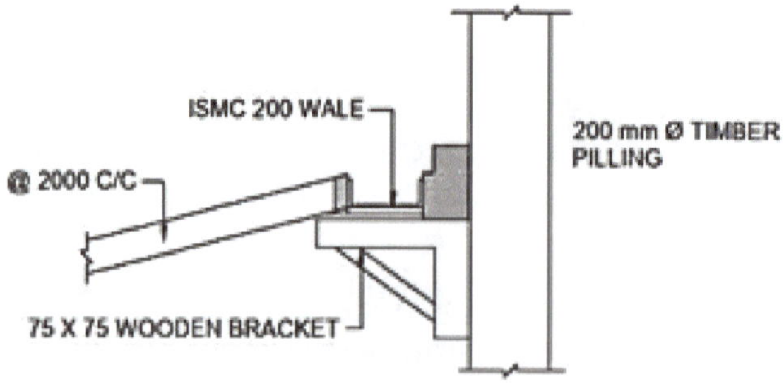

Fig. 5.46A: Enlarge View of Strutting

The lesson learned from this job is that site-specific constraints call for innovative design. Solutions to typical problems can be achieved by creatively applying locally available materials. Assessing the risk involved and understanding the site conditions are prerequisites for working out these solutions.

Fig. 5.47: Concreting of RTS Box in Tank

Fig. 5.48: Concreting of Substation near Park Street

Concreting for this section for the RTS corridor and Sub-Station **(Figs. 5.47 & 5.48)** was carried out in situ using man-heads. Using transit mixers or dumpers was also not feasible due to the soft ground in the tank area. Concrete was produced using an electrically operated Pan Mixer of 30 cu.m. Capacity. Though the production of concrete by Pan Mixer satisfied the quantity required, as pouring was done manually, the concreting period for the RTS box of 21m long, and the amount of approximately 200+ cum could be completed in 18/20 hours.

Before the concreting process, the reinforcements and water bars were placed for construction and expansion joints, and MTP engineers checked the placement of the water bars. The concrete was done in three phases: Base slab with construction joints, side walls with water bars at the junction of walls and the top slab. Expansion joints were placed in dumbbell shapes across the complete box at an interval of 21m. The station, which has a double-story structure in height, **Figure 5.49** with three slabs, Base, Middle & Roof, i.e., the roof of RTS, being a thick slab, took a longer time than the RTS box section **(Fig. 5.50)**. The construction of the station was carried out in four phases.

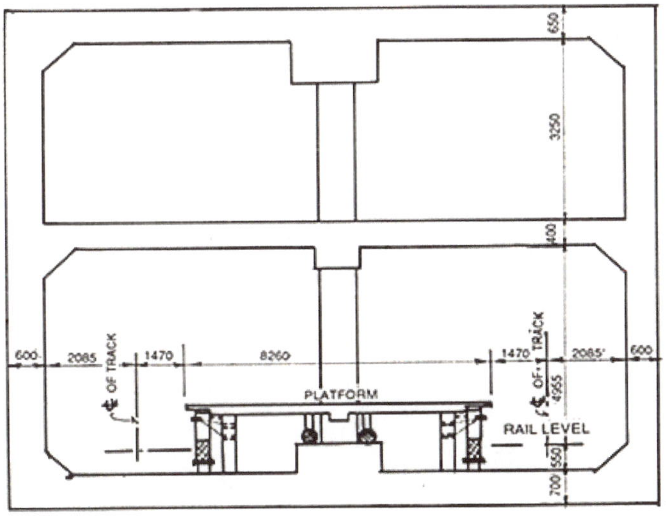

Fig. 5.49: Cross Section with Mezannine

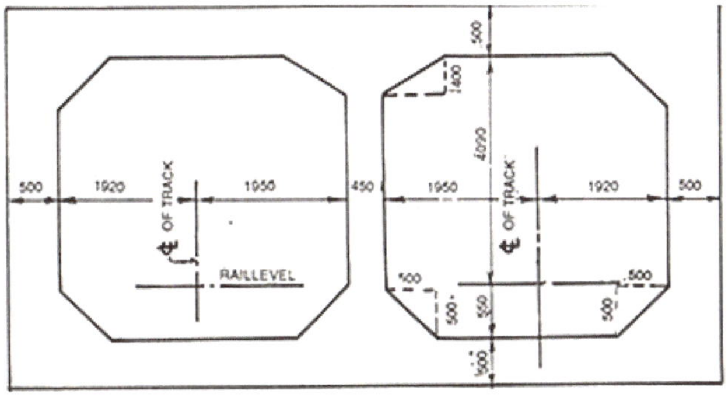

Fig. 5.50: RTS Cross Section

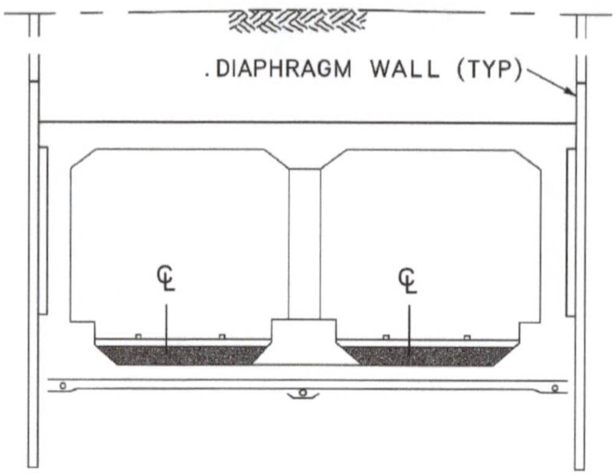

Fif.5.51: RTS Section with DW – Semi-Integrated

5.15 Excavation for RTS Corridor and Stations Using Shoring Walls

Metro construction is complex, requiring the application of several cutting-edge technologies in civil, electrical, signaling, and telecommunication engineering. Indian engineers, backed by their own experience and supplemented by references to systems abroad, adopted advanced technologies in the following fields for the first time in India.

The basic mode of construction applied was cut and cover, using mostly diaphragm walls and other approaches in sections. Extensive temporary decking kept the traffic flowing smoothly over the cut while construction work was in progress below.

The stability of the excavation, ground movement, control of water into the excavation area, effect on adjoining structures, effect of overburden **(Fig. 5.51),** and design of structural elements such as diaphragm wall and struts are distinct yet interrelated features in the design of braced excavation. The depth of penetration of the supporting wall below the bottom of the cut was determined by considering stability against the bottom heave. For this purpose, the client specified a safety factor of not less than 1.5. It is particularly important in brown field areas with surcharge from heavily congested roads and abutting Structures and where the diaphragm wall terminates in clay.

Fig. 5.52: Traffic Snarls in Kolkata Roads

5.15.1 Underground Corridor (RTS) for Running Section

The MRTS finally adopted by Kolkata is a double-track system with steel wheels on steel rails (with 8-car trains), mostly underground **(Fig. 5.54)** shows the train in an underground station. Station locations have been chosen based on the commercial and population concentration along the route, taking the guide from the tram and bus stops. Construction of the underground running box used a) the Cut-and-Cover technique, b) the UG corridor using a Tunnel Boring Machine (TBM), and c) the new Austrian Tunnel Method (NATM). The excavation was carried out mainly using the Cut-and-Cover method. The method comprises three approaches: i) Shoring walls as an earth retaining structure, ii) a Wall as Semi-integrated with a running box, and iii) an integrated wall with the box. The shoring walls again were of different types, like a) Concrete Diaphragm Wall, b) Steel Sheet pile walling, c) made with H-piles and Timber lagging, and d) Contiguous bore piles.

5.15.2 Cut-and-Cover Technique

The cut-and-cover technique consisted of putting up two vertical retaining walls to retain the ground abutting excavation and the structures at adjoining area of excavated trench.

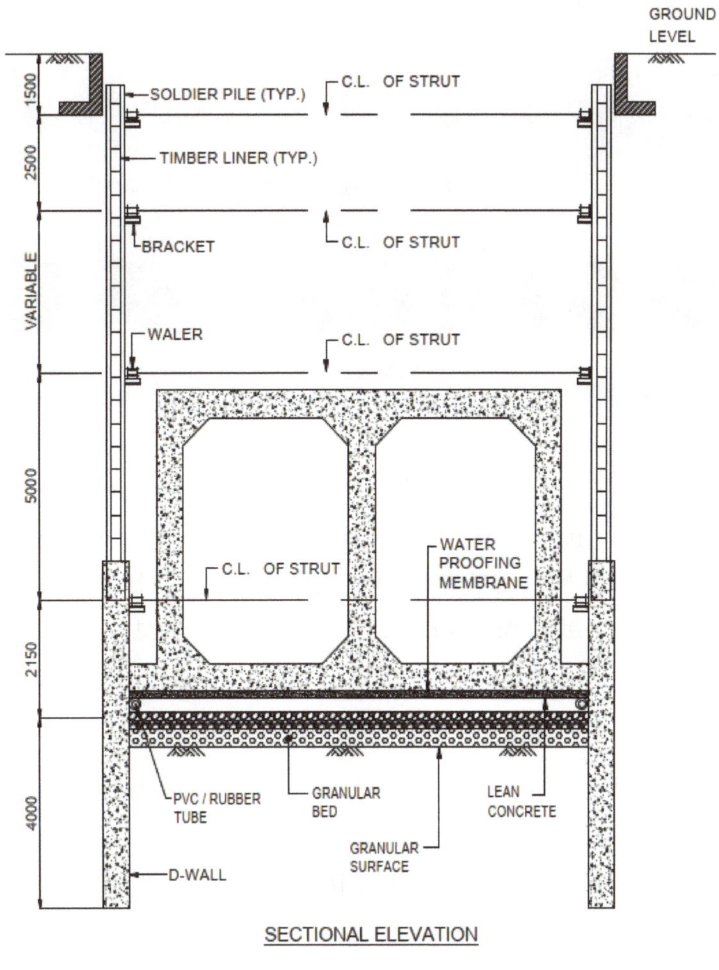

Fig. 5.53: RTS Box with Cut-and-Cover Excavation using Combination of DW & H Pile & Lagging

Note: In sections where H-Piles & Timber Lagging were used in N-S Metro, H piles were embided near top slab. In addition, Excavation using only H-Piles & Lagging were not used as depth of excavation was comparatevly more. The method is suitable for shallow to medium depth excavation.

Trenches of required width are to be made for boxes to accommodate the station structures and the running line between the stations. The walls were typically separated by 10 meters on running sections and 19 meters at stations. The design was such that the ground and structures outside the excavation area were least affected during the implementation. In this technique, the supporting structure comprises a rectangular box constructed in RCC that carries two tracks. In heavily built-up areas, like Chitpur yard (included in the Tunnel Section later part), a single tunnel carries a single track. For double tracks, two tunnels were required.

Fig. 5.54: Metro Train at Underground Station

5.15.3 Strutted Excavation with Diaphragm Walls

The Diaphragm Wall (DW) is used for underground Highways, Railway tunnels, Multistoried Basements and Foundations, and Metro Corridors and stations. It is designed to carry vertical loads, where required, and earth and water pressures. It is generally cast 3-6m long and between 0.6 -1.5 m thick, with end profiles, concave and convex shapes or ends having grooves. End profiles help DW panels cast side by side to make the geometry of a continuous wall.

DW panels in the North-South metro used concave and convex end profiles and steel tubes of 600mm diameter on both sides/one side, depending on the location, to construct 0.6 m-thick. DW for the East-West Metro, grooves of the designed profile were used for ends fabricated from steel sections. In both cases, panel joints are made watertight. The design-build tender was submitted by CPPL (India) with Phillip Holzmann (PH), Germany, as advisor. The tender was submitted with Dr. BC Roy (Author) as team leader. Excerpts from the Tender Document floated by MTP (R) (Metropolitan

Railway Projects, under the Ministry of Railways, Government of India-GOI) have been included in **Annexes 1 & 2** and included in **Figures 5.55, 5.56 and 5.57** show different stages of construction as included in the tender submission. These included a typical plan of DWs, proposed stages of DW construction, including DW casting, and the sequence of RTS box construction, including installations of intermediate struts.

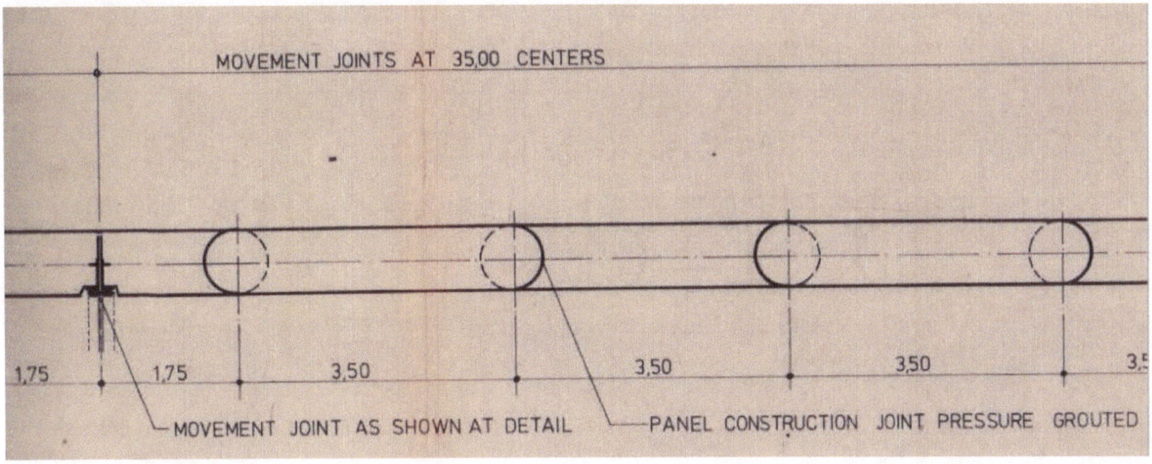

Fig. 5.55: Plan Showing Typical Arrangement of Diaphragm Panels

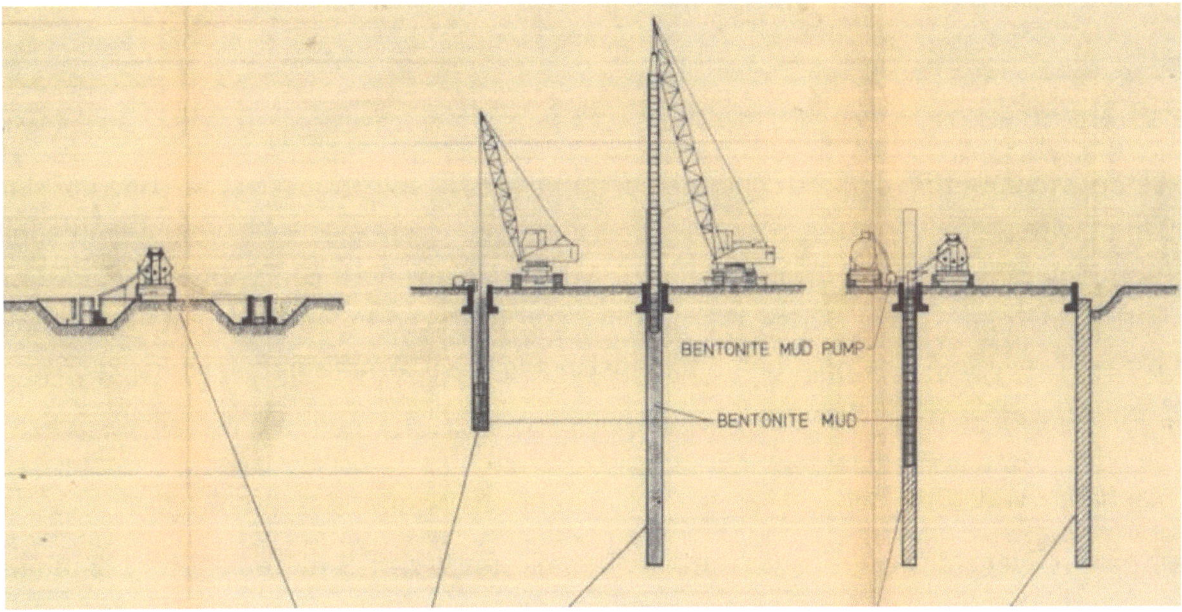

Fig. 5.56: Typical Section Showing Stages of Diaphragm Wall Panel

The construction of DWs is done step by step. Approximately one-meter-deep Guide Walls **(Fig. 5.56)** are constructed parallel with a gap between two parallel lines to maintain the gap required for DW design (600mm & 800mm in the N-S metro corridor)

Step two is to excavate between two parallel sidewalls using mechanical Grabs (manually to obtain a smooth surface) and hydraulically/mechanically operated grabs. During excavation, the excavated pit, with spoils, continuously removed, is filled with bentonite slurry of the required specific gravity to maintain consistency and prevent the pit from collapsing. Where the diaphragm wall terminates in clay, a safety factor against the bottom heave may be determined. During excavation, the excavated pit, with spoils continuously removed, was filled with a bentonite slurry of the required specific gravity (1.04-1.05) to maintain the necessary consistency and prevent the pit from collapsing.

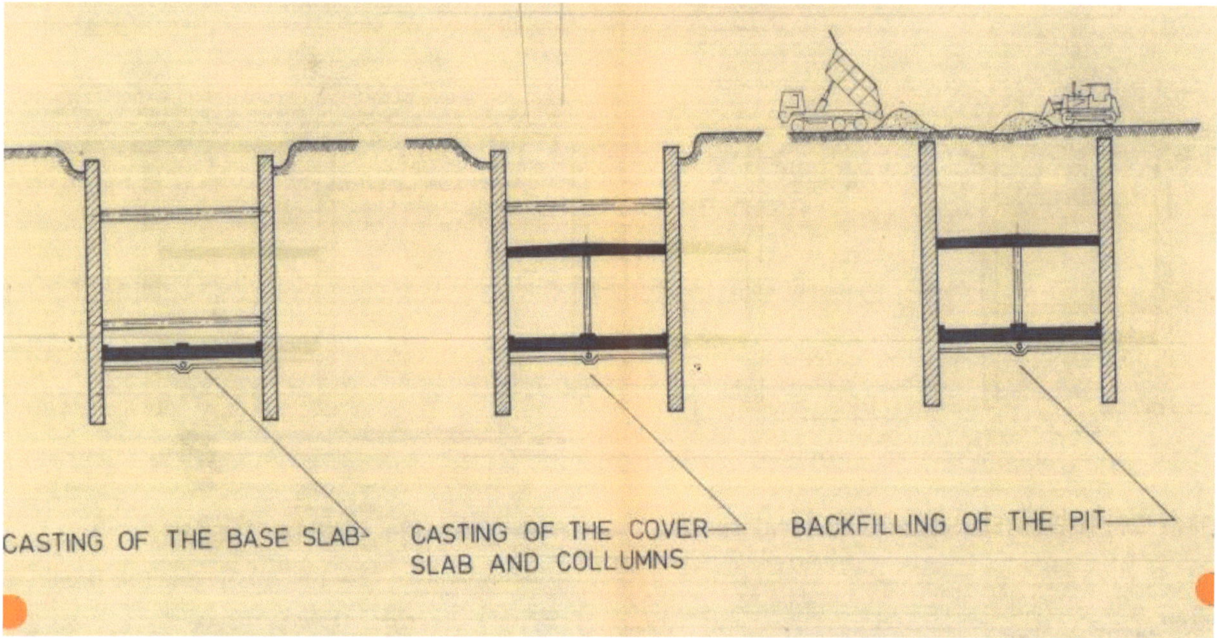

Fig. 5.57: Stages of RTS Box Construction

The next step is to pour the required concrete per approved specifications (tremie concrete slumping 150-160 mm) with the M20 per the design. For N-S metro, concrete of 1:2.4 of volume mix converted to mix (nominal) by weight for strength of M20/M25. High-strength concrete was used. Concrete displaces the bentonite slurry during the pouring and is stored in the bentonite tank through a flexible pipe for recycling. The next panel to construct is the alternate panel, which keeps a gap of one DW length.

5.15.4 Diaphragm Wall (DW) Panels

DW wall panels were designed as continuous beams, mostly of 3-meter width, commensurate with the size of the Kelly used for excavation, commonly supported vertically over a series of struts (supports) for retaining earth on either side of the cut. Additional horizontal reinforcements were provided at strut locations to act as built-in Wales (or Cross beams) between which diaphragm walls were considered as span. The maximum earth pressure was adopted from Terzaghi-Peck's empirical formula in the design of the N-S metro UG corridor. Besides earth pressure, lateral pressure due to surcharge loads from traffic and adjoining buildings was also considered. **Fig. 5.58** shows the mechanical grab of 4t by weight imported from Germany and used for the excavation of DW as shown in Fig. **5.59.**

Fig. 5.58: Clamp shell Bucket- Mechanical Grab

Temporary traffic decking was used at most locations for the NS Metro as a temporary earth retaining DW. The details of the DW technique are further discussed in Chapter 6, Part 3, East-West Metro. The trenching for DW was used with a Kelly-grab (grabs in use are included in **Fig 5.58)** attached to the crane and a mechanical grab (Clamp shell bucket, 4t in wt., imported from Germany). Tube form articulation was done at panel joints. The thickness of DW ranged from 600 to 800 mm, penetrating to a depth that satisfies safety against bursting and heaving. A round PVC water bar was used at the box-box joints, constructed between two parallel DWs. The general grade of Concrete adopted was M20/M25. (Earlier, concrete of lower grade using volumetric mix was used, but the advent of mechanical and chemical admixtures has raised the grade of concrete to M20 and Above). Concrete grade M40 was also used in some places. For the last two decades, concrete design mixes have adopted only weighting lines with the output of the appropriate mix design, mixing in line with the production of a proper mix design.

5.15.5 Construction Approach

The approach to the design and construction of diaphragm walls is of two types: (i) the **Top-down method** and (ii) the **bottom-up method**. The bottom-up procedure was mainly adopted for the N-S Metro, and the top-down method was used for a small section.

Fig. 5.59: Two Dumpers in action: Loading Excavated Earth and pouring Concrete in DW

5.15.5.1 Top-Down and Bottom-Up Construction

A unique method, viz., **top-to-bottom construction**, constructs a single box tunnel. The scheme adopted is as given below:

i. After excavation and strutting (top layer) up to the bottom of top slab, the top slab with side haunches was cast first. This was supported on specially designed steel brackets welded to the steel liner of each pile (at 1.5m intervals) on either side.
ii. Lay water/elastomer tube at the junction of wall (to be cast later)/DW.
iii. Fbricate and erect top slab reinforcements in position over a base made farm.
iv. After casting the top slab, including haunches on both side at lower side, which would also act as a strut for the diaphragm walls, excavation was continued, and the second layer of strutting was installed and preloaded.
v. Further excavation was done up to the final level, and mud matting was completed.
vi. The bottom slab with haunches was cast. The sidewalls continued up to the 2nd layer of struts. The bottom slab was extended till the DWs on both sides designed for bending and axial forces.
vii. Struts were de-loaded, and further casting of sidewalls was done. The sidewalls were cast up to 1.0m below the top haunch level.
viii. The gap between the sidewalls and the top slab was filled with concrete through specially kept holes of 400mm width at 3.0m intervals in the top slab on either side.
ix. The junction between the sidewalls and the top slab was filled with concrete through a specially provided space 400mm thick.
x. After removing the shuttering, the junction between the top slab and the sidewall was grouted with epoxy.
xi. To facilitate muck removal from below the slab, two openings of 5m width each were kept in the top slab. These openings were later closed after the conventional method of completing the box.
xii. Lay the water proofing layer and rubber/elastomer tube at the junction of base slab and wall/ DW as applicable, above mud mat before placing the reinforcement for base slab.
xiii. Fabricate and laying of base slab reinforcement, leaving the appropriate dowels for walls on either side and upper part of reinforcement of longitudinal beam connecting the central columns.
xiv. Cast the base slab leaving water bars at the appropriate locations.
xv. Cast the walls to a height as per approved drawings.
xvi. The next step is to remove horizontal props in sequence, satisfying the design requirements.
xvii. Complete walls on both sides to the bottom of the top slab and ensure joint rigidity per the drawing.
xviii. Backfilling above the roof up to the final formation level as per approved detail.

The factors which led to the choice of this method of construction in this reach are:

i. Minimise the layers of strutting, which were enormously time-consuming and delayed the box construction.
ii. Due to the acute space constraint at the site, the cutting spoils were to be carried by trucks to the dumping ground, which was far away from the work site. By casting the top slab first, at least for some period, the cut spoils from the bottom half of the cut could be lifted and dumped on it. In other words, this meant simultaneous backfilling on the box top for part depth.
iii. The box construction was quickened as elaborate centering and shuttering for the significant portion of the top slab were avoided as the slab was cast on earth.
iv. Decking for construction activities could be minimised as the top slab space was available.

However, the major disadvantage of this scheme was the concrete of the joint between the top slab and sidewalls, which had to be done unseen through chutes and holes from the top. The structural continuity in the reinforcement was, however, ensured. It was seen that the casting of such joints was entirely satisfactory, and the few minor defects were made good from the inside by proper grouting.

The **activities of the bottom-up method, primarily used in the North-South Metro Corridor, are like** those of the top-down method.

In the bottom-up method, excavation and putting horizontal struts at required spacings are the first two steps, followed by casting of bottom slab, remove horizontal struts per the required sequence, cast the walls and intermediate column to support the top slab in the middle, or, as per the design, complete the box casting of the top slab.

The top slab over-staging was used in N-S Metro for bottom-up construction, and the details are included in **para 5.14.2.**

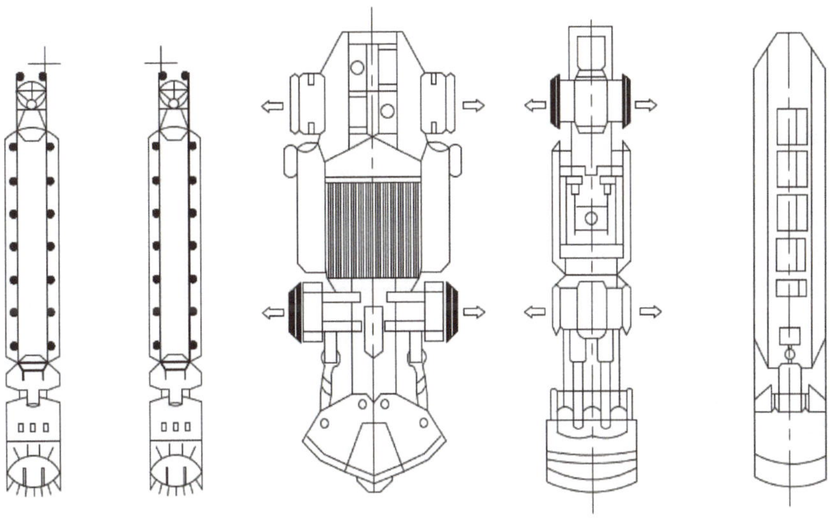

Fig. 5.60: Types of Diaphragm Wall Grabs generally in use

Indian Standard published a code of practice for the Design and Construction of Diaphragm walls, IS: 9556-1980, which was reaffirmed in 2003. The design of the North-South Metro was carried out before 1980, and the Diaphragm wall was constructed around 1977 in the Maidan area before the publication of IS-9556. However, the method used for the North-South metro is like what the IS code recommended, using steel tubes at the end/ends to facilitate the construction of a joint between the abutting joints. Different types of DW Grabs are shown in **Fig. 5.60** and possible joints between consecutive DWs are included in **Figures 5.61,5.62 & 5.63.**

5.15.6 Excavation

The Excavation was carried out between two rows of DWs after strutting or anchoring both sides outside the box. The design of compression struts accommodates earth pressure on the diaphragm wall. The excavation was done manually and mechanically. Manual excavation was carried out, in a significant part of the metro corridor and tank areas Like, at General's Tank, besides the top 3m for the RTS and Station locations.

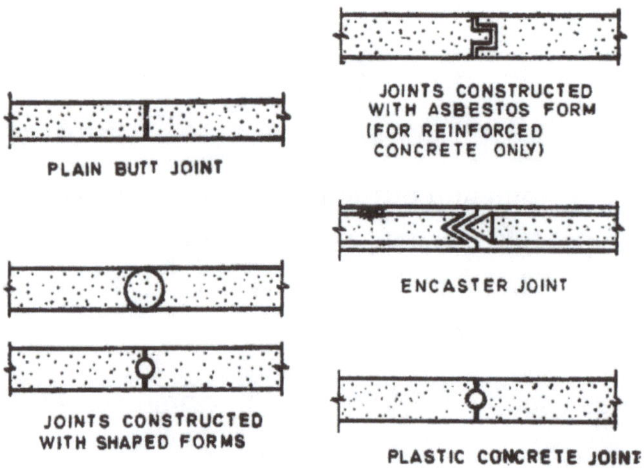

Fig. 5.61: Joint of Abutting DWs (IS: 9556)

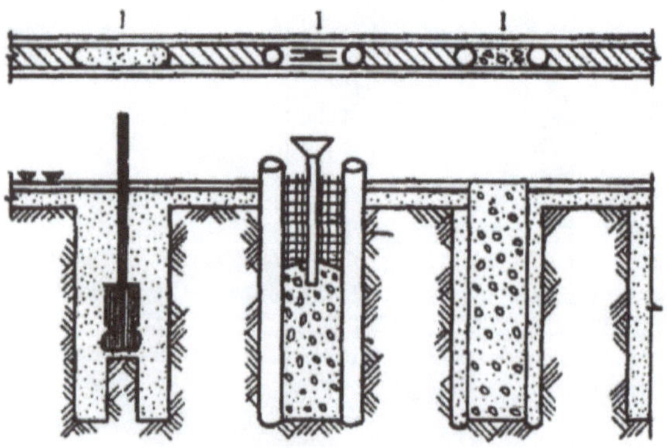

Fig. 5.62: Construction of Alternate Panels using Tremie Pipe (IS:9556)

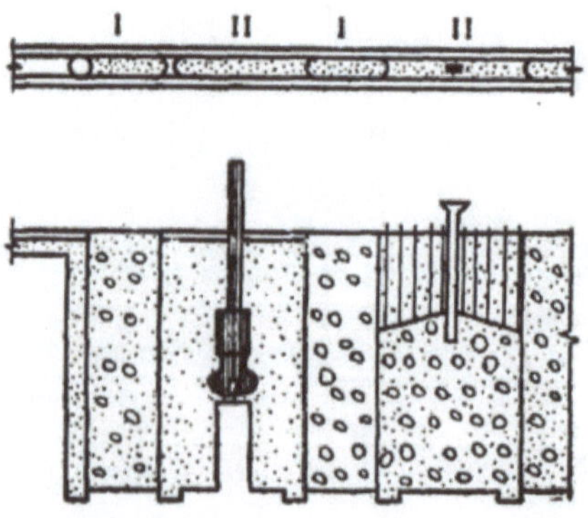

Fig. 5.63: Construction of Panels Left Between Alternate Panels (IS:9556)

The excavation was carried out for the remaining portion using mechanical and hydraulically operated grabs/mechanically operated clamshell buckets, 3m below the top of the founding level and blinding concrete, including the waterproofing layer, with an average depth of approximately 10 m. The grabs were operated from the crawler-mounted Cranes. Gantry-operated cranes were also used in the Maidan Area. However, it had restricted use in areas in the city due to limited space and heavy traffic on the city roads.

Manual excavation from the top of the ground level was used to prevent DW from yielding due and the presence of underground utilities. Manual excavation was also resorted to at the final level to avoid over-excavation and make the grade level as smooth as possible. Though the machines were used in substantial areas, the excavation was carried out in stages with the installation of struts at the appropriate levels until the final level was reached. The excavation for more than 14m cuts between DWs included a 2.5m berm on either side to provide adequate support to the earth. In top-down construction for integrated single cell boxes, chutes were used on both the shoring walls on both sides **(Ref.Fig. 5.64).**

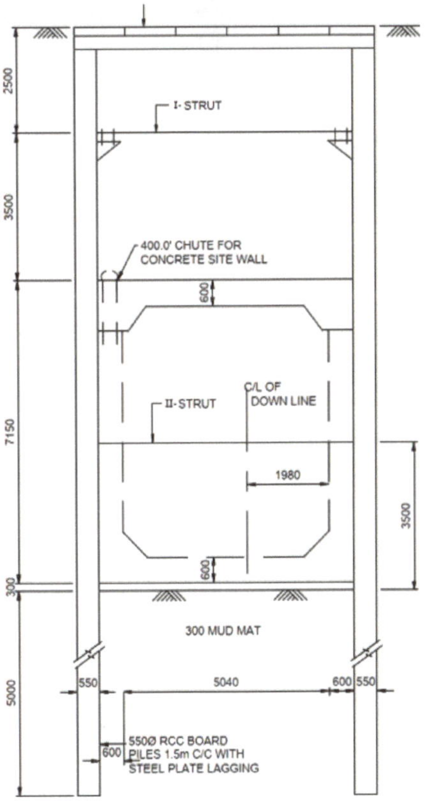

Fig. 5.64: Concreating of Integrated RTS below the top with slab with Chutes on both Sides

Fig. 5.65: Removal of Spoils Using Bucket Lids Opening Downwards

Gantry or crawler cranes removed spoils from the excavation, and where possible, buckets **(Fig. 5.65)** with a bottom opening and one cu.m. capacity were used to dispose of the spoils. Crawler cranes were 30/40t capacity. Head loads were primarily adopted for manual excavation in tank areas and the top 3m in other cases to dispose of the spoils. Spoils were dumped over the deck in case of top-down construction and removed at night in dumping areas like the Maidan and Noapara Car Depot areas. For bottom-up construction the spoiles temporarily stacked at a predetermined location and removed at night.

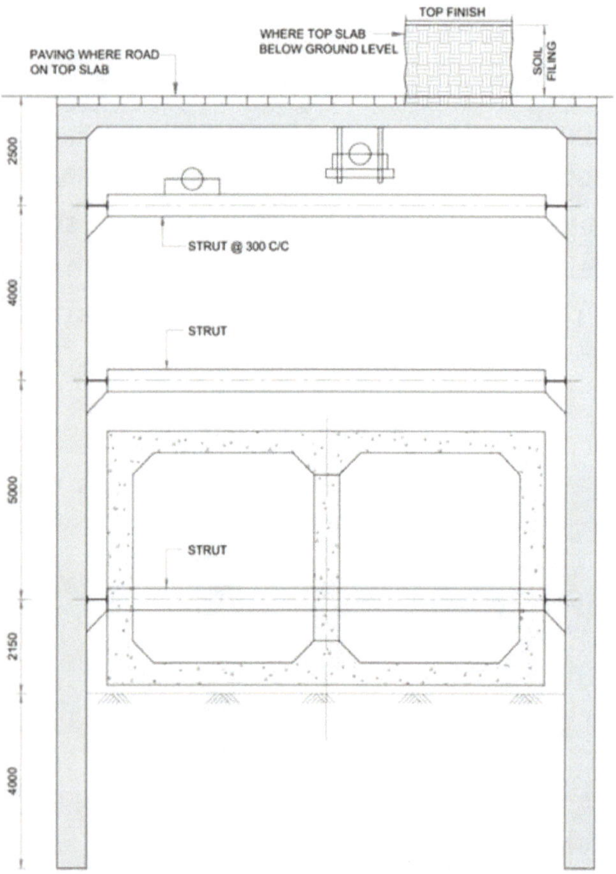

Fig. 5.66: Structural Elements in Position incl. RTS Box for Top-Down Construction

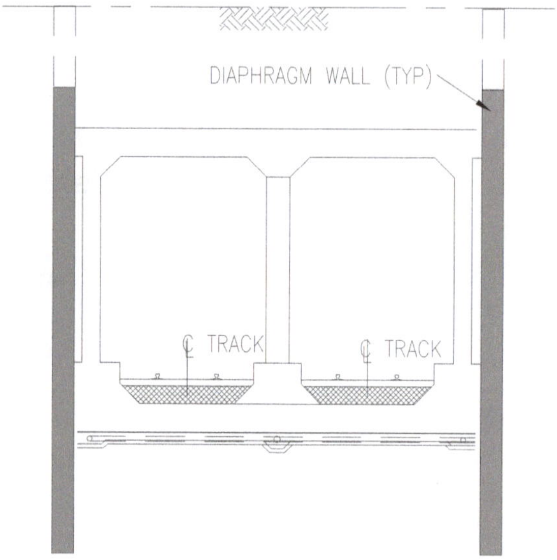

Fig. 5.67: Section of Semi-Integrated DW

Figure 5.66 includes strut locations in addition to the top deck, for top-down construction, at locations where the RTS corridor crosses an important road, with or without backfill and provision of a traffic deck. **Figure 5.67** shows a section of the RTS corridor during operation. **Figure 5.68** includes a view during excavation in J.N. Road, and **Figure 5.69** includes a view where a Hoe-excavator was possible to use.

Fig. 5.68: Strutted Excavation in JN Road

Fig. 5.69: Excavation Using Excavator

Part of the section was planned to be excavated using a Drag Line **(Fig. 5.71),** where a site of sufficient length and encompassing free area was found. Available space in Chitpur area was limited due to operating railway infrastructure, including multiple tracks **(Fig. 5.70),** and excavation was carried out using special construction equipment.

Fig. 5.70: Special Unit Used for Excavation in Chitpur Yard

Fig. 5.71: Excavation Using Drag Line

5.15.7 DW with Struts

Panel widths of the diaphragm wall were generally limited to 3.0 m to restrict the construction time for each panel to less than a day. So that it minimises disturbances to adjoining structures due to construction activities. The 3.0 m panel width also facilitates the adoption of a single strut system for each panel. The excavation time for each 3.0 m wide and 16.0 m to 23.0 m deep panel was 4 to 5 hours. Bentonite slurry of specific gravity 1.07 to 1.10 was used. The Bentonite slurry was recycled and reused after being pumped back to vats. 3-6 hours of continuous concrete work were required for each panel. The end of adjacent panels was interlocked by forming semicircular concave and convex ends in alternate panels using cylindrical form tubes 600mm in diameter. Panels were cast using a concrete tremie.

The struts were in the middle of the 3.0 m panel. The Wales were built into the diaphragm wall panels. However, Wales was used for options with Sheet Piles and H-piles and timber lagging. The vertical spacing of the tiers in the northern section, towards Shyambazar, was limited to 3.0 to 3.5m, whereas in the southern section, towards Tollygunge, it was up to 4.5m. The reasons for such close spacing of struts were:

1. Cuts were more profound in north than the South Section and varied between 13.0 to 14.0 m.
2. Cuts were broader to provide longitudinal traffic and construction decking.
3. A series of old buildings were located near each other on either side of the diaphragm walls, and accordingly, the ground settlement had to be minimised.

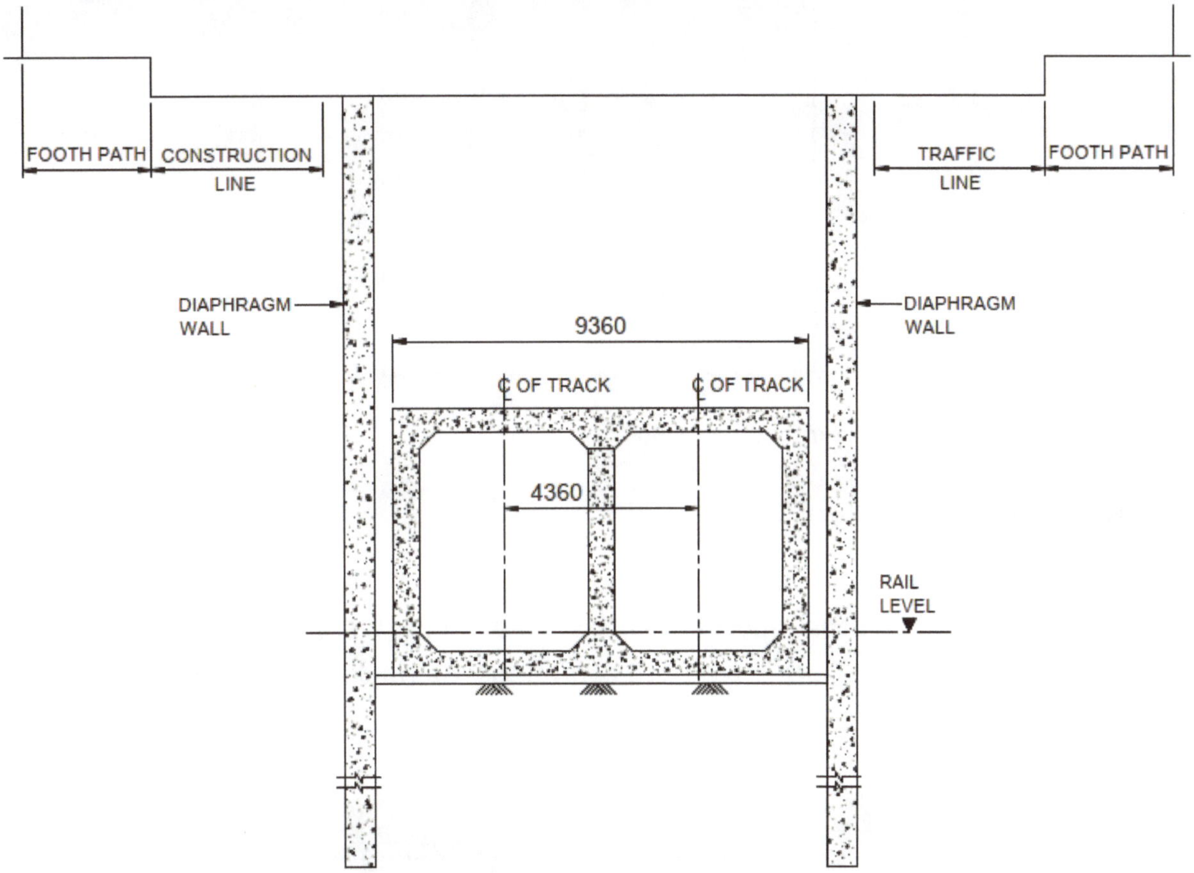

Fig. 5.72: RTS Box Showing Construction Lane & Traffic Lane abutting Excavation on either Side
Note: RTS box independent of DW and Safety Parapet on either side of trench not shown.

In the original planning, the first layer of temporary struts was to be de-loaded and reused for the second layer of temporary struts below the second-level permanent struts. However, practical difficulties arose at the site in de-loading and reusing within the confines of the cut. Therefore, these were also left as permanent struts and removed at the time of box concreting. The design of compression struts accommodates earth pressure on the diaphragm wall. The strut sections were standardised and made with two nos. ISMB 600 joined by batten plates. This enabled reuse. The struts were preloaded using hydraulic jacks to fit in position along with timber/steel wedges.

Figure 5.73 includes the pressure as per tender provision for cohesionless soil (excluding Water pressure) and for deep excavation.

Figure 5.74, a photo taken in JN Road where DW extended up to Road Level and top layers of Struts in position.

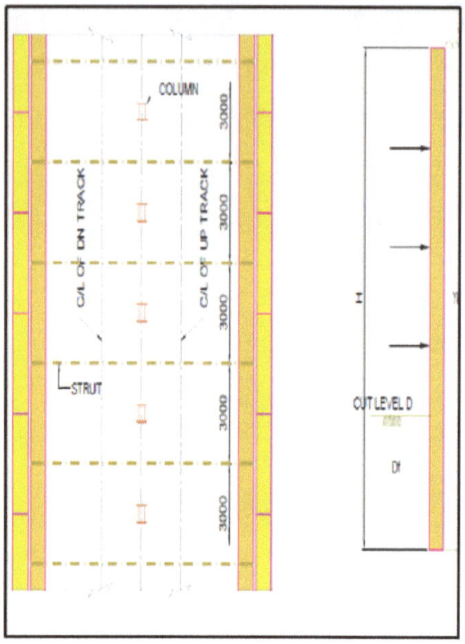

Fig. 5.73: Diaphragm wall pressure diagram for Granular Soil

5.15.8 Walls using Steel sheet piles

DW was mainly used in the North-South Metro as shoring walls, except for a short section integrated with the RTS Box. In addition, other approaches, like Sheet Piling, were adopted in a few sections of the N-S metro corridor commensurate with site conditions. All these systems were further detailed in the section covering East-West Metro in Chapter 6 of the book.

Fig. 5.74: Excavation in Progress with Struts in Position

(Note: Shoring wall is made of DW extended till Road Level where required and Struts are in Position)

The sheet piling substructure/foundations retain soil and are subjected to soil pressure. They were used in the construction of temporary and permanent works. The sheet piling may act as a cantilever wall, with or without bulkheads. A tie bar anchored back into the stable soil volume holds the sheet pile at the top. Anchors can be one or more anchors to help the excavation for a deep cut using strutted excavation.

In strutted excavations for N-S Metro, back anchors in the soil were replaced by struts for each Sheet pile. This approach was used between two parallel rows of Sheet piles. The excavation would have been unencumbered in case of use of back anchors. Total excavation could have been done mechanically, making it faster and cheaper. However, the same could not be used as the size and strength of sheet piles required to satisfy the design requirement besides the necessity of long anchore required to satisfy the coil strength. The sheet piles were used only, which were available and satisfied the design and site requirements.

For N-S Metro, Russian Engineers prepared the concept, and sheet piles were proposed for shoring. During that period in India, the available length of sheet piles was 7 to 8 meters, and the section modulus needed to be better used in shallow to medium excavation. To use anchors, the required length of anchors was too long to satisfy soil properties and use of multi-anchor design was found to be messy. And so, use of Anchors was replaced by Struts. Sheet piles required, mostly, were to be imported with the required section modulus. However, for the N-S metro, a small section for which sheet piles used were available in India and was good for shallow depth. The proposal for imported sheet piles was discarded. Different views of Strutted excavation with Sheet Piles are included in **Figures 5.76 and 5.77.**

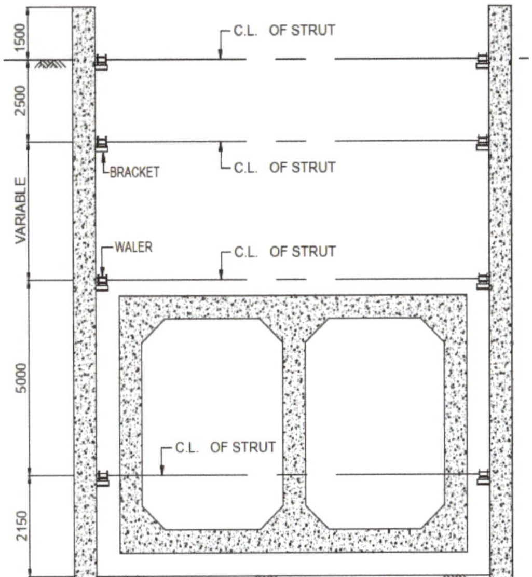

Fig. 5.75: Sheet Piles with Struts and RTS Box

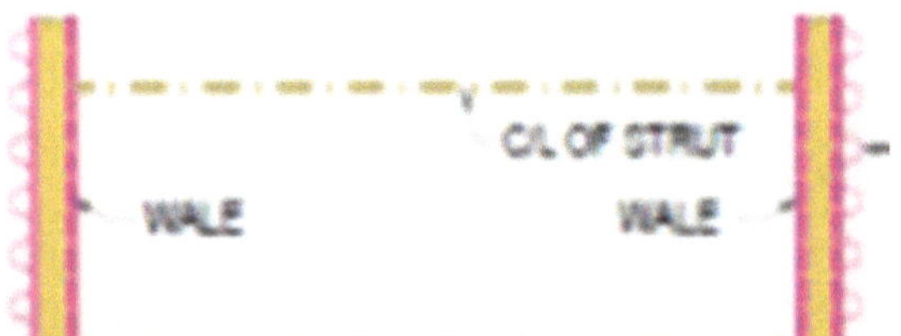

Fig. 5.76: Plan of Sheet Piles including Wales and Struts

Fig. 5.77: Sheet Piles with Struts in Position

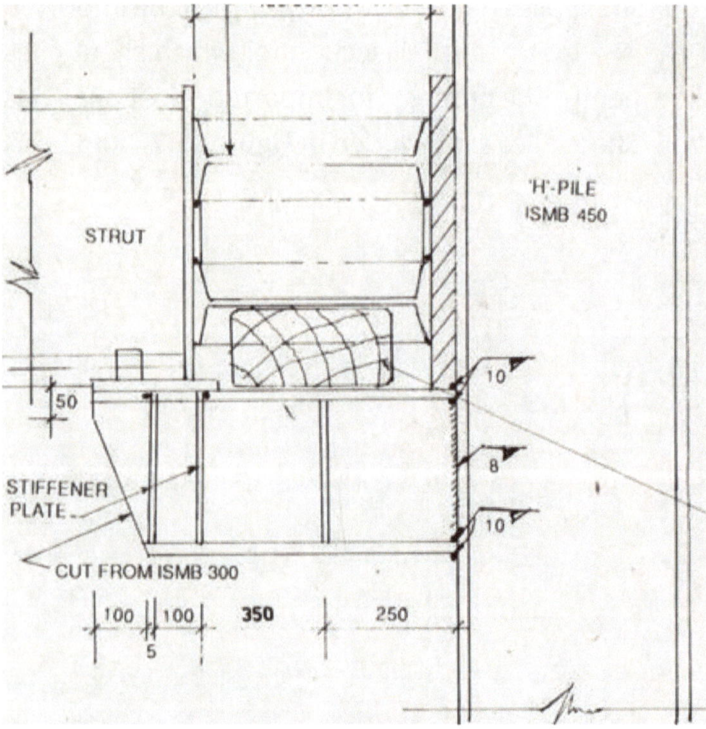

Fig. 5.78: Junction Between H-Pile and Wale

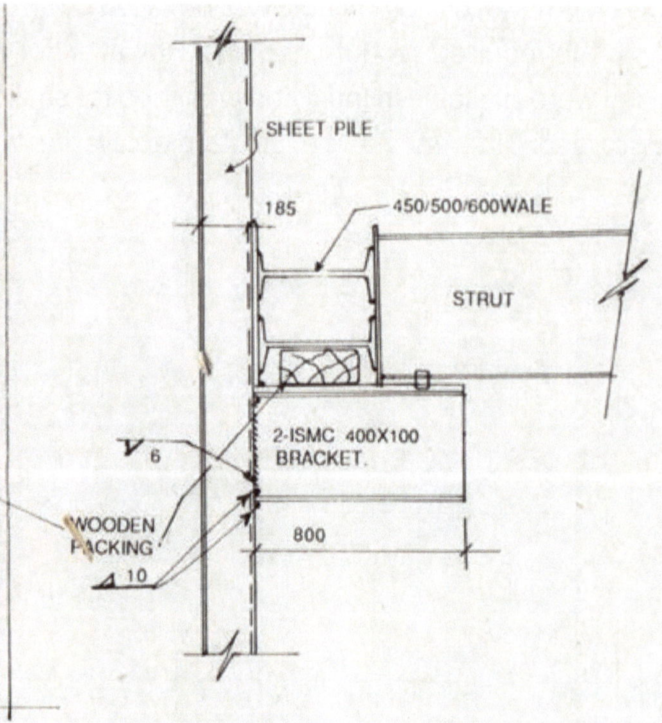

Fig. 5.79: Junction with Sheet Pile

The junctions between the Struts and the Sheet/Soldier piles were like **Fig. 5.78 and Fig. 5.79.**

But without Wales. Struts were fixed directly with DWs using steel brackets to facilitate the erection and jammed with DW's inner side using seam plates or timber wedges.

5.15.9 Steel H-piles with timber/steel lagging

The excavation was carried out in sections, mainly in the Maidan Area, free of chartered utilities, using Steel H-piles for the uppermost part and supported on diaphragm walls are cast below. The strut sections were standardised and made with rolled sections **(Fig. 5.83),** joined with battens as per the approved design. These were designed as continuous walls over Wales/Struts for maximum earth pressure in soft, medium clay soil with other surcharge loads. External steel Wales were provided to act as cross beams for steel strut loads, and the continuity of the sheet was ignored in the longitudinal direction.

At the top, above DW, one Strut was used and supported in Wales. The Wales spanned longitudinally between H-piles. Wale and Strut junctions were designed as semi-rigid Junctions. Lower struts were fixed against the DWs.

This method was adopted, in general, for sandy soils, shallow cuts, and at ground levels, to avoid damaging the existing utility and to provide future utility. Timber lagging spanning between steel H sections at intervals of 1 to 1.5 meters, driven/or embedded in DW cast approximately 3m below the grade level for the transfer of earth pressure for shallow to medium depth. A maximum earth pressure intensity based on Terzaghi-Peck's empirical formula was adopted. Other lateral pressures due to surcharge loads were considered. **Figs. 5.80, 5.81, and 5.82** include views of steel H-piles with timber lagging for the North-South Metro. H-pile and timber lagging were used for the top portion of the excavation and embedded in DW starting 3 meters below. This approach was used to save costs and keep the layer above DW free of encumbrance.

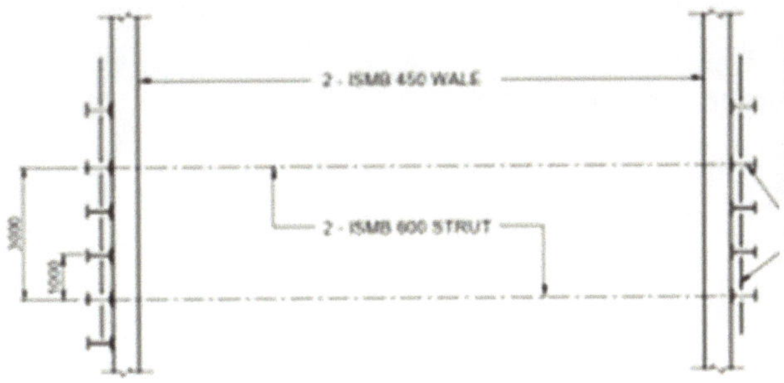

Fig. 5.80: Sectional Plan-H-Pile & Timber Lagging

Fig. 5.81: View During Construction

Figure 5.77 is a plan view and includes the spacing of soldier (H) piles and timber lagging. **Figure 5.78**: View During Construction, and **Figure 5.79** shows the RTS Box, and the approach was to lay concrete in the bottom slab. **Figure 5.80** includes standard steel built up sections used as struts for all types of Shoring Walls.

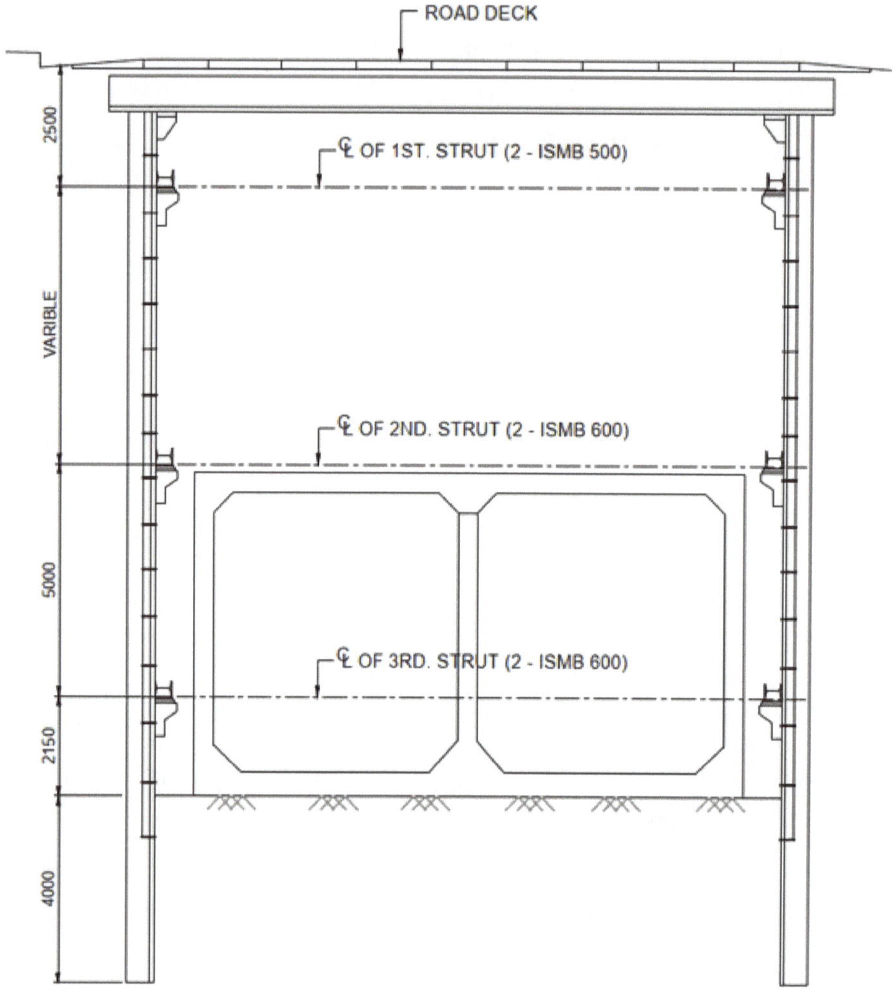

Fig. 5.82: H-piles, Lagging, Struts & RTS free from Shoring Walls for Bottom -Up Construction

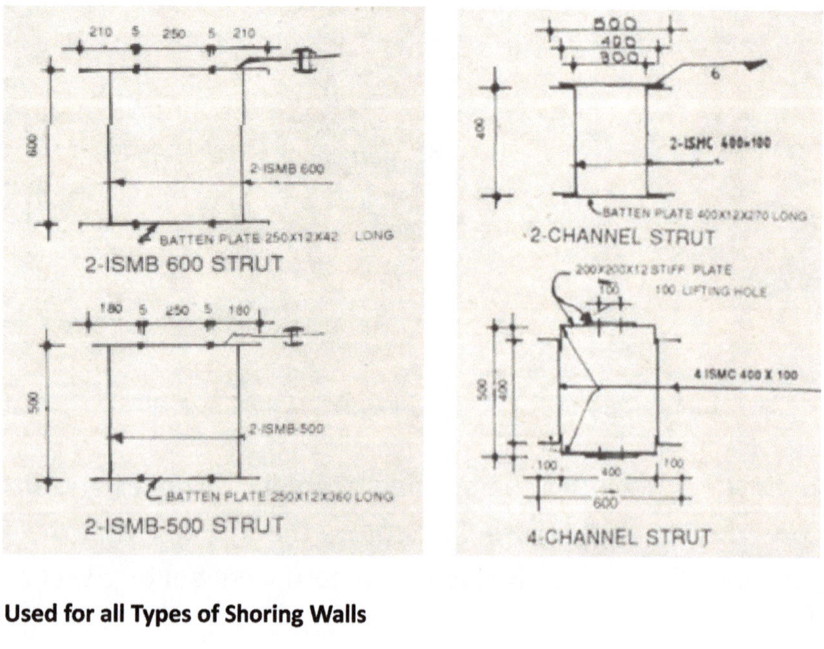

Fig. 5.83: Strut Sections Used for all Types of Shoring Walls

5.15.10 Contiguous Bore Piles

Contiguous pile retaining walls were constructed using the abutting bored piling approach. They are an ideal solution for urban areas as they can be built without encroaching on adjacent properties. Contiguous pile walls were formed by constructing a line of circular piles abutting each other. This approach was used for a small section, where the construction of DW or Sheet-pile walling was not possible. The construction of a single box (a length of 187 meters) between Belgachia and Chitpur yard adjacent to the completed circular tunnel, using compressed air technique, was done by 550 mm dia. cast-in-situ RCC piles driven continuously in line of Diaphragm wall. As was anticipated, the standard technique of diaphragm wall of 3-meter length might affect the already completed circular tunnel. The strut sections were standardised and made with rolled sections, joined with battens as per design. While standardisation, there were two types, one for the top strut (for solder pile and lagging using Wales) and the second at deeper levels with heavier sections. In this case, Struts were erected against steel Wales in the longitudinal direction. Reference may be made for the sequence included **in Figures 5.84, Fig. 5.85 and 5.86.**

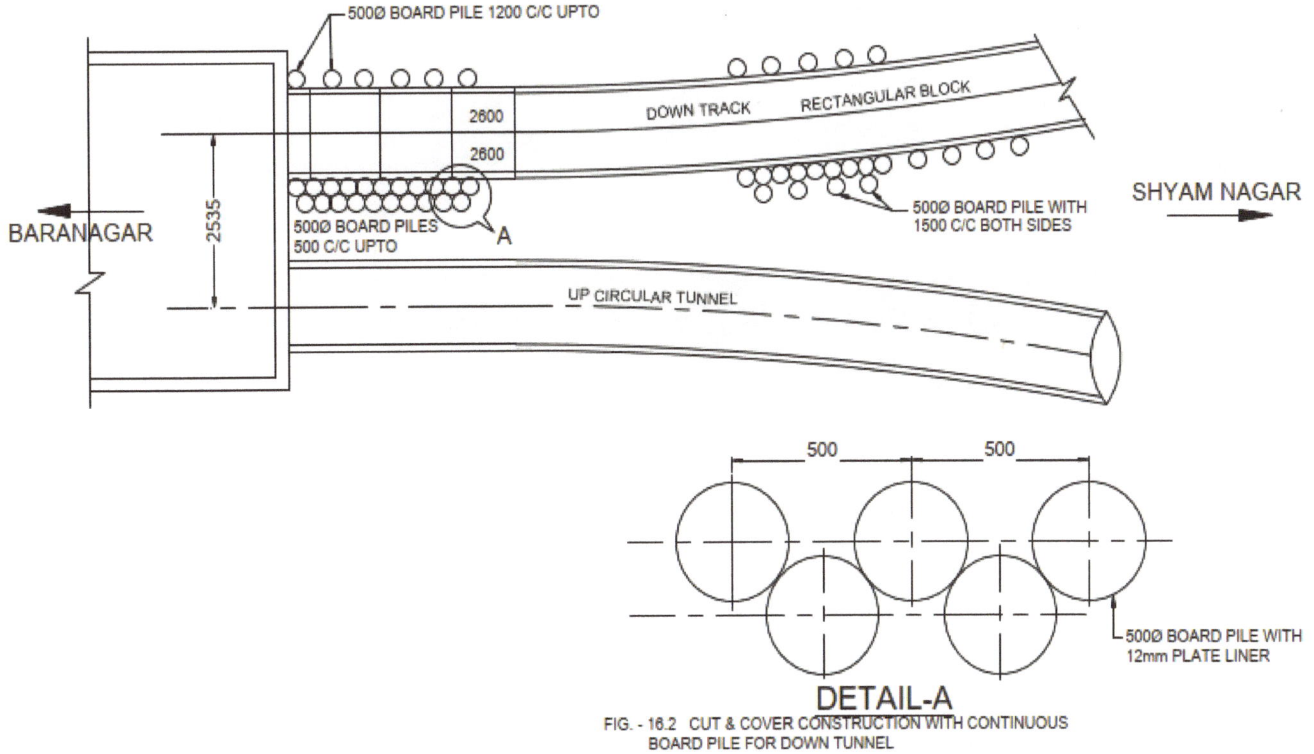

Fig. 5.84 & 5.85: Retaining Wall made of Bore-cast in Situ Piles

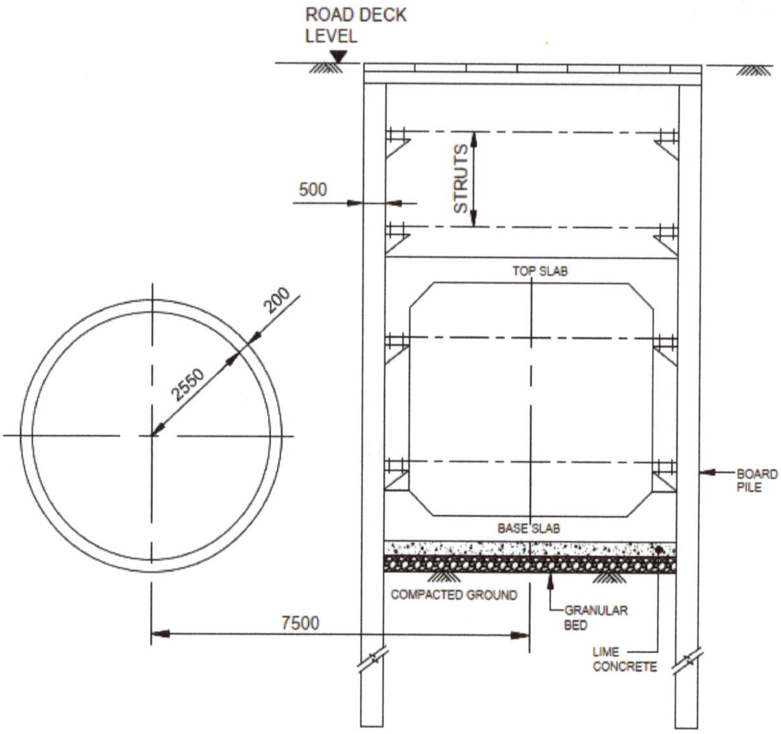

Fig. 5.86: Bore-Pile wall Abutting Tunnel

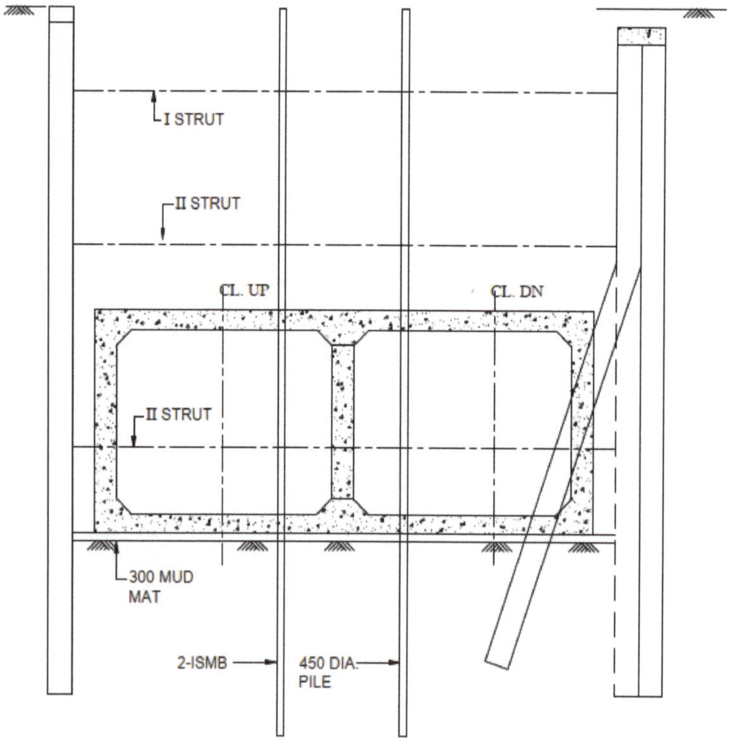

Fig. 5.87: Tilted DW Intruding on the UG Corridor (Exaggerated View)

Figure 5.87 includes the number of bored piles used in RTS, with wider sections, to reduce the slenderness ratio of struts. In a few locations, DW intruded on the RTS box. The intruded DWs had to be dismantled to clear the corridor, stations and/or sub-stations. And it could be done only after constructing DWs encircling the distorted section, as shown in **Figure 5.88**. It needed to work out a proper design and construction approach, including precautions, so that new walls did not intrude on the corridor.

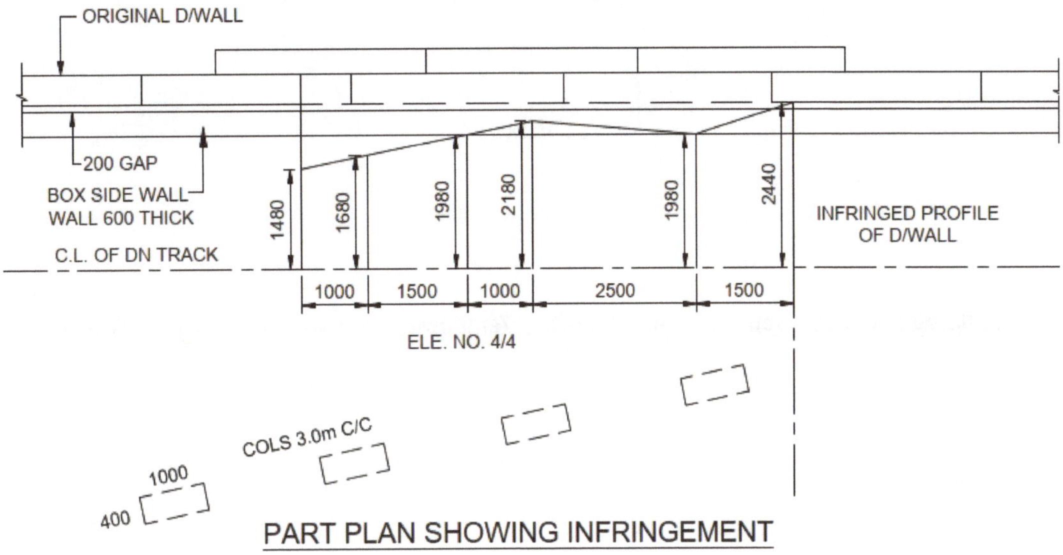

Fig. 5.88: Additional Diaphragm Walls to replace tilted DW (Schematic)

Figure 5.88 also includes the installation of bored piles in wider sections of the corridor, such as at Stations or Substations, as discussed in earlier paragraphs. The design needed one or two piles, depending on the width of the excavation. These piles were also used as piers to support a temporary road deck, where applicable and to reduce the slenderness ratio of struts as per design drawing. It needs preplanning, a construction approach, and the details between the junction and the bored piles to be used at the final stage of construction. Station Box or in other areas as required, ceased their use, or the construction of the Top slab was completed for Bottom-Up construction, and the temporary struts were removed. In case of Top-Down construction, in a few sections after the top slab was completed, Struts were removed, and the piles were converted to piers after appropriate junction details were completed to take the design loads. **Figure 5.90** includes the reinforcement details of the Bored Piles.

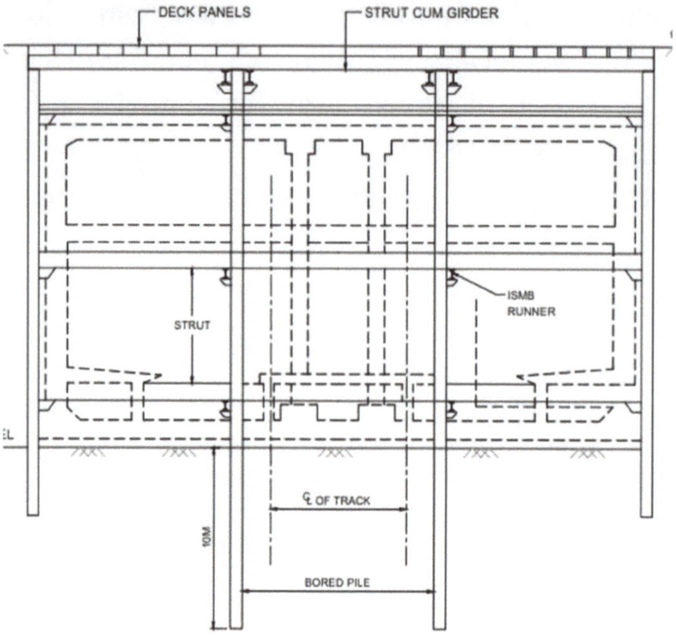

Fig. 5.89: Temporary Bored Piles at Station Locations Supporting Temporary Deck and Reduces Slenderness of Struts

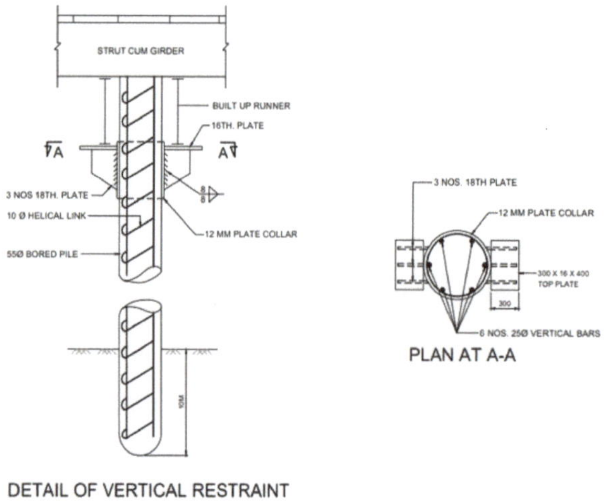

Fig. 5.90: Details of Bored Piles

5.16 Special Diaphragm Wall

5.16.1 Around Shiv Kali Temple

A case of Special Diaphragm Wall construction was undertaken around the Shiv Kali Mandir **(Fig. 5.91)**. This Temple is in the middle of the road, north of the Sova Bazar Metro Station.

The temple structure could not be shifted from its existing location, so the alignment of RTS corridor was modified to suit the site condition and was constructed on either side of the temple in the available space. To protect the temple structure from distress during construction, DWs were around the temple, and access to the temple was provided through decking on all sides. Special supporting arrangements for diaphragm walls **(Fig. 5.92)** were designed, stood well during construction, and had the least disturbance to the temple structure and activities.

Fig. 5.91: Shiv Kali Temple

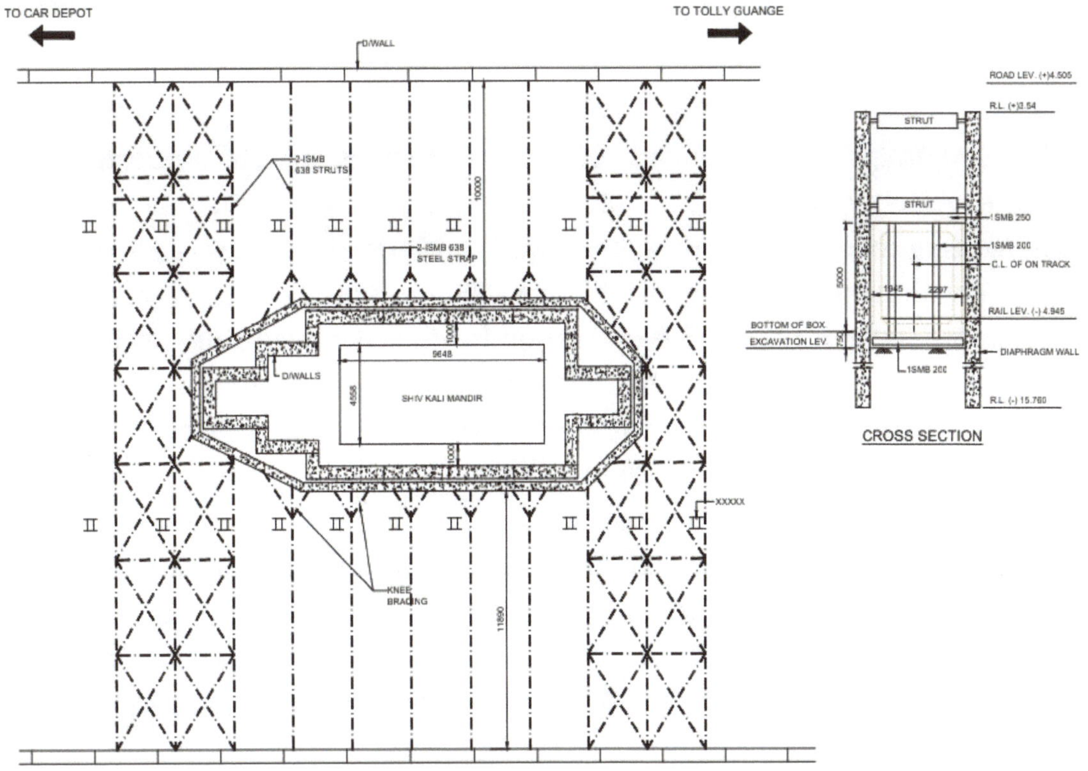

FIGURE 30 TIMBER PILE ARRANGEMENT

Fig. 5.92 a & 5.92 b: Construction of DW Around the Mandir and Cross Section

A temporary deck was installed in front of the temple and removed after the access road was reconstructed. **Figure 5.93** includes the single track up/down on either side of the temple for operation of the metro corridor.

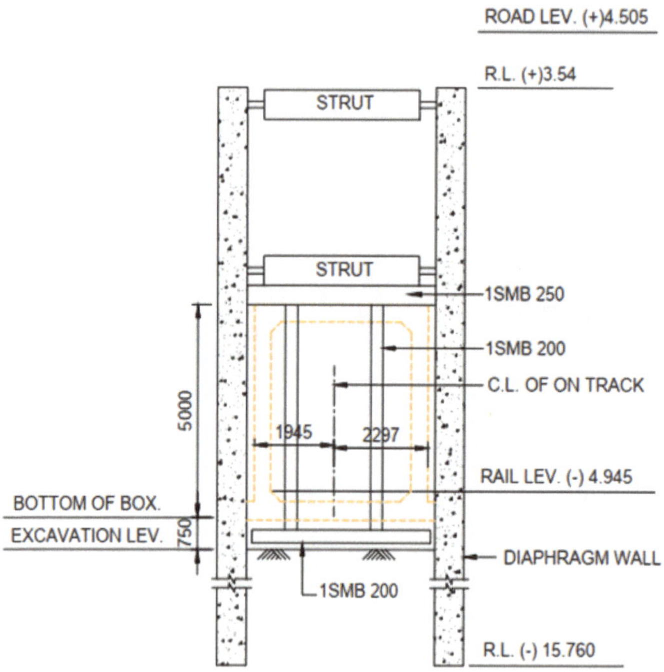

Fig. 5.93: RTS location shown for One Side of the Temporary Decking

5.16.2 Construction Under Existing Tollygunge Railway Bridge

The Existing Railway Bridge at Tollygunge is of two short spans, simply supported, with girders supported on one Masonry pier in the middle and two masonry abutments on either side. The Bridge is on an existing road and of low height. The N-S Metro continues as an underpass below the existing road, dividing the corridor with separate RTS boxes, and each box was implemented in each span, as detailed in **Figure 5.94.** The excavation for the DWs was done using specially made equipment due to space constraints and two diaphragm walls for each box. Excavation between the DW s in the section was made with equipment below the first strut level. The excavation was then completed using grabs up to the full depth underwater. The U frames were positioned, and the bottom plug was concreted by tremie concrete underwater and at levels of the bottom plug. Water in the enclosed DW pit was pumped, and second layer of the strut, erected. The water in the pit was then pumped out to full depth after providing a counterweight with sandbags on the U frame to neutralise the uplift pressure below the bottom plug. The box was then constructed under dry conditions. Figure **5.95** shows the plan of the shoring wall used in combination with bored piles, cast using a steel tube.

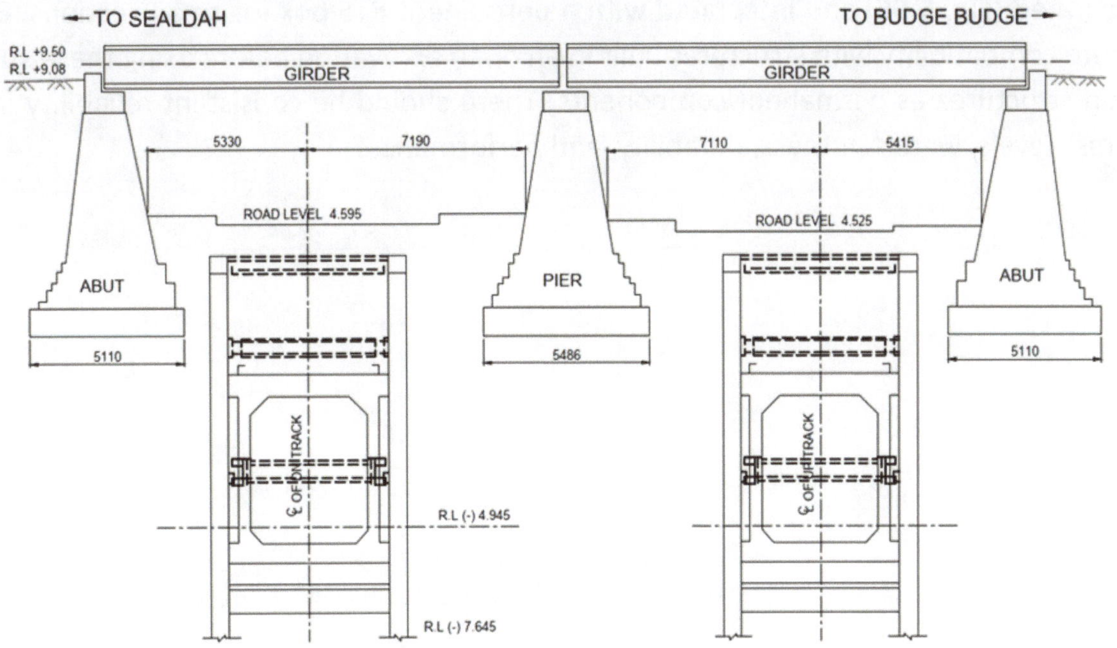

Fig. 5.94: RTS in Two Parts Under the Existing Railway Bridge

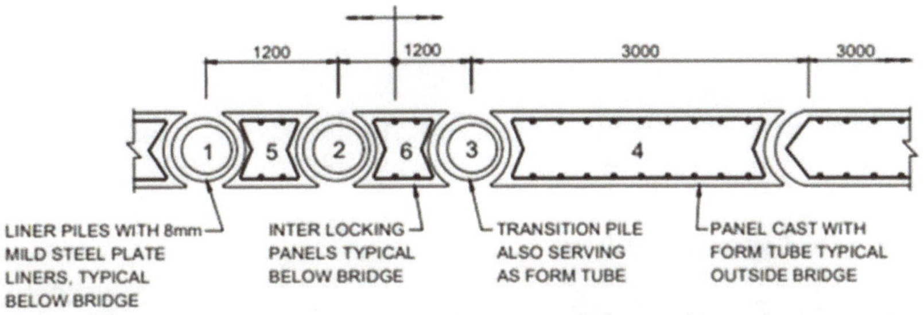

Fig. 5.95: Plan of Short-length DW Integrated with Bored Piles

5.17 Design of Diaphragm Wall (DW)

(Typical Calculation of N-S Metro DW included in Annexe 2)

This section discusses the design basis development achieved in India during the last four decades concerning Indian rail-based structural systems. Design and construction methods adopted in the first Indian metro during the 1970s and being adopted today are included in the relevant paragraphs in East-West Metro, Chapter 6, Part 3, and Mumbai Line 3 in Part 4.

In the section on the East-West Metro, an overall approach to metro design is compared between the North-South and East-West Metro to get a clearer picture of the paradigmatic changes, boosted by technological development, over the period.

Diaphragm walls constructed using the slurry trench method achieve their best economy when used as part of a permanent underground structure. In the North-South metro, the diaphragm wall was semi-integrated **(Fig. 5.96)** and integrated with a permanent RTS box for small sections. Engineers must assure compatibility with structures built in an open excavation before using these support-of-excavation structures as permanent components. There should be consistent reliability in applied loads, stress levels, watertightness, durability, and performance.

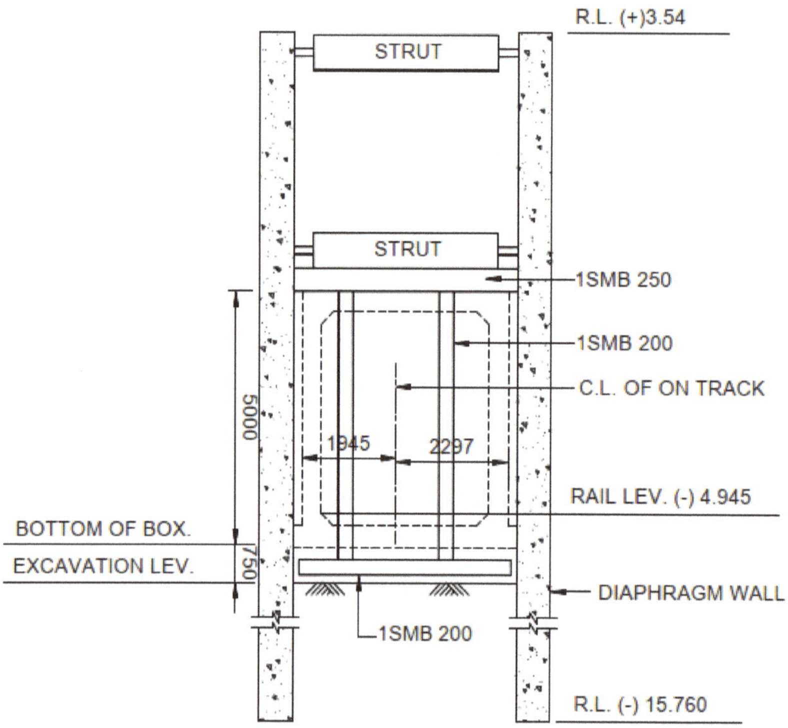

Fig. 5.96: Semi-integrated Single Cell Box for Single Track

Soil mechanics is an art rather than a science; this definition is inherent, like soil. It is usually a heterogeneous material, sometimes artificially deposited, subjected to varying degrees of consolidation, surcharge loads, and pore-water pressure, and all these conditions are subject to considerable variation from one point in the ground to another. Even the best available methods for determining the parameters that define soil properties give results that might vary by 100 per cent from actual conditions. When the variations of properties from point to point are considered, assessing average soil properties for design purposes becomes an art and not science. Assumptions for earth pressure and its distribution should be consistent with those used for the soil properties' parameters. Field measurements of earth pressure behind braced retaining structures show considerable variation. Actual performance cannot be correlated with sophisticated assumptions, and their use has little justification. Earth pressure and its variation are materially affected by the excavation method, the construction sequence, and the installation and possible prestressing of bracing.

In the case of soil parameters that define the properties of the soil, simplifying assumptions are warranted in establishing design earth pressures on the diaphragm wall. Goldberg, Jaworski and Gordon have comprehensively studied earth pressures on lateral support systems. The recommendations are shown in **Fig. 5.97**, where the first three cases are from Terzaghi and Eck's empirical data for flexible walls.

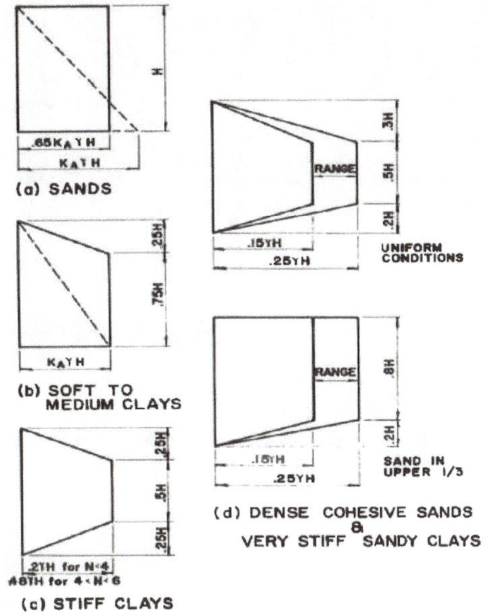

Fig. 5.97: Pressure Diagram as per IS 9556

North- South (NS) Metro, Kolkata, the first metro in India, with Internal funding, mostly underground, started in 1974; following experiences gained across the world during those days, analysis, particularly for rigid structures like diaphragm walls (DW), was adopted. DW was used mainly for shoring both sides of the metro corridor, adopting strutted excavation and considering earth pressure as pressure at rest. Strut spacing was based on earth pressure and DW rigidity. Like Peck's apparent earth pressure distribution, pressure models were similar for cohesive and cohesionless soils. **Fig. 5.98.**

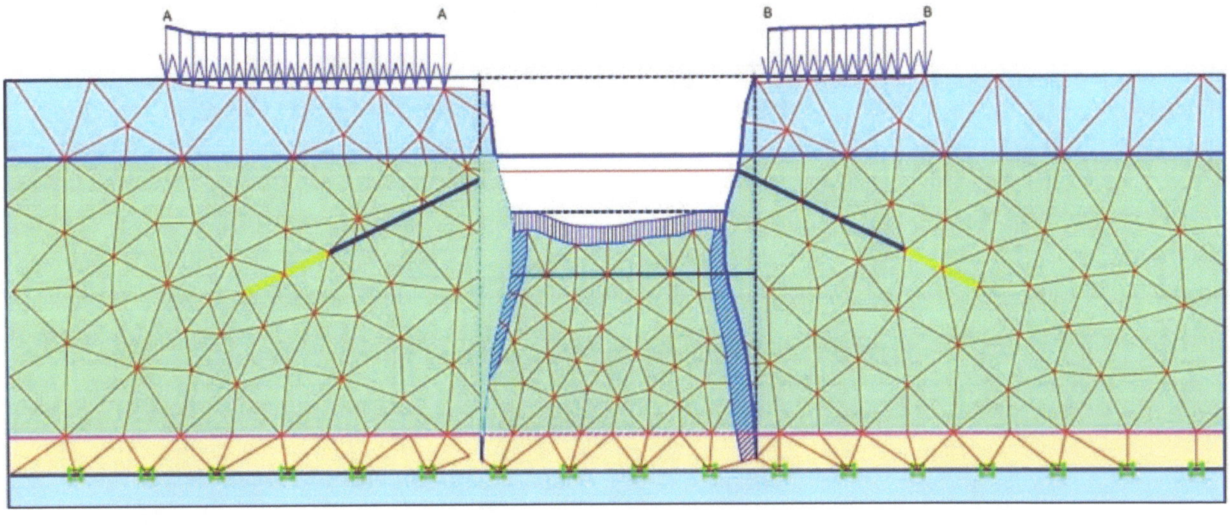

Fig. 5.98: Earth pressure distribution.

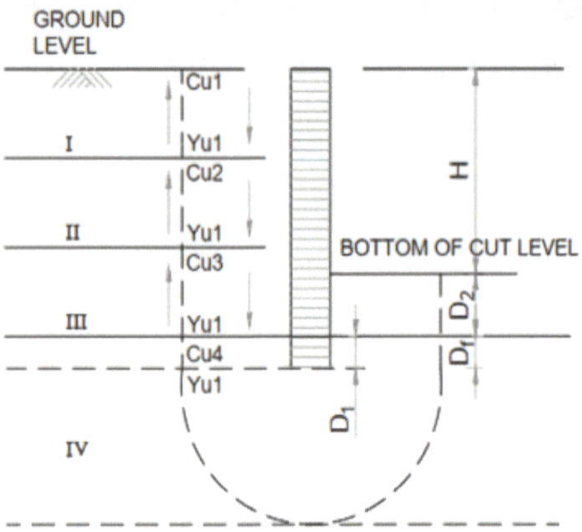

Fig. 5.99: Stability Against Bottom Heave

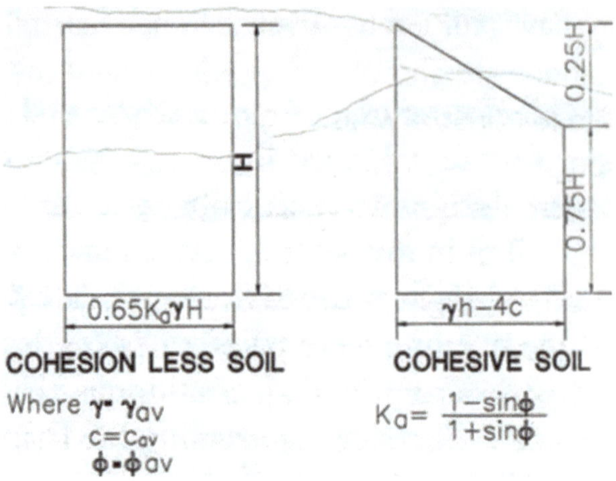

Fig. 5.100: Peck's Apparent Earth Pressure Diagram

$$F.O.S = \frac{Cu_4 Nc + r_3 D_2 + r_4 D_1 + \sum Cu(H + D_2 + D_7)/D_1}{\sum r(H + D_2 + D_r)} \geq 1.5$$

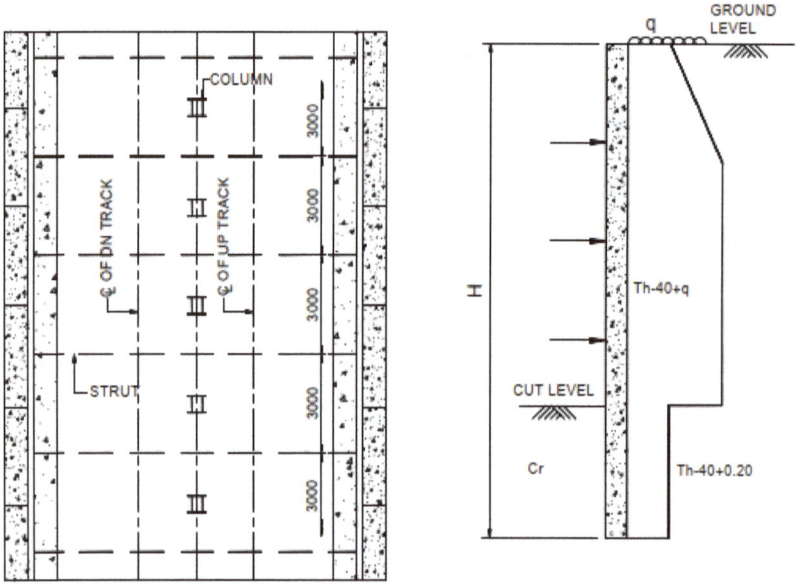

Fig. 5.101: Pressure on Diaphragm Wall as per KMRC

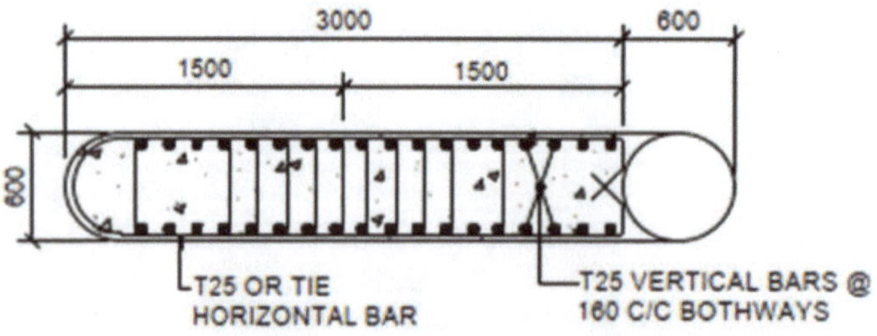

Fig. 5.102: Typical DW Section in Plan with Reinforcement

Figures 5.99, 5.100 and 5.101 show Bottom heavier and the typical earth pressure distribution, the empirical formula adopted, and typical plan with strut locations. Whereas **Figure 5.102** shows a sectional plan of DW with typical reinforcement. Engineers now have more sophisticated tools to perform multiple analyses using different parameters.

The underground corridor structure for the N-S metro is an RCC box of 500 mm thick side walls, with 500-600 mm thick top and bottom slabs, and a row of columns or two equally placed at the center from the center of the alignment, in case the gap between two parallel DWs is more than, say, 9m, such as in Substation and Station areas (mostly 19m in width) to reduce the slenderness. Substation roof at General's Tank is 2000mm, intermediate slab & Base slabs were substantially thicker than RTS box section. A gap of 500mm was kept between the diaphragm wall and RTS box wall, filled with earth. This gap takes care of thefilledin the diaphragm wall construction and increases the overall construction width. As applicable, the box was constructed between two parallel DWs of 600/800 mm thick. Measurement of diaphragm wall movement, strut load and ground settlement had been made at different sections of the metro construction. **Figure 5.104** gives the diaphragm wall deflection, corresponding to strut locations and soil properties, shown in **Figure 5.103**, at different stages of excavation. These measures helped to control the effect of ground movement on nearby buildings on the highly congested Chittaranjan Avenue in Central Kolkata. These buildings were mostly of RCC framed construction or load-bearing construction. Detailed instrumentation was carried out in the Maidan area and a few other places along the corridor.

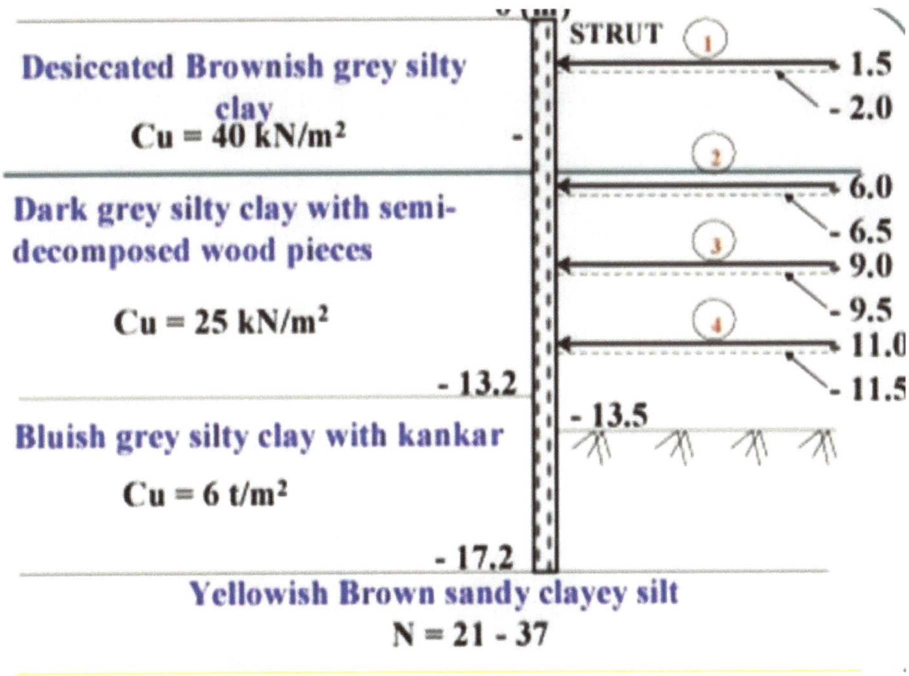

Fig. 5.103: Excavation Using Bottom-Up Method with Struts and Typical Soil Profile

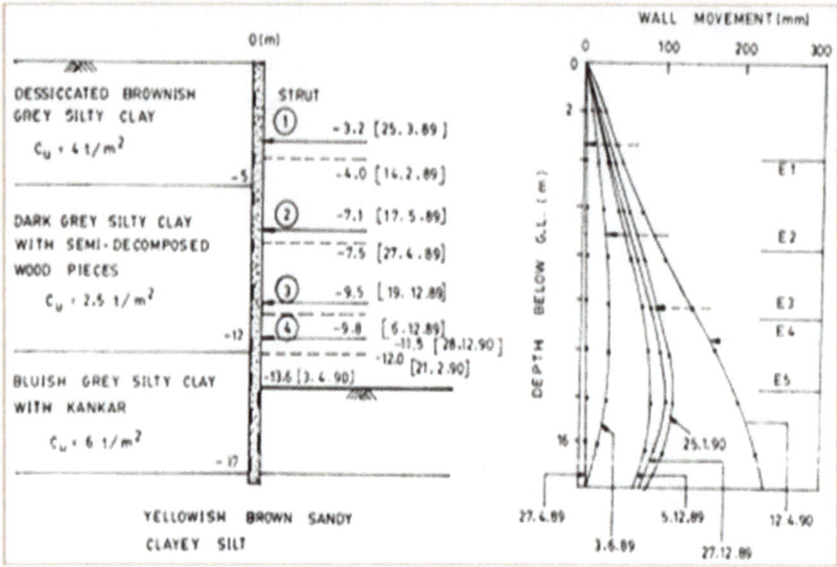

Fig. 5.104: Deflections of DW During Different Stages of Excavation

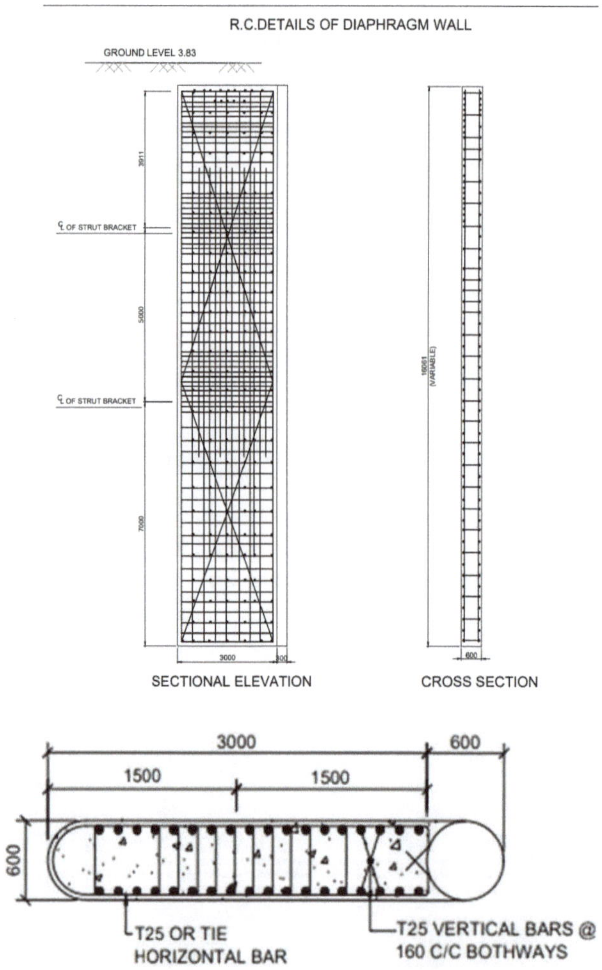

Fig. 5.105: Typical RC Details of Diaphragm Wall

Typical reinforcement details for Diaphragm Walls are included in **Figure 5.105.**

5.17.1 Cross Diaphragm Wall

Cross diaphragm walls were constructed along B.B. Ganguli Street at the junction of C.R. Avenue and B.B. Ganguli Street, with their cut-off level kept considerably below the floor slab level of the Central Station box. This was done to facilitate the future construction of the East-West Metro line between Ramrajatala and Salt Lake as was planned earlier. Proposed Central Station of East- West metro was on Bowbazar Street near the intersection of Chirraranjan Avenue and Bowbazar Street. The design of N-S metro in the area kept a provision of integration/Transfer station with E-W metro. E-W metro was planned below N-S metro corridor.

5.17.2 Remedial Measures

Cases of 'leaning diaphragm walls towards the cut occurred in both the north and south sections. The incidences were comparatively less in the north. The maximum infringement was to the order of 800 mm. As there was no scope to shift the affected downline inwards, the only alternative left was dismantling the infringing diaphragm wall panels to obtain precise moving dimensions. To facilitate this, enveloping diaphragm walls were introduced at these locations, outside the affected diaphragm wall. The tilted diaphragm walls were dismantled during the excavation.

5.17.3 Strutting in deeper cuts

The struts were in the middle of the 3.0 m panel. The Wales were built into the diaphragm wall panels. The vertical spacing of the tiers in the northern section was limited to 3.0 to 3.50 m, whereas in the southern section, it was up to 4.5m. The reasons for such close spacing of struts were:

i. The cuts were more profound in North than the South Section and varied between 13.0 to 14.0 m.
ii. The cuts were broader for longitudinal traffic and the construction deck.
iii. A series of old buildings were situated near each other on either side of the diaphragm walls.

In the original planning, the first layer of temporary struts was to be de-loaded and reused for the second layer of temporary struts. Practical difficulties arose at the site in de-loading and reusing, in a few cases, within the confines of the cut. In such cases, Struts were left as permanent struts and removed at the time of the concreting of the box. A uniform strut section with ISMB 600 or other sections was adopted for all the struts and was fixed between the D-wall for the road girder and the girder. **Figure 5.83** includes standard built-up sections used. Decking was designed and erected with possible reuse. The sequence of Strutting, facilitating Mechanical Excavations, after placing the struts in position for a width of 14m in the case of Stations were worked out and included **in Fig. 5.106** and **5.107.**

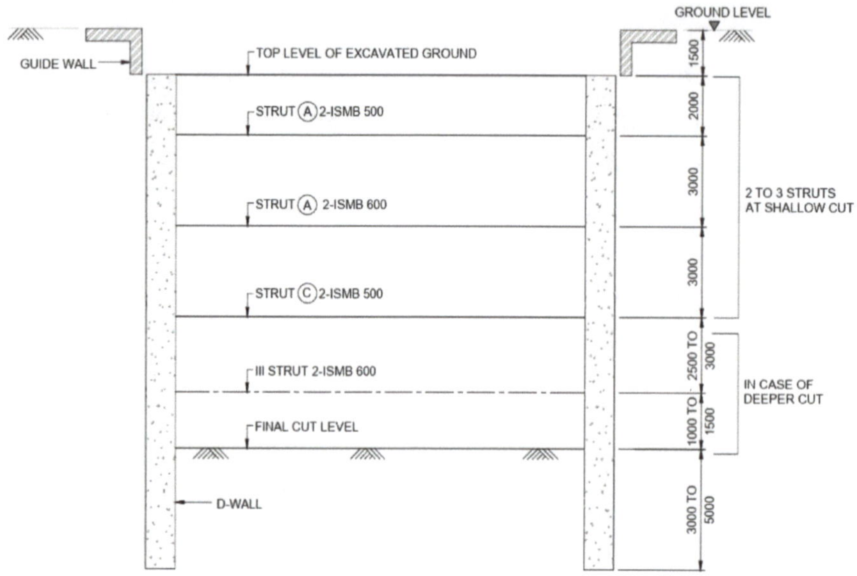

Fig. 5.106: Strut Level for Corridor Towards Shyambazar

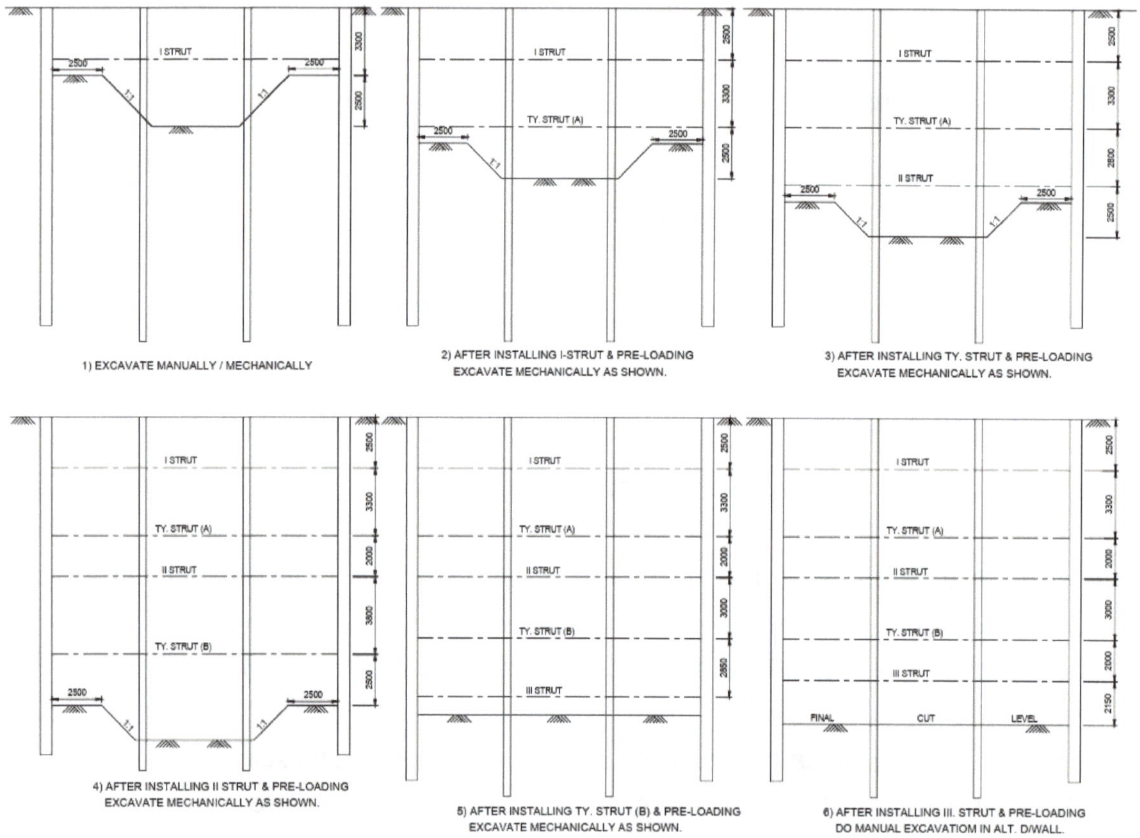

Fig. 5.107: Sequence of Strutting and Excavation

The excavation was done below after struts were erected and pre-loaded in stages. Mud Mat was cast at the final level. The box was constructed in phases, and the struts were removed. Waterproofing was done, including in the latter part of the book. After the RTS boxes were done, backfilling was carried out on the top of boxes. Road decking and girders were removed, and the road was completed and restored. The projected U-bars, left on the bottom slab for integration of concrete plinth **(Fig. 5.108).**

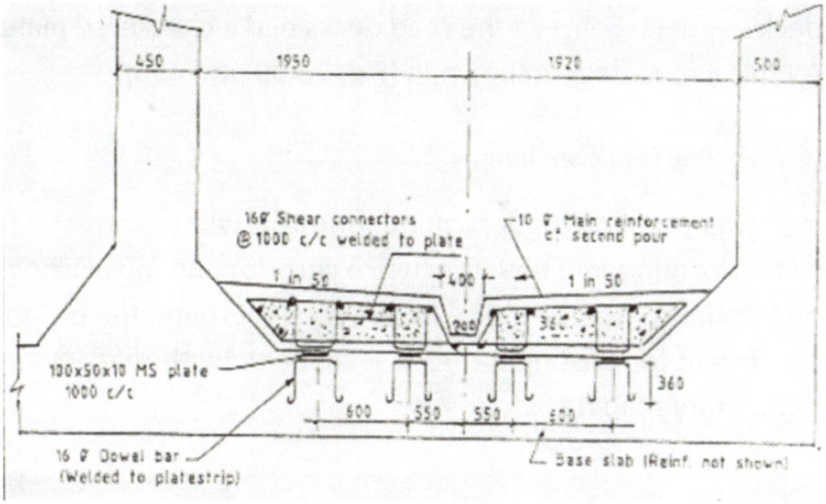

Fig. 5.108: Part Section through One Track of RTS

5.17.4 Road Deck

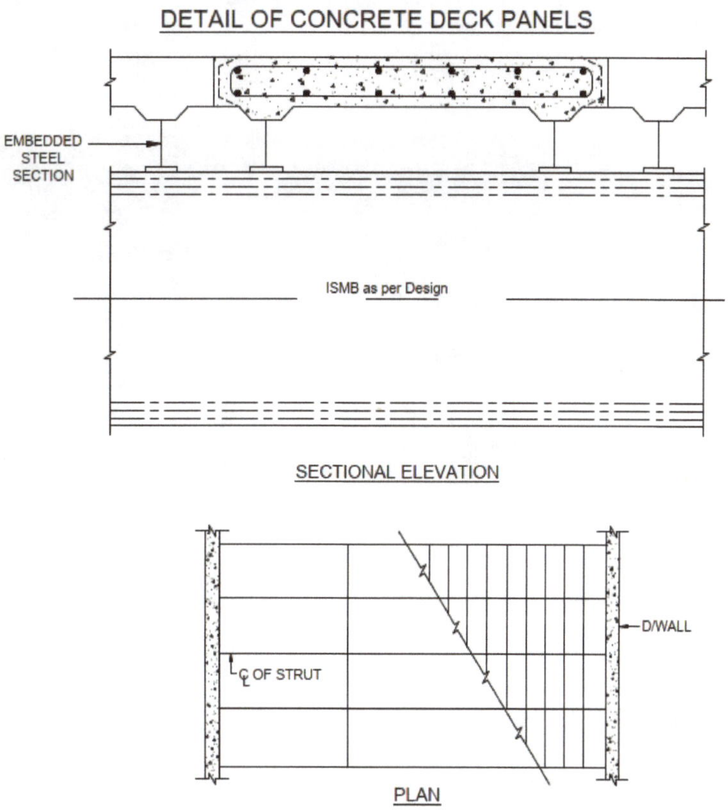

Fig. 5.109 & 110: Plan and Section of a Typical Deck at Road Level

Most of the south section was constructed without a continuous road deck. The road deck was restricted to serve during construction, and the traffic deck primarily was at the crossing. Llarge-scale road decking was done in the north section between Esplanade and Shyambazar. Road decking was completed using composite precast road decks supported by top-level struts, which in turn supported the road deck's girders. Some of the road decks had a checkered plate topping instead of a concrete topping. A typical road deck is shown in **(Fig. 5.109 and 110).**

5.17.5 Temporary Traffic Decking for Diversions

The N-S Metro corridor was primarily underground, and the width of existing roads was limited, which was inadequate to accommodate both the metro corridor and the existing roadway for traffic. It was worse at station locations. In some areas' road network around the proposed metro corridor, with narrow roads, insufficient for existing traffic, was made up with road widening where possible and/or constructing temporary road deck.

Fig. 5.111: Diversion of Tram Line near Birla Planetarium

In such cases accommodating diverted traffic was possible, adding a sufficient temporary road deck parallel to road on the cut for the metro **(Fig. 5.111). Figure 5.112** includes traffic diversion plan where existing road widths are wide enough.

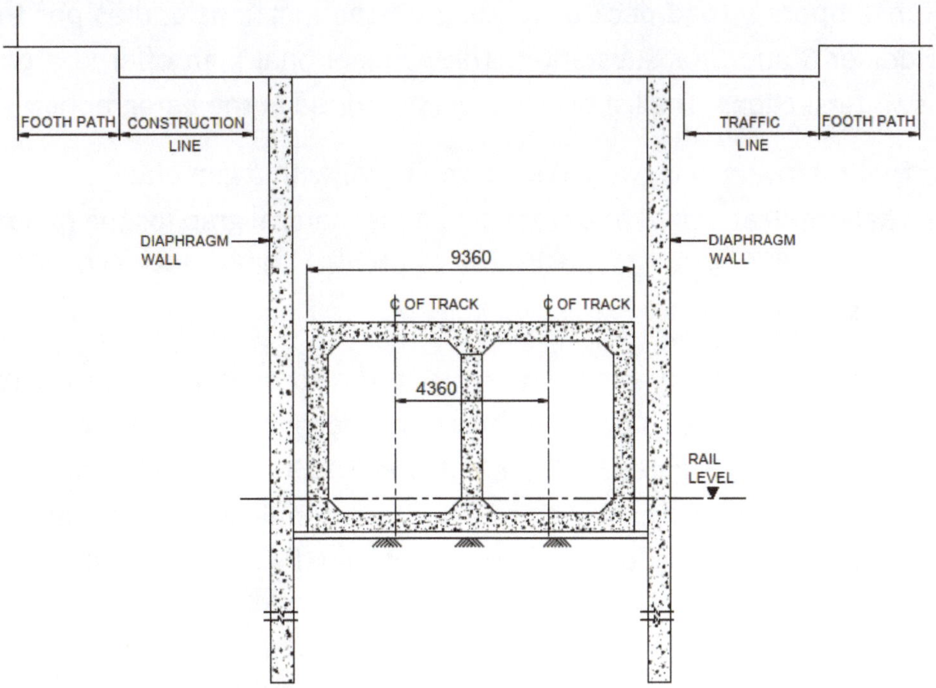

Fig. 5.112: Traffic Diversion with Two Lane Carriageway on one side

Decks were built parallel to/over the excavated section **(Fig. 5.113)**, the tram tracks were clubbed where needed, with road width/decking as applicable. Part footpaths, carriageways, and a temporary road were integrated with the existing roadway **(Width of the Construction Lane and Footpath reduced to Minimum Required)** wherever possible.

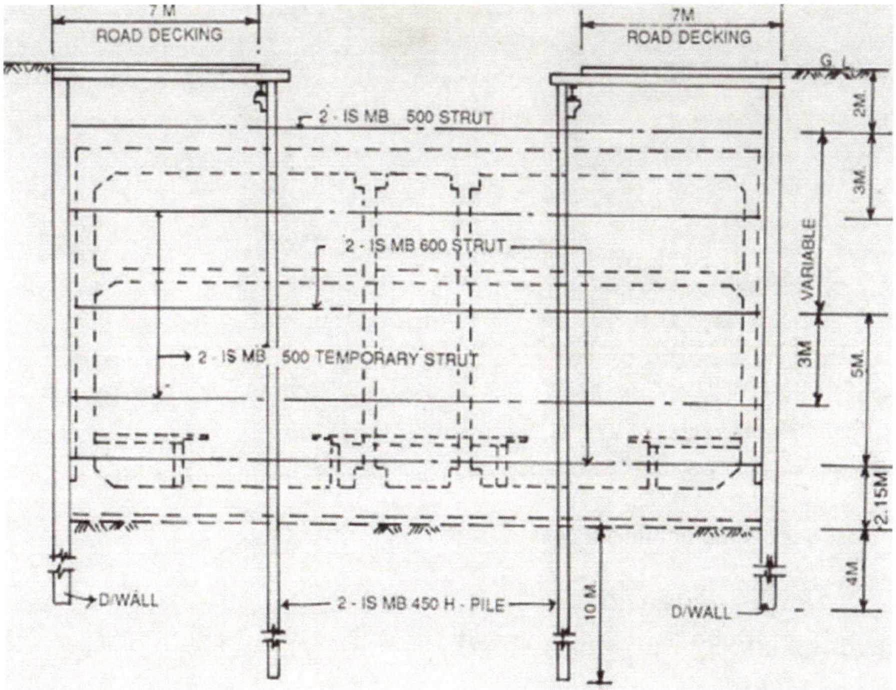

Fig. 5.113: Temporary Road Deck on Both Sides - Where full road width is required for a Station

Different plans for temporary road deck depending on the locations and as per the metro width needed for Corridor or Station, or Substation. The conventional plan offered a cross-deck for all essential and busy intersections. The list of crossovers is added in the paragraphs given below.

The system used for crossovers was two DWs, on the road, with a gap equal to the metro width to cross over, using a Kelly grab attached to a crane and a mechanical grab for the maidan section. And Kelly Grab in other areas. PVC water bars were used between consecutive DW joints. The grade of concrete adopted was M20/M25.

As such, in addition to road crossovers, a requirement was to build a temporary road deck along the corridor. Widening the road was done with the geometry as per site requirements. Types of Road widenings are included in, Figures **5.111, 5.112, and 5.113** show the temporary deck for traffic parallel to the metro corridor on either side. Cross traffic perpendicular to the metro across decks, as included in **Figure 5.114.** were provided at all important junctions and busiest intersections, which are as follows:

1. AJC Bose Road and Chowringhee Road
2. Elgin Road and AT Mukherjee Road
3. Chakraberia Road and AT Mukherjee Road

Fig. 5.114: Temporary Cross Deck at Major Intersections

4. Hazra Road and SP Mukherjee Road
5. Rashbehari Avenue and SP Mukherjee Road
6. Kake Avenue and SP Mukherjee Road
7. ratapaditya Road and SP Mukherjee Road

Fig. 5.115: Temporary Road Along the Corridor on one side

Figure 5.115 includes road widening on one side of the metro corridor for both road traffic and tramway.

5.17.6 Subway by Box Pushing Method

Dumdum Metro Station was being constructed abutting the suburban station at Dumdum. To facilitate the interchange of commuters between the Metro station and the Suburban station, a passenger subway was constructed **(Fig. 5.116).** The subway crossed three platforms and six running lines of the Suburban Railway. The top of the subway was 1.6 m below the rail level. The size of the subway is 8m x 3.6m externally, with 48m, length. The construction of the subway was with rectangular precast box sections and pushed into the embankment by high-powered jacks, taking reaction from an anchored thrust bed. The face of the embankment at the entry was protected using sheet piles cut to size as required.

Fig. 5.116: Underpass Below Rail Lines

5.17.7 De-Watering during Excavation

The groundwater table in the city is high. There is no appreciable difference in level throughout the year. However, the level is different between the north and south parts of the city. The water table in the Belgachia area is as high as (-) 2 m, whereas in central and south Kolkata, the water table is around (-) 4.5 m. The bottom of the RTS corridor was at minimum (-) 9 m, level.

Fig. 5.117: Dewatering using Well Point

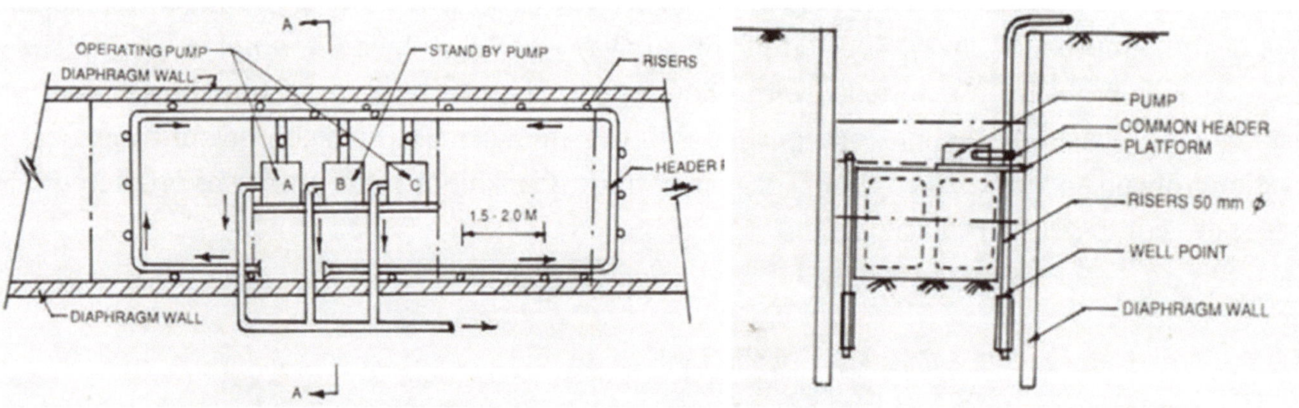

Fig. 5.118: Long and Cross Section of Well Point System

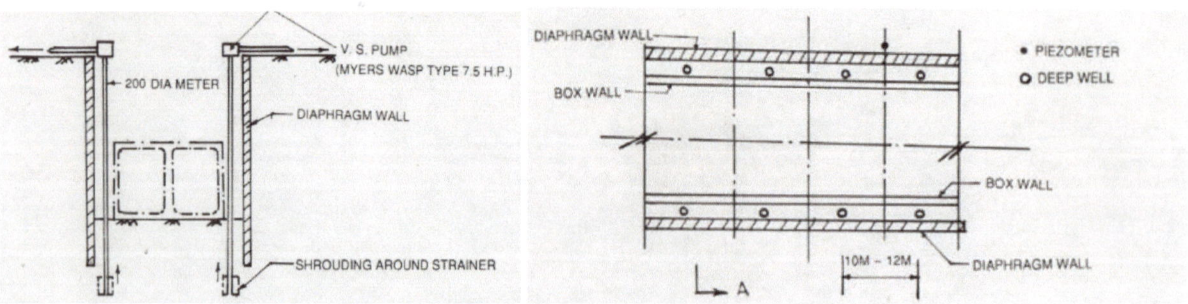

Fig. 5.119: Section and Plan of Deep Well System of Dewatering

De-watering during the excavation had to continue till the excavation reached the final excavation level. As such dewatering must continue till a sufficient level of construction was done to neutralise the uplift of the constructed section. The tank areas were dewatered using the required capacity of the pump and drained into the city's stormwater system.

Two systems, a well point **(Fig. 5.117)** and a deep tubewell **(Fig. 119),** were adopted in other areas. The well point system was adopted in the Maidan area and in the Tollygunge area. The system **(Fig. 5.118)** consisted of small diameter stainers, 50mm and 75 mm in diameter, and 1.5 m long brass pipes were used with closed ends. Well points were installed at a spacing generally of 1 to 4 m and attached with a header pipe of 150 mm to 300 mm in diameter. The header pipes were connected to a combined vacuum and centrifugal pump of 50 HP.

The deep tubewell system consisted of 150-200 mm diameter tube wells with a spacing between 10 to 12 m and with a 5 to 10 m long strainer, depending on the area to be covered. Deep wells were used in areas where large quantities of water needed to be pumped out. The location of deep wells was either outside or inside the excavation area, depending on the space available **(Fig. 5.119).**

5.18 Instrumentation of Diaphragm Walls

(As presented by the Author in Lucerne)

As the design was carried out using an Empirical Formula, during the construction of the Metro corridor in the Maidan area, it was decided that post-construction (diaphragm wall) investigations would be carried out to collect data and compare them with theoretical calculations based on the empirical method to study the structure's performance. The study was carried out by Dr. BC Roy, the Author and project in charge, and Prof. Nitin Som of Jadavpur University during the late seventies after the struts were put in position. Later, Prof. Som and his team continued the instrumentation in other sections and briefed the latter in the section.

5.18.1 Field Measurements

For Test Section 1 in the Maidan area, the investigations were carried out on panels 18, 19 and 20. Simple instruments could only be installed to measure the strut loads, ground settlement and diaphragm wall movement due to the late decision to monitor the performance of the structures. For the Test Section between Girish Park and Mahatma Gandhi Road Station, extensive instrumentation was done at three subsections as shown in **Fig. 5.139,140,141** (included towards the end of this section) before execution. The test sections were monitored to measure earth pressure on the diaphragm wall, pore pressure variation in the soil, ground settlement, wall movement, and strut loads. Measurement instruments included magnetic settlement gauges for ground settlement, Demac gauges for strut loads, an inclinometer (vibrating wire type) for diaphragm wall movement, earth pressure cells (vibrating wire type) for lateral pressure and both Casagrande and vibrating wire piezometers for pore pressure at different depths. However, in both the test sections, strut loads were measured using a mechanical extensometer with an accuracy of 0.001 mm, manufactured by MV Mohr & Federhoff, West Germany.

5.18.2 Approach and Assumptions

i) Assumptions

Design of braced excavations is an empirical science, and no established analytical procedure is still available. There are, in general, four possible modes of failure

- excessive movement of walls
- yielding of supporting struts
- bottom heaves in cohesive soil and
- piping in granular or bursting in cohesive soils

Accordingly test sections were instrumented to measure earth pressure/pore pressure variations/ ground settlement/wall movement/strut loads.

ii) Approach

- Before start of excavation earth pressure on either side corresponds to Ko - pressure at rest
- First stage of excavation allows DW to deflect as cantilever, however rigidity of DW does not allow earth pressure to come down to active pressure
- Thereafter lateral movement is restricted due to introduction of struts
- Stress concentration occurs in the vicinity of struts as excavation progresses
- Earth pressure develops depending upon the strut spacing & DW rigidity. However, pressure never comes down to active pressure
- Peck's apparent earth pressure is generally considered

5.18.3 Results and Observations

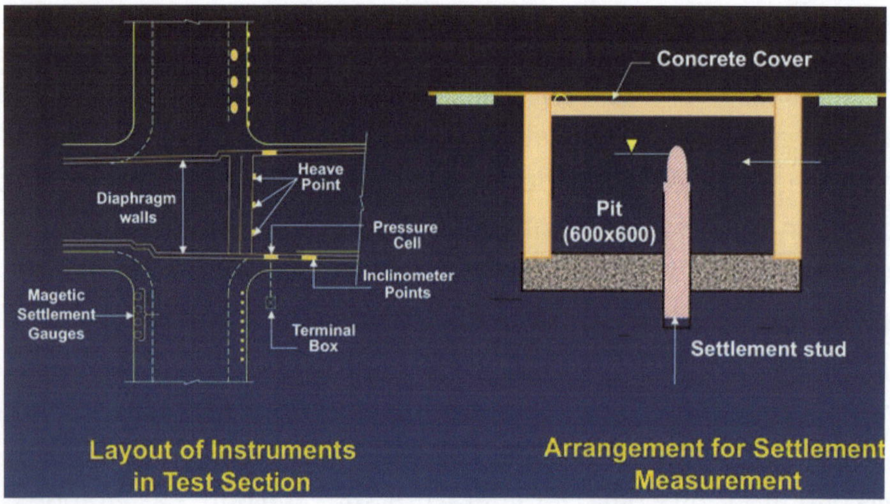

Fig. 5.120: Lay out of Instruments

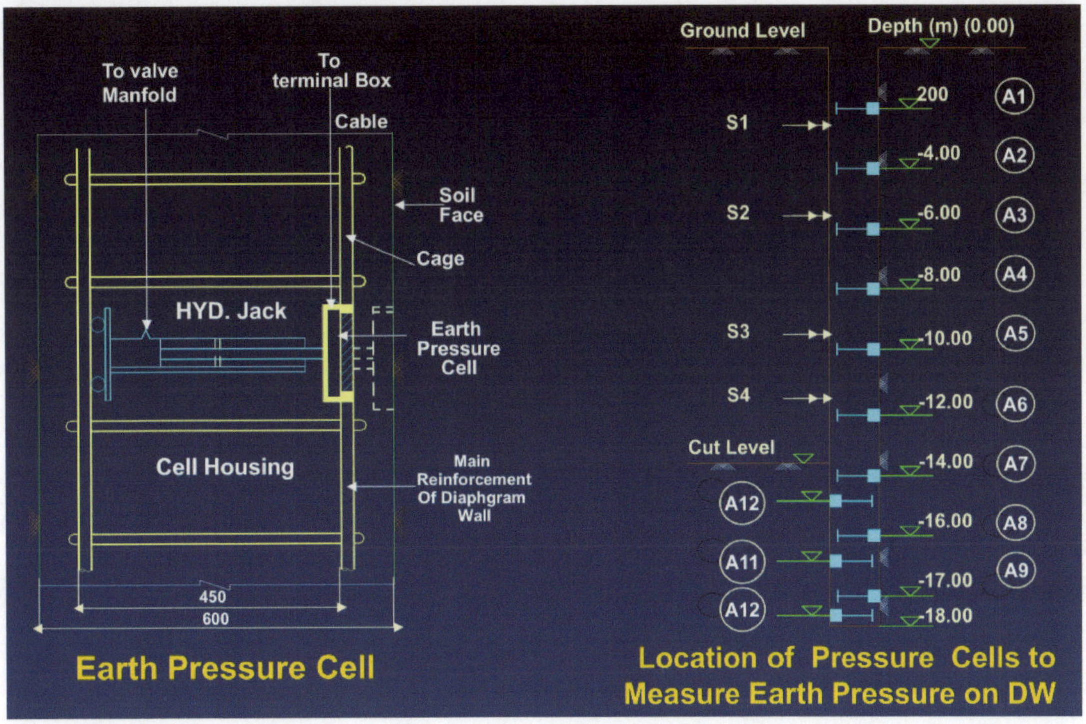

Fig. 5.121: Location of Pressure Cells

From the distribution of measured pressure, it is observed that

- Concentration of pressure at strut locations*
- Earth pressure above cut, increased with depth and is always greater than corresponding active pressure

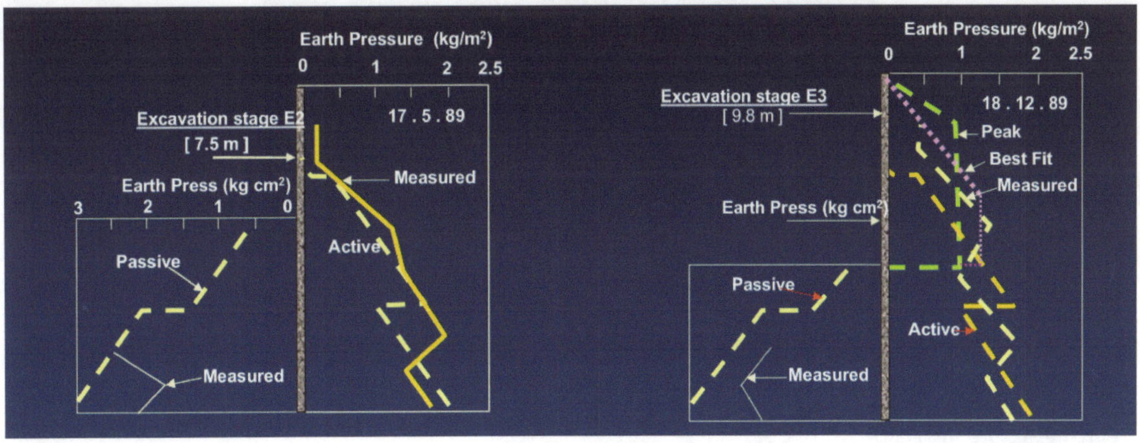

Fig. 5.122: Earth pressure

- At any stage of excavation earth pressure increases in depth but remains essentially constant beyond a certain depth
- Earth pressure below cut trends to approach active pressure on cut side & passive on other side

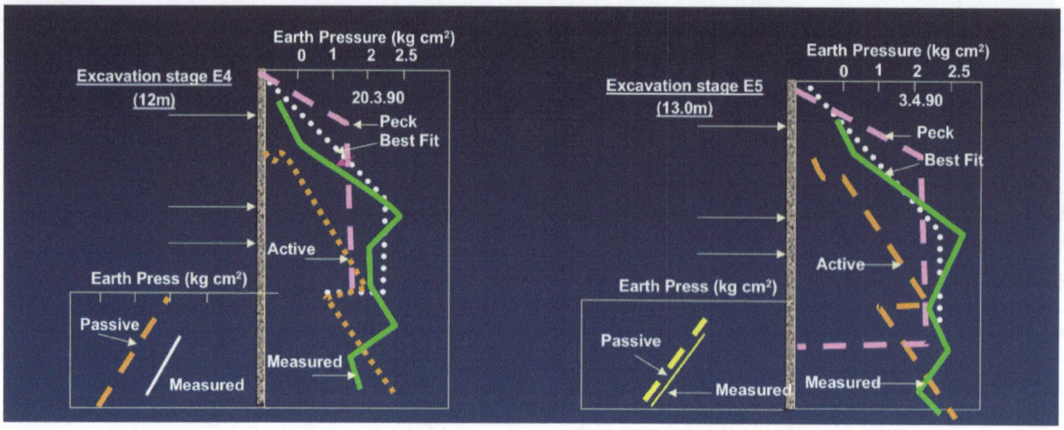

Fig. 5.123: Excavation vis-vis Earth Pressure

- During different stages of excavation pressure increases with depth up to a stage and then becomes essentially constant
- Earth pressure approaches active below cut on cut side & passive on other side
- Best fit plot from measured observation resembled Peck's envelope but varies in magnitude

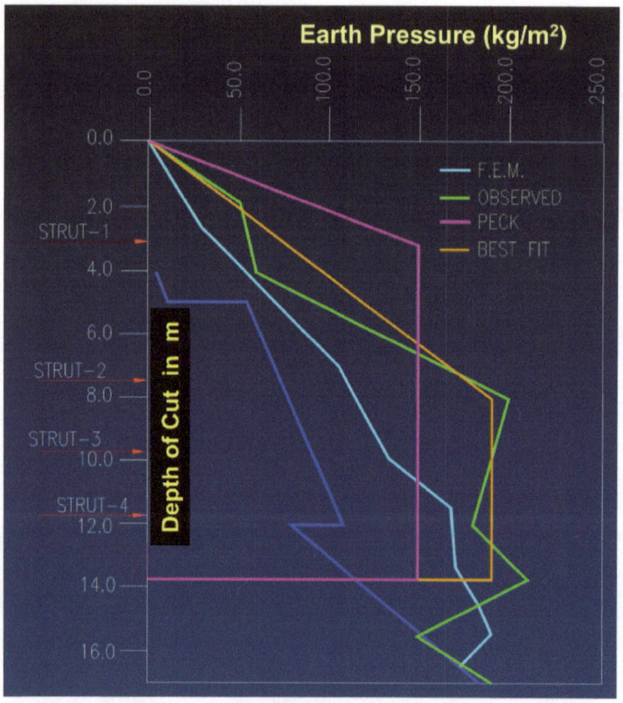

Fig. 5.124: Earth Pressure vis-vis Depth of Cut

- Sequence of construction involving successive excavation and fixing of struts determines the deflection pattern
- At any stage of excavation, wall movement is restricted by placement of struts
- Shallow over-burden on the cut side (passive) allows movement of DW below cut - greater depth of penetration increases fixity & reduces deformation below cut

Cut Level & Struts **Strut Loads** **Wall Movement in mm**

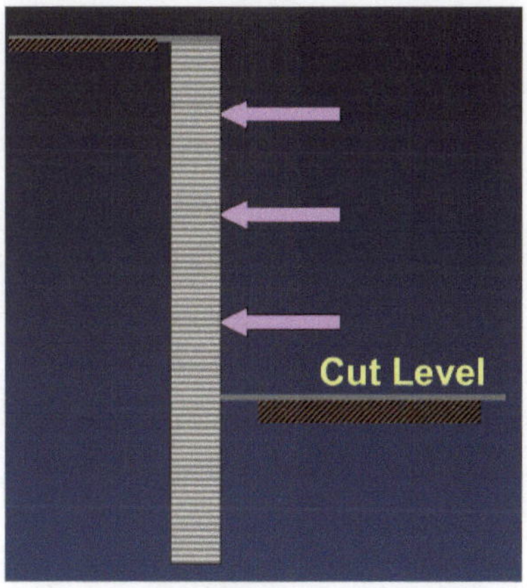

Fig. 5.125: Cut Level

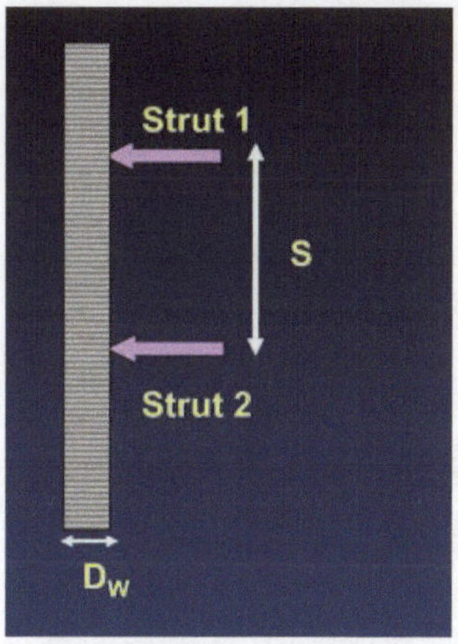

Fig. 5.126: Strut Level

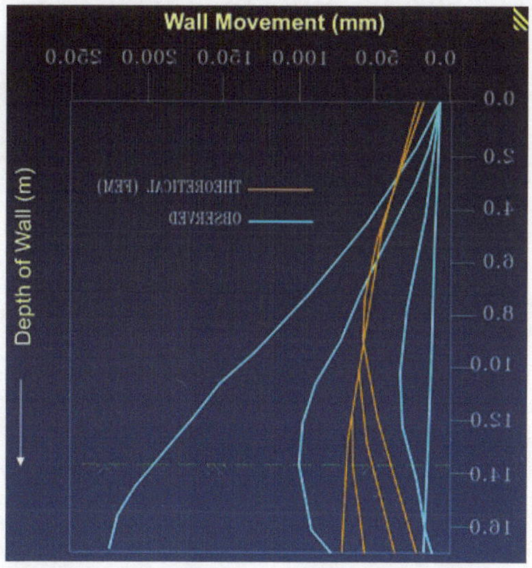

Fig. 5.127: Corresponding Wall Movement

Strut loads depend on the rigidity of DW, spacing of struts & soil parameters and varies with stages of excavation

- Peck recommended apparent earth pressure distribution for the estimation of strut loads
- Actual strut loads also depend on construction methodology and speed of construction. For initial estimation of strut loads, Peck's approach still can be considered

- **Deflections of DW were recorded at different stages of construction and observations were**
    - Movement above cut was restricted due to presence of struts
    - Rate of increase was more as the depth increased
    - Large movement noticed (increased from 100mm to 225mm) for last stage of excavation between 12 & 14m
    - With the increase in depth of penetration the movement will be restricted

- **Strut loads recorded with time and variations Strut Loads in the loads with depth of cut**

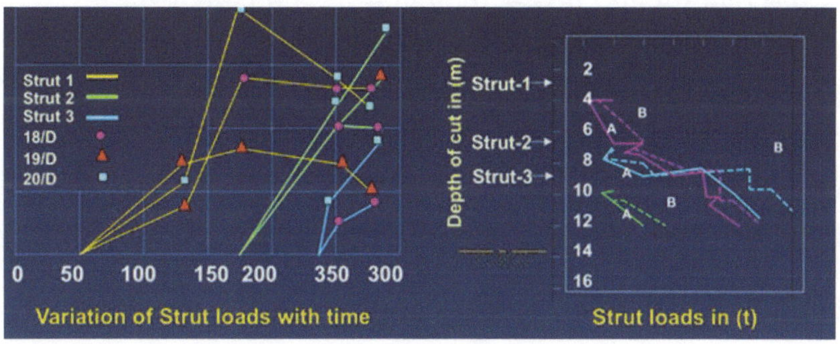

Fig. 5.128: Variation of Strut Loads with Time

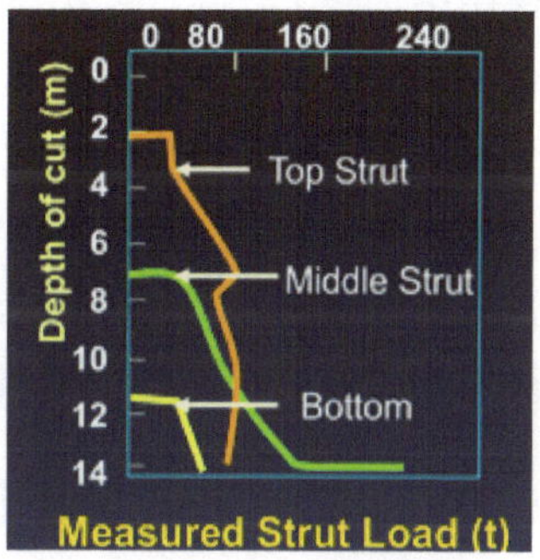

Fig. 5.129: Strut Loads with Depth of Cut

The maximum load in Strut A is reached before installation of strut at level B and after installation load at level A strut either reduced or remain unchanged

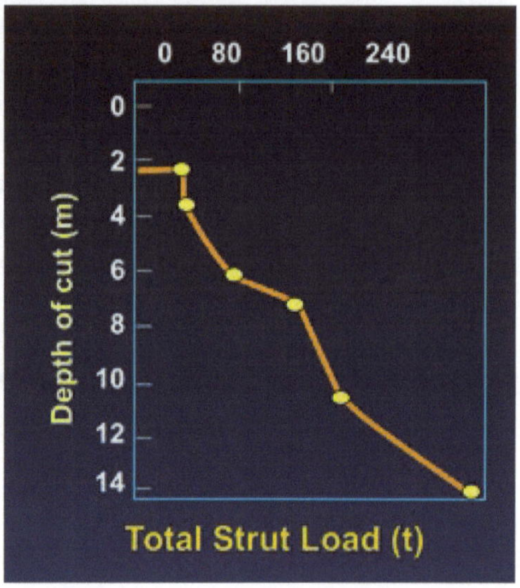

Fig. 5.130: Strut Load at Location-1 with Depth of Cut

Depth in (m)	Measured in (t)	Peck's apparent in (t)
10	73.6	79.3
12	108.3	111.6

- However, this was not always true. There were several cases where load at 1st (Say A) strut went on increasing despite Introduction of 2nd (B) and 3rd (C) struts.
- Total strut load however increased slowly for shallow depth say up to 7m, beyond which load built up much faster with increase in depth.
- Strut loads measured/calculated by Peck's approach & obtained from best fit trapezoidal diagram compares very well (variation of 5-15%).

Strut Spacing/Ground Settlements

- Strut spacing affects the lateral load carrying capacity of DW in flexure and deflection
- Larger the number of struts more is the rigidity and lesser movement results in leser ground settlements
- Spacing to optimise between increase of rigidity and construction facility
- To study the effect of strut spacing, ground settlement occurred during the time gap between successive installations were measured
- Settlement increases rapidly for unsupported depth of cut of more than 4m

  Optimum strut spacing is found to be 3-4m for deep excavations in soft clay for thickness around 60cm.

Ground Settlement

- Ground settlement is the surface manifestation of sub-soil deformation
- Both magnitude of settlement and the zone of influence are important
- Figure shows variation of maximum ground settlements with depth of cut using normalised value of settlement/depth of cut in percentage. Results show straight line variations with increase in value of settlement with increase in depth

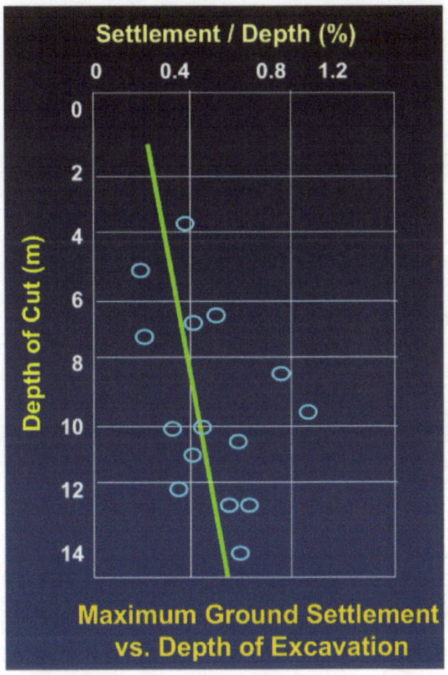

Fig. 5.131: Maxm. Settlement wrt Depth of Cut

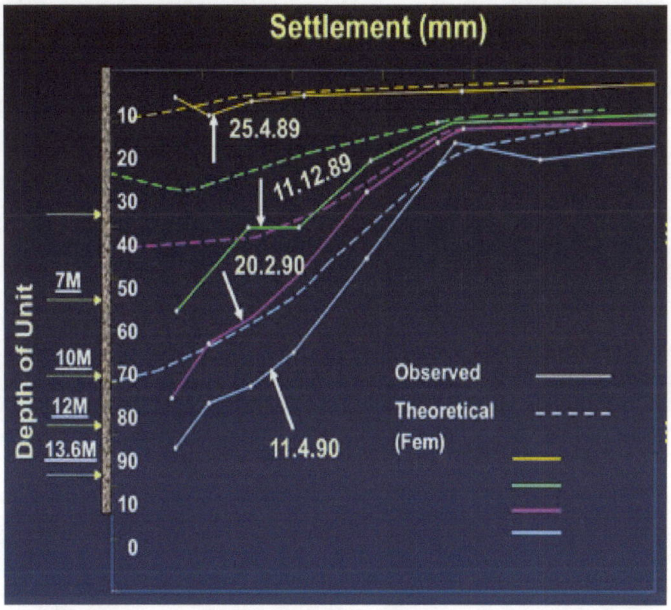

Fig. 5.132: Settlement wrt Depth in Unit

Ground settlements were more pronounced within an influence zone, in this case, 10m from the excavation areas

- The observed values of settlements were less than the calculated values using Peck's approach
- Wall deflection and ground settlement increases with increased cut width. This may be due to larger zone of plastic deformation in case of wider cut

Effect on Adjacent Buildings

Type of observations undertaken were

- Settlement observation on the plinth/floor and building
- Observation of the tilt of the building
- Development of cracks in walls/floors

- Around 60 buildings had shown small and minor damage. 28 buildings suffered moderate to severe damage
- Settlement of buildings varied between 70mm – 856 mm
- Rotation in building measured to vary between 1 in 36 to 1 in 475

Mathematical models were evaluated to study the parametric performance of the Calcutta Metro excavations. The sequential excavation and strut installation models in the stratified soil were simulated for field conditions by removing the elements row-wise in front of the wall and making the excavated surface stress-free. The parametric comparison between theoretical and observed results indicated the excavation's performance.

Compared to **Fig. 5.133**, the theoretically predicted earth pressure is well within Peck's envelope but far from trapezoidal. The best-fit plot of the earth pressure resembled Peck's envelope, but the magnitude varied in places.

Fig. 5.133 includes the theoretically predicted earth pressure is well within Peck's envelope But not trapezoidal. However, the best-fit envelope resembled Peck's envelope.

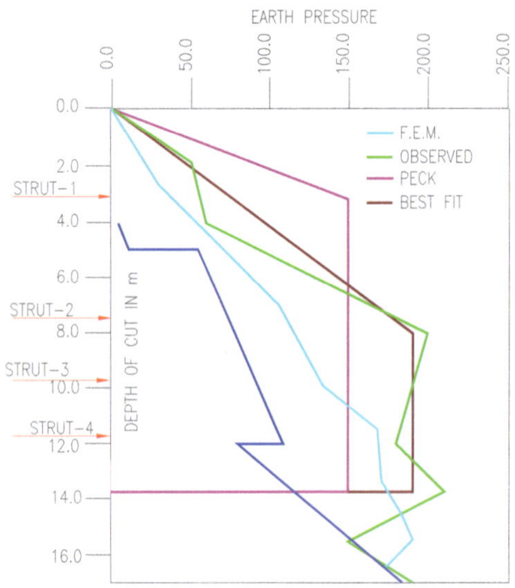

Fig. 5.133: Comparison of measured stress with Peck's Apparent Earth Pressure

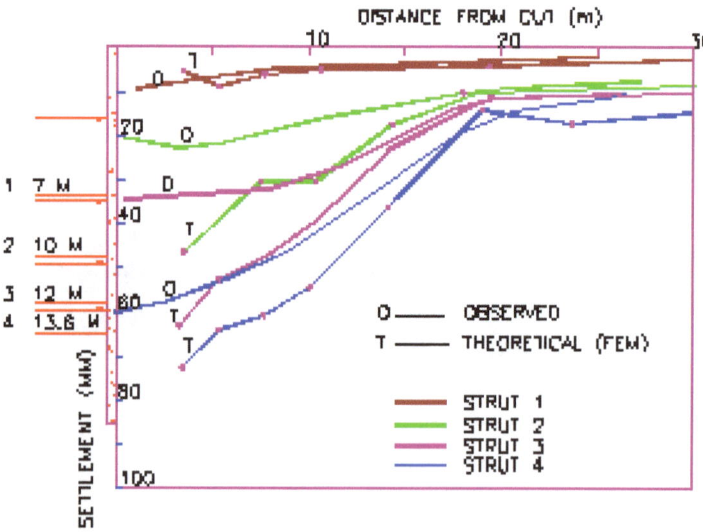

Fig. 5.134: Ground settlement profile in test section 2C

Fig. 5. 134 represents the observed and the theoretically predicted ground settlement profiles on different dates during the excavation in Test Section 2C. The magnitude of the settlement and the zone of influence are essential in establishing the adverse effects of the construction on adjoining buildings. **Fig. 5.135** shows the arrangement for measuring the ground settlement in Test Section 1 for panels 18, 19 and 20. The comparison between the measured and designed strut load is illustrated in tabular form in **Fig. 5.132** represents the measured settlements against time for different test locations illustrated in **Fig. 5.133**. For Test Section-1, the variation of ground settlement with time at settlement points L10 to L14 had generally been more than that on the other side at settlement points L5 to L9, presumably because of moving vehicles and surcharge.

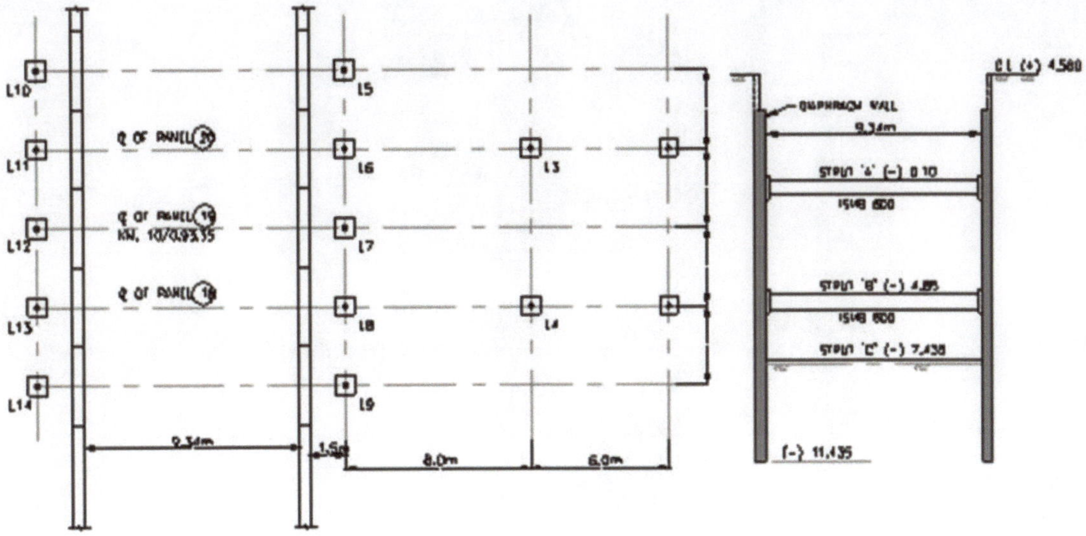

Fig. 5.135: Instrumentation for ground settlement measurement in test section 1

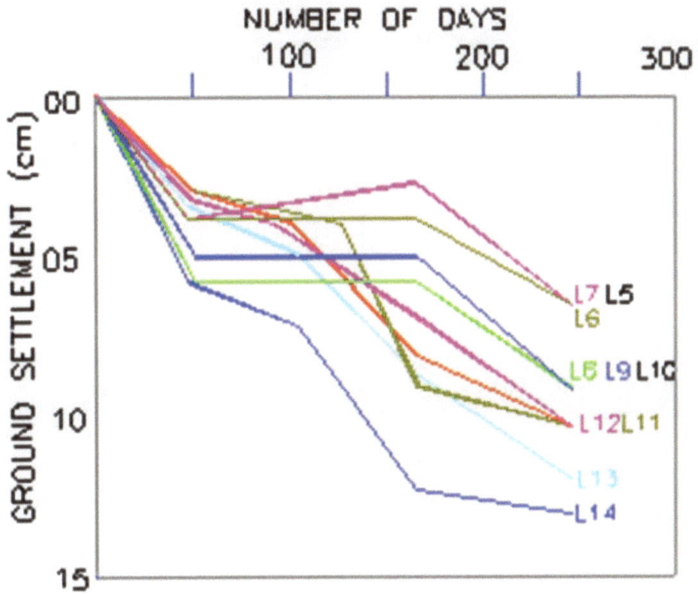

Fig. 5.136: Ground Settlement in Test Section 1

Fig. 5.136 represents the measured ground settlement against time at different locations illustrated in **Fig. 5. 133**, while **Fig. 5.137** illustrates the simulated deflected shape. The absolute wall movement at points below the top of the diaphragm wall could not be assessed.

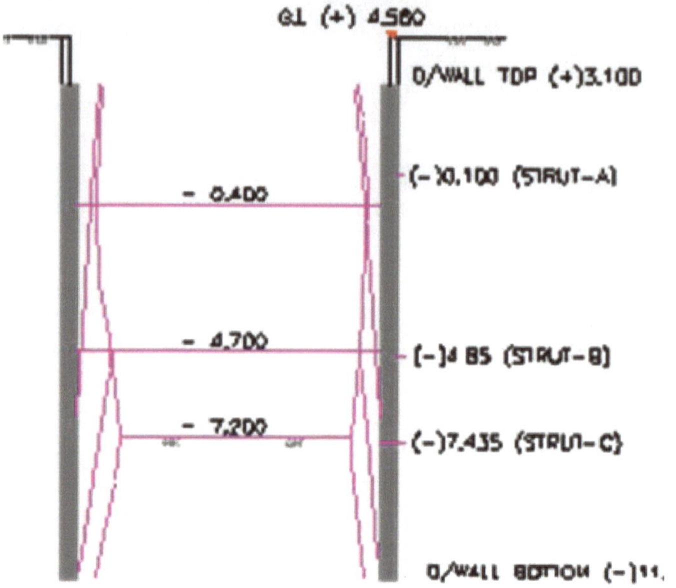

Fig. 5.137: Deflection of Diaphragm Wall

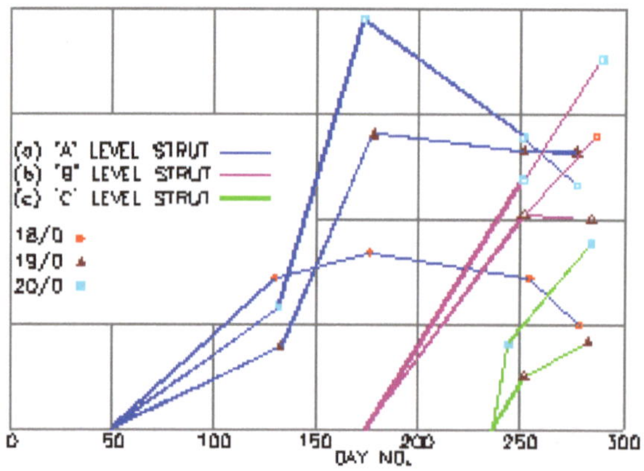

Fig. 5.138: Strut loads in test section 1

However, the points could only be obtained after excavation up to the desired level had been completed, and by that time, the points had already moved by a certain amount. **Fig. 5.138 shows** the maximum load at the level **A** were reached immediately before **B** struts were installed. There after the load of **Strut A** went on decreasing as the **level B Struts** started taking load.and **Fig. 5.138** show the variation of strut loads with time at Test Section 1 and the variation of strut loads with depth of cut at Test Section 2, respectively. The loads in 'C' level struts were relatively small, as would be expected, because there was little excavation below these struts. Moreover, it has been observed that only 50% of the total ground settlement occurs during the excavation, 40% during the construction of underground structures, and 10% at the time

However, the points could only be obtained after excavation up to the desired level had been completed, and by that time, the points had already moved by a certain amount. Show the variation of strut loads with time at Test Section 1 and the variation of strut loads with depth of cut at Test Section 2, respectively. Moreover, it has been observed that only 50% of the total ground settlement occurs during the excavation, 40% during the construction of underground structures, and 10% at the time of backfilling. Thus, to reduce the settlement, the construction time for underground structures shall be optimised to the bare minimum.

The thickness of diaphragm walls does not contribute much to reducing ground movement; instead, the penetration depth is essential in a semi-integrated type of structure. Integrated structures account for the increased diaphragm wall thickness based on bearing considerations. The panel width of the diaphragm wall should be limited to facilitate a single strut system and to avoid complications in strut loading and wall movement.

In addition to instrumentation in the Maidan area, later the study continued in a limited way to measure the strut loads, ground settlement, and tunnelling stresses in the tunnel section between Belgachia and Shyambazar and the Cut & Cover section in the north of Esplanade in three locations. The Instrumentation was carried out to measure the i) earth pressure developed on the Diaphragm Wall (DW), ii) ground settlement during excavation and placing of struts, iii) DW movement, iv) Strut loads to check the stability and compare the theoretical calculations. Three cut-and-cover excavation sections measure earth pressure in DW, pore pressure variation, ground settlement, DW movement, and Strut loads.

The study used Pressure Cells **(Fig. 5.137)** with a Vibration wire in the Maidan Area. The Maidan section also used piezometers, Magnetic settlement gauges, inclinometers, and Demac gauges. Similar instrumentation gadgets, among others, are being used even today.

Table.5.5: Comparison of Strut Loads between Measured and Designed

Strut Level	Diaphragm Wall Panel	Maximum Strut Load Measured (t)	Design Strut Load (t)
A	18-D	170	165.75
	19-D	280	
	20-D	375	
B	18-D	265	64.0
	19-D	208	
	20-D	322	
C	18-D	NO STRUT	132.03
	19-D	80	
	20-D	158	

Fig. 5.139: DW with Pressure Cells to Measure Earth Pressure

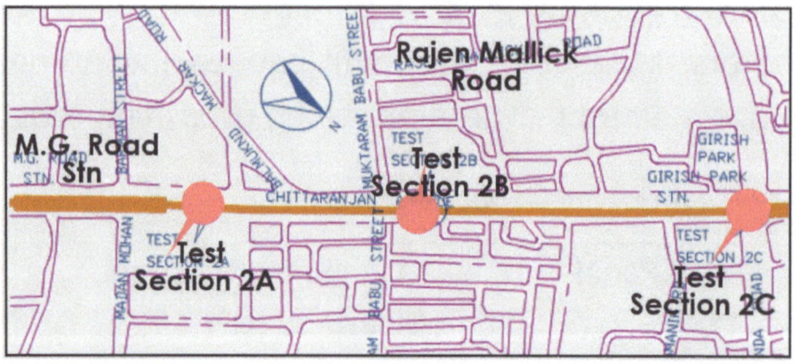

Fig. 5.140: Location of Test Sections-A

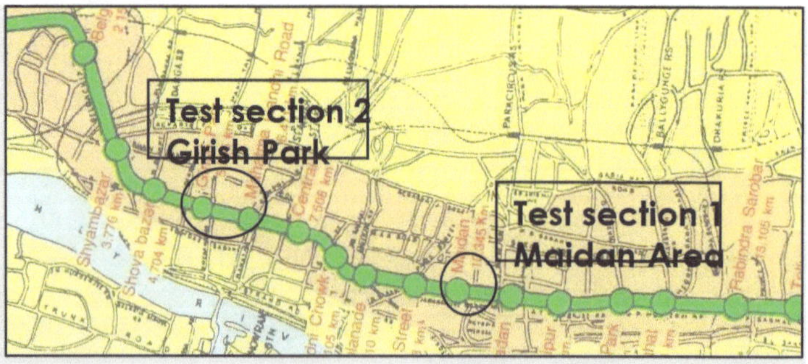

Fig. 5.141: Locations of Instrumentation--B

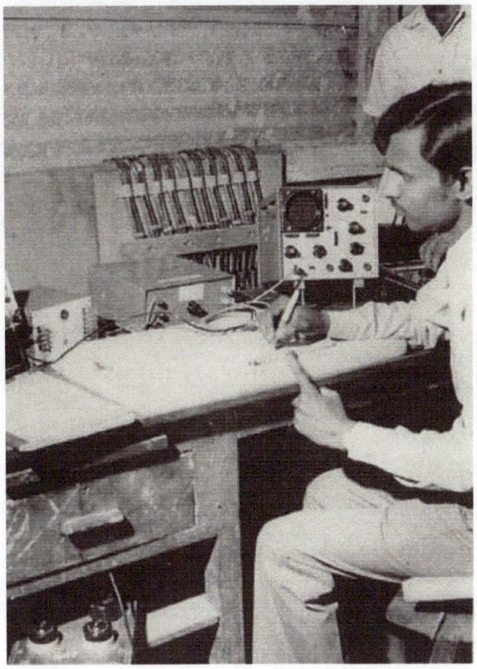

Fig. 5.142: Measuring Soil Pressure in Laboratory.

structures shall be optimised to the bare minimum.

5.18.4 Interpretation & Conclusions

- Wall thickness does not affect significantly the magnitude of wall deflection
- Increase in wall thickness does not have significant effect on ground settlement.
- From the parametric study different factors, it is observed that strut spacing and penetration of DW are two significant factors for the performance of braced cuts
- Settlement observed at different stages of construction
    - Excavation stage - 56% of the total settlement
    - Box construction stage - 28% of the total settlement
    - Backfilling stage - 16 % of the total settlement

It is observed that settlement increases with time and Strut Loads.

5.19 Tunnelling

The first foreign technology adopted was precast concrete tunnel segments, successfully developed and used in Germany across the River Main in the early 1970s. The author had the opportunity to work with Philipp Holzmann (PH), the designer and builder of that tunnel. This technology was introduced in India for a small section of the North-South Metro across the Chitpur railway yard. However, the implementation team faced challenges during segment erection, and the remaining portion of the section was completed using the older method of cast iron segments with the help of Russian engineers. PH was also associated with CPPL in bidding for the first tender floated for the Chitpur railway yard, which planned tunnel construction using EPB-TBM.

5.19.1 Shield Tunnelling at Chitpur Yard

Kolkata soil is soft and unstable in places, making regular tunnelling impossible. Hence, tunnelling was carried out using Compressed Air.

Tunnelling by compressed air started with a double D shaft sunk by a well-sinking method. It was made airtight by providing a steel frame cover with a rubber lining and held down by bolts embedded in the concrete of the well-steining top. After the tunnel had advanced, the air lock shifted inside the already completed tunnel, and the working locations of the air locks were arranged.

Air Pressure

The stability ratio at the tunnel's face is a critical factor in determining air pressure during construction.

Stability ratio Nc = Pz/Su

where Nc = Stability Factor

Pz = Vertical Soil Pressure at the depth of the axis of the tunnel

Su = Undrained Shear Strength of soil at the face of a tunnel

A stability ratio Nc of less than 5 is recommended for safe working. Where there are more than 5, compressed air brings it down, i.e., Nc = (Pz-Pa) Su, where Pa is the air pressure required. Air pressure between 0.5 to 1.0 atmospheres was used in N-S Metro construction.

Air Pressure Regulation

Close monitoring is essential during construction as excess pressure leads to accidents, and lower pressure causes the tunnel face to become unstable. Hence, the pressure must be continuously regulated to ensure the tunnel face's stability.

5.19.2 Design of Lining

The circular tunnel in the Shyambazar—Belgachia Section and in the underpass leading to the Car Shed was designed for full overburden pressure and horizontal pressure due to soil at rest on the rigid circular ring. Because of the soil condition and the fact that the depth of the tunnel was not too large compared to the Diameter of the Tunnel, no relief due to the arch action of soil was taken.

A detailed study was conducted to explore the possibility of using a lining with hinged joints, and it was found that the ring's deformation was excessive, making the tunnel unusable.

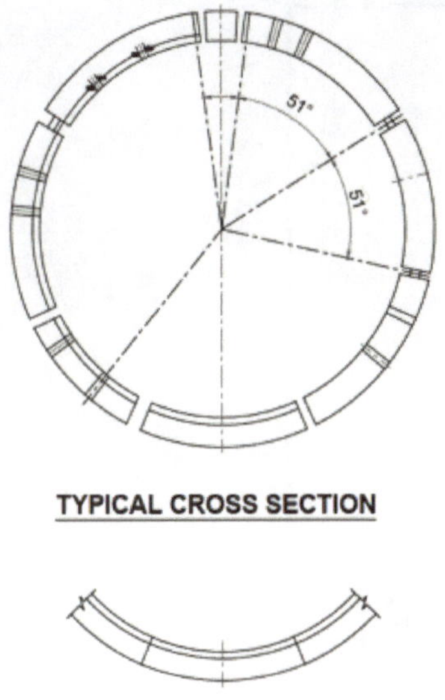

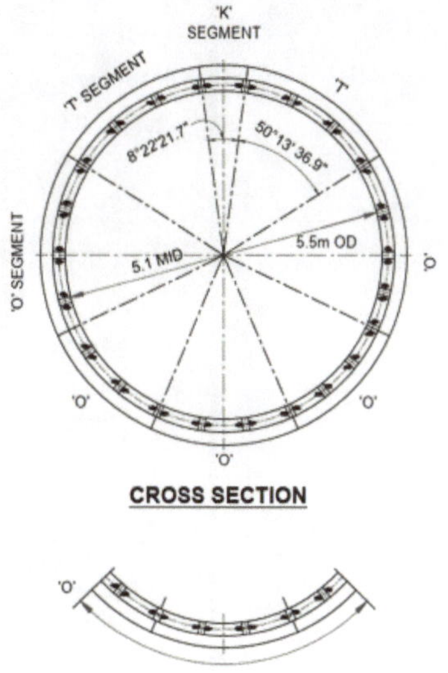

Fig. 5.143: Cross Sections of Concrete Segment

Fig. 5.144: Steel Segments

High-tensile bolts were used to join cast iron ribbed-type lining segments to form the ring **(Fig. 5.143 & 144)**. Curved bolts could be used in the case of solid slab RCC lining, but they were not convenient to use at the site. A special arrangement was made to substitute the bolting arrangement with steel rods fixed in holes by a specially designed adhesive. The top Key Segment could not be bolted by this method.

Loads on the tunnel segments included in Figs. **5.145.** Access Shaft is included **in Figure 5.150. Fig. 5.146** includes cutter wheel of TBM**, Fig. 5.147** shows TBM assembly**, Fig. 5.148** includes segments at rear of TBM and **Fig. 5.149** includes installation of Cutter Wheel**.**

The lining was designed as a ring with one hinge at the crown. The possible determination due to soil was assessed. Extensive tests, both in the lab and field, were made to prove the efficacy of the joining arrangement before adoption. A full-scale prototype ring was tested before the design was adopted. The C.I. lining was used in the starting portion and critical locations, such as under the canal location, while Precast Concrete Lining was used in the remaining portion. The concrete grade used was M 40 to keep the thickness (200 mm) as minimal as possible. The concrete was produced by using mechanised vibratory compaction and steam curing.

Shield equipment plan and longitudinal sections are shown in **Fig. 5.151 & 152,** Shaft reinforcement in **Figure 5. 153** (representative) and **Figure 5.154** for Twin Tunnel at Chitpur.

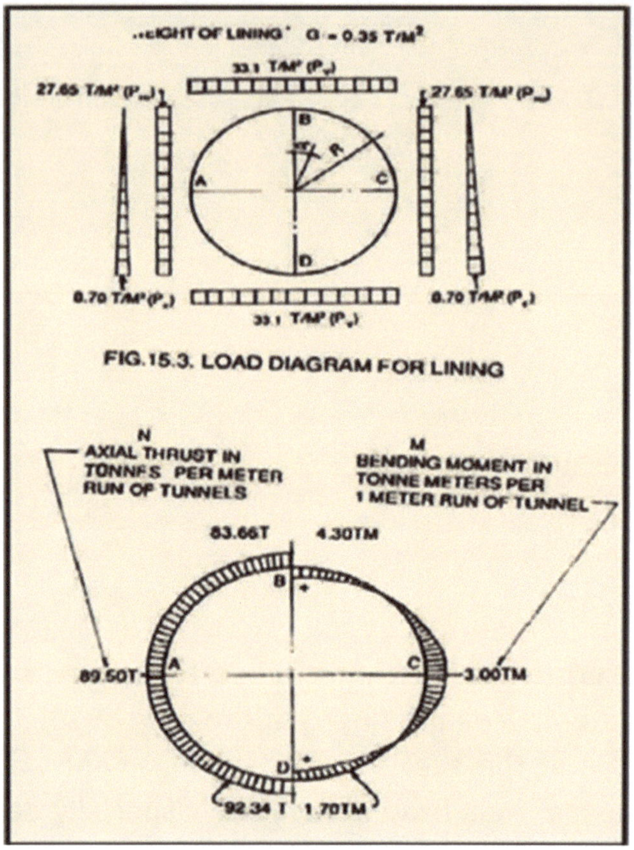

Fig. 5.145: Load on Shield Tunnel

Fig. 5.146: Lowering Cutter Wheel at Shaft

Fig. 5.147: Assembly of TBM in Shaft

Fig. 5.148: Segments at rear of TBM

Fig. 5.149: Installation of Cutter Wheel to TBM

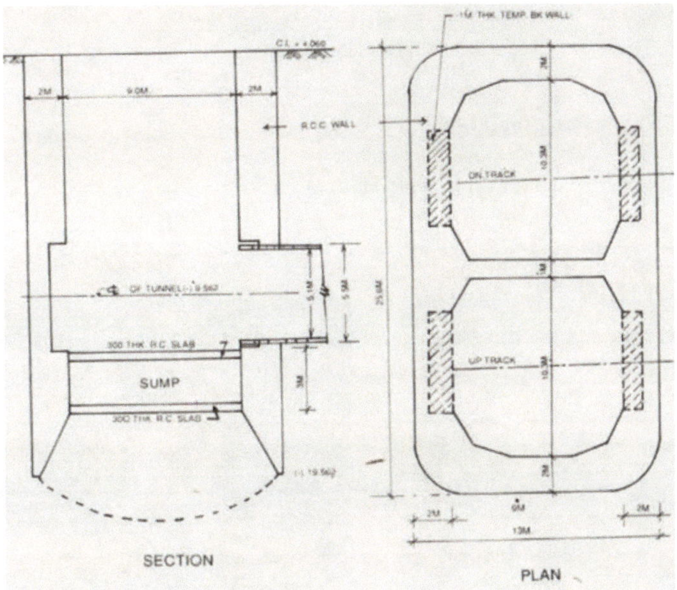

Fig. 5.150: Section of Access Shaft

The access shaft was sunk to start with by the well-sinking method. The shaft was like D-type. Steining thickness of two meters and was made of RCC. The intermediate wall was one meter thick. The overall dimension was 25.6mx 13.0m, and the bottom plug was 4.5m. A drainage sump had been provided below the plug and the floor. The shaft was used as a Compressed Air Zone. The steining was designed to withstand the air pressure of 1.25 atmospheres. The bottom plug had to withstand the uplift pressure. As the shaft was meant structure to withstand air pressure, the top had to close, and the design was done using a steel structure.

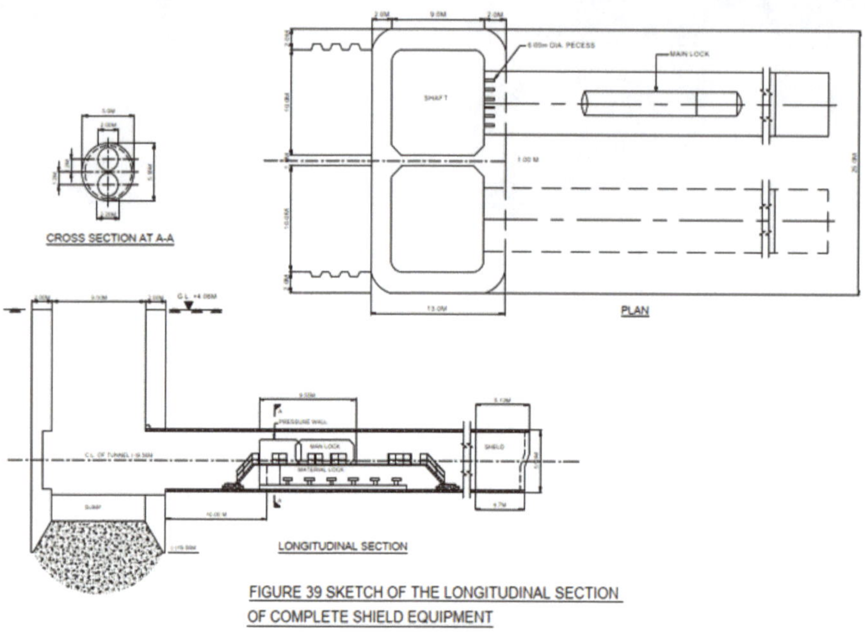

Fig. 5.151: Access Shaft and Equipment Position (PLAN)

Fig. 5.152: Shield Equipment – Longitudinal Section (Lower Figure)

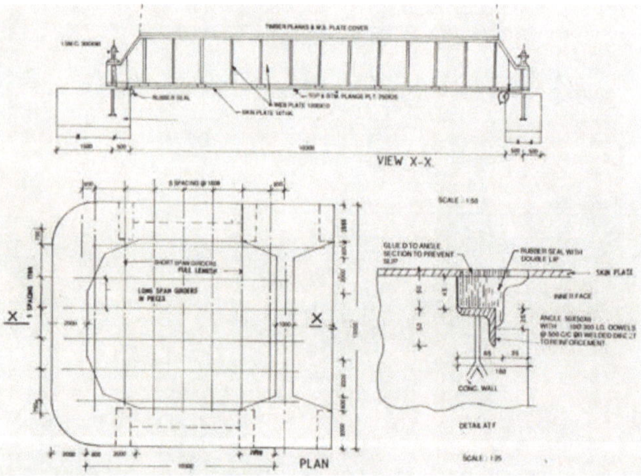

Fig. 5.153: Shaft Reinforcement

Fig. 5.154: Twin Tunnel at Chitpur Railway Yard

Tapered Segments and Settlement Study

To introduce curvature in the tunnel alignment and correct the error in straight alignment, solid cast iron tapered linings varying from 40 mm to 20 mm were designed and manufactured. The requirements of such linings are 10 per kilometre in straight and one after two rings for curves of 400 m. A detailed study of the likely settlement and its effect on buildings in the influence zone was made. Accordingly, necessary precautions for evacuating or fixing observation points on the building were adopted during construction. Steel mould for concrete segment included in **Figure 5.153.**

Fig. 5.155: Steel Mould for Precast Concrete Segments

5.19.3 Tunnelling by Blade Shield

The 5.10 m diameter tunnels in the underpass were constructed using the Blade Shield Tunnelling method without using compressed air. The blade shield consisted of a 5.70m outer diameter cylindrical steel shell with cutting edges consisting of 23 nos. of trough-shaped blades in front and a tailpiece for casting RCC lining rings.

On completion of the preparatory works in the entrance pit, the face of the diagram wall was punctured, and the down-tunnel work of the underpass started. After the tunnel had advanced up to 30m. Thereafter, the shield nose started showing a downward-dipping (nose diving) tendency. Attempts were made to correct the vertical alignment and arrest the nose-driving tendency by adjusting the jack pressures. However, this did not yield results, and a rapid nose-dive took place. The total dipping was about 967 mm from the desired level.

It was found during the investigation that a weak pocket of soft/weak medium silty clay was encountered below the tunnel at this point. The value of N of this portion was 4-5 at the Dumdum end and 5-9 at the car shed end. While tunnelling, the entrapped water below the embankment caused a loss of confinement of clay particles, resulting in a loss of shear strength and bearing capacity. This led to the nose-dive of the heavy shield machine at this location. Attempts were made to strengthen the base below the shield by micro piles of 150 mm diameter G.I. pipes driven by reverse jack pressure, taking reaction from the jack, but this was unsuccessful.

A specially designed drilling rig did the inclined drilling for piles. They are to work from within the 5.1 m diameter tunnel. The plan consisted of i) the drilling was carried out with mud circulation, and ii) an automatic mixing plant starting from water and dry products, bentonite, and cement, which was supplied for drilling and injection grouting. iii) The injection equipment consists of an automatic batcher and high-pressure pumps. iv) Water and compressed air circuits.

A high-pressure pump (over 20 MPa) was provided for the water jet. For the cement grout an injection medium-pressure pump (up to 12 MPa) was provided. The air compressor could deliver air of 24 cum/min.

The jet grouted piles by the Rodinjet technique. In this method, seven nos. of inclined jet piles of about 2.0 m dia. were done below the tunnel bottom in each sequence. After the jet grouping of the base below the shield, the shield was lifted by erecting a specially designed semicircular lifting frame within the shield. The reaction was transferred through hydraulic jacks to a specially cast RCC load distribution beam at the base above the jet grout column bed. After lifting and correcting the shield to properly align the voids created at the base grouted with concrete. Once the lifting was achieved, further progress was smooth till the completion of the Down Tunnel. For the up-tunnel, the jet-grouted column base was created from the beginning, and the tunnel was completed in the sequential operation of advanced jet grouting, shield advancement, casting concrete lining, etc.

5.20 Station Planning

In the Kolkata N-S Metro, two types of station platforms have been used: island and side platforms.

Fig. 5.156 shows a view of a UG station platform during operation.

Fig. 5.156: UG Station During Operation

Fig. 5.157: UG Station (Netaji Bhawan)

Fig. 5.157 Includes station with island platform. Island platforms have the advantage of requiring a minimum duplication of facilities, like stairs, escalators, lighting and other services. The total platform width will generally be less for island platforms since the same can be used to accommodate both the morning and evening peaks. Each platform must be of sufficient width for side platforms to accommodate peak traffic. Side platforms, on the other hand, are more adaptable to direct access from sidewalks. Most stations were designed with central island platforms. Side platforms were planned only at stations where local topography does not permit an island-type platform except at a high cost. In Kolkata Metro, all stations have been provided with an island platform except the overground stations of Dum Dum and Tollygunge. In contrast, Park Street is the only underground station with a side platform.

Fig. 5.158: North-South Metro during Construction

Whereas island platform stations were designed with multiple-level arrangements, the side platform stations were constructed using single or multiple-level arrangements. In single-level stations, ticketing, fare collection, and loading operations take place at the same level. A mezzanine is provided for ticketing and fare collection in two-level stations, while train loading and unloading are accomplished at a separate level.

Two-level stations allow for controlling the number of passengers waiting on the platforms at any time. When the number of passengers on the platform becomes too large, scheduled train loading times can be maintained using the mezzanine as a holding zone.

Fig. 5.158 includes a photograph of underground corridor during construction in JN Road during 1980.

5.20.1 Station Mezzanine

A mass rapid transit system has its platforms only for the ingress and egress of passengers to and from trains. All other connected activities are dealt with at station mezzanines, generally at a different level. The principal function of the mezzanine is to provide for the movement of people between platforms and road access and for the activities associated with that movement. The main activity is paying the fare so that the mezzanine will house the booking offices, ticket vending machines and barriers to control entry into and exit from platforms. The mezzanines also accommodate several other facilities. Its size, shape and layout vary from station to station depending on the volume and nature of traffic and facilities provided. For mezzanines, a headroom of 3 meters has been used, but in some places, this has been reduced to 2.3 meters due to the constraints of vertical space.

The other facilities in the mezzanines are divided into two main categories: those required in connection with system operation and the stations, particularly those provided for convenience and as amenities to the users. Of the first category, the most important is the Station Master's office. This houses the public address equipment for the magazine and the platforms and means of communication with the Control Centre. CCTV monitors and AFC control are also located in that room.

At all stations, space has been provided for escalators, which were proposed to be installed in the beginning and for those to be provided in the future. All stations have also provided space for air conditioning & ventilation equipment and electrical substations. Facilities required for the convenience of and as amenities to the passengers include illuminated direction signs, train indicators, information booths, counters for selling newspapers and other selected commodities, restrooms and first-aid facilities. Although the primary design considerations of the mezzanine are the efficient operation of the rapid transit system, all of them serve a secondary but essential function of a pedestrian subway. In Kolkata Metro, the Mezzanines have, therefore, been designed wherever possible so that there will be free movement between all access points.

5.20.2 Access to Station Mezzanines

The underground stations have been designed with mezzanines under significant roads at the level of the subway roof. Access to the mezzanines has been provided from both sides of the road.

Where there is a critical road junction above efforts have been made to provide access to the mezzanine from all corners of the intersection. Access from surface streets to the underground Mezzanines are by stairways, except in a few stations where escalators have been provided. **Photo 5.159** of Shyambazar Station showing platform. The station is 1,625 m south of Belgachia Station, the first UG station of the corridor. Shyambazar is the hub of commercial activity in the northern part of the city. The commanding area of the station is very large, and has a large population from the suburbs emanating, travelling by many Bus routes. **Fig. 5.160** includes Entry/Exit of Shyambazar Station and **Fig. 5.161** shows combining concourse and platform.

Fig. 5.159: Shyambazar Metro Station

In many areas of Kolkata, the pavements are not wide enough to accommodate access to the mezzanines. In such cases, adjoining buildings have been acquired, providing access to the mezzanines. At several locations, the metro service buildings have been used for this purpose.

Fig. 5.160: Entry/Exit of Shyambazar Station

5.20.3 Access to Station Platforms

Access from the mezzanines to the station platforms is by stairway or escalators, depending on the vertical separation of the two levels. Escalators have been provided for upward movements of more than 5 meters. At stations with huge passenger loads, escalators have been proposed even where the ascent is less than 5 meters.

Fig. 5.161: Station with Concourse and Platform

5.20.4 Platforms & Passageways

Platforms: The arrangement of entrances and exits to station platforms considerably affects the time required to load and unload trains. For a given train length and number of passengers to be packed, loading time depends upon the distribution of passengers along the platform, the ratio of total door length to car length and the relative size of movements into and out of the train. The loading time is minimised when the number of passengers using each door is equal. Obtaining a uniform distribution of passengers along the platform is, however, difficult, and passengers tend to locate themselves at points that provide them with a favourable exit upon arriving at their destination or points they feel will be less crowded than other points.

The location of platform entrances affects the relative loading of each car in the train. If entry to the platform is at its center, the number of passengers and cars tends to be less.

Similarly, passenger loading in the central vehicles tends to fall off if there is access to the platforms at the ends. If two stairways are provided to access the station platform, they should ideally be located at the quarter points. This arrangement has been primarily adopted by the N-S Kolkata Metro.

Platform widths depend primarily on the maximum number of passengers expected to be on the platform at any time. Given this number, acceptable loading densities (passenger concentration per sq. meter of platform) can be applied to determine the required width of the platform. The selection of an appropriate loading density for the platforms must be done carefully after considering the nature of the movement on the platform and the distance the passengers walk on the platforms.

Fig. 5.162: Esplanade Station

Photographs included in **Fig. 5.162** belong to the Esplanade Station. The station is 704.5 m south of Chandni Chowk and 8.81 km from Dumdum Station, and the first station at the northern fringe of Moidan. It is the focal point of the old splendour and affluence of old Kolkata, which, in the early days of British Rule, happened to be the capital of the Country. Internal view of Interchange station at Noapara included in **Fig. 5.163.**

5.20.5 Length and width of Platformss

The platform's length depends on the train's length. The length of an 8-coach train is 162.4 meters in the N-S Metro and allowing a margin for the inaccuracy of train stopping a length of 170 meters has been provided. The width of the platforms, depending on the intensity of traffic, has been provided as follows:

Island-type platforms	–	10 meters standard
	–	8 meters absolute minimum
Side type platform	–	4 meters minimum to be increased to suit the traffic.

The platform levels have been kept equal to the level of the metro car's floor under loaded conditions, 1080mm above the top of the rails. The platform headroom has been kept between 4 and 4.5 meters, so clearance below the platform signage is 2.8 to 3m.

In calculating the width of the platform, a minimum safety zone of 65 cm on the train side and 25 cm on the offside has been kept. A column on a platform is at least 150 cm away from the platform's edge. In calculating the width of the platform, a density of 1.5 people per sq. meter was considered.

Passageways: This term refers to all means of access to and from station platforms, including corridors, stairways, and escalators.

In determining the desirable capacity and arrangement of passageways, the following figures have been used in Kolkata Metro to determine widths.

5.20.6 Capacity of Passageways

Type of passageway	Capacity per meter width of passageway per hour	
Level corridors and platforms	For movement in one direction	4000
	For movement in both directions	3500
Staircases	For the downward journey	3500
	For the upward journey	3200
	For both upward and downward journeys	3000

5.20.7 Entries and Exits at Stations

The location and position of entries and exits start and terminate at the road level and lead to the circulating hall (mezzanine/concourse) through staircases and corridors, which suit local conditions and needs. All entries and exits are provided with rolling shutters to prevent outsiders from entering the subway system when the trains do not run at night.

Fig. 5.163: Noapara Transfer Station

5.20.8 Staircases

In the N-S Metro, most stations have escalators to handle large volumes of passengers with ease and comfort. A conventional staircase is also offered to escalators. Only staircases are considered adequate where the heights of ascent and descent are less. In some stations, only staircases have been provided for the present, but space has been left in the designs for the installation of escalators in future. In N-S Metro, the staircases are generally provided with steps of 32 cm tread and 14 cm rise.

5.20.9 Facilities for the Guidance of Passengers at Stations

Platforms are provided with Train Indication Boards, which give the train's destination or the name of the terminal station for passengers' guidance.

All platforms, waiting places, and mezzanines should have loudspeakers to maintain continuous communication between station train dispatchers and passengers. The system also allows the Train dispatcher at Central Control to make announcements directly when required. A stand-by battery-operated UPS has been provided so that the system is available in case of power supply failure, avoiding the confusion that would otherwise be created just at the critical time.

Fig. 5.164: Baranagar Station During Construction

For the guidance of detraining passengers, indicators have been provided showing the exit direction.

Figure 5.164 includes different views during the construction of Baranagar Station.

5.20.10 Barriers and Escalators

Barriers: Suitable barriers are provided at entrances from mezzanines to platforms. Initially, these barriers were manned and operated manually. Now, manual operation has been replaced by automatic barriers. The capacity of the automatic obstacles is 25 passengers per minute.

Escalators: Escalators have a standard slope of 30°. To cater to passengers' changing demands, the escalators have arrangements for changing their direction of movement. The escalator requirement has been calculated based on a standard carrying capacity of 8000 people per escalator at a 1-meter width per hour.

The stations' all-access and egress arrangements, including escalators and staircases, have been designed for peak traffic, and the escalator tunnel has been integrated with the station structure.

5.21 RTS Box Corridor and Design

Based on practices followed in different countries, experience gained across the World, analysis, particularly for rigid structures like DW, was adopted in Kolkata N-S Metro.

The distribution of earth pressure corresponding to the soil pressure at rest (k_o) on either side of the diaphragm wall is one of the most important design considerations. Strut spacing is influential for the earth pressure development and rigidity of the diaphragm wall itself. The earth pressure models for both cohesive and non-cohesive soils which was like that of Peck's apparent earth pressure diagram. An instrumentation program, concurrent with the execution, with real-time feedback, fixed and validated the lateral pressure due to building surcharge. It was considered as 5.5 t/sqm at a minimum distance of 2 meters horizontally and 1.5 m vertically below the ground surface. This pressure was presumed to act continuously along the side of the diaphragm wall wherever applicable. Lateral pressure due to road surcharge had been divided into two mutually exclusive cases: during construction and service. Stability of the excavation, ground movement, control of water into the excavation area, effect on adjoining structures and design of structural elements such as diaphragm wall and struts are distinct, yet interrelated features in the design of braced excavation. The depth of penetration of the supporting wall below the bottom of the cut was determined from the consideration of stability against bottom heave, as shown below. The factor of safety against bottom heave was kept at not less than 1.5, as per the formula below

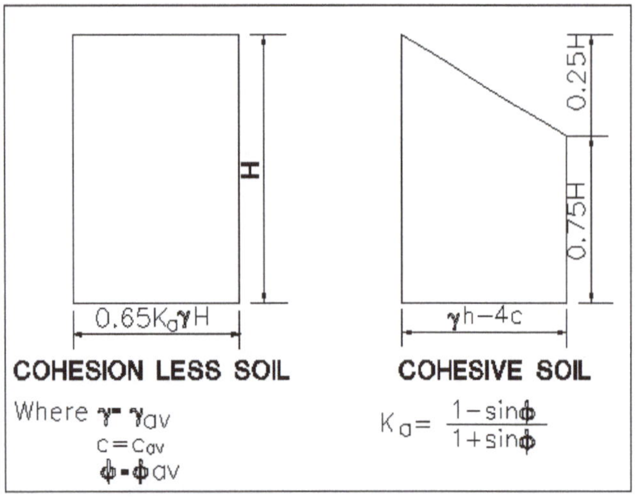

$\gamma_{av} \rightarrow$ Average unit weight of soil

$c_{av} \rightarrow$ Average cohesion

$\phi_{av} \rightarrow$ Average angle of internal friction

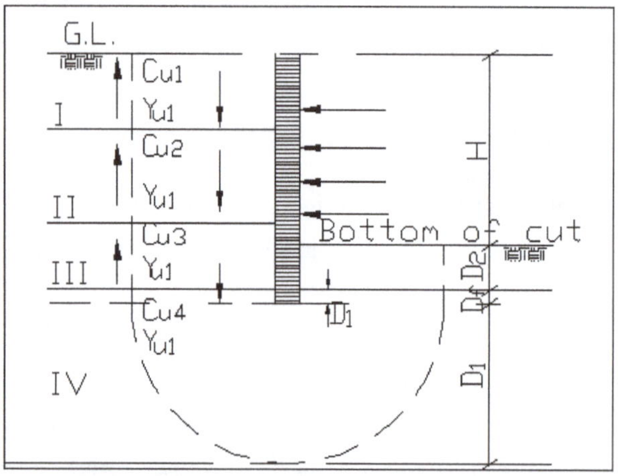

$$F.O.S = \frac{Cu_4 Nc + r_3 D_2 + r_4 D_1 + \sum Cu(H + D_2 + D_7)/D_1}{\sum r(H + D_2 + D_r)} \geq 1.5$$

600mm thick (800 mm in some areas) was cast mostly in 3.0m panels into bentonite slurry trenches. The depth of penetration is in general is 3-4m. In certain cases, depth increased to 7-8 meters, due to soil condition or to carry additional loads of Flyover in future or for planned high-rise buildings in future over the Metro Section.

Panel widths were also increased in places up to 6 meters with two rows of Struts, as per site requirements. The figures shown above have also been discussed for convenience.

Struts were located at the middle of each Panel, except for wider panels and three or four layers, depending on depth/as per site requirement.

In N-S Metro, different design approaches were adopted depending upon Site/Design condition/ requirements, like Cross DW, Bored Cast in situ RCC pile walling, Top-bottom Construction, first time in India.

Remedial Measures

Cases of leaning of Diaphragm walls towards the cut side occurred in many sections, and maximum infringement was in the tune of 800 mm. As remedial measures enveloping DWs were built outside and tilted DWs were broken.

Where the sides of the DW trench collapsed during concreting (due to mechanical failure of cranes), enveloping the DW system was introduced. In places of considerable honeycombing and exposure of reinforcements, steel plates were welded.

To retain a Temple in the middle of the Road, enveloping DW were built around the Temple, and an entry was provided through temporary top decking. Under the Railway Bridge across a narrow road special design was adopted to construct the underground Corridor in several sections and keeping the safety arrangement like grabbing under water by filling the excavation, using sandbags for protective measures, underwater grouting and others.

The N-S metro between Dumdum and Tollygunge has all stations other than Dumdum and Tollygunge underground. Dumdum station was elevated, and Tollygunge was at grade.

All stations are two stories, with the base slab supporting platforms, tracks, and a third rail for traction; the upper floor between the base and roof is the mezzanine, popularly known as the concourse level, which accommodates all passengers' facilities and the facility's required metro operations. The roof is below ground, with 2/3m of filled-up soil at the top, duly relaying the diverted utilities and for new utilities as needed. However, before filling, the waterproofing layer was laid and tested.

Fig. 5.165: Park Street Junction at grade with station underground

Fig. 5.166: Park Street Station During Christmas (Telegraph India)

Park Street Station is located under Park Street Junction (Fig. **5.165).** The stations at Esplanade and Park Street, a focal point of a business district. Last two decades, this area has seen an expansion of business activities along with an increased population. These two Stations are busy throughout the year and become overcrowded during festivals and sports events.

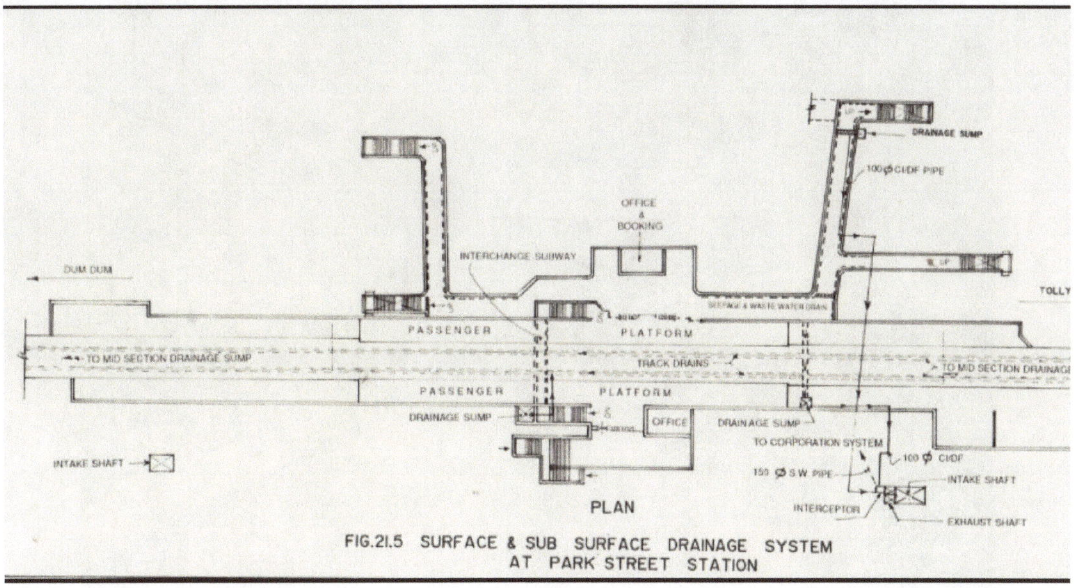

Fig. 5.167: Park Street Station Plan

Fig. 5.168: Park Street Station During Morning Hours (www.telegraphindia.com)

Figure 5.166 is a view of Park Street station during the Christmas period. The station becomes overcrowded every year during Christmas, and the Metro management team, the State Transport Department and Police authorities must take appropriate measures to have optimal flow for passenger traffic while all types of road traffic are closed except emergency vehicles. Similarly, many stations along the corridor must be provided, similar management during the Durga Puja and Kali Puja/Diwali festivals. **Figure 5.167** shows the Park Street Station plan, and **Figure 5.168** shows the station mezzanine during the off-peak period.

Fig. 5.169: Tollygunge Station At-Grade – Last Station for Original Alignment

Tollygunge Station **(Fig. 5.169)** is the southern terminal of the Dumdum-Tollygunge metro corridor. The station is the only station at grade, acquiring a strip of land from the Tollygunge Club, which has a Golf Course and other sports facilities. A Car shed has been sited here besides a staff colony with 128 units of quarters.

5.21.1 Design of RTS Box

RCC structures for corridors, stations, substations, and all underground structures were designed per the relevant provisions of the Indian Code of Practices, IRC, and IRS. During the design process, the international practices where the Indian codes were silent were referred to, particularly for the design of strutted excavation and referred internationally to the design process adopted for Metros.

The underground structure is an RCC box with a thick 500 mm side wall, top and bottom slabs, and a row of columns spaced at 3 m with top and bottom longitudinal beams. Subsequently, a gap of 500mm was kept between the diaphragm wall and the to-be-filled-with-earth. This gap prevents inaccuracy in the construction of the diaphragm wall, though it also increases the overall width. The box was constructed between two 600 mm-thick diaphragm walls A stretch in one of the sections (section 11) was constructed as a semi-integrated construction, where the top and bottom slabs were integrated with the diaphragm. A gap of 75 mm was kept between the side wall of the box (a structural element) and the diaphragm wall. While this has the advantage of a less construction width, it imposes stricter control on diaphragm wall construction. A small section was constructed as an integrated structure with DW. Reinforcement details are shown in **Fig. 5.171.**

The RTS box and Substation in the tank area were analysed as independent structures using Kani's Method and designed as per IS:456. **Fig. 5.170** includes reinforcement details during placement.

Each unit of the box is 21m long in the longitudinal direction between two consecutive Expansion Joints. 225 mm wide dumbbell-shaped PVC water bars were used.

Fig. 5.170: Reinforcements of Base Slab During Laying for RCC Box

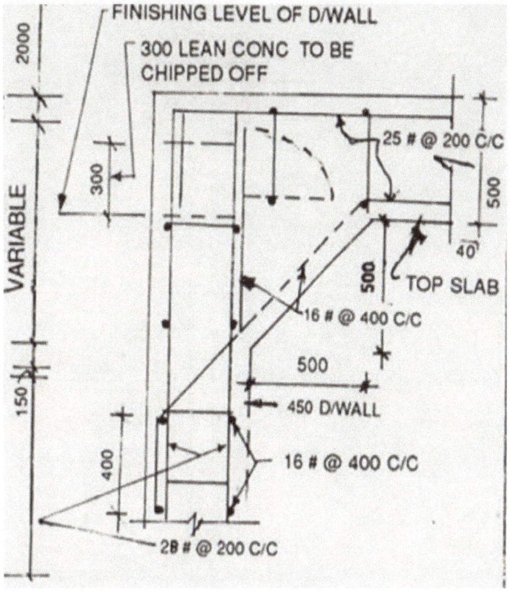

Fig. 5.171: Reinf. Details Between Wall & Top Slab

As mentioned earlier in a few sections RTS box was designed as semi-integrated and integrated structure. For a semi-integrated structure RTS box wall was designed for water pressure only, besides other loads. And DW was designed for soil and water pressure, besides its dead load & surcharge where applicable. Whereas for an integrated structure, DW must carry all loads vertically and horizontally. All joints for independent, semi-integrated, and integrated were designed as rigid joints. **Figures 5.170 & 5.171** include reinforcement at joints.

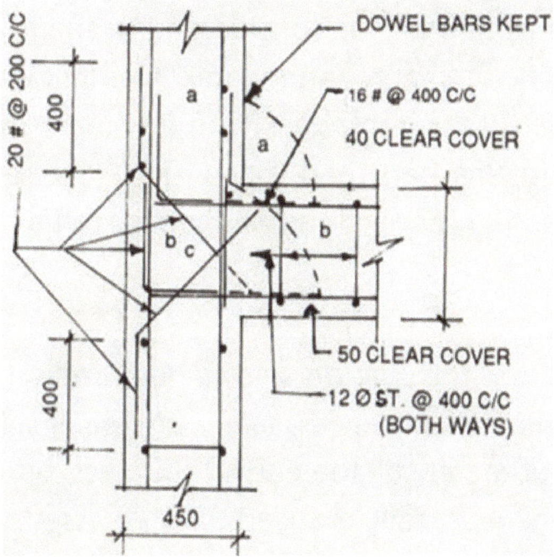

Fig. 5.172: Reinforcement Details Between Wall and Mezzanine Floor (Stations and Substations)

Figure 5.172 shows the reinforcement details between DW and Concourse Slan. **Figure 5.173** includes reinforcement details of integrated DW and RTS top & bottom slab, in integrated design. In this case, the DW & RTS box wall jointly carries the load.

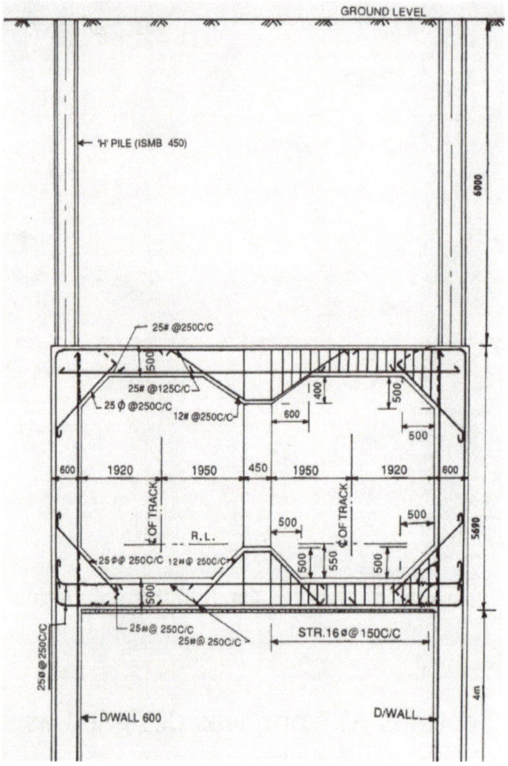

Fig. 5.173: Reinforcement in Integrated Box

The analysis and design of RTS were done using conventional methods. The study was done using Kani's technique, and the joints were designed to be rigid.

Three types of Structural forms were used. The first one independent RCC box within the Parallel shoring walls, like DW and free from DW, was considered. As an independent embedded box, wall pressure was considered for soil, surcharge, and water. As mentioned earlier in the section, the concrete generally used was M20/M25 controlled with a volume mix (1:2:4/1:1.5:3/1:1:3), MS & Cold Twisted High Tensile reinforcing bars, and rolled and Plate sections manufactured in India. Fabrication using welding and bolted connections was carried out as per the Indian Codes.

5.22 Temporary Structures

Temporary structures required for the Station and other auxiliary buildings were essential for technological upgrades, cost & time effectiveness, safety, and quality. It was necessary for scaffolding to be of high quality and safe not merely till the end of a project, but to continue for projects in the pipeline so that the cost could be drastically reduced through asset depreciation. In other words, in the language of corporates, these must be treated as capital goods. In India, bamboo scaffolding was considered the dominant construction type. Even today, it is primarily used for smaller & non-repetitive structures. Though bamboo scaffolding was not safe enough, most of these temporary structures were used due to their low cost, ease of handling, and availability. As such, in India till yesterday, scaffolding and formwork were low-technology and labour-intensive. With global markets/sites opening, there have been radical changes in construction technology for over a decade. The predominant material used in scaffolding changed from bamboo to timber posts for low height & steel posts with specially made joineries. As of today, Indian Company L&T has developed with DOKA (L&T and DOKA joint-venture) for custom-made formwork. The form works changed to resin-bonded plywood sheets, customised steel formwork, plastics, and aluminium (Alu form) for buildings or FRP for dedicated finishes.

Formwork contributes to one of the most critical functions in any significant project, particularly in quality, cost, and time. Reliability and risk assessments are being done to effectively reuse temporary structures in cases with similar loads, which reduces time and expense. **Figure 5.174** shows the staging for a substation using pipe scaffolding. **Figure 5.175** shows timber scaffolding. **Figure 5.176** includes completed frame of Park Street Station.

Fig. 5.174: Staging used for Substation

Fig. 5.175: Staging and Shuttering in Stations

Fig. 5.176: Completed Frame at Park Street Station

5.23 Water Proofing

Waterproofing underground metro corridors, substations, and stations is a challenging job. The water table in Kolkata is generally high at 5-6m, and the metro's excavation depth was around 11m. During the construction of the underground structures, the concrete was mixed with waterproofing compounds and water bars were used for every construction and expansion joints. However, for the N-S metro corridor, the concrete used was of grade M20/M25, less dense than the concrete used on the day. A general method evolved, which, in turn, was believed to take care of all combinations of situations concerning the inflow of subsoil water in the box. In the initial stage of construction, two layers of polyethene films were laid with bituminous coatings laid over lean concrete and wrapped around the box **(Fig. 5.158)** and **Figure 5.177** includes water bars across box joints and includes dumble shaped water bars at the expansion joints. The Dumble shaped water bars with a 25mm central bulb were used along the periphery and on two layers. For construction joints, water bars were serrated on both ends. **Fig. 5.178** shows water proofing layers around the RTS box.

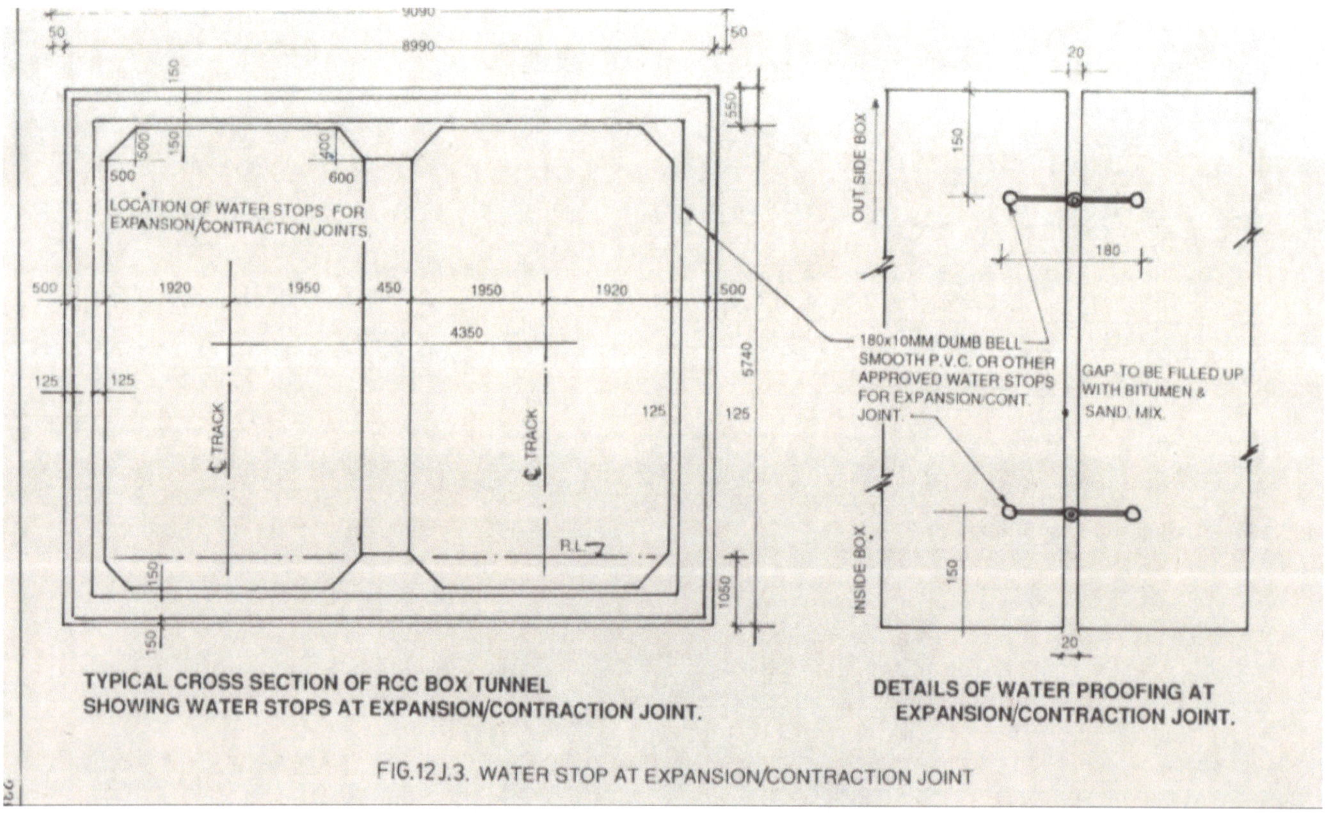

Fig. 5.177: Water Bar Across Box Joints

In the construction joints, locations were preplanned, and a fixing arrangement is included in **Fig. 5.179** and reinforcement details used are shown in **Figure 5.180.** In both cases, water bars were 225 mm long.

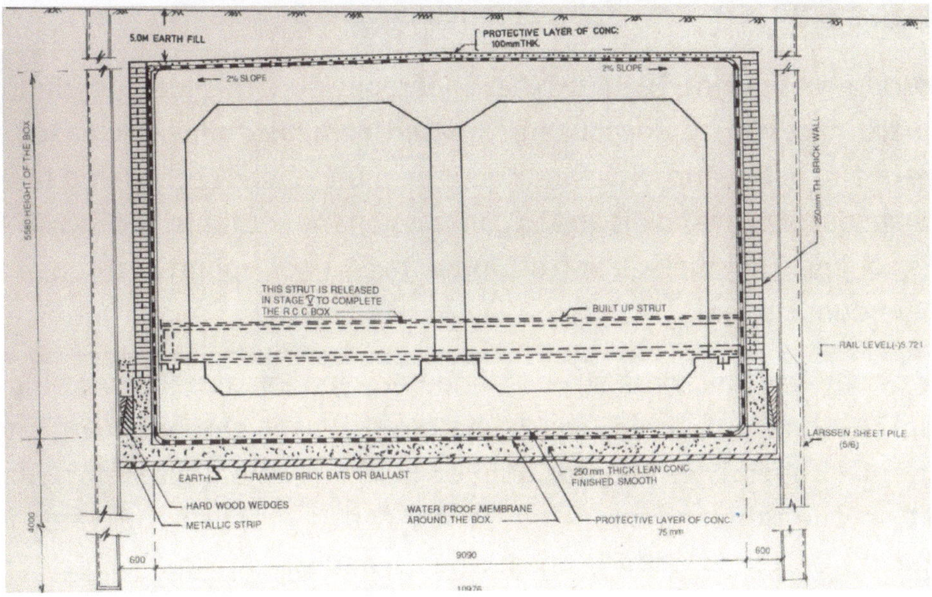

Fig. 5.178: Waterproofing layer around the RCC Box

The wrapping of the waterproofing membrane around the RCC box was not feasible all along the corridor, particularly for semi-integrated and integrated sections. Waterproof layers were laid on the roof and the specifications were followed, as described below.

i. The roof surface was finished smooth, 12mm thick, with 1:6 cement mortar,
ii. A layer of bitumen 80/100 grade conforming to IS:73-1961 was laid at the rate of 1 Kg/sq.m
iii. A layer of black polyethene film 1000 gauge, with cut bitumen adhesive in overlaps, was laid over a hot bitumen surface.
iv. Laying of 100-gram brown Craft Paper laminated with a layer of bitumen,
v. Laying of a hot layer of bitumen as detailed in item (ii),
vi. Laying of 1000 gauge black polythene film as included in item (iii),
vii. Laying of the second layer of Craft Paper as given in (iv)
viii. Applying Hot Bitumen at 1kg/per Sq.m.

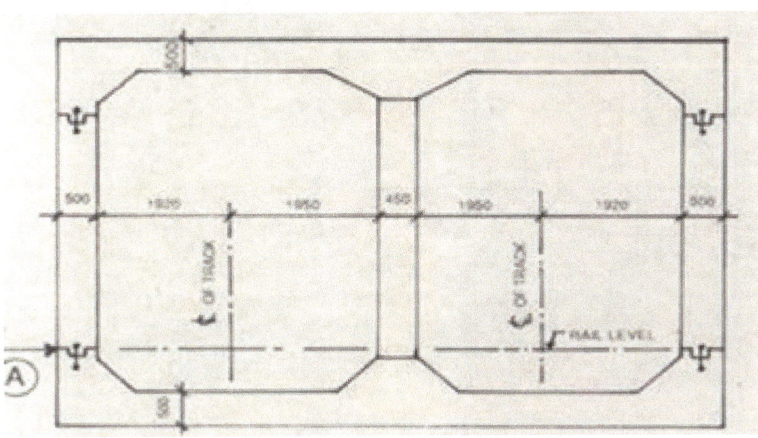

Fig. 5.179: Water Bars at Construction Joints

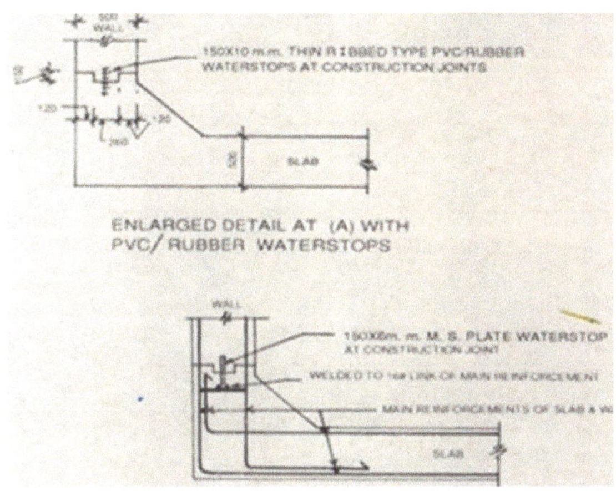

Fig. 5.180: Reinforcements at Construction Joints

5.24 Depot and Car Shed

The N-S Metro corridor was planned with two Car Sheds, one at both ends, Dumdum and Tollygunge. Two car sheds were necessary on account of phased commissioning. However, the depot was planned at Noapara and in the end, Noapara Depot cum Car Shed is scheduled to be the largest in India. The Tollygunge Car Shed was built first to commission the southern section of the corridor, i.e., Tollygunge and Esplanade. The small section at Dumdum End was operated through a temporary car shed built near Dumdum Station.

A temporary car depot was built on a strip of land between Eastern Railway lines and RTS lines entering Dumdum Station. The lines in the car shed were on the embankment, and the dead end provided a substantial buffer. The shed was built on 50m x 70 m of land. It includes one line with an inspection pit on one side, and one fixed Gantry with a manually operated hoist to lift the bogie underframe. It transferred underframes, wheels, and axles to load vehicles below between RTS lines.

The car shed at Tollygunge End was built on 1.613 acres of land, 174.8 m x 37.35 m, and 682m from the station. This area was used for stabling, car washing, a wheel lathe, and signal cabin. The 9.2 m wide shed is made of north light trusses as roof material supported on columns at a spacing of 16.2 m and 20.2 m. The Inspection pits and floor were made of concrete M15. The Yard facilities were i) a Stabling Yard, ii) a Pit-type wheel lathe, iii) a Battery loco shed, iv) a Car washing plant, and v) Washing lines. The shade included 2nos electrically operated 30-ton capacity supported on Gantry Girder.

Fig. 5.181: Temporary Car Shed Near Dumdum RTS Station

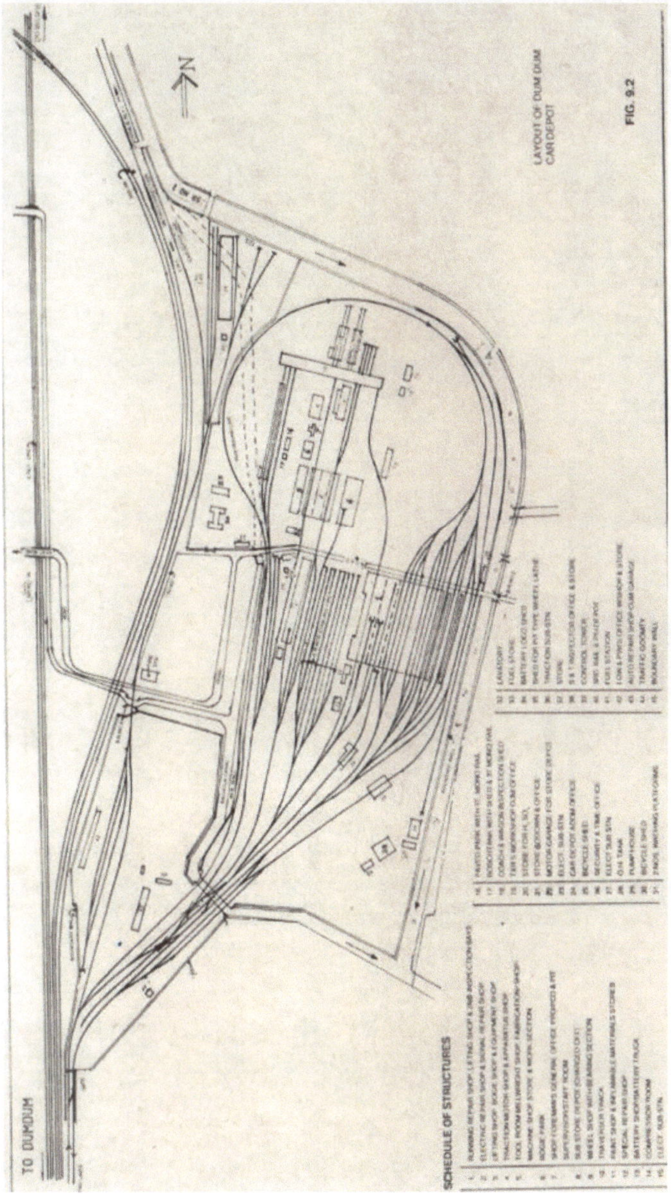

Fig. 5.182: Plan of Noapara Depot

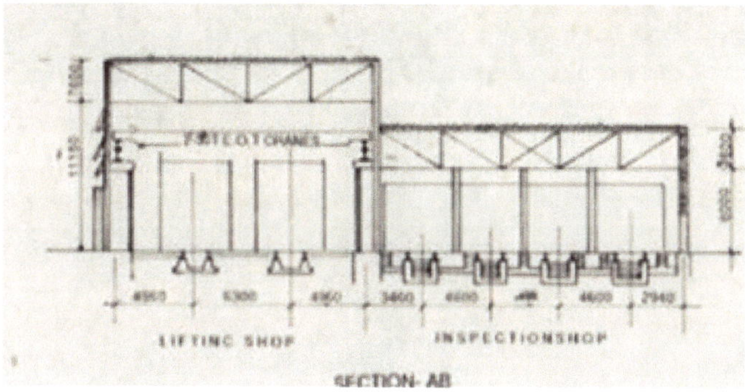

Fig. 5.183: Cross Section of Depot

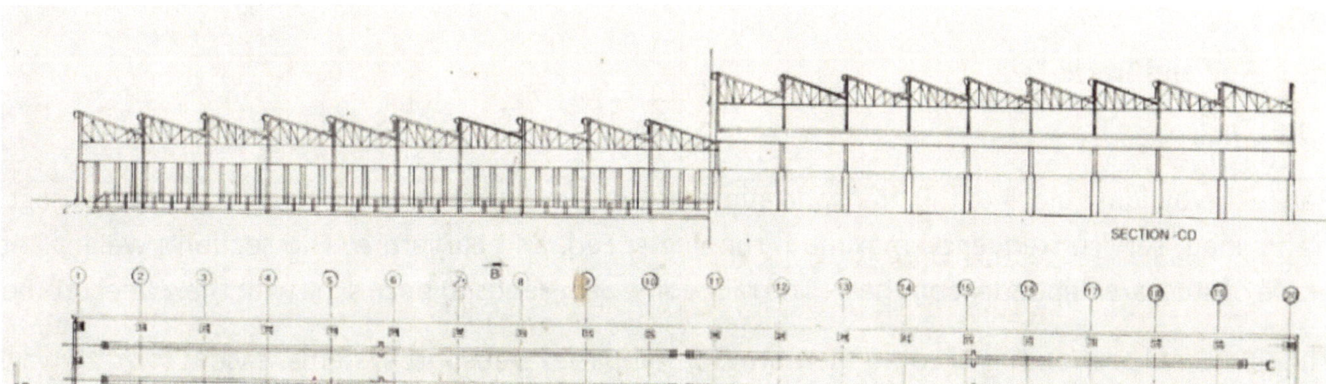

Fig. 5.184: Noapara Covered Car Shed – Longitudinal Section

Fig. 5.185: Noapara Car Shed

Fig. 5.186: Internal View of Tollygunge Car Depot

Figures 5.181 to 5.186 include different views of Depot and Carshed.

5.25 Rail and Tracks

5.25.1 Rail

Rail is the building block of this transportation system. The rail section is selected based on usage, traffic load, service frequency, maximum running speed, and curvature. The section's weight and grade of steel are dependent on these. The rail profile also needs to be in sync with the wheel profile.

The 60 KG rail section has become the norm for rail-based transport systems, except for tramways with distinct rail sectional profiles. This is only for track on embankments; track on depots is often on level land without banks. Sometimes, tracks can also be in cuttings.

5.25.2 Track Formation

In a general way, it refers to the layers comprising the blanket, the subgrade/top layer of formation. The Formation Top is the boundary (interface) between the ballast and the top of the blanket or subgrade (where the blanket layer is not provided). The track foundation consists of ballast, sub-ballast, blanket and subgrade, which are placed/exist below the track structure to transmit load.

Cess is the portion at the top of the formation level that extends from the toe of the ballast to the edge of the formation. Ballast is made of crushed stone, and the desired specifications are placed directly below the sleepers.

A sub-ballast is a layer of coarse-grained material provided between blanket/subgrade and ballast and is confined to the width of the ballast section only. However, sub-ballast is not in vogue on Indian Railways.

A blanket is a layer of specified coarse, granular material of designed thickness provided over the full width of the formation between subgrade and ballast.

The upper part of the subgrade is formed into a prepared subgrade layer, which generally has a cross-fall. The subgrade is the upper part of an embankment or cutting provided above subsoil by borrowed soil of suitable quality up to the bottom of the blanket or ballast. For embankments, subgrade may be imported soil, whereas in cuttings, it is the naturally occurring soil of sufficient strength.

Cohesive Subgrade: A subgrade constructed with soils having cohesive behaviour, i.e., shear strength is predominantly derived from the soil's cohesion, is termed a cohesive subgrade. Normally, soils having particles finer than 75 microns exceeding 12% exhibit cohesive behaviour. As per the IS classification, all fine-grained soils and GM, GM—GC, GC, SM, SM—SC, and SC types of soils exhibit cohesive behaviour.

Cohesionless Subgrade: Subgrades constructed with cohesionless, coarse-grained soils, i.e., where shear strength is predominantly derived from internal friction of the soil, are termed cohesionless subgrades. Typically, soil particles are finer than 75 microns; less than 5% exhibit cohesionless behaviour. As per the IS Classification, GW, GP, SW, and SP soil types fall in this category.

The purpose of having all or some of these layers below the sleeper, like ballast, sub-ballast, blanket, etc., is to progressively reduce stress to a level that can be borne by the parent formation below without any permanent deformation. The ballast also serves the additional purpose of helping maintain track geometry and rigidity and providing resilience. Cohesionless behaviour as per IS Classification, GW, GP, SW & SP types of soils fall in this category.

The purpose of having all or some of these layers below the sleeper, such as ballast, sub-ballast, blanket, etc., is to progressively reduce the stress to a level that can be borne by the parent formation below without any permanent deformation. The Ballast also helps maintain track geometry, rigidity, and resilience. Dispersive clayey soils normally deflocculate when exposed to water with low salt content. They are generally highly erosive and have high shrinkage and swelling potential. Crumb, Double Hydrometer, Pin Hole, and Chemical Tests can identify these soils.

The subsoil is the soil of the natural ground below the subgrade and embankment fill.

5.25.3 Other Types of Track Structure

In addition to the conventional ballasted track type, other track structures exist for specific purposes. Some are briefed below:

Direct Fastening Track is typically used on a viaduct/tunnel with a concrete deck. The rail is typically laid on a plinth/slab connected by the fastening system with anchorage. The plinth is connected to the deck through shear connector and friction forces. **The floating slab** is not structurally connected to the deck slab. Generally, a resilient mat is provided between the deck and slab, holding the rail in position.

Embedded Track is a system in which the rails are embedded in concrete, with only the running edge and gauge face exposed to support rolling stock wheels. This track type is generally used on at-grade sections with a high interface with roads. The Embedded track is applied at a location in the depot where staff are walking safely on the track. The embedded track in the workshop is carried out for the following locations:

Lifting line: Tracks will be embedded through the entire length of the track in this workshop area.

Wheel Lathe: The underfloor wheel lathe is installed in a pit inside the workshop floor. The rails will be covered only up to the pit edge. The Pre-Pit and Post-pit tracks will be laid as embedded tracks.

Bogie drops: The bogie drop equipment is also installed in a pit inside the workshop floor. Like tracks in the wheel lathe shop, the Pre-Pit and post-pit tracks will be laid as embedded tracks.

The stabling track is generally a **ballasted track.**

Ballast-less Track with Sleeper is a track system in which the rails are fixed on concrete sleepers with specific fasteners and supported over ballast. This track type is generally used on at-grade sections with a low interface with roads.

Special Track in Depot/Maintenance Yard: The entire depot area will be at the same level. The longitudinal profile of tracks will be flat, and the top rail level will be the same. Different types of track applications will be used in the Depot area according to functional requirements.

Washing Apron: This type of track shall be used for the washing area. It is a Plinth Track with no derailment guard and special drainage scheme provision. It has an Auto Coach Wash, Internal Heavy Cleaning line, and Bogie Wash Facilities.

Track on Columns: This type of track inspects bay lines and the pit wheel lathe (pit area). It is supported by metallic poles (steel columns) installed vertically beneath rails with a unique fastening arrangement. The metallic poles will be fixed in a concrete foundation. The purpose of this track is for the inspection personnel to examine the underneath portion of the coaches ' embedded Track.

5.25.4 Track Gauge Used in India

Track gauge is the distance between the inner edges of both rail heads of the track, measured 14mm below the top surface of the rail head. Historically, different gauges were used in various parts of India. Cargo transhipment from one gauge to another was always done by manually unloading and reloading the goods from one train to another.

i) Broad Gauge

The decision to use a gauge wider than the one in Great Britain was made with an eye towards economies in freight movement and to ensure stability in the face of Indian weather and the perceived threat of cyclonic winds. This is now found nationwide; all major passenger and freight routes are broad gauge of 1676mm.

ii) Meter Gauge

This is still found in many places despite the push to convert everything to broad gauge. Four people could sit comfortably alongside, which would have been 1000mm to the metric system, or 1 m.

iii) Narrow Gauge-1

This gauge, of 762 mm, was adopted in various parts of the British Empire. The most well-known line is probably the Kalka–Shimla route. The rationale for the narrower gauges was economy in building the lines — they could be laid much faster than broad gauge lines and in more challenging terrain. It was envisioned that narrow-gauge lines would act as feeder lines to the broad-gauge and meter-gauge lines.

iv) Narrow Gauge-2

A few places in India have the even narrower 2–foot gauge: New Jalpaiguri-Darjeeling, Neral Matheran, and the Gwalior branch lines, which include Gwalior-Sheopur Kalan, Gwalior-Shivpuri, and Gwalior-Bhind. The gauge is 610mm

v)Standard Gauge

This has been used in a few places in India. The Calcutta tram lines were probably the most well-known before the various metro systems in India introduced standard-gauge trains Indian Metros mainly use the Standard Gauge of 1435 mm.

vi) Other Indian Gauges

From an interoperability perspective, having a uniform gauge for the entire network is always beneficial. A broader gauge increases traffic capacity as the coach's width increases. Also, with a broader gauge, the wheelbase allowed is generally more, resulting in more excellent stability at higher speeds, which imparts more safety during operation.

In 2009, the Task Force proposed setting up a Special Purpose Vehicle (SPV) with an initial corpus to implement a wide-ranging integrated multi-modal transport system for suburban commuters. This led to balanced and sustainable growth across the region. The NCRTC, the nodal agency for implementing this multi-modal transport system for the NCR, was formed.

5.25.5 Track Structure in N-S Metro

The track structure used for Kolkata N-S Metro consisted of three ballasts and fewer track types (M-1A, M-6, and M-7). Type 1A with 60 kg/m rails fastened to cast iron bearing plates, including 4.5mm thick grooved rubber between rails and bearing plates, besides 12mm thick grooved rubber pad below bearing plate. As included above, all were fastened to a concrete bed anchored with studs. It is fixed in high-density polythene inserts, with holes left for pouring the second layer of concrete.

For track type M-6, similar rails used in M1A were fastened to wooden sleepers. The wooden sleepers were fixed to the reinforced concrete bed using steel holding-down bolts. A grooved rubber pad was placed between the rail and the wooden block. The M6-type track structure was used on the elevated section between Dumdum and Belgachia.

60 Kg/m wear-resistant rails were laid over 10mm thick, with designed, continuously grooved rubber pads and fastened to second-pour concrete. A levelling layer of epoxy and silica-sand mortar was laid over the concrete bed below the rubber pad to ensure correct levels. MCI inserts were fixed to the second pour of concrete. For grouting and fixing down the holding down bolts and MCI inserts, Araldite GY 257, Hardener HY 840 and quartz-sand mix (100:50:350) by weight were used.

Track for Yards and Sidings: In stabling yards, conventional 90R track on wooden, cast iron, or steel sleepers was used. Details of Track Structure are included in **Figs.5.186 to 189.**

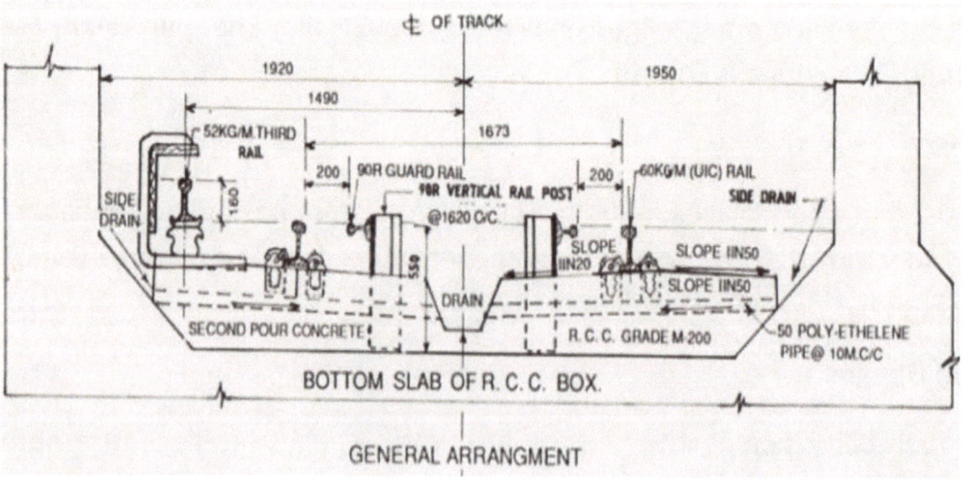

Fig. 5.187: Plinth Detail of Single Track

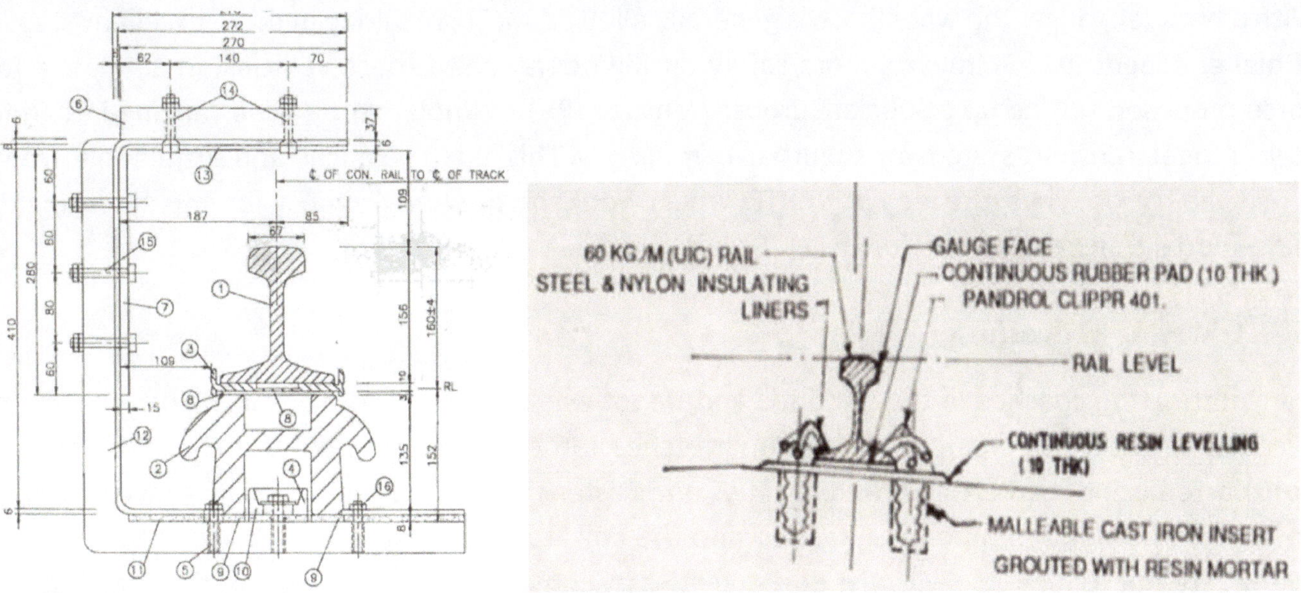

Fig. 5.188: Fixing of Rail

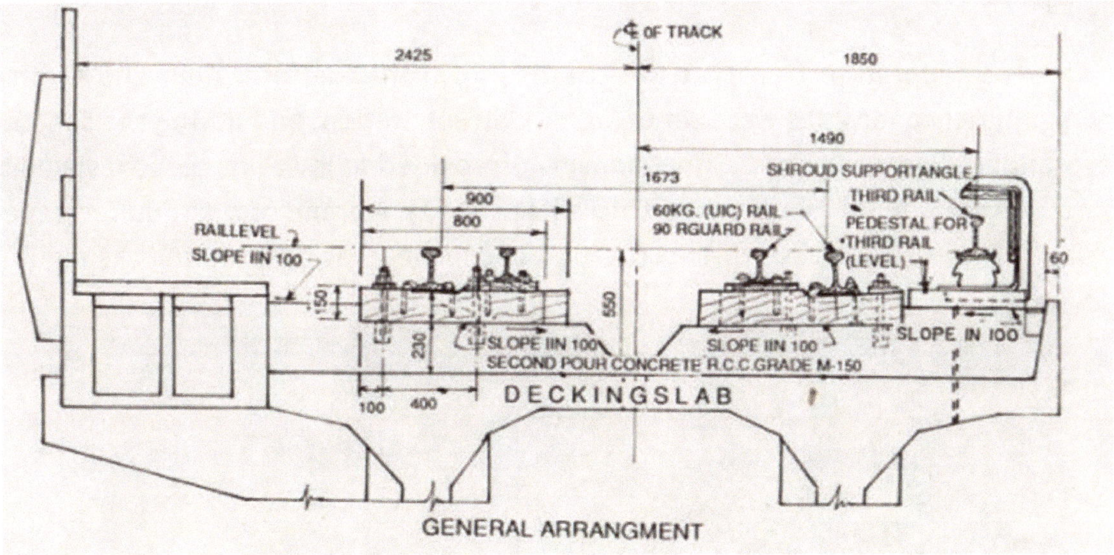

Fig. 5.189: General Arrangement of Track on Concrete Deck

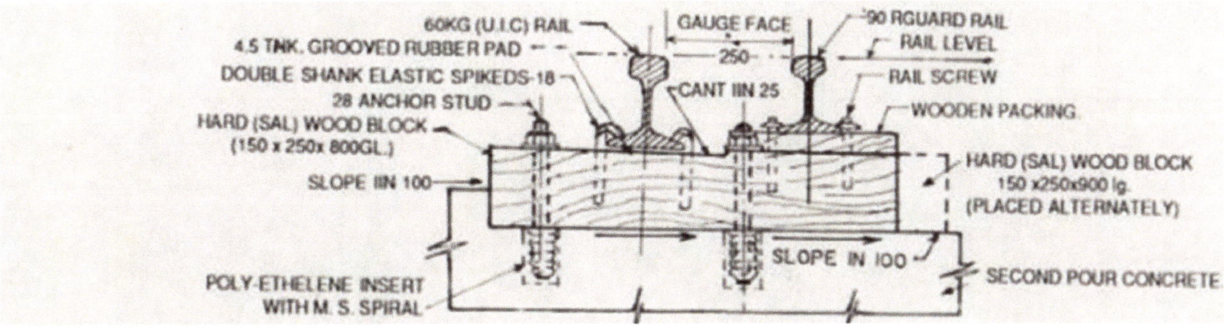

Fig. 5.190: Enlarge Detail of Track with Wooden Block

With a broader gauge, the wheelbase is generally allowed more, resulting in more excellent stability at higher speeds, which imparts more safety during operation of these vehicles. In 2009, the Task Force proposed setting up a Special Purpose Vehicle (SPV) to implement a wide-ranging integrated multi-modal transport system for suburban commuters. This led to balanced and sustainable growth across the region. Thus, the idea of NCRTC, which would be the nodal agency for implementing this multi-modal transport system for the NCR, was formed.

5.26 Lowering of Coaches

Transporting the coaches to the Car Shed and Depot was one of the most challenging jobs, mainly because construction was carried out in stretches rather than continuously. The coaches were transported using the Port lines and facilities and Eastern Railway lines. The coaches were brought down from Dumdum temporary Car Shed into the Kolkata Port Trust Railway lines on Strand Road, taking care not to interfere with the Port activities by constructing short siding at the site chosen in construction sidings of Hooghly River Bridge Commission. The site was selected to load the coaches in Road Trasport. The end of siding ends with a ramp and a solid platform of 22m long, 4m wide and 1.4m high.

The Vehicles were adopted to transport the coaches from the Platform built and Metro Tunnel. Temporary maintenance facilities were set up at Park Street Station, and during the construction of Park Street Station, a 23m long and 2.3m opening was provided to lower the coaches. Vehicles used for transportation were multi-wheel trailer trucks **(Fig. 5.191).** A crane of the required capacity **(Fig. 192)** was used to lower the coaches. Transportation of coaches was done during the night.

Fig. 5.191: Metro Coach Brought to Depot by Multi-Axle Trailer Truck (Reference: "The Kolkata Metro)

Fig. 5.192: Crane is Used to Lower the Coaches System

5.27 Power Supply

The power supply system was designed to maintain reliability, economy, flexibility, and safety. It incorporated the optimum utilisation of equipment types. It ensured the continuity of the entire/ partial supply to maintain safety. Power supply is the most vital element of any MRTS's expenses, and costs constitute approximately 40% of working expenses. Underground stations warrant lighting, ventilation, and air conditioning. Signaling had to function, tunnels and stations were used to prevent flooding using electric pumps, and passenger movement was done using escalators and lifts for disabled people, all of which needed electric power.

Power and energy for RTS requirements are in three major components: power for traction, Power for equipment, and auxiliary equipment and services for the Car Shed, Repair shed, Staff Quarters, Administrator buildings, canteens, etc. Power for the Dumdum-Tollygunge section was estimated, and Traction is 40 MVA, Auxiliary Power is 23 MVA, and for others—2 MVA. After the corridor extension to Dakshineswar and New Garia, the power required more than 65 MVA, which was needed for 16.45 km between Dumdum and Tollygunge.

CESC made the 33 KV power supply available at four North-South corridor receiving stations. The 33 KV was lowered to 11 KV through a 16 MVA 33/11 KV transformer with two no's for each receiving station. The power supply for auxiliary services was from two auxiliary transformers connected to the two sections of the busbars of the traction substation.

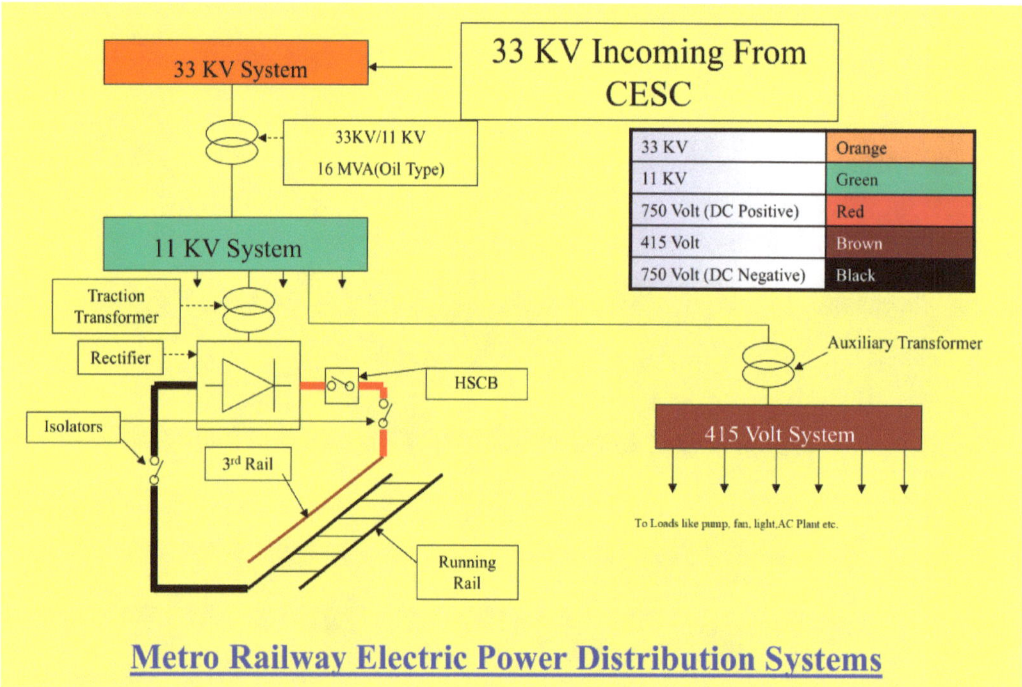

Fig. 5.193: Power Distribution System

As such, the principle adopted for designing the power supply distribution was such that the first degree of failures would not affect the running of services, including other auxiliary services.

For the original alignment, 11 **Traction Substations** were designed along the Metro route, four of which were planned to combine with receiving stations; one was at the Depot, and the remaining six were at different stations. At the traction substations, the 11KV supply was converted to 750v DC through transformer rectifier sets having silicon diodes with high Repetitive Peak Inverse Voltage.

The number/capacity of rectifier sets satisfied the 90-second headway with chopper-controlled cars with regenerative braking.

Fig. 5.194: Underground Auxiliary Substation

Most of the N-S Metro corridor stations (for Original Alignment) are underground, so emergency lighting was essential, and batteries were provided at both ends of the stations.

To ensure reliability and timely isolation of the fault and faulty equipment/sections, protective arrangements were provided, which are:

i. 33 KV/11KV transformer
ii. 11 KV Bus section
iii. 11 KV AC/750V DC Rectifier Transformer Protection
iv. 750 V DC Rectifier
v. HSCB and Isolator Panel
vi. 11 KV/415 V Auxiliary Transformer
vii. 11 KV & 415 V Feeder Protection
viii. Battery Protection
ix. Indicating & Recording Instruments

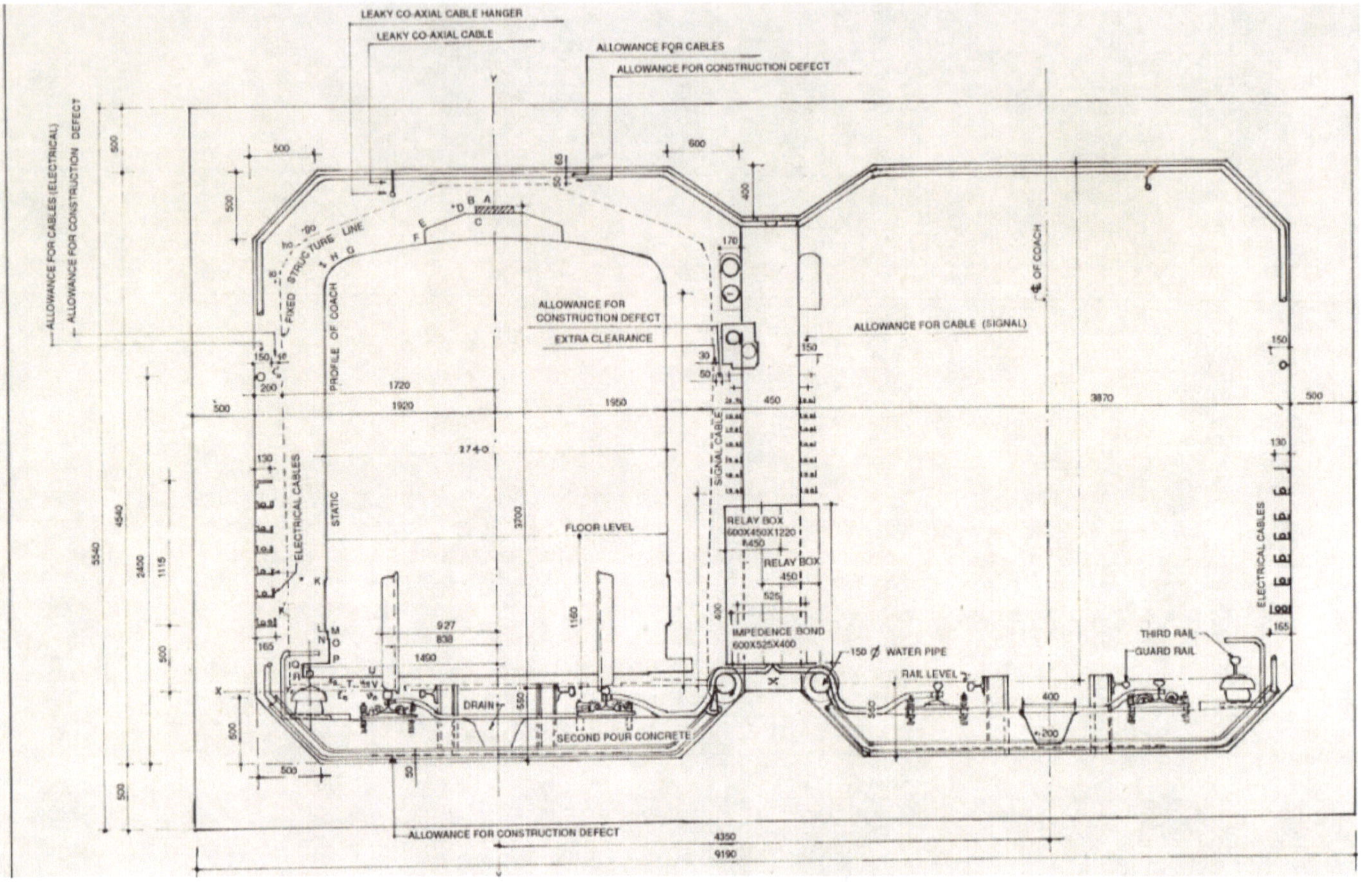

Fig. 5.195: RTS Box with Cable Profile including Location of Third Rails

The Power supply system in the North-South Metro corridor was divided into two major categories: i) Traction Control System & ii) Auxiliary Control System. Two remote-control panels, including tele-control and tele-signal, are housed in the Administrative Building Control System at the center of the route.

5.28 Third Rail

A third rail, also known as a live rail, electric rail or conductor rail, provides electric power to a railway train through a semi-continuous rigid conductor placed alongside or between the rails of a railway track. It is typically used in a mass transit or rapid transit system, with alignments in its corridors, fully or almost fully segregated from the outside environment. Third-rail systems are usually supplied from direct current electricity.

In the case of the North-South Metro, the power supply to the operating is fed through the third rail, which is located near the station. The traction substation is located at either end of the platform, commensurate with the direction of operation. **Fig. 5.195 & 5.196** show a typical view of the third rail at the station. The third rail is powered by a DC high-speed circuit breaker from two adjacent substations.

For the safety of passengers and the maintenance team, the third rail is located away from the platform side, with fibreglass shrouds painted red. The inter-tripping facility was made between the DC high-speed feeder breakers of two adjacent traction substations feeding the same third rail section to ensure.

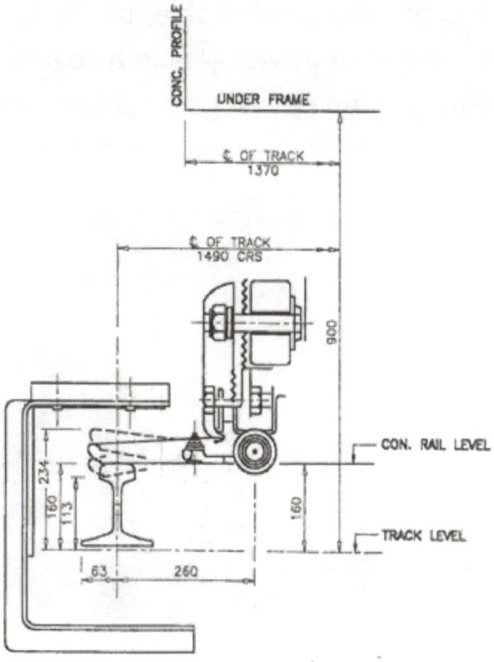

Fig. 5.196: Current Collection System

5.29 Rolling Stock

Kolkata Metro's rolling stock consists of five classes: three classes for the blue line (line 1), one of which is also used for the purple and orange lines, and two classes for the green lines. All trains receive 750V DC using a third-rail electrification system. Trains can operate up to eight for blue, purple, and orange lines and six cars for the green line.

Kolkata Metro was constructed in the late seventies; there were technical limitations. Due to tunnel dimensions and being part of Indian Railways, Blue Line, the first metro line in India, opted for a meter gauge shell, 22.7m with broad gauge, 1676 mm tracks. However, the track is a standard gauge for the green line, i.e. line 2, like international standards.

First, the N-S metro was planned with a four-coach train manufactured in I CF Chennai. Soon after, due to traffic demand, the train module was changed to eight coaches. As such, over the years, the metro line has been extended in both directions, North and South, increasing the elevated and underground.

Rapid transit coaches were designed to satisfy i) Reliability, ii) Traffic needs, iii) Noise reduction, iv) Fire hazards, v) Lightweight, vi) Quick reversibility, vii) Automatic operation, viii) Equipment's layout, ix) Scheduled speed & x) platform height from rail top is 1080 mm. Basic dimensions like length, breadth and height were worked out based on traffic characteristics, coach & tunnels and station costs, operational cost, and the headway. Coaches were designed with 2400 passengers per train (made of 8 coaches), and dimensions are considered in the following paragraph. Cares were taken so that it should be pleasant and not suffocating. Sitting arrangements were in the transverse direction, as sitting in the longitudinal direction is uncomfortable, particularly during starting and breaking. To accommodate the maximum number of passengers in a coach, the ratio of standing and sitting was 5/7 to 1. The occupancy density is eight passengers per sqm, which is a crushing load. Axle load calculation was considered based on 60 kg per person. Coaches of a train are vestibuled.

Fig. 5.197: Rolling Stock – North-South Metro (railwaygazette.com)

For N-S Metro, nine rakes were manufactured by ICF, Chennai, and N-S Metro started operation between Esplanade and Bhavanipur in 1984. The colour of the coaches was primarily green and later changed to white with blue stripes. By around 1988, 9 more rakes were delivered to Kolkata and used for the extended portion. The length of each rake was 156m, coach dimensions were 19.5x 2.74 x 3.55m, maximum speed 55 km/h for the first lot & the second lot was 80 km/h. Average operation speed of 30 km/h. The first two lots' ventilation system was forced with washed and cooled air. Later, all the coaches supplied were AC and started operating on the extended elevated section. ICF built the Coaches in collaboration with BHEL and Knorr-Brennecke, Germany.

The electric traction equipment developed in collaboration with RTS authorities developed the DC Chopper system (development was principally in Japan, Germany, Sweden, Switzerland, and Italy) using thyristors that enable stepless variation of the voltage of the incoming supply for applying to the traction motors. The system uses DC motors, and the power supply is 600 V/1500 V. By chopping, a low voltage is applied to the traction motors to start, and the voltage is varied continuously during the train's running. The chopper used to have regenerative break down to a low speed during braking.

The length of the couplers is 20.3 m. The car's width was determined by considering benches with two people sitting on either side with a gap of 690 mm to facilitate quick alighting and boarding even if the standing passengers were leaning against the transverse benches. Each car has a driving space (cabin) of 1400 mm. The driving space is provided by one end of a driving unit that covers 1400 mm in length. The Power supply is through a stainless third rail, 750V DC. There are 24 stations in line 1, with 17 power sub-stations receiving power from CESC (Calcutta Electric Supply Corporation).

A third rail system with 750V DC is a traction for all lines. Metro Rail, Kolkata's operator for line 1, plans to change from Stainless Steel to Aluminum. Upgrading the signaling system is essential to increase passenger per rake and line capacity by reducing the headway, i.e., reducing the interval to 90 seconds. The Railway Board of India has approved the proposal, cleared the invitation for Tender, and advised RVNL to take care of it.

The implementing authority is taking the necessary steps to install the CBTC rakes from China for additional rakes.

The important details of the Rolling Stock are reproduced:

i. Gauge: 1.676,m Broad gauge (Indian Gauge)
ii. Rake: Consists of 8 coaches. Driving cabs: Each Rake (Train)- two driving cabs
iii. Coach Dimensions: 20.4 x 2.74 x 3.70 m
iv. Distance between Bogies & Wheelbase: 13.515m and 2.450m, Wheel Diameter: 860 mm/780mm
v. Traction Voltage: 750 V DC – Third rail.
vi. Coach body: Corten Steel. Inside panels: anodised aluminum
vii. Automatic Door closer system & vestibuled
viii. Radiotelephone and Communication facilities
ix. Ventilation: Roof-mounted 230V single-phase Ventilation fans.

5.29.1 Maintenance of Rolling Stock

An essential ingredient in the successful running of a railway is a well-maintained system, and train maintenance is the most important. Trains are built of several mechanical and electrical systems with several moving parts. Rolling Stock is the most maintenance-intensive, vulnerable and cost-intensive. However, replacement is costlier.

5.30 Signaling and Telecommunication (S&T System)

The signaling and Telecommunication System of the Kolkata North-South Metro Corridor was designed to include

Signaling

Continuous Automatic Train Control System (CATC) comprises

i. Automatic train protection (ATP) and Cabin signaling.
ii. Automatic Train Operation (ATO)
iii. Automatic Train Supervision (ATS)
iv. Centralised Traffic Control (CTC)

Panel Interlocking was provided at pre-selected stations. Automatic Route Setting was designed for two terminals of the original route, Dumdum and Tollygunge, while other stations, namely Maidan, Esplanade, and Girish Park, were designed for manually controlled CTC operators. The standby system on the block worked in the remaining stations, while a Route Relay Interlocking system was adopted at Car Depot.

Telecommunication

The telecommunication system adopted for the first Metro corridor includes

i. Train/Traction, Maintenance control telephone System
ii. Emergency Control Telephone System
iii. Duplex VHF Train Radio Communication

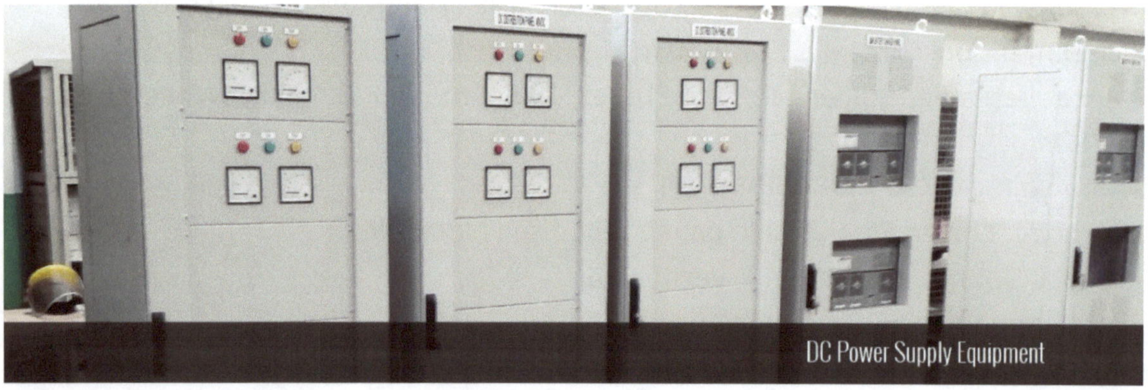

Fig. 5.198: DC Power Supply Equipment

iv. Automatic Telephone Exchanges at Metro Administrative office, Dumdum and Tollygunge depot
v. Passenger assistance telephone system
vi. Miscellaneous P & T (Post & Telephone) facilities
vii. Passenger announcement system
viii. Continuous passenger supervision and surveillance are done using CCTV.
ix. Centralised time system
x. Automatic Fare Collection System.
xi. Train Destination Board etc.

5.31 Ventilation System

Need

1. To ensure the supply of fresh air for the Passenger (O2)
2. To dilute obnoxious and hazardous gases
3. To bring down the humidity
4. To dispose of a large quantity of heat by Traction Eq., Auxiliary appliances Human Body, etc.
5. To dispose of the heat leaked inside the Subway from the earth

Air Ventilation Design Criteria

Fixing Up Ambient Condition (90% of day of any season) from last 5 Year Record from Metrological Department at 10 Hrs. when peak Traffic Occurs.

Season	Temperature	Relative Humidity
Summer	36	55
Monsoon	33	70
Winter	27.2	57

Because of pollution and Dust, Air washery had been provided at every fresh air intake. Now at Esplanade station, dry air filters have been provided & air washery closed to save electricity consumption.

Fig. 5.199: Air Ventilation Criteria

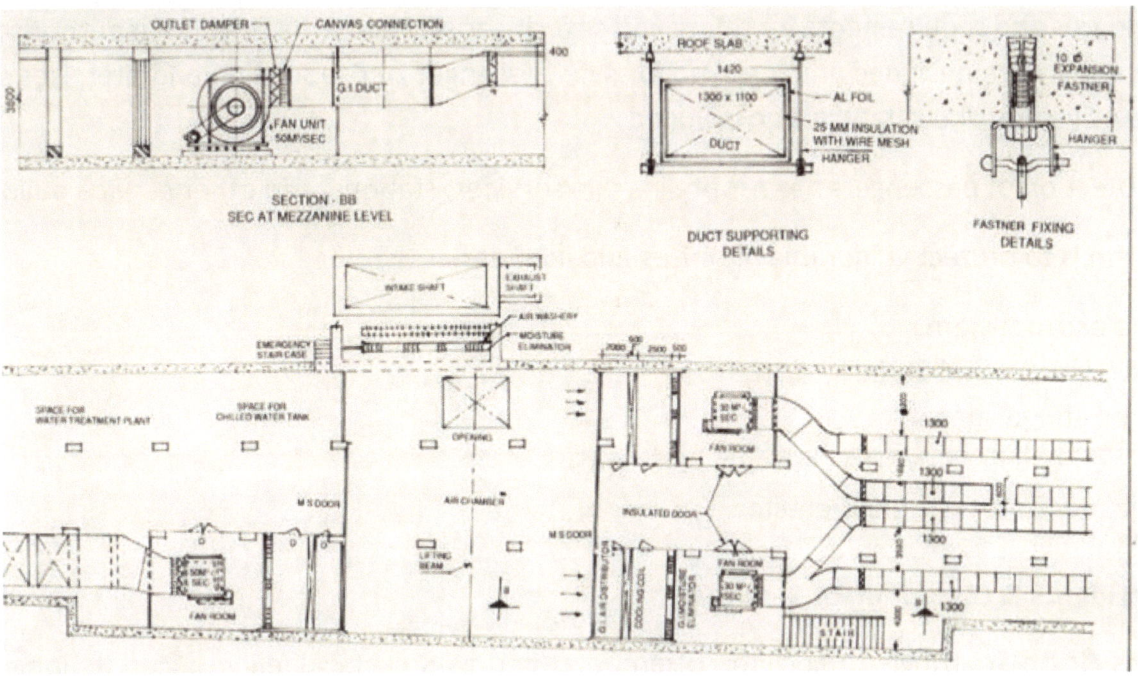

Fig. 5.200: Ventilation System and Equipment Layout at Ventilation Mezanine

5.32 Schedule of Dimensions (SoD)

North-South Metro was India's first metro, so it was decided to frame the Schedule of Dimensions. To start, a static coach profile was decided based on the proposed dimensions, like the length of the coach from buffer to buffer, center to center bogie, and wheelbase, in discussions with Metro Authority and RDSO-Lucknow, and it was decided to follow the same for future Metros. The design of the static coach profile was based on

i. Coach dimensions based on a study of dimensions of Metros abroad.
ii. Literature Survey.
iii. Survey of EMU stocks.

The Dynamic profile was finalised concerning the static profile based on the following criteria

iv. Vertical Movement
v. Horizontal Movement
vi. Rotational Movement.

The Kinematic Movement affected by

vii. Straight Track
viii. Curved Track – Superelevation, Middle & End throw and Additional sway.

Besides the safety margin between fixed structures & moving trains, and the guiding principle of EMUs.

5.33 Safety & Security Systems

The standards and requirements for safety and security arising from the fire and unauthorised entry into premises were designed and installed for safe passenger transportation and premises safety in the Metro System. The safety design included

- Protection of passengers against fire in train services, stations, and other service buildings.

The system is to protect vulnerable premises and includes,

- Fire alarm system.
- Fire Hydrant and Sprinkler System.
- Fire Extinguishers.
- Closed-circuit television with video analytics.
- Security Gates – Metal Detector.

5.34 Accidents & Disruption of Services

The city's North-South Metro corridor, often described as Kolkata's lifeline, faced its longest ever disruption of 13 hours after **floodwater entered the tunnels** between Rabindra Sarobar and

Tollygunge, **as per a TOI report**. Tracks were submerged, and the third rail that powers trains had to be switched off to allow pumping.

Even as trains managed to run on the Maidan–Dakshineswar stretch, it wasn't until 5.38 pm that services could resume on the Maidan–Bansdroni section, **once safety trials confirmed the rails were dry enough.**

An underground retaining wall of the North-South Metro corridor was damaged during civic pipe-laying activity on SP Mukherjee Road, prompting safety concerns for the city's busiest mass transit line. Metro Railway engineers confirmed that a section of the diaphragm wall—critical for structural stability and flood prevention—was compromised during ongoing installation of a 1,400mm water pipeline by city authorities. In response, the Metro administration requested to repair and a shift in pipeline alignment, on priority to safeguard tunnel integrity.

The incident occurred near Mudiali, where workers laying the steel water line unintentionally breached a 3-metre stretch of the diaphragm wall, about one metre below ground level. **(Urban News)**

5A. Modernisation of Existing TVS, ECS of Metro Railway, Kolkata, Line 1 (Ayesa)

Underground Section Incorporating Energy Efficiency and Reliability During Normal Operation, Congested Operation, Emergency Mode Demand.

A.1 Background

The Kolkata Metro line 1 was commissioned, with its underground section in the 1980s. The existing Tunnel Ventilation System (TVS) and Environmental Control System (ECS) were designed based on the standards, technologies, and operational requirements prevailing at that time.

With decades of service, these systems are now ageing, less energy efficient, and face frequent reliability issues, leading to challenges in maintaining safety and comfort conditions for passengers, especially in emergency situations such as fire, smoke, or congested operations.

Metro Railway, Kolkata, awarded the Detailed Design Consultancy for the modernisation of the existing TVS, ECS of the Metro Railway Kolkata underground section, incorporating energy efficiency and reliability during normal operation, congested operation, and emergency mode demand to M/S **Ayesa Ingenieria y Arquitectura S.A.U.**

A2. Details of Existing System

The VAC system provided in the Kolkata Metro is generally known as the Open Type System with tunnel cooling. The stations are provided with an air conditioning system (mechanical cooling) to remove the heat generated by the passengers, train propulsion or braking, and the heat released by other equipment such as lighting, signage, escalators, fare gates, etc. The system is also required to supply necessary fresh air for the biological needs of passengers and staff inside the subway. The train coaches were not air-conditioned. Therefore, the tunnel section is also supplied with cool air. This is achieved by means of high-capacity fans installed at the station ends, which deliver cool air in the tunnel sections.

A3. Scope

All the above fans are centrifugal fans, and it is not possible to rotate them in both directions. However, the VVVF drive is being installed for rotation at a selected speed. The train operation in Metro Railway is being done using AC and 12 Non-AC rakes (which shall be phased out by 2025). Till 2010, only non-AC coaches were in operation. Since its inception, there was UPE to remove heat from SBR, but no OTE is present even though Metro Railway shifted to AC rake.

To accomplish this objective, Detailed Design Consultancy for modernisation of existing TVS, ECS of Metro Railway, Kolkata including Smoke extraction system design for a typical underground station with associate tunnel including provision for heat extraction from Coach AC unit.

A4. Object of consultancy contract

The design of HVAC was carried out by Kharagpur IIT in 70s. The calculation was done manually. In 70s there were hardly any Metro Railway in tropical countries except Hong Kong Metro. Hong Kong Metro went ahead with close loop system which was later followed by other tropical countries. Metro Railway already modified the design from open loop system to semi closed system for energy saving in station and tunnel cooling.

Now Metro Railway has decided to get the state-of-the-art TVS and SCS system integrated with SCADA controlled fire alarm system in tunnel and stations, to have state of the art smoke extraction system. Metro Railway Kolkata decided to engage Detailed Design Consultancy (DDC) of international repute to have cost-effective, reliable and state of the art TVS, TCS, fire extraction system redesigning the existing system.

The purpose of engaging DDC is to study existing semi-closed system of TVS, ECS system using simulation software to know the adequacy/inadequacy of existing system compared to state-of-the-art TVS, ECS & emission system to address the demand for normal, congested and fire situations in tunnel & amp, stations. Metro station and tunnels. DDC must validate the design using software, and prepare the tender document covering specification, sizing of equipment's, BOQ, likely sources of supply and AFC drawing including Client's approval.

DDC shall ensure complying with the relevant standards of ECS, TVS for metro railway including fire safety provision in NFPA 130 and National Building Code. It may happen that required design changes require compromise of structural design, in that case optimal design changes shall be proposed without structural design alteration.

3D Computational Fluid Dynamics (CFD) Analysis and 1- Dimensional analysis for all stations and tunnels to meet NFPA 130 (latest) requirements for tunnel ventilation system. The DDC shall also conduct value engineering optimising tunnel ventilation system capex and opex costs.

Need for Modernisation

1. Energy Efficiency
   a. Existing systems operate on outdated designs, with high energy consumption.
   b. Modern equipment (EC fans, VFDs, smart controls) can reduce energy demand by 20–30%.
2. Reliability in Different Modes
   a. Normal Operation: Maintain temperature, humidity, and air quality in underground stations and tunnels.
   b. Congested Operation: Handle peak passenger loads with enhanced ventilation and cooling.
   c. Emergency Mode: Efficient smoke extraction and pressure control during fire or breakdowns to ensure safe evacuation.
3. Compliance with Latest Standards
   a. NFPA, ASHRAE, and international metro safety standards now demand higher redundancy, reliability, and fail-safe operation compared to the original design.
4. Passenger Comfort & Safety
   a. Improved air circulation and temperature control for daily commuters.
   b. Enhanced smoke control and evacuation safety in case of fire emergencies.
5. Scope of Modernisation:
   a. Replacement of old fans, dampers, and motors with energy-efficient and reliable equipment.
   b. Upgradation of control and monitoring systems with SCADA/IoT-based automation.
   c. Integration of Variable Frequency Drives (VFDs) and demand-based operation logic.
   d. Reconfiguration of emergency smoke extraction system in compliance with latest fire safety codes.
   e. Upgradation of ducting, sensors, and air distribution network.

6. Expected Benefits:

   a. Energy Savings: Significant reduction in operating costs through energy-efficient technology.
   b. Reliability: Enhanced system availability with minimum downtime.
   c. Safety: Improved emergency handling capacity ensuring passenger safety.
   d. Passenger Comfort: Better air quality, temperature, and humidity management.
   e. Sustainability: Reduction in carbon footprint aligning with green metro initiatives.

7. Conclusion:

   Ayesa Ingenieria y Arquitectura S.A.U. has been providing detailed design consultancy services for the modernisation of Tunnel Ventilation System (TVS) and Environmental Control System (ECS). While executing these services there are also some modifications required for existing dation to accommodate new requirements for which Ayesa has been providing comprehensive design consultancy for architectural planning, E & amp; M works (including electrical, plumbing, firefighting and M& amp; E sssSCADA).

PART 3

CHAPTER 6

East-West Metro, Line 2

6.1 Introduction

Moving from one place to another is an inseparable part of modern city life. Movement is the essence of business and livelihood, of growth and development, indeed of life itself. In cities today, with their teeming millions, the overflowing traffic, the uncertainty of travel duration, the prospect of travel is daunting. Travelling in a city is as problematic as it is necessary, as tiresome as it is inevitable. Achieving unhindered travel may seem utopian in cities today, but it is possible to make it significantly better. The aim of the East–West Metro corridor project was to make travel in Kolkata a reliable and pleasurable experience.

Indian underground metro evolves in such a way that it adopts different designs and construction techniques based on site conditions, projects, materials, construction equipment, and engineering solutions. Complexity and land availability for executing a megacity like Kolkata have an expanse of urban conglomeration around its core, which houses millions of people.

Running through the heart of Kolkata **(Fig. 6.1),** bridging the city's most populous areas, the corridor has been planned to reduce travel costs and time. Normally, one would take at least an hour and a half and use multiple modes to travel between the extremities of the city. Users would find themselves reaching their destinations in air-conditioned comfort in a shorter time than they take today and with unbelievable reliability.

Kolkata, a major metropolitan city of India, has a population of more than 20 million living in the Kolkata Metropolitan Area (KMA in 2024), which covers an area of more than 1876 sq. km and has a population density of 24000 persons per sq. km, which is one of the highest in the world. The rapid growth of Kolkata has made it increasingly difficult to maintain the minimum acceptable level of urban services, and most of them are showing visible signs of deterioration. Transportation services are one of the worst affected. Subsequent efforts to solve the problem by augmenting public transport vehicles barely touched the fringes of the problem. The average weekly passenger volume increased by 100% from 1960 to 1970, and more in the early twenty-first century, which rose by more than 190% in bus travel, while tram travel remained stationary due to capacity saturation. Meanwhile, the number of suburban rail passengers for Sealdah and Howrah stations increased by

323% between 1960 and 1970 and continued to rise rapidly. Kolkata tops India with 2448 cars per Km.

The Green Line, Line 2, has a planned eastern extension from Sector-V to Teghoria. It will cover 22 km and have 17 stations from Teghoria (Haldiram) in the east to Howrah Maidan in the west, of which 11 will be elevated and 6 underground stations.

Green Line's Extension Under Construction Stations List

Teghoria/Haldiram, Raghunathpur, Baguihati, Dumdum Park and Keshtopur

Fig. 6.1: East-West Metro Corridor During Construction

6.1.1 Connecting Two Major Terminals

Two major rail terminals, Howrah on the west of the River Hooghly and Sealdah on the east, were designed to serve millions of suburban passengers daily. An efficient connection between Howrah and Sealdah stations, which between them cater to about 24 lakh passengers a day, is a dream for every commuter for whom the daily trip into and out of the city at these two stations is a nightmare. The East–West Corridor will finally realise this dream. And there is yet another advantage to the new metro that would cause the whole city to breathe a sigh of relief. No needless sprinting from the station to the bus stop to the auto stand, lumbering from the auto stand to the share-taxi stand. The East-West Metro Corridor is expected to provide the Kolkata commuter a seamless way to travel.

This automatically means moving people and goods across an expanding landscape. And, as it expands, the things to be crossed loom ever more significant. Metro lines are no exception. In a congested city like Kolkata, the crossings may be short but are complex. Since the sixties, Kolkata has been spreading in all directions. The same is for the City of Howrah, across a major river like the Hooghly. Consequently, the East-West travel patterns are becoming more intense. At the same time, rail-based systems are under different planning and implementation stages, besides other developmental initiatives, particularly in the road sector, on both banks of the River Hooghly. As the cross-river traffic increases continuously, water transport is taken care of.

6.1.2 Crossing the Hooghly

When one steps out of Howrah Station and heads towards the city, one is faced with the daunting task of crossing the Howrah Bridge. Any attempt to revive its long-faded use is difficult. Teeming with every kind of vehicle possible, the bridge has become a dreaded prospect for every Kolkata–Howrah–bound traveller. This is where the East–West metro, which combines an almost dreamlike appeal with uncompromised efficiency, travels more than a 10-storied building below the water surface for more than 500 meters across the Hooghly. The reinforced concrete segmental tunnel linings have a thickness of more than a quarter of a metre, caulked with a composite gasket of neoprene and hydrophilic rubber to seal the tunnel from ingress of water. This marvel of modern technology is a resounding first in India and a rare venture worldwide.

An Engineering feat and the technological precision and expertise that is required for a project that makes it possible to travel under the Hooghly in a state-of-the-art metro line. Recognising the precarious situation of travel in Kolkata, KMRCL, which was authorised to build the metro corridor, adopted technologies at the planning stage to minimise discomfort. The underground portion of the corridor has adopted the latest boring process, with the top–down cut–and–cover process sparing citizens inconvenience and traffic snarls to a great extent. The boring of twin tunnels **(Fig. 6.2)** was planned with "Earth Pressure Balance" Tunnel Boring Machines, to be procured from abroad.

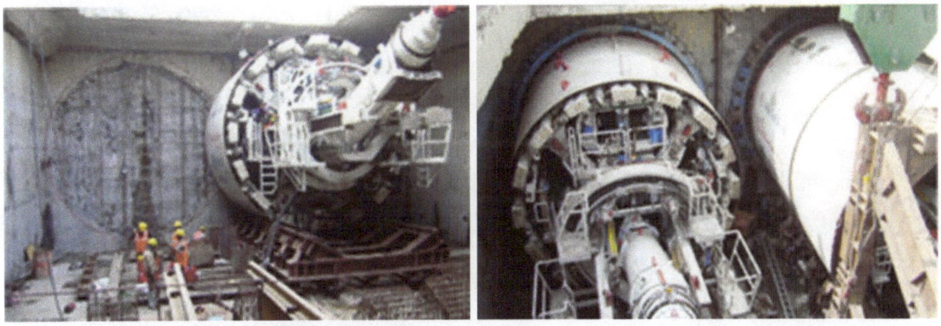

Fig. 6.2: Tunnel Below River Hooghly (Ref. KMRC Report)

Fig. 6.3: View of Completed Entry to a Station

The Kolkata East-West Metro project involves numerous challenges. Twin tunnels of the East-West (E-W) Metro pass below the river, which is 520m wide and approximately 14m below the riverbed. Tunnels are 5.55m in diameter and made of precast concrete segments. **First, precast concrete segments were introduced in the mid-seventies under the River Main in Germany.**

Fig. 6.4: Completed Entry to Howrah Metro Station

Fig. 6.5: Completed Entry to Sealdah Metro Station integrated with Sealdah Main

The new engineering involved meticulous planning of the alignment and experience in Tunnel execution under the River Main in Germany, which helped the author use his experience gained in Germany in the implementation of the East-West Metro.

The East-West Metro, detailed in Part 3, holds the distinction of being India's first metro to make an underground crossing under a major river, the Hooghly. This corridor is partly elevated **(Fig. 6.3)** and partly underground **(Fig. 6.10)**. The elevated sections utilised precast concrete segments and in-situ construction, while the underground corridor employed TBMs and precast concrete segments. Underground stations were constructed using the cut-and-cover and top-down methods. Instrumentation played a crucial role in both the North-South and East-West Metros.

The most important aspect of the East-West Metro is that the line (No.2) connects the two major terminal stations **(Figs. 6.4 and 6.5)** among the busiest stations of the world and crosses a major river. Besides the two major corridors of Kolkata Metro, the East-West & North-South Metro integrate at Esplanade **(Fig. 6.6).** Both corridors were planned for extension. The extension of the North-South Metro has already been completed and is in operation. The planned extension of the East-West Metro, from Howrah Maidan to Santragachi, is shown in **Fig. 6.7.**

For the East-West Metro, KMRC (Kolkata Metro Rail Corporation) appointed **MYCEL** as the General Consultant. The MYCEL consortium comprised **AECOM (Maunsell), Yachiyo, CES, EGIS, and LHPC,** where the author served as Joint Project Director, and Project Director for a significant period, **reviewing RITES' report on alternative alignments** between BBD Bag and Sealdah. The original alignment was approved during the implementation of the Kolkata Metro network plan and incorporated provisions for a transfer station at BBG (Bowbazar) Street & Chittaranjan Avenue for integration with the East-West Metro, during the design and construction of the North-South Metro. The subsequent change in alignment significantly delayed the project's completion.

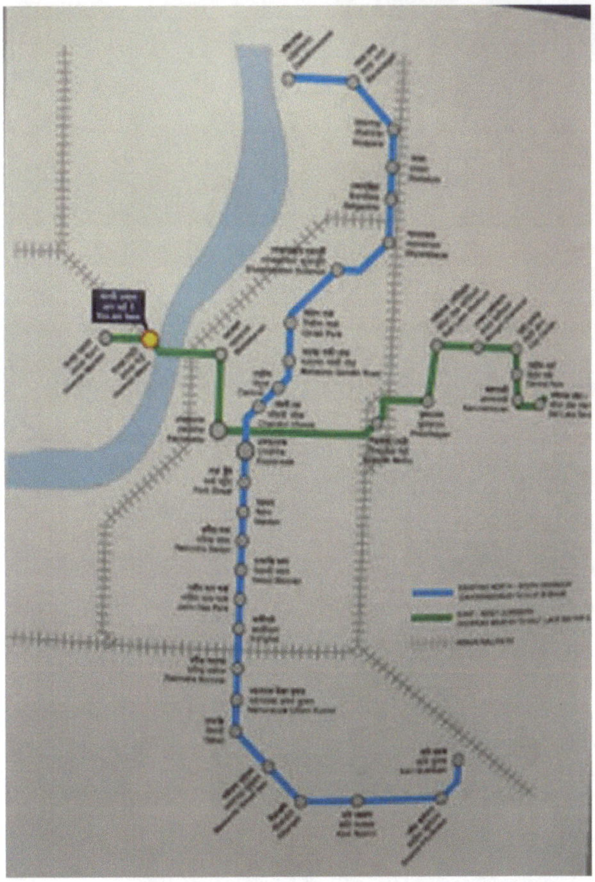

Fig. 6.6: **Integration of East-West & North-South Metro**

Fig. 6.7: Planned Extension of East-West Metro

The Indian underground metro industry evolved in such a way that it adopts different construction and design techniques based on site conditions, project type, material availability, and engineering solutions. Methods and techniques currently used in cut-and-cover tunnel and station construction have evolved environmental factors at the work site. Complexity and land availability for execution of the work are also significant factors in establishing the construction and design methodology. Contractors commonly use cut-and-cover techniques for stations, ramps and or Corridors. The construction sequence is planned after the identification and relocation of utilities, installation of earth retaining structures, installation of temporary traffic decking, excavation to the bottom, and construction of the station box. The cut-and-cover technique is still being continued, depending on the availability of sufficient space. Several Metro corridors are planned in this city and suburban region, totalling 137 + Km of new lines. This Line 2 corridor is already in operation in two sections, and the Hon'ble Prime Minister of India has inaugurated the corridor **(Figs.6.8 & 6.9).**

Fig. 6.8: Hon'ble Prime Minister of India is inaugurating the Line 2, Green Corridor

Fig. 6.9: Hon'ble Prime Minister of India visited the East-West Metro

Fig. 6.10: One of the underground Station Entrances of the East-West Metro

A very challenging Metro corridor has been implemented in the East-West direction, with a 16.6 Km **(Fig. 6.11)** corridor running underground through densely populated and built-up parts of the city. A future eastern extension (Other than an extension in the west beyond Howrah Maidan) from Sector V to Teghoria is planned. It will cover 22 km and consist of 17 stations, ranging from Teghoria in the east to Howrah Maidan in the west, comprising 11 elevated and six underground stations **(Fig. 6.10).** There are currently two operational sections, one from Howrah Maidan to Esplanade and the other from Sealdah to Salt Lake City, Sector V. At present, more than 100,000 passengers commute every day, and this figure is expected to be one million every day by 2035 when an entire metro line will be operated, connecting two of the biggest terminals of the world, Howrah and Sealdah.

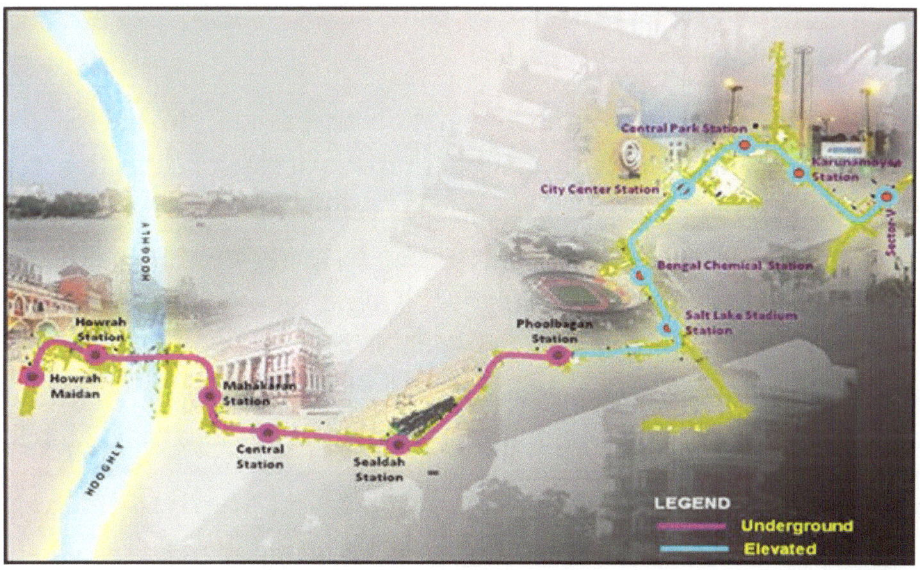

Fig. 6.11: Original Alignment of E-W Metro

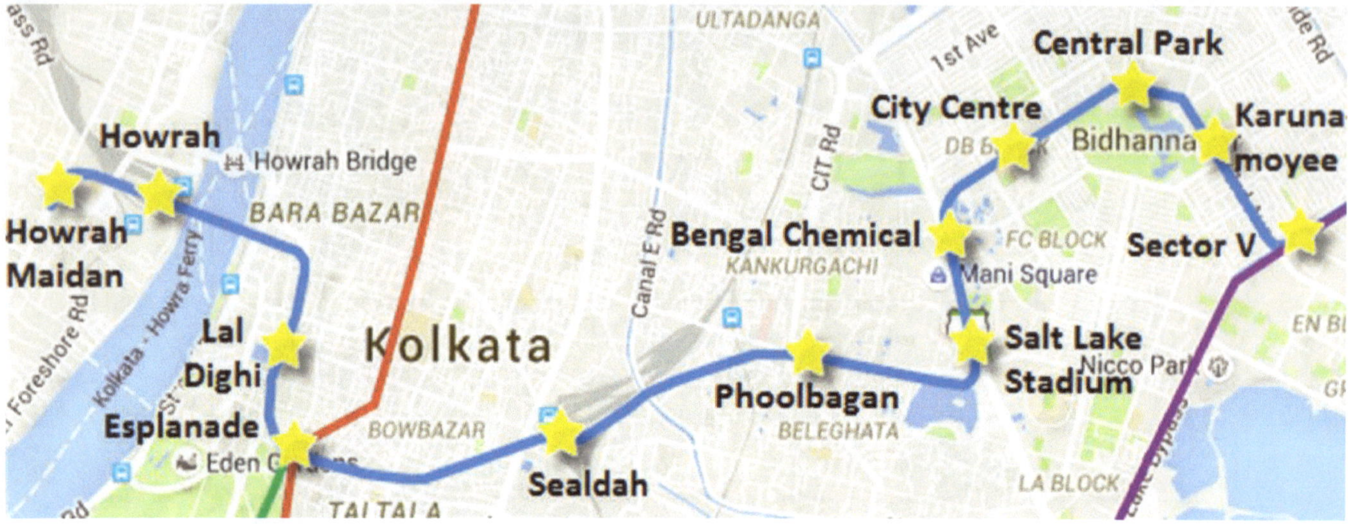

Fig. 6.12: East-West Modified Alignment

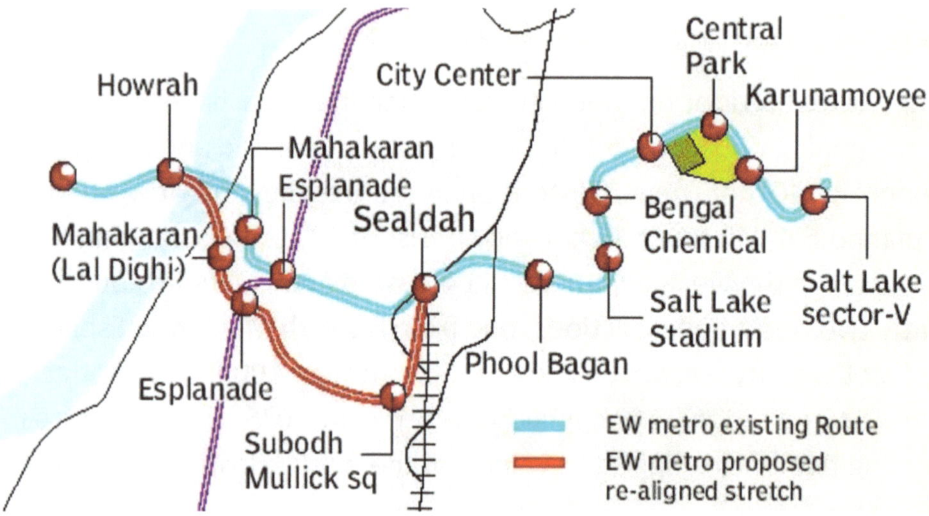

Fig. 6.13: Plan Showing both Original and Executed Green Line

As per the UN-Habitat report, the global urban population increased from 220 million in 1900 to 8025 million in 2023. Such fast urbanisation demands the futuristic design of cities to accommodate mega-populations while creating a high quality of urban life. Tunnels provide an effective transportation corridor below the Earth's surface with minimal impact on busy city life at the surface level. The advent of Urban Rail projects in the last century has changed the dimension of travel in densely crowded cities, and the success of such metro projects largely depends on effective tunnelling in busy districts of the cities. A cross-river facility primarily to connect the two busiest rail terminals on the Indian Railway network. The catchment area of this section sports a population of more than 3.5 million.

The construction of Line 2 started in March 2009. The East-West Metro is to run from Salt Lake Sector 5 in the east to Howrah Maidan in the west. **Figure 6.11** includes the original alignment passing through the Central Station of the North-South Metro. **Figure 6.12** includes the modified alignment that has been executed, and **Figure 6.13** includes both the alignments. Line 1, implemented and is being operated by Indian Railways, Line 2 is being implemented by a new organisation, Kolkata Metro Rail Corporation (KMRC). The 16.6 km long MRTS Route comprises six underground stations, six elevated stations, and a depot on 40 acres of land at Salt Lake's Central Park.

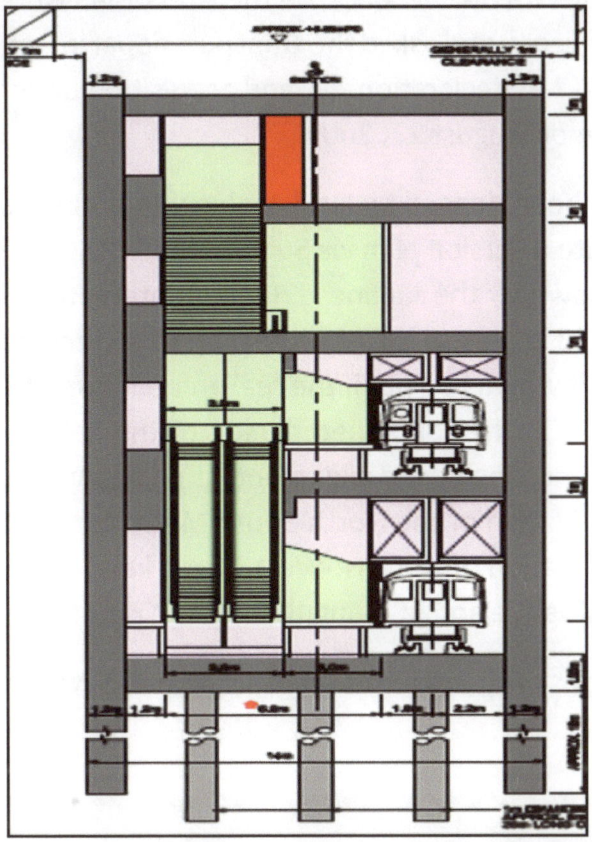

Fig. 6.14: Stacked Station, way to go for limited Right of Way **Fig. 6.15: Internal View of Howrah Maidan UG Station**

At the Central Station, the East-West metro alignment could have intersected the existing North-South metro corridor, providing an interchange between the two underground metro corridors. The space constraints at Central Station and mined tunnelling (NATM) below the N-S line faced the challenges during construction. While designing the North-South Metro, provision was kept for the East-West Metro to cross under with provision for integration between the two stations at Central. During the implementation of the Line 2 corridor at the instance of the three funding agencies, an alternate scheme had to be evolved to avoid acquiring a few buildings (which were in an advanced stage of acquisition), with a plan to accommodate tracks at different levels **(Fig. 6.14).** However, one of the three funding authorities did not accept the solution. It was decided to reroute through Esplanade, which was implemented, resulting in Station spacing, the longest between Esplanade and Sealdah, causing substantial delay and an increase in cost. The developmental benefits of this population because to the new metro facility. **Figure 6.15** includes the first underground station and terminal at the west at Howrah Maidan of the planned corridor.

i) Land Issue (Ref. Times of India)

The original alignment of Line 2 was through the BB Ganguly (Bow Bazar) street, and one of the causes of the long delay of the East-West Metro corridor was a land hurdle. The Calcutta (Kolkata) High Court had ruled that land acquisition at Bow Bazar for the proposed Central Station (as per the original alignment) on the Saltlake – Howrah Maidan route was invalid because the land losers were not compensated within the stipulated timeframe of two years, and asked the transport department, which took the possession of the land on January 24, 2012. A declaration for land acquisition under section 6 of the Act was published in the Kolkata Gazette on August 27, 2009.

The order spelt trouble for the metro corridor that had been dogged by land acquisition problems and was running about two years behind schedule. A route diversion plan via Subodh Mullick Square and SN Banerjee Road to reroute the corridor turned down by the Japanese Bank of International Cooperation (JBIC), due to cost escalation from Rs. 2,253 crores to 4,874 crores, and the funding agency wrote to the State Government in August, saying it would pull out if the realignment is made. The Author had the opportunity to discuss the matter with the concerned person of the Railway Board, and it appeared that they were also not in favour of changing the alignment, as provision was made during the design and construction of the Central Station of the North-South Metro corridor, and at the same location, a provision of an interchange Station of the East-West Metro kept below the North-South corridor at the Junction of Chittaranjan Avenue and BB Ganguly Street at Bow bazar.

However, later, the realignment was agreed by all parties with a transfer station with the North-South corridor at Esplanade.

ii) Land for Extension

For the extension of Kolkata's East-West Metro corridor from Salt Lake Sector V to Haldiram (Teghoria), Metro Railway has issued a land acquisition notice, a key step toward expanding the proposed line by 6.5-km stretch. The route will pass through Kestopur, Baguiati, Raghunathpur, Dumdum Park and Teghoria/Haldiram ending near VIP Road at Haldiram. The Metro authority marked 171 land parcels along the proposed alignment and officially announced the start of the acquisition process. As per the Metro Railways (Construction of Works) Act, 1978, the notice restricts any construction within 20 meters on either side of the route for five years from the date of publication. "No building or other construction shall be made or done above the proposed Metro alignment or within the 20m width on either side," the notice states. If completed, the extension will greatly benefit commuters in densely populated areas like Lake Town, Kestopur, and Baguiati. Additionally, it will provide a key interchange at Teghoria, enabling passengers to switch to the New Garia-Airport Metro line. This would facilitate easier access to the airport from Howrah, Sealdah, and Salt Lake without detours through New Town.

iii) Rehabilitation of Project Affected People.

Any large infrastructure project is expected to displace or affect in other ways people within its alignment. KMRCL has factored in this possibility from the stage of planning. It has not only provided for the resettlement of people affected by the project between Howrah Maidan and Sector V but has also aimed to improve their condition. A few of the resettlement activities

- A modern five-storey building has been constructed at 10 Lu Shun Sarani to accommodate tenants occupying shops in the Central station area.
- An entire indoor stadium was relocated in the Subhash Sarovar area.
- One hundred and twenty-nine (129) shops in the well-known Sree Market have been relocated after the construction of temporary shops close by to facilitate the construction of the Howrah Maidan Station
- A set of 56 pucca quarters has been built close to the Subhash Sarovar area to Provide Accommodation to KMC staff living in Swabhumi.

iv) Green Project

One of the prime concerns of the project has been to restore the road and greenery after construction is completed. About 800 trees have been uprooted so far for the construction, and they have been replaced by planting 4200 saplings. Walkways have been reconstructed as soon as possible. To minimise disruption of traffic as well as inconvenience to pedestrians, construction has been largely carried out along the median of existing roads.

The depot and the operational control centre being constructed at Salt Lake Central Park were designed with a state-of-the-art green facility, which will have a minimum of a silver rating under Leadership in Energy & Environmental Design (LEED). The details of the Depot have been included later.

The East-West Project is designed with technological expertise and equipment harnessed from across the world. The Metro Corridor project was planned with the hope, fulfilment of the dreams and demands of the citizens of Kolkata.

v) Accident During Tunnelling at Bow Bazar

Besides the delay due to the land acquisition process, delays occurred in surveys and investigations that had to be carried out along the revised route, within a minimum period, abutting structures, and above the realigned corridor.

In addition, a major delay occurred during the tunnelling operation at Bowbazar Area. An unprecedented collapse happened in Bowbazar area. The KMRCL authorities termed the damage caused by the tunnel boring work of the East-West Metro as "unprecedented" and said that experts from Singapore and Hong Kong are being called to assess the situation. On Sunday night, several buildings collapsed and developed cracks in the city's Bowbazar area, reportedly due to underground Metro corridor construction. "It's such a congested area that the machinery inside the area could not be taken in. We will try from the esplanade. It's a difficult task, but we fell short of a few 700 meters to reach Sealdah underground," authorities said.

According to reports, "several houses **(Figures 6.15 & 6.16)** collapsed following seepage of water in the ground under their foundation, owing to a tunnel being bored. More than 350 residents are reportedly affected and have now been evacuated and shifted to nearby hotels".

Fig. 6.16: Collapsed Building

Fig. 6.17: After Resuming Construction

The figure shows a typical collapse of buildings. As a result, tunnel construction at the Metro project had been brought to a complete halt. Cracks appeared in around a dozen houses in the Bowbazar area following water seepage during the construction of a passage to connect the two tunnels of the East-West Metro on Friday. The seepage of water had been arrested, but all underground work had been put on hold.

6.2 The Project

This is the second of the two Corridors, the east-west metro corridor **(Fig. 6.11)** conceived in the 1971 Master Plan. Implementing it now is urgent because of the developments in the metropolis that have reached further east, away from the river. While the road area percentage in locations such as Salt Lake City is higher than in the metropolitan area, traffic chaos is just waiting around the corner if the facilities are not augmented.

The East-West Metro alignment brings the main lines of Indian Railways closer to the eastern fringes of the metropolis. A spur to the airport is also planned to enable airport connectivity. As such, an underground transfer station at the Airport is in the advanced stage of construction.

The alignment has two distinct parts: one stretch runs elevated for 5.6 km **(Fig. 6.17)** and the other underground for 8.9 km, including the section running beneath the river. The depot and cross-over lines account for 1.2 km of tracks. The total length is 16.6 km. The maximum design speed for the metro is 80 km/hr, while the average travel speed reckoned as 35 km/hr. The general platform dimensions are 140 m long x 12 m wide. All underground (UG) stations were planned for cut-and-cover construction, and the UG corridor Tunnel section **(Figs. 6.18 & 6.19)** was built using EPB-TBM. The system was designed to operate with Standard Gauge.

Fig. 6.18: View showing the Elevated Corridor with Curve in Plan

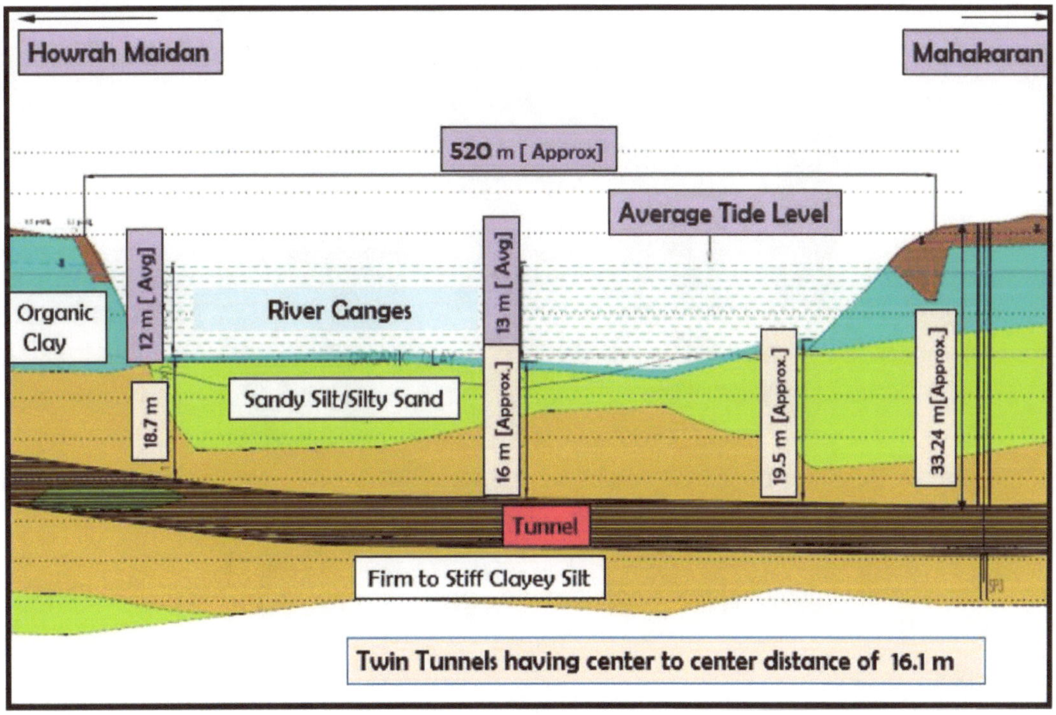

Fig. 6.19: Tunnel alignment below river vis-à-vis geotechnical profile

In initial years of operations, the plan of Line 2 was to have 11 rakes, each consisting of 6 cars, with headway was planned to be 5 minutes and be reduced gradually to 1½ minutes as demand increases.

The ruling gradient is 1 in 25 (Compensated). The minimum radius of the horizontal curve for the underground section (UG) is 200 m; for the elevated section, 120 m; in the depot, 100 m; for stations,

1000 m; and the minimum radius of the vertical curve is 1,500 m. The rolling stock will consist of AC Car Rakes. The carrying capacity of each rake is 1,500 and 1,800 passengers, at 6 and 8 persons per sq.m density, respectively. Traction power is by 3rd rail, 750 DC. The incoming power at 33 kV is from Calcutta Electric Supply Corporation (CESC) at two locations—Traction Rectification Substations at eight stations and Auxiliary Substation at every station to supply all other electrical loads. The Rolling Stock will consist of AC Car Rakes with a 4M + 2T formation. A 3-phase Induction Motor drive with VVVF Control and compatibility with the signalling system with Regenerative Braking, a "Green" feature, is used.

The signalling is a Continuous Automatic Train. Control (CATC) with ATP for System Protection; ATO for Controlling Train movement under the overall Control ATP; ATS System for overall supervision from the central control; and stations with points to have line-side signals to protect points. The Communication System planned for mobile communication between trains and the Control Room, including mobile maintenance staff. A Public Address System in trains and stations, Telephone Communication among staff through local small exchanges, a Control telephone with a central controller and Station master, a CCTV Surveillance System, a Public Information Display System, and a centralised clock System are part of the Line-2.

Fig. 6.20: Tunnel Section

Ticketing has been designed through a Closed System with subsystems: Entry Gates to be operated with Token Tickets or Smart Cards; Exit Gates; Ticket Vending Machines—ticket Office Machines/ passenger-operated Machines; add-on Fare Machines at each station; and a central computer to acquire data on passenger and financial transactions.

The East-West corridor is now in operation except for the section between Esplanade and Sealdah. Hon'ble Prime Minister of India has inaugurated the section.

6.3 Issues and Desirable Transport Characteristics

6.3.1 Issues

General Problems or issues related to the transportation system are i) High population density, ii) Unplanned Development, iii) Minimal Right of Way, iv) Mixed traffic & Congestion, iv) Road area- Kolkata's road area was 6.2 % of the city area in compared to Delhi 26% & 22% in Mumbai, and v) Presence of old and heritage buildings. **Figure 6.20** includes the traffic congestion during the eighties, and **Figure 6.21** includes the NMTs on Kolkata roads. This is why the North-South Metro was built underground during the eighties.

Fig. 6.21: Mixed Traffic

Fig. 6.22: NMT Traffic in Kolkata Streets

6.3.2 Desirable Transport Characteristics

Affordability: Must be affordable considering both Money & Cost.

Accessibility: Comprehensive Mobility Plan for spider-webbed network. Last-mile connectivity for door-to-door service.

Convenience: Reduced pollution, less congestion, stress relief and more comfort.

Reliability: Punctuality, frequency of services and interchange facilities.

Safety: Reduce accident frequency. Provide a disability-friendly system and pedestrian safety.

Vehicle Technology: Modern and attractive vehicle technology.

Dedicated facilities: For non-motorised vehicles.

6.4 Salient Features

Total Length: 16.6 Km

Elevated Section: 5.6

Underground Section: 8.9

Elevated Stations: Salt Lake Stadium, Bengal Chemical, City Center, Central Park, Karunamoyee, and Salt Lake Sector V.

Underground Stations: Howrah Maidan, Howrah Station, Mahakaran, Esplanade, Sealdah, Phool Bagan.

Stabling & Depot: Central Park, Salt Lake City

Platforms: 140mx12m

Rake: Comprises six coaches per Rake (Train)

Maximum Design Speed: 80 Km/hr.

Average Speed: 35 Km/hr.

Gauge: Standard Gauge-1,435 mm

Headway: 5 mts/1.5 mts.

Design PHPDT: Year 2021 – 25000

Train Composition: 6 Coaches (2MC + 4TC)

Carrying Capacity : Motor Coach (MC)- 252 Pax (42 + 210)
: Trailer Coach (TC)- 260 Pax (50 + 210)

Rolling Stock: 2021 year-19rakes

Fare Collection: Automatic Fare Collection

Passenger Integration : Existing N-S Metro Line- Esplanade
: Suburban & Mainline- Howrah and Sealdah
: Bus Depots- Howrah and Sealdah
: Ferry Passenger- Howrah
: International Bus Terminal- Salt Lake City
: Suburban Rail System- Sealdah

Traction: Third Rail-750V/DC

Fig. 6.23: Inner view of Metro Coach

Fig. 6.24: View of Concourse of Esplanade Station

6.5 Traffic

Fig. 6.25: Passengers waiting for Boarding (East-West Metro-Late Nineties)

Fig. 6.26: Typical View of Sealdah Terminal

Infrastructural facilities (2004):

Features	Units
Total Road Length (in KMC area)	1870 km
Total Railway Length	29.87 km + 8.5 km added in 2010
Metro	16.45 km + 8.5 km extended till New Garia in 2010
Circular Rail	13.42 km

	2000	2011	2021	2025
Estimated no. of vehicles (1) in KMA				
range from		1,538,000	2,348,000	2,800,000
		1,662,000	2,586,000	3,018,000
Total transit trips per day (2)		23,500,000	29,900,000	32,200,000
Freight (million tons/annum)		124.71	144.82	169.70
By road		92.81	103.00	123.00
By rail		19.90	28.82	33.70
By river (3)		12.00	13.00	13.00
Trans-river traffic (vehicles/day)	142,000	203,000	277,000	312,000
Goods vehicles entering/leaving KMA (4)	41,000 (1998)			
Persons entering/leaving KMA	1,200,000			2,000,000

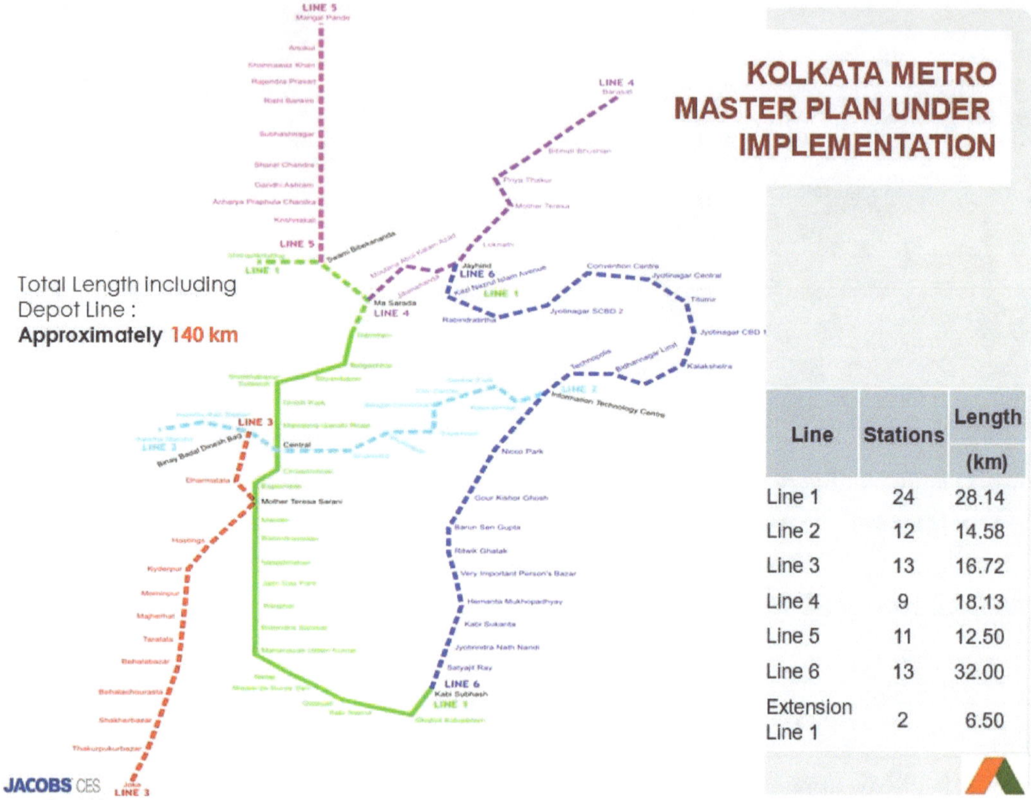

Fig. 6.27: Master Plan Kolkata Metro

Fig. 6.28: Platform Screen Door for Elevated Station

6.6 Recent Trends in Transport Structures

In recent years, segmental construction has been a preferred solution, primarily for elevated metro and highway corridors. In the case of the precast segmental construction method, the precast elements are cast in the casting yard through specially designed moulds, can be pre-tensioned or post-tensioned and transported to the site. This applies to the Span-by-Span Method of Construction

(SSMC) with concrete beams/girders or steel beams, a composite structure with concrete for the deck. The limitation is that the span lengths and lingths are 18/20 m, which can be transported from the pre-casting yard to the site through urban corridors.

In case of other forms of structures, such as concrete U-Girder or Box Girder, with span length preferably between 25 -40m, which can be extended up to 50 m. The small segments are made transversely, which contrasts with precast girder bridges.

Generally, the joints between two consecutive segments are either glued by a specially designed epoxy formulation or dry joints, that used in Nivedita bridge approach viaducts. Segments are orthogonal to the longitudinal axis of the bridge. The reinforcement is not continuous through joints unless it is required for structural reasons. The designs with transverse segments are dictated by the shape of segments, weight, arrangement of pre-stressing and methods of casting, transportation and erection.

The two principal methods of construction generally being adopted are span-by-span and the cantilever method of construction. For both constructions, the precast segments are cast in the casting yard either by the long line casting method or the short line method of casting, and in span construction, the erection or launching of the segments can be done by overhead launching or by underslung erection method, or the balanced/free cantilever method of construction can be done.

Segmental construction offers the advantages of repetitive construction procedures, minimum impact on traffic and the environment during construction, and economical and durable structure. Segmental construction can be used even on small-radius structures, which is common in urban environments and useful for curved access ramps. The slender piers and long, shallow superstructures make concrete segmental construction aesthetically pleasing. Precast concrete segments have the advantage of superstructure being erected faster. Segments are built using either a short-line or long-line method, briefed later in the chapter.

There are three ways to erect segments.

a. **Span-by-Span:** In the span-by-span method, an entire span is assembled, erected and is suitable for span lengths up to 30m.
b. **Balanced Cantilever:** The superstructure is erected by cantilevering out from opposite sides of the pier. The segments are added simultaneously or alternately to each cantilever to maintain a balanced system. The balanced cantilever method is economical for medium span lengths between 50m to 150m.
c. **Using Progressive Placement:** For small and medium spans, corridors can be erected using full-span-length precast RCC, Pre-tensioned and post-tensioned Girders, U-girders, and Box girders. However, for larger spans, erection starts from one end and erects segments sequentially. Progressive placement usually requires the placement of temporary piers in the middle of each span.

Most Indian metro projects use precast segments of the same design form, length, width, and height. Various Structural forms, such as single-cell boxes, voided slabs, girders, and slabs, are currently used for urban flyovers, elevated corridors, and metro corridors, as shown in **Fig. 6.28.**

The single-cell box girders tend to be uneconomical for short spans and pose problems transporting them using the existing handling infrastructure in an urban environment, except where sufficient space is available or across a river (such as the Bridge across Airoli in Mumbai). The voided deck system negates some disadvantages but needs extended construction involvement (in case of left-in shuttering for construction) and equipment for large size and weight. At the same time, voided deck is too heavy to handle.

Precast RCC (comparatively heavy), post-tensioned and pre-tensioned I girders with span lengths up to 20/25 m can be handled at the site, erected, and transported from the Casting yard. The casting yard may be at the site where space is available to set up a casting yard or at a suitable location that can handle transport for several sites, in case city road geometry is ideal for transporting the girders within a reasonable distance from the sites. Some sites in Delhi were provided with precast pre-tensioned girders cast in a Centralised Casting yard. Segmental boxes were also delivered at the site from a typical Casting Yard, like the Second Vivekananda Bridge, East-West Metro and several metro corridors in Kolkata and many other urban areas across the country, like Delhi, Mumbai, Bangalore, Chennai, etc. The precast elements were for short to medium-span length RCCs and precast girders, Segmental Box sections, and U-girders span by span.

Structures made of steel prefabricated, transported & erected at desired locations are also used in projects with span lengths up to 20/25 m & structural form as steel girders. The structural form is a single-cell box (with a segment length of around 2.5/3.0 m) for span lengths up to around 30/35m, but more than one as per design requirement. For fabrication, it can be centralised in a fabrication yard or can be fabricated from existing facilities. Even fabrication can be done in established workshops like those in Kolkata, Mumbai, and Chennai, where advanced technology for fabrication is available.)

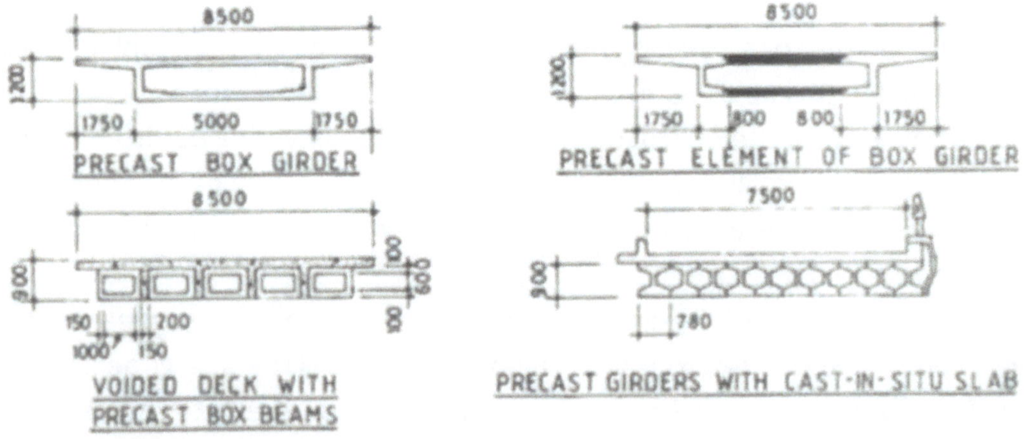

Fig. 6.29: Possible Forms of Precast Concrete Elements

For spans using Precast segments, erection is done using a single span and then deck continuity can be done using around 150 mm-thick reinforced in situ concrete for single spans or continuous span configurations, as done for the number of flyovers in Delhi, using precast-pre-tensioned segments **(Fig. 6.28)** and steel-concrete composite sections.

Precast segmental construction grew out of the mass production concept to prefabricate off-site for the elevated deck, for different project sites within an urban area and has advantages like:

a. Enhanced quality control of casting in factory conditions.
b. Ability to adjust the rate for girder casting to suit the rate of erection by increasing the number of moulds and erection tackles for the project requirement.
c. Reduction of disruption to the existing users of the site.
d. Flexibility in casting the deck in parallel with the construction of foundations and sub-structures.
e. Overall reduction in the construction programme due to possible parallel activities and hence cost savings

6.6.1 Other Forms of Precast Sections

Other than Reinforced Beams, pre-tension/post-tension girders, the following forms **(Fig. 6.29)** are being used in India. **Figure 6.30** includes typical cross section of Segmental Box.

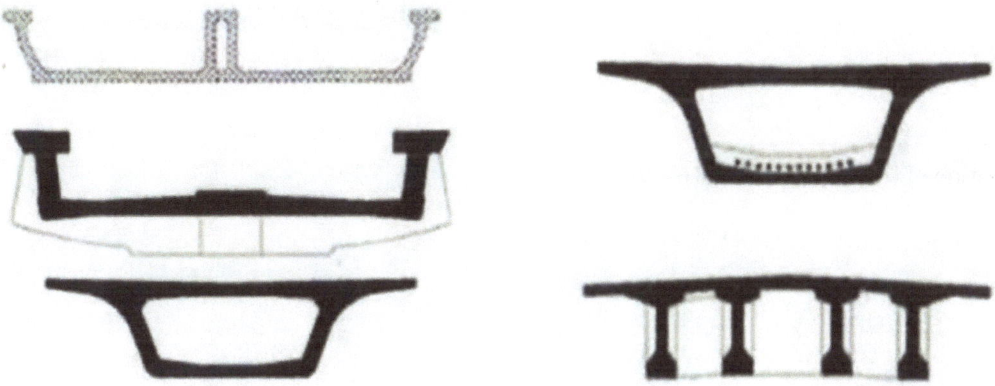

Fig. 6.30: Other Forms of Precast Segments

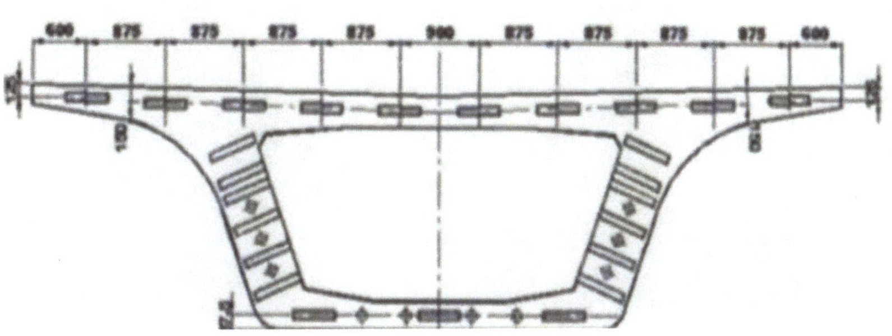

Fig. 6.31: Typical Cross Section of Segmental Box

6.6.2 Methods of Construction

Pre-cast methods of construction have a process of quality assurance:

i. Pre-casting of segments **(Fig. 6.31)** -U Girder
ii. Transportation & Handling of precast segments
iii. Erection of precast Segments

A few photographs of erections used in Indian Metros are included for reference in **Figs.6.32, 6.33 and 6.34.**

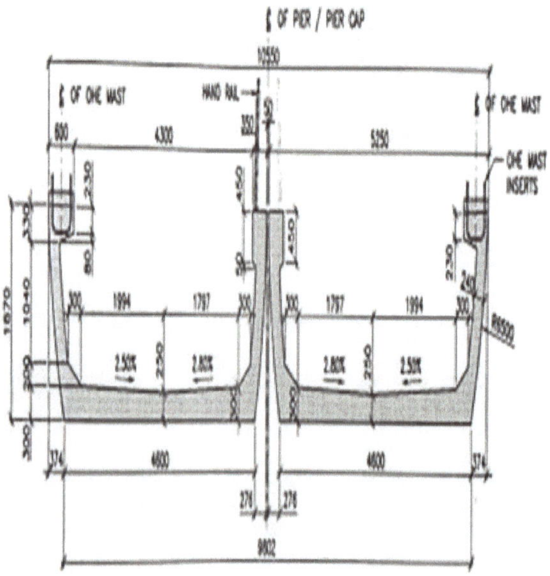

Fig. 6.32: Cross Section of U-Girder (Ref. Delhi Metro)

Fig. 6.33: Erection of U-Girder (Delhi Metro)

Fig. 6.34: Precast Girder Lifted & Placed in Position

Fig. 6.35: Precast Girder during Lifting to Place

6.6.3 Pre-casting technology

Match casting is the essential technology used for precast segmental bridges. Each segment is individually cast using already-cast adjacent segments as a face shuttering form. Thus, the system ensures a perfect fit between adjacent segments in their final position. So, a span in SSMC or a cantilever in CCM is incrementally cast segment by segment.

6.7 Design Parameters and Material Specifications

6.7.1 Metro Corridor

The Corridor is 16.67km long, with 9.842 km underground and 6.828 km elevated sections, with the Depot at grade. It includes six underground and six elevated stations.

The **superstructure** is 8.2m wide, supported by a precast, prestressed segmental box girder, other than in the Duttabad stretch, where the superstructures are three continuous/two spans continuous and supported by steel-concrete composite box girders. A minimum 550 mm Walkway is provided for the safe passage of passengers. As per the design, needed to transfer vertical load, bearings are supplied with Elastomeric bearings and POT-PTFE bearings.

And the shear key for the transfer of in-plane forces. Spherical bearings have been provided in the Duttabad area. Minimum 5.5m vertical clearance from the road level and 0.475m minimum.

Horizontal clearance from the inner edge of the pier head has been maintained.

The **substructure** is generally made of single-cast in-situ circular piers with a hammerhead pier head, gradually widening at the top to support the box at its web location through appropriate bearings.

Cantilever piers and RCC Portals have been used to suit the horizontal layout of the corridor.

Crash barrier and Parapet: A New Jersey-type crash barrier has been provided at the road level against the pier to protect the same from vehicular traffic.

Precast counter-balancing type, 3m long, top width varying from 550 mm to 750 mm to accommodate SOD, **RCC parapet** with the provision of stitching beam (250 mm x 400 mm) at the bottom has been provided.

Pile Foundations: Cast-in-situ vertical bore piles have been used. Pile Design & Capacity–per IS 2911, Pile Estimated Settlement -IS 8009 (Part-II) IRS CBC.

Concrete: 24/25 kN/cum for RCC & PSC & 23 kN/cum for PCC, as per IS-875 and modular ratio as per IRSCBC.

Minimum grades of structural concrete: PSC-45, RCC-M35, PCC-M30; Piers & Columns

Cement: The cement grade used is 43/53, OPC, conforming to IS-8112 and as laid down in -M50/ M45/M35,

Thermal coefficient: As per the IRS Bridge rule for concrete & steel, **Poisson's ratio**-0.15

Structural Steel: Hollow Section- IS:4923, Structural Sections: IS-2062. Steel Tubes-1161,

Stainless steel is not used; steel conforms to IS-2062 & IRC-24. The design is as per the IRS Steel Bridge Code & BS 5400, and steel concrete composite structures are as per BS 5400 (Parts 1 & 3).

Structural Steel for Miscellaneous Structures: Design as per IS-800

Reinforcement: Wt. 77 kN/cum., TMT bars, grade Fe 500 (min), conforming to IS 1786, seismic as per Zone 3 and with elongation of 14.5% (IRS Concrete Bridge rules).

Concrete Grades: Piers & Columns-M50/M45/M35.

Pre-Stressing: Steel conforming IS-14268, class-2 low relaxation uncoated stress relieved strands. Pre-stressing cables are 12/19 -15.2 nominal diameters, seven-ply class 2 strands.

The design is based on nominal diameter. Of each strand-15.2mm, nominal area-140 sq.mm., Breaking strength-260.7kN, UTS-189.94Kg/sq.mm.

Pre-stressing Details: As per IRS CBC, Jacking Force-195.5 kN, 75% of Breaking Strength, Prestress loss, Sheathing, Anchorages- IRS CBC.

Loads: DL as per actuals.

Superimposed DL- Plinth-3.5MT/m, Rail & Fastening-0.565MT/m, Parapet2.1MT/m based on the actual section adopted, Railway track-0.08MT/m (as per Rail Section), Cable Trough-0.5 MT/m, Third Rail including Pedestal-0.275 MT/m: **Total-7.5MT/m**

Live Load: As per Rolling Stock adopted – Axle Load: 16 tons, Number of Coaches-6, Length of Coach-21.3m, Overhang on either side-2.15m Bogie Wheel Base2.3m Distance between Axles-12.40m

For Design: Critical locations and calculation of maximum forces due to moving Loads are to be taken when calculating BM, Shear, & Longitudinal forces worked out. Foot path load was taken as 490 kg/sqm, including the dynamic effect as per the IRS bridge rule

Coefficient of Dynamic Effect/Braking Traction/Centrifugal Force/WL/Seismic

Load/Temperature Effect: As per the IRS Bridge Rule

Temperature Effect: Overall, Differential and Temperature Gradient-IRC 6

Differential Settlement: 12mm for long-term and 6mm for short-term

Buffer Load: Design considered (not considered separately)18% of Tractive Force (for the fully loaded Train, i.e.3840kN./**Racking Force**- 4kN/m per track on rail top.

Long Welded Rail: Guidelines vide BS report" RDSO Guidelines "

Derailment Load: Applied horizontally 85.33kN/m for 3 m, i.e. 256 kN

Load Combinations SLS & ULS -5 primary load combination groups as per the IRS CBC

6.7.2 Elevated Station Buildings

Cement: Plain and reinforced concrete, Clause 5.1 of IS: 456

Concrete: IS-456

Structural Steel: Hollow Section- IS:4923, Structural Sections: IS-2062. Steel Tubes-1161

Reinforcement and Cover: Wt. 77 kN/cum., Characteristic strength, fy 500 kN/sq mm, Es-200 000 kN/sq mm, and IS-456

Concrete Grades: Piers & Columns-M50/M45/M35, Floor Beams, Slabs, Stairs & suspended slabs-M35, Foundations, Slabs on grade & others- M30.

Fire Resistance: Resistance period of 2 hours both for Concrete (IS456) and Steel (IS 800)

Earthquake: IS-1893 (Including drift limitation) and detailing as per IS-4326 (for Strength) & IS-13920 (for Ductile)

Temperature, Shrinkage, Creep, Other Forces & Effects: IS-456

Earth & Water Pressure: IS-875 (Part 5)

Settlement: IS-1904

Ultimate Load Combinations: Load Combinations and Factors for RCC-IS 456, for Steel-IS 800, Load Combination- IS-1893 and as per IRS CBC & RDSO Guidelines

Serviceability Load Combination: As per IS 456, IS 1343, IS-800 & IRS CBC

Deflection: IS-456

Lateral Sway: H/500, H-Height of Building

Fatigue: RCC Structures- IRS CBC, Steel Structures- IRS Steel Bridge Code, Other Cases- IS 800.

Foundations: Types- Spread, Raft or Pile foundations

* Foundation not to fail in Shear, Settlement should not exceed as per Table 1 IS-1904 Safe Bearing Pressure- IS-6403
* Pile Design & Capacity–IS 2911, Pile Estimated Settlement -IS 8009 (Part-II)

References: IS, IRS, IRC, BS/Euro Code, AASHTO & Specialist Literatures (Where applicable)

6.7.3 Cut-and-Cover Construction

The underground Stations were designed using cut-and-cover method of construction. The Corridor was designed with two approaches: i) Open Excavation in Tank areas, as per design ii) Using shoring walls along the corridor, with two walls, one on each side, across the designed RTS Box, Station, and Substations, as per approved drawings. Different types of Shoring walls were considered to select the right one. Those were Diaphragm Wall, Sheet Piles, H-Pile and Timber Lagging, and Abutting Bored walls to form a wall.

Design with Shoring walls, two approaches were considered, one using the Top-Down and the Bottom-Up method for the other. The RTS corridor and Wider sections, like Stations, substations, were designed with DW, using three approaches: i) Independent RTS box, ii) Integrated, and iii) Semi-integrated with DW. For Shoring walls other than DW, designed as an independent structure along the RTS corridor.

The maximum width of the trench was measured between external faces of shoring walls of both sides. Minimum width left between shoring wall's inner side and the outer face of RTS was 600 mm, so that one man space is available for construction needs. For the road crossings, the design needed special attention and was based on site-specific solutions. The trenches on roads were backfilled, and roads repaired with utmost priority, taking all safety precautions during road crossing operations

6.8 Elevated Corridor

6.8.1 Basic Features

The elevated Corridor was designed and built by Gammon India, and the Stations by Simplex Infrastructures.

The total length of the elevated part, from Phool Bagan to Salt Lake Sector V, is 5559.451 m, including the six stations but excluding the depot and crossover lines. The length, including the cross-over length and Depot lines, is 6828.009 m.

1. The elevated corridor length is 4719.451 m.
2. The total length of the six stations is 840 m (each station is 140 m long).
3. The depot is at Salt Lake's Central Park on 40 acres of land, with a total depot line length of 809.517 m.
4. The cross-over length is 459.041 m.

Provision has been kept for future extension to the airport from the Central Station in Salt Lake.

A line diagram of the elevated corridor at Salt Lake is included in **Fig. 6.35**.

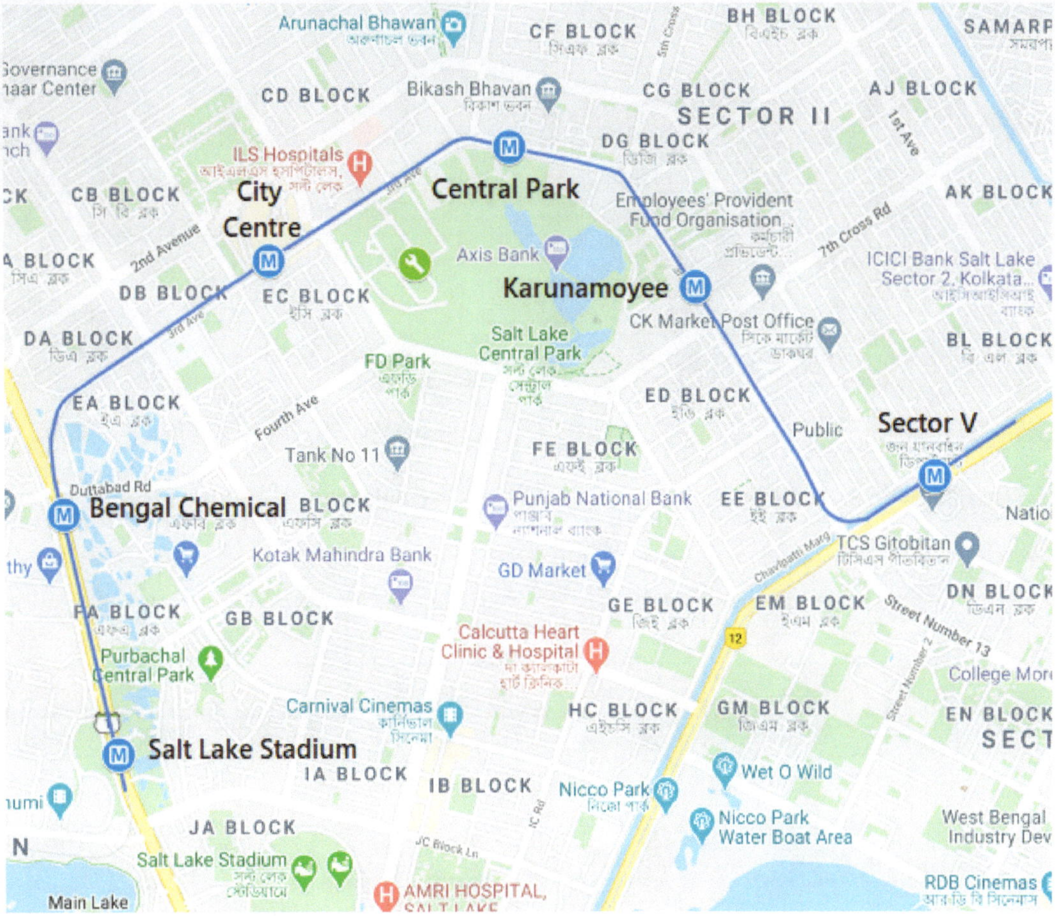

Fig. 6.36: Line Diagram Showing Elevated Corridor at Salt Lake

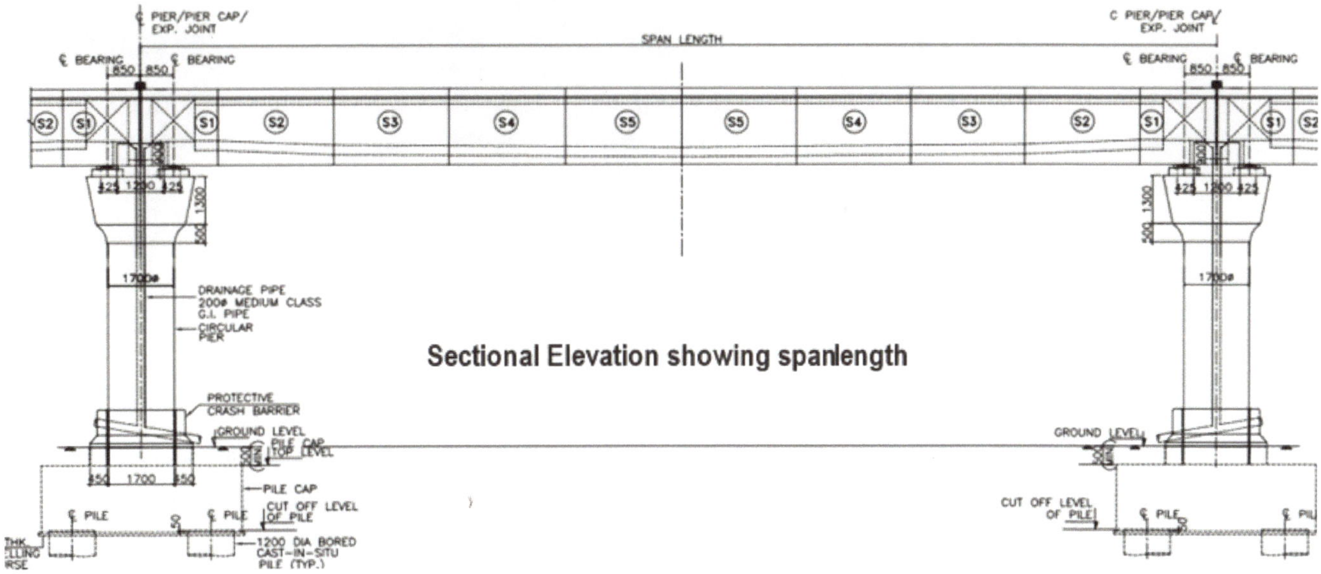

Fig. 6.37: Sectional Elevation of a Typical Viaduct Span

Figure 6.36 includes a typical viaduct span**, Fig. 6.37** shows after erection of segments and precast parapets, **Fig6.38** shows side view of the corridor.

Fig. 6.38: Erected Corridor with Precast Parapets

Fig. 6.39: Enlarged view without Precast Segments

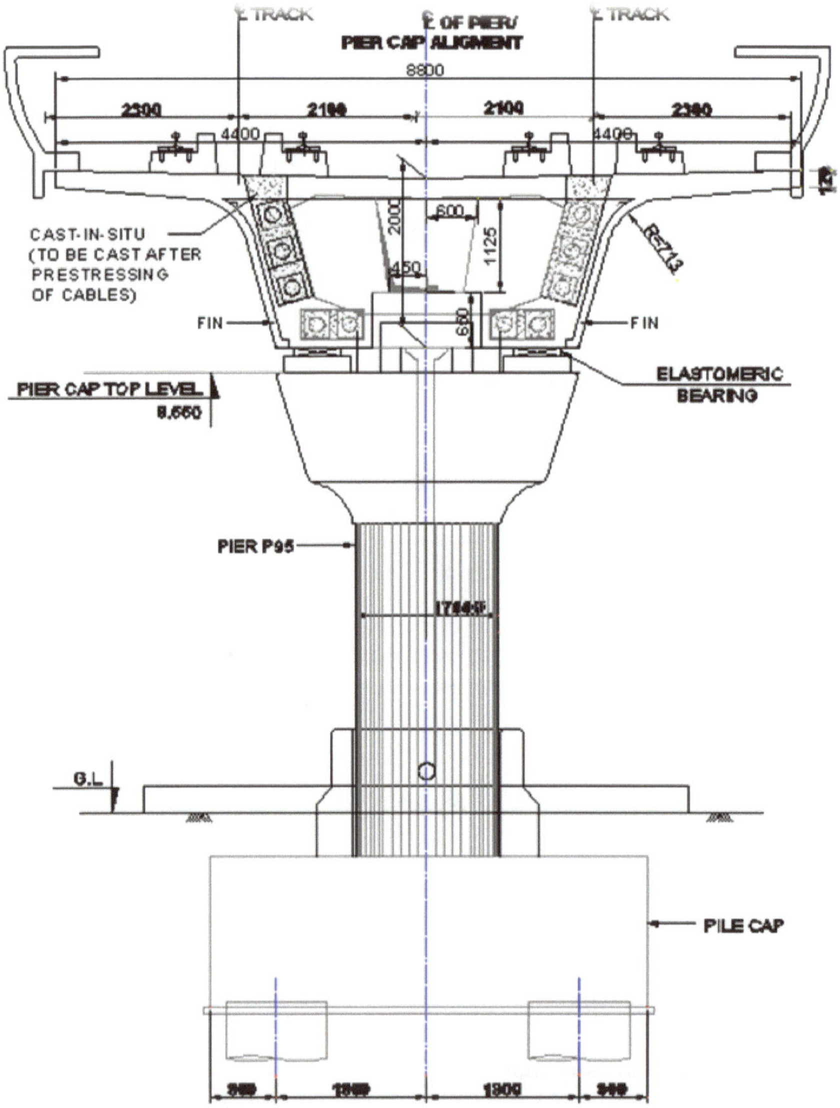

Fig. 6.40: General Arrangement with Piles, Pile Caps, Pier & Superstructure

Fig. 6.41: Top Layer of Pile Cap Reinforcement and Starter Bars for Pier, and Pier Protection.

Fig. 6.39 includes Cross section of the corridor and **Fig. 6.40** includes reinforcements of pile cap top layer and pier protection.

6.8.2 Foundation and Substructure

Pier foundations were designed with bored in-situ piles at pier locations (between 19m-45m), depending on the suitability of the soil strata encountered. The pile cap was kept at least 500 mm below the road level to facilitate necessary drainage from the viaduct and utility crossing. Circular piers, as per design, including rectangular pier caps, were cast in a single lift to give a good finish without joints, except for very tall piers. To protect the piers against collision with moving vehicles on the road, concrete crash barriers (Pier Protection) were provided up to a height of 1.0 m above road level around the piers.

Pile capacity was worked out based on results obtained from the final set of Geotechnical Investigations and verified by conducting Initial Load Tests, the number commensurate with IS codes. The acceptability of driven piles was based on routine tests. The grade of concrete and reinforcements was based on design parameters included in the earlier section.

The foundation consists of a pile foundation, which has a group of 4 to 6 piles with a diameter of 1200mm and a length of 19 to 45m. The total number of pile foundations is 257, with 1081 piles. An initial pile load test has been carried out at four different locations, and a routine pile load test has been carried out on 1% of the total pile to ensure the capacity of the pile. Also, integrating pile testing has been carried out on 25% of the total piles.

6.8.3 Pile Cap

The elevated corridor is mostly in the Salt Lake area, except for a portion on the EM bypass. The Salt Lake is filled up, the area at the top layer of soil is very loose, consisting of silty soil, and the water table was above the bottom of the pile cap. This caused major problems during excavation for casting pile cap and utility shifting. Pile caps were designed for a size of 5.2mx5.2mx1.8m. For protecting slopes and providing a proper dry surface for casing, the shoring arrangement was provided. This shoring arrangement consisted of 8mm plate driven in between ISMB joist. **Figure 6.41** includes piers and pier caps at EM bypass.

Fig. 6.42: Pier & Pier Cap in Progress at EM Bypass

Fig. 6.43: Routine Load Test on Piles

6.8.4 Pile Load Testing

Four initial pile load tests and 1% of the total pile numbers were tested as per the contract and IS 2911. Integrating pile testing was on 25% of the total number of piles.

Figs 6.42, 6.43 and 6.44 include the photos of pile load tests as they were carried out.

Fig. 6.44: Initial Load Tests of Piles

Fig. 6.45: Photos During Lateral Load Tests

6.8.5 Substructure and Superstructures

The substructure includes 1.7m standard piers, 201 circular piers, and 32 portal piers. 32 Portal piers are 20 to 25m wide and 24 cantilever pier caps of 3.6x1.8m size **(Fig. 6.46).** Piers have been cast in single pour, formwork included in **Fig. 6.45.** Superstructures for Span lengths between 19-28m were made of precast segmental box. For span lengths of 25-31m, precast segmental and 22-25 m superstructures were made of in-situ box girder between Central Park Station to Karunamoyee Station. As such, for a deck width of 8.8m precast segmental box, for a deck width of 4.4m (Single track) and a width varying between 8.8m and 14.187m, with a span length of 25m, cast in situ RCC box girders were used.

Fig. 6.46: Pier Formwork Casting by Single Pour

Fig. 6.47: Casting of Pier Cap

6.9 Pre-cast Segment

The casting of precast segments was carried out using a long-line casting method. Normal segments have an 8.8m deck width and 3m length, whereas pier segments were 1.9m long. Four nos. of long line beds had been set to cast 1700 segments. The maximum weight of a segment was 52 tons, which governed the handling and erection equipment capacities. Holes were kept in the deck slab and soffit slab for the lifting frame and temporary pre-stressing.

6.9.1 Design for Elevated Structure

The basis of horizontal and vertical alignments was based on basic design parameters as per the Contract. The standards for geometric design, general features, specifications, and other requirements for the detailed design were based on detailed topographic and geotechnical investigations along the corridor. The elevated corridor was designed in line with the following:

a. For an Axle Load configuration based on Coach design and Train configuration.
b. The ballast-less track structure used continuous welded rails. The resultant forces from rail temperature variation were considered.
c. The Indian Railway's Concrete Bridge Code was followed to design RCC sub-structure and foundations and reinforced prestressed concrete segments with external prestressing in line with IRC codes of practices, specialised International Litertures and experience from Indian Projects of similar nature.
d. Segmental Box Section was recommended for all spans. The design was to provide a suitably designed sound barrier all along the corridor.
e. The superstructure was supported on single columns erected mostly along the road's median. Portal frame support was considered at a few locations where single-column support was not feasible.
f. The overall width of the superstructure was 8.8m, with the track centre at 4.2 m.
g. The minimum clearance above the road level was 5.5 m.

The superstructure was chosen considering its ease of construction, fast track, and maximum standardisation.

6.9.2 Precast Segmental Construction

In the East-West Metro corridor, precast segments have been used extensively in contrast to the method used in the North-South Metro of the same city. Segmental viaducts/bridges offer the advantages of repetitive construction, minimum interference to traffic, and an economical and durable structure. Segmental construction can traverse curvature with a small radius. For segmental design and construction, piers can be on small footprints, and slender piers supporting long and shallow superstructures make concrete segmental structures aesthetically pleasing. Concrete superstructures can also be coloured to blend into the environment, like the Metros in Kolkata.

In recent years, Indian engineers have used the precast segmental construction method of medium-to-long viaduct structures, particularly in urban infrastructures like elevated urban corridors and metro viaducts. However, the country's first Metro used in-situ concrete for elevated and underground corridors and stations. The concrete used was of nominal mix, and high-grade concrete was rare. The shuttering and scaffolding used locally available materials like Bamboo/wood for scaffolding and wooden planks for shuttering. Though the form-finish construction was slowly gaining momentum, the use of shuttering made of Steel/timber ply glued over boards began due to the architectural requirement. The builders were used to labour-oriented constructions and avoiding the use of mechanical construction plants & tools. However, the construction of high-rise buildings started using precast elements like building facias and composite precast reinforced concrete with RCC in situ concrete topping. The author had the opportunity to introduce precast facia and planks for floors in one of the high-rise buildings in Kolkata during the mid-seventies and later used in the Road building in Delhi. During the early nineties, due to the availability of modern construction plants and equipment, the construction of comparatively large projects with reduced construction periods, multiple uses of mould, and use of the precast structural/non-structural elements started gaining momentum.

At present, both design and construction engineers, supervising and the client's engineer, are all preferred for the use of the precast structural elements in a project, cast at the site or in the casting yard. The system provides easy quality control, ensuring no deviations from the design section for all similar sections and maintaining the scheduled time frame. The specially designed moulds practically ensure zero variations are transported to the location and post-tensioned together. As such, precast segmental construction grew out of the mass production concept to prefabricate off-site, which has advantages like:

- Enhanced quality control of casting in factory conditions.
- Ability to adjust the rate of casting to suit the rate of erection by increasing the number of resulting flexibilities to control the construction period. Reduction of disruption to the existing users of the site.

  *The flexibility to run the deck casting parallel with constructing foundations and sub-structures.

- An overall reduction in the construction time resulted in a decrease in Cost, Time, and early operation of the facility.

Fig. 6.48: During Construction - EM By-pass

Fig. 6.49: During construction of Station on EM By-pass

The construction method can be adopted by various types of available erection equipment, such as overhead, underslung or falsework systems. Depending upon the equipment chosen, the delivery of each segment can take place over or below the deck, which needs to be considered during the equipment selection process. Overhead gantries are designed and fabricated by the Contractor having enabling structures. However, the contractors can procure from specialised suppliers as per requirement.

Regarding size and weight of segments, as per project requirements like in the metro system, it depends on the width of the segments (for both box & U-sections being used in Indian Metros) to support single track, double track or three tracks (where provision of limited length to stable a Metro Rake). The pier segments are stiffened with diaphragms for anchorage, deviation of longitudinal external post-tensioning cables, and transfer of the loads to the bearings. The large, concentrated forces act on the pier segment during erection. The front leg is supported in the pier segment when the overhead launching system is used. The weight of segments is an essential parameter for segmental viaducts as segment transportation, the capacity of cranes, the size & capacity of trolley trucks and the winches of the launcher, etc.

6.10 Casting Yards for Precast Box Segments

Three casting yards have been used in the East-West (E-W) Metro. The first one is to cast segments of the elevated corridor (precast box section) at Salt Lake, the second is for the tunnel segment for the contract between Sealdah to Saltlake at Rajarhat, and the third one is at Howrah for the section between Howrah Maidan to Sealdah, for tunnel segments.

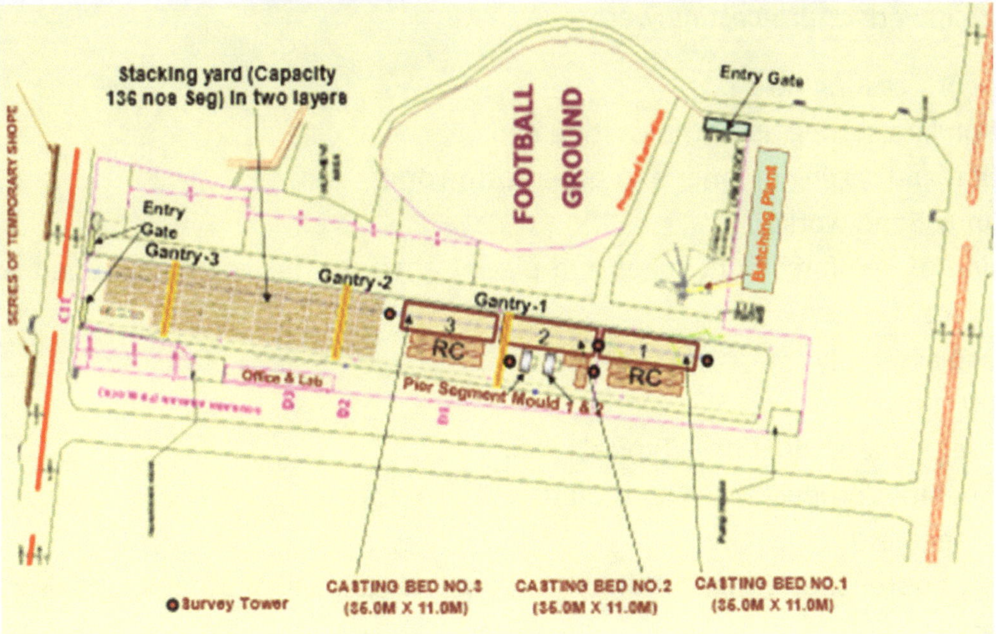

Fig. 6.50: Casting Yard- Segments for Elevated Section

The casting yard for the box section was designed for the elevated corridor between Phool Bagan and Sector V, Salt Lake.

The planning and design of the casting yard **(Fig. 6.49)** for the viaduct section was carried out, considering the steel yard, QC lab, stacking area, entry and exit gates, storage and office area. The casting yard was planned and designed based on the following operations:

a. Planning and designing the casting yard based on geometry, number, size, and type of segments
b. Working out the casting bed to satisfy the project schedule, number of segments, type, and geometry-wise casting bed. The planning should consider the environment, such as rainy seasons, foggy weather, etc.
c. The casting yard should be near the site to reduce transportation time, away from dense residential areas and areas with available utility facilities nearby. The capacity of the casting yard would depend upon the Sufficient areas required for storing raw materials, testing laboratories, batching plants, water and electric supply, working hours (considering work during the night), required segments, temporary stacking area, etc.

All activities related to the construction of reinforced concrete precast segments were carried out in the precast yard, located in Salt Lake, including:

i. Rebar storage & rebar jigs area (rebar cage fabrication area)
ii. Segment casting area (moulds)
iii. Stores and Workshops
iv. Segment storage area
v. Site offices & basic facilities for staff

The activities followed for pre-casting were:

a. Layout of pre-casting yard
b. Fabrication and setting up of the Casting bed.
c. Fabrication and setting of inner and outer formwork.
d. Rebar Cut & Bend works
e. Fabrication of Reinforcement cages and placing them at the appropriate bed with the inner form in place
f. Rebar cage assembly
g. Geometry control
h. Pre-cast segment production – Typical mould
i. Precast segment production - Pier mould
j. Curing of segments
k. Stripping of outer shuttering, continuing curing and removal of inner shuttering as per design specification
l. Segment marking
m. Segment Finishing works
n. Segment remedial works
o. Segment stacking
p. PT duct fabrication
q. Segment Transportation

The planning and design of the casting yard for the viaduct section was planned and designed based on the following operations:

a) Planning and designing the casting yard based on geometry, number, size, and type of segments

Fig. 6.51: Aerial View of Casting Yard for Precast Box Segment

Gammon India was the design-build contractor for the elevated corridor, and the plan of the Casting Yard (**Fig. 6.50**) was based on the data included below.

6.10.1 Data on Segment Casting

i.	Total Segmental Spans	: 179 Nos.
ii.	No. of Segments	: 1630 Nos
iii.	Pier Segments	: 358 Nos
iv.	Segments Cast in Long Line	: 1272 Nos
v.	Time Taken	: 385 days
vi.	No of Beds	: 3 Nos
vii.	No of Moulds	: 10 Nos
viii.	Weekly Production	: 8 Nos
ix.	Monthly Production (both Long & Short)	: 96 Nos

Future extension up to the Airport is planned from the Central Station.

The following Photographs **(Fig. 6.51 to 6.55)** show the activities carried out from the casting yard

Fig. 6.52: Segment Mould

Fig. 6.53: Outer Segment Mould

Fig. 6.54: Total Formwork

Fig. 6.55: Fabrication of Reinforcement Cage

Segment Reinforcement Fabrication in Progress

Fig. 6.56: Fabrication of Steel Reinforcements (Slide 2)

6.10.2 Overhead Launching Girder used for East-West Metro

Spans having a radius curvature greater than 800m were erected by launching the girder. The total length of the launching girder was 64.75m (7 boxes of 9.25m length each). The maximum lifting capacity of the launching girder was 450 tons. The Segments for maximum span length that could be erected was 31m. Launching girder was able to erect span up to 300m radius of curvature and span of 3% gradient. Launching Girder has a hoist of 80-ton lifting capacity and a lifting height of 18m. The main box of the launching girder was 2m wide by 3.5 m deep. It consists of front support having a telescopic leg for adjustment of height in gradient, middle support was designed for 500-ton force, and it was provided with 4 nos of 200-ton jack for height adjustment. The launching girder was provided with a trolley having rollers for smooth movement of the launching girder during the forward movement of the launching girder.

Spans were erected employing an overhead launching girder. The segments were brought over trailers of suitable capacity from the casting yard to the location required. The launching girder was erected over the piers. Segments were lifted with the help of a fixed mounted 80-ton capacity hoist. **Figures & Phtos from 6.56 to 6.61** include stacking, lifting and transporting of Segments.

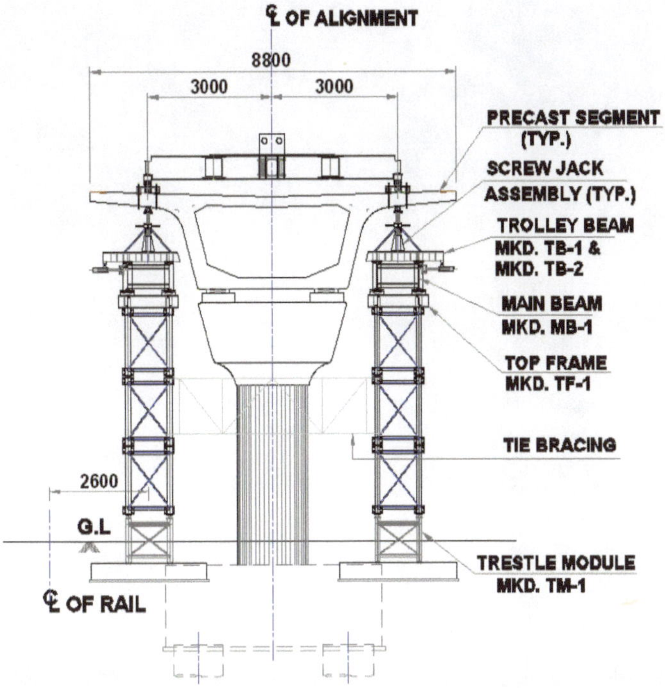

Fig. 6.57: Temporary Structure for erection

Fig. 6.58: Segmental Erection-Using Crane

Fig. 6.59: Lifting of Segments in Dutta bad

Fig. 6.60: Lifting Segment from Trailer Truck

Fig. 6.61: Launching Truss Moving to Next Pier

Figs. 6.62: Stacking Yard for Precast Box Segments

6.11 Erection of Segments

6.11.1 Equipment Used

The 250 t Crane used for the erection of the Steel Bridge and Segments is used under Slung meyhod of erection.

Fig. 6.63: Equipment Used for Lifting of Segment

Fig. 6.64: Viaduct with Pre-Cast Segments in Progress

6.11.2 Launching Sequence

i) Salient Features of Launching Girder

1. Type of superstructure to be lifted: - Pre-cast, single cell box girder
2. Type of construction: Simply supported
3. Maximum span length: - 36.00 m

4. Minimum span length: - 19.00 m
5. Maximum height of lift from ground: - 18m
6. Maximum Longitudinal Gradient: - 3%
7. Maximum span wt. to be erected: - 525 t
8. Minimum Radius can be negotiated: - 280m

ii) **Major Steps for Erection of Precast Box Segments**

Using Overhead Launching Girders, precast concrete segments, transported from the Casting, and stacked at the site, span by span.

Launching Girder over two erected spans, ready to move to the next span. The following sketches to erect the segments in two stages. The first one is to erect the Launching Girder. In case it is already erected on an earlier completed span, then it has to be moved to the following span, where the segments are to be erected. The stages for launching of Girder and the sequence of erection have been included in the following sketches.

Launching Girder: Position at Different Stages

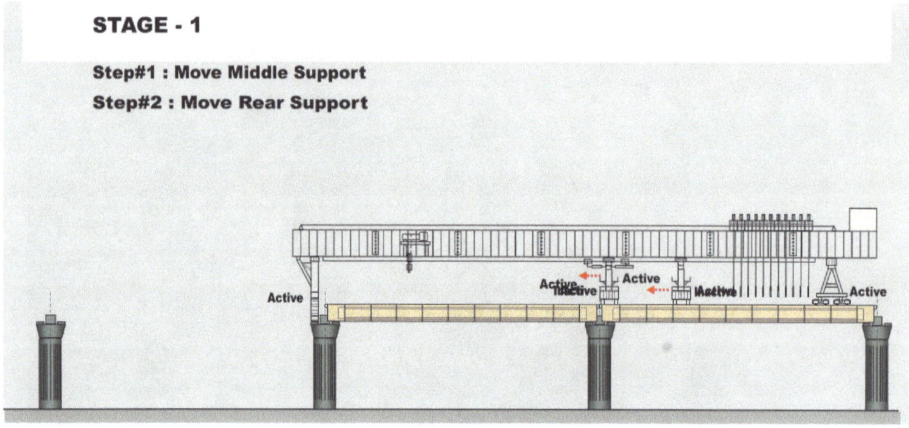

Fig. 6.65 A

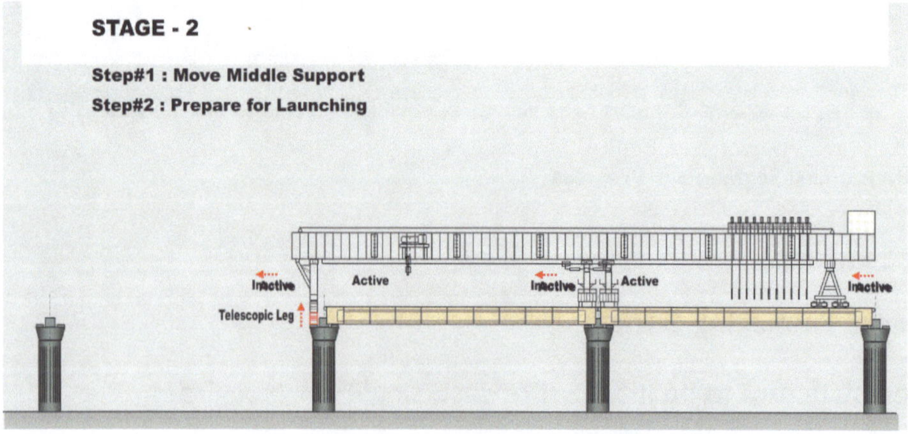

Fig. 6.65B

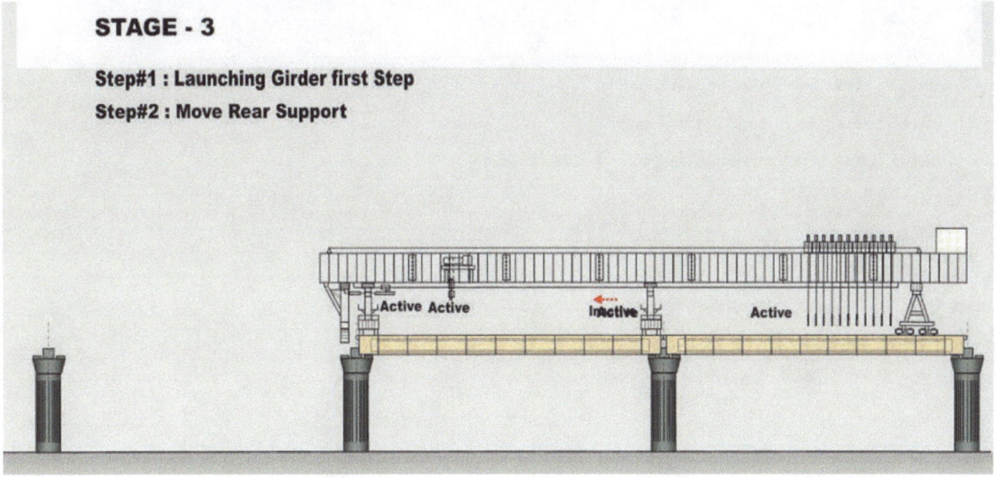

Fig. 6.65C

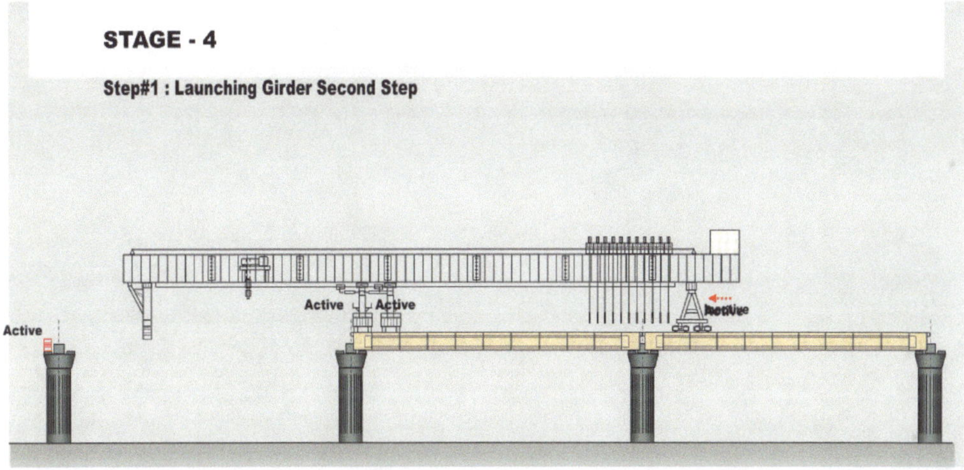

Fig. 6.65D

Stages of Segment Erection

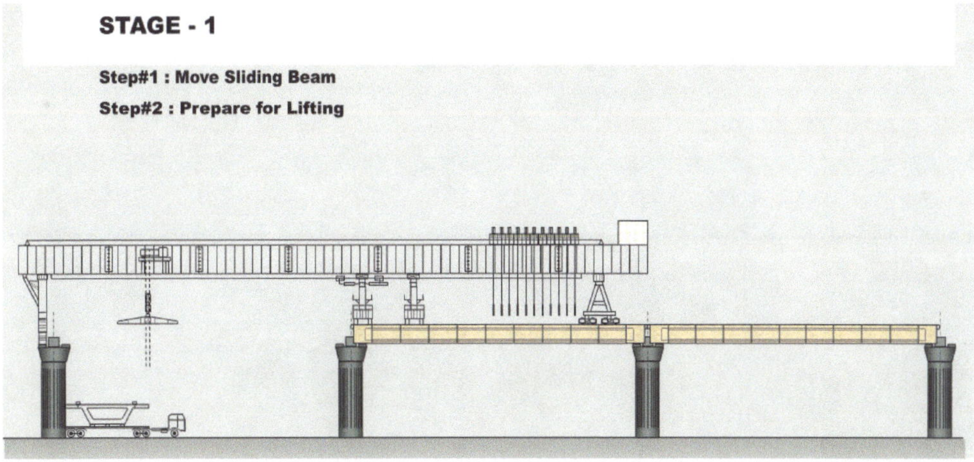

Fig. 6.66A

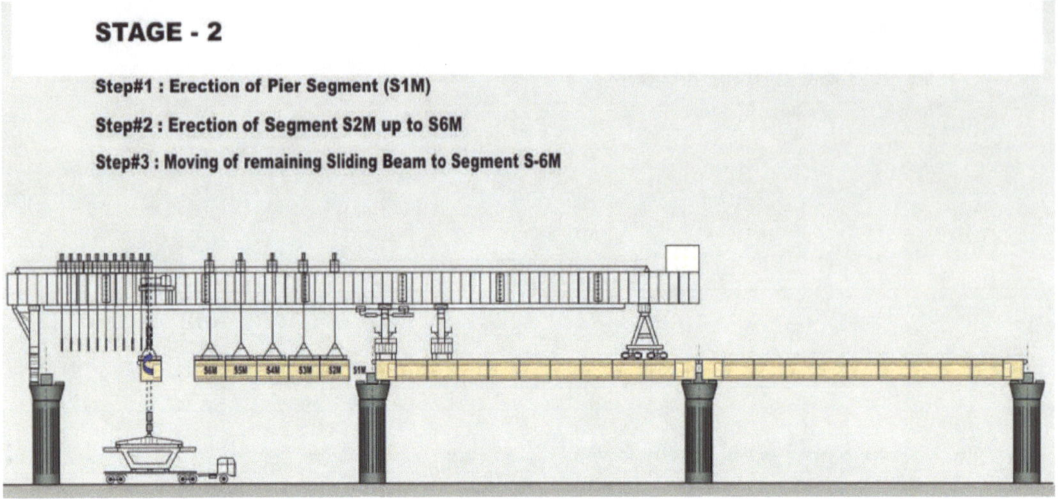

Fig. 6.66B

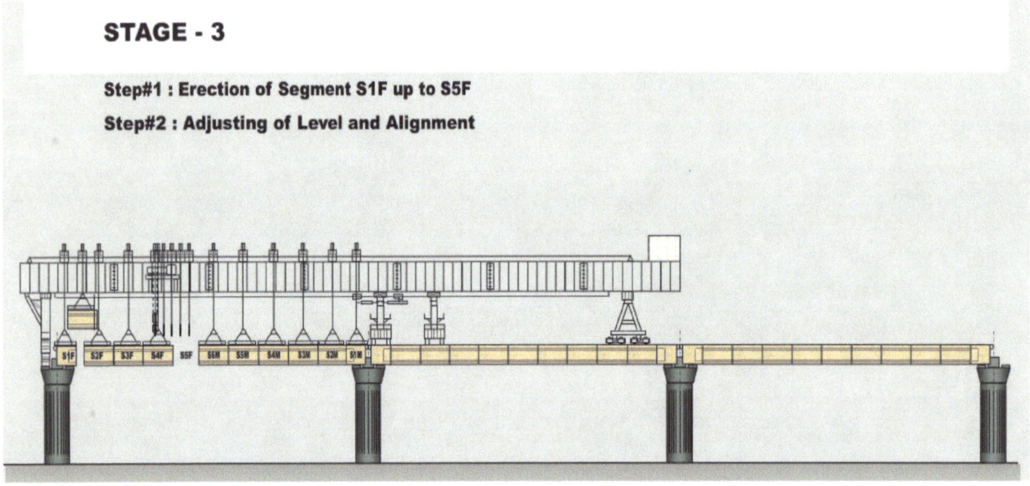

Fig. 6.66C

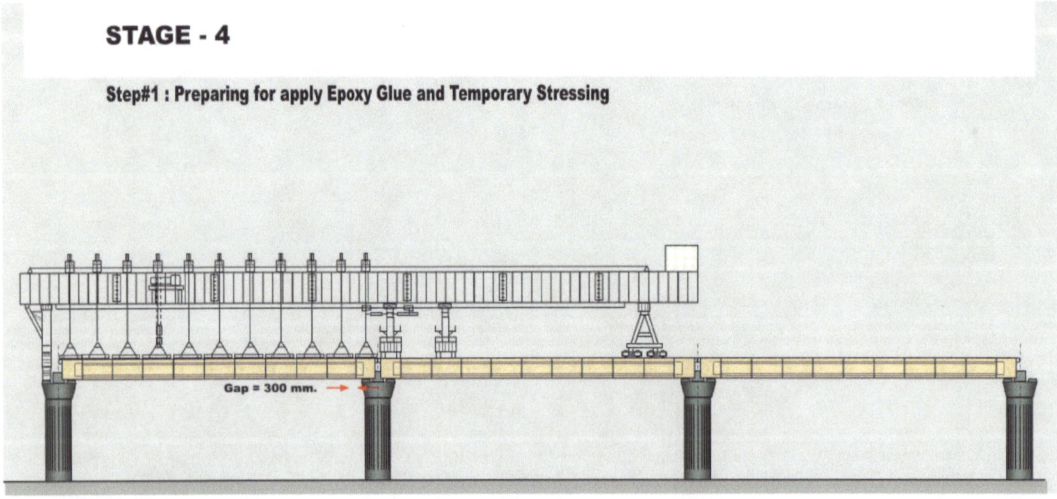

Fig. 6.66D

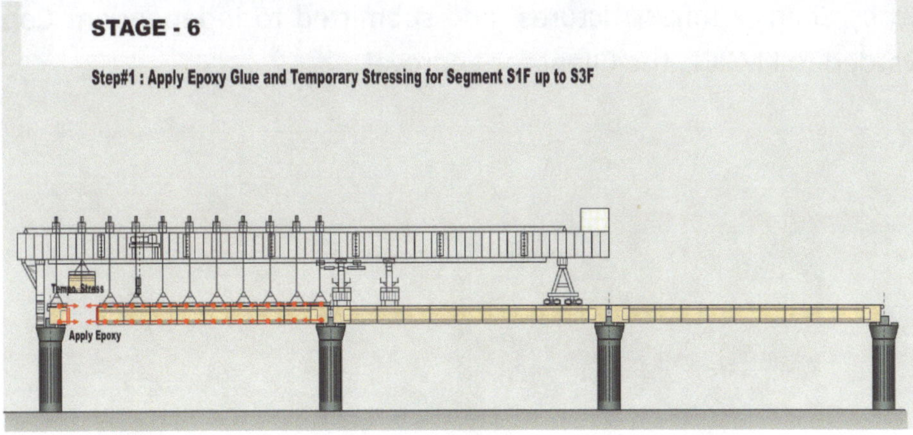

Fig. 6.67A

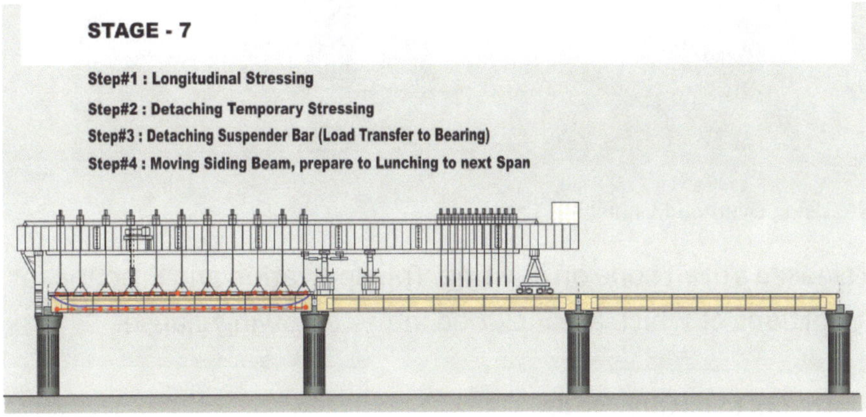

Fig. 6.67B

Fig. 6.68: End Section of Precast Girder supported on Bearing (Stage 5 not included due to repeatation)

The layout plan of the Casting Yard, material storage, concrete batching plant capacity to produce required quantities including wastages, the form and design of moulds, the casting bed, the sequence of activities, shop drawings, equipment to handle prefabricated reinforcement cages, curing arrangement, method to be used for de-moulding, handling of precast segments and stacking in dedicated locations in layers (2 or 3) as per the approved design encompassing all activities from start to finish was done by Gammon India, the Contractor for the elevated section (Elevated Stations were constructed by Simplex Infrastructures) and submitted to Independent Consultant (MYCEL) who in turn forwarded to KMRCL, the Client for approval.

Fig. 6.69: Erection at Site Using Overhead Launching Truss

Later, in 2022, IRC released a draft code on 'Casting, Transportation and Erection for Pre-Cast Concrete Segment', relevant portions of which are included in the following clause.

6.11.3 Shifting of Segments to Casting Yards

Segments shifted to Stacking Yard using Goliath gantry and stacked in layers as shown in **Figs.6.70 & 6.71**. Segments were transported using multi-axle Trailer Truck **(Fig. 6.72)** and erected in position using over head Launching Girder **(Fig. 6.73).**

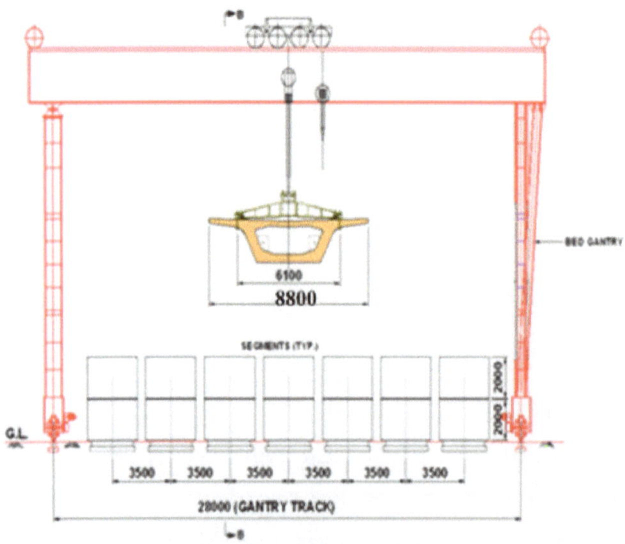

Fig. 6.70: Goliath Gantry Shifting Segments

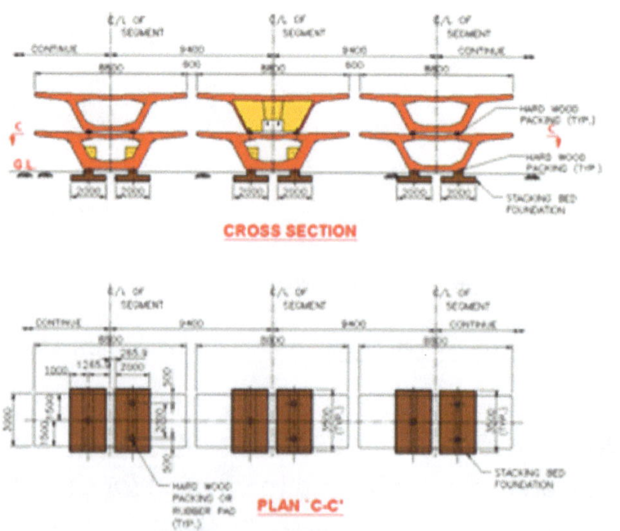

Fig. 6.71: Segments at Stacking Yard

Fig. 6.72: Segment is on Trailer to move to Site

Fig. 6.73: Launching Shaft Lifting Segment

The Indian Road Congress (IRC) published a draft for Design Methodology and Criteria Planning and Operation (IRC 87-2022). The approach used for the East-West Metro is like draft IRC 87 and reproduces the relevant portion for ready reference hereunder.

6.12 Design Methodology & Criteria Planning and Operations (IRC 87-2022/Draft)

Casting Yard Planning: Basic requirements of pre-casting activity shall be planned in the casting yard. Various activities like bar bending, casting, stacking and handling of pre-cast elements, and storing & sorting of raw materials shall be properly studied before finalising the layout. Methods for segment casting include the layout of the casting yard, set up and operation of the casting cells, movable rain and sun shades, geometry control stations, the storage and handling of rebar cages, the preparation of as-built geometry data, placing and finishing concrete, curing of concrete, form stripping, bond breaking, and other similar items shall be planned efficiently. Equipment for segment casting, including details of the forms and casting cells for the manufacture of the segments, surveying the segment, lifting and transportation of the segment in the yard and other similar items.

The following are the key points regarding the establishment of a well-planned casting yard:

i. A schedule of materials for segment casting, including concrete, reinforcing steel, prestressing steel, duct filler, and other similar items.
ii. Each segment number and the direction of erection.
iii. Segment dimensions including widths, lengths, thicknesses, tapers, fillets, radii, working points, post-tensioning, clearances, rebar dimensions and spacing, embedded items, holes, anchorage positions, and other similar items.
iv. Post-tensioning requirements: Check post-tensioning for consistency with pre-approved post-tensioning hardware and provide part numbers for department pre-approved systems on the shop drawings.
v. The volume of concrete, the weight of reinforcement, the weight of post-tensioning in each precast segment and the total weight for reinforcement and post-tensioning for the superstructure shall be summarised and tabulated on the shop drawings.

vi. Details and calculations for any localised strengthening for concentrated supports and loads or reactions from any special erection equipment placed in locations not already allowed for in the Plans.
vii. Details and supporting calculations for any modifications to segment geometry, cross-section dimensions, or segment length including any required changes to reinforcing and post-tensioning.
viii. Details of permanent and temporary embedded items, including inserts, blackouts, temporary openings, holes, and other similar items; any localised required strengthening and the materials and methods to fill and finish the holes.

Casting of segments shall not be started until the Engineer approves the relevant shop drawings, calculations, casting manuals, concrete forms, concreting and the post-tensioning system components and layout Plans. Stacking yard Planning Segment storage including layout of the storage area, method of supporting the segments, single or double stacking, placing erection marks and segment identification, and other similar items.

Calculations and details for lifting, storage and stacking of segments.

Additional strengthening of the segments to accommodate stacking was done as required at the casting yard, or arrangements were made for safe transfer of the load to the base without affecting the behaviour of the segment geometry. Segments shall be handled using only the devices shown on the shop drawings.

Stacking was at level in the upright position. Segment stacking and shipment shall be on the approved three-point bearing system, which does not introduce a twist under self-weight. Stacking superstructure segments one upon another shall not be done unless the Engineer approves.

Provide a **storage area** of suitable stability for the segments to prevent differential settlement of the segment supports during the entire period of storage. **Transportation** of Segments/Girders before shipment: The engineer shall thoroughly inspect each segment for damage. Thoroughly cleaning the faces of all joints of laitance, bond-breaking compound and any other foreign material by light sandblasting before shipment shall be done. Make no repairs to minor spalls or chipped areas on the joint surfaces until after the segment's erection.

6.13 Erection Using Underslung Gantries

Many precast segmental box-girder bridges have been erected with underslung self-launching gantries. The gantry supports the precast segments for the span to be erected under their side wings with adjustable carts that roll along the main girders. The main girders have front noses and rear tails to control overturning during launch. The underslung gantries are simple to design, assemble and operate. They do not require hangers, and spreader beams to suspend the segments, like in overhead launching girders. It allows fast loading of the segments on the gantry and facilitates access to the segments with working platforms applied to the main girders. The underslung gantries are typically less expensive and labour-intensive.

The gantries support the deck segments under the side wings and are designed to erect precast segmental box girders. Precast segmental bridges and precast girders are generally placed using ground-based cranes to lift each segment or girder. Ground access to the spans may not be feasible by the presence of existing infrastructure or the height to which the segments must be raised can exceed the reach of ground-based cranes The geometry of these gantries prevents their use for the simultaneous erection of adjacent bridges by side shifting the gantry from bridge to bridge, and these gantries are therefore addressed at linear erection from abutment to abutment.

The most visible feature of a launching gantry is the twin parallel girders, which can either be above or below (underslung) the bridge deck. However, a single beam can also be used, typically in an upper-beam configuration. The launching gantry machine is usually sized for the construction project, with the length of the twin main girders being more than the distance between spans. This length enables the launching gantry to span the gap between two adjacent bridge piers while providing allowances for the distance required for launching to the next span and flexibility of movement to accommodate curved paths between piers.

In some cases, hinges have been inserted into the gantry girders to allow tighter curves. The launching gantry girders are supported at each pier by braced frames, which have a limited range of movement to facilitate placement of bridge segments or bridge girders; the launching gantry does not generally contact the bridge deck.

For the East-West Metro, both Overhead and Underslung erection methods were used.

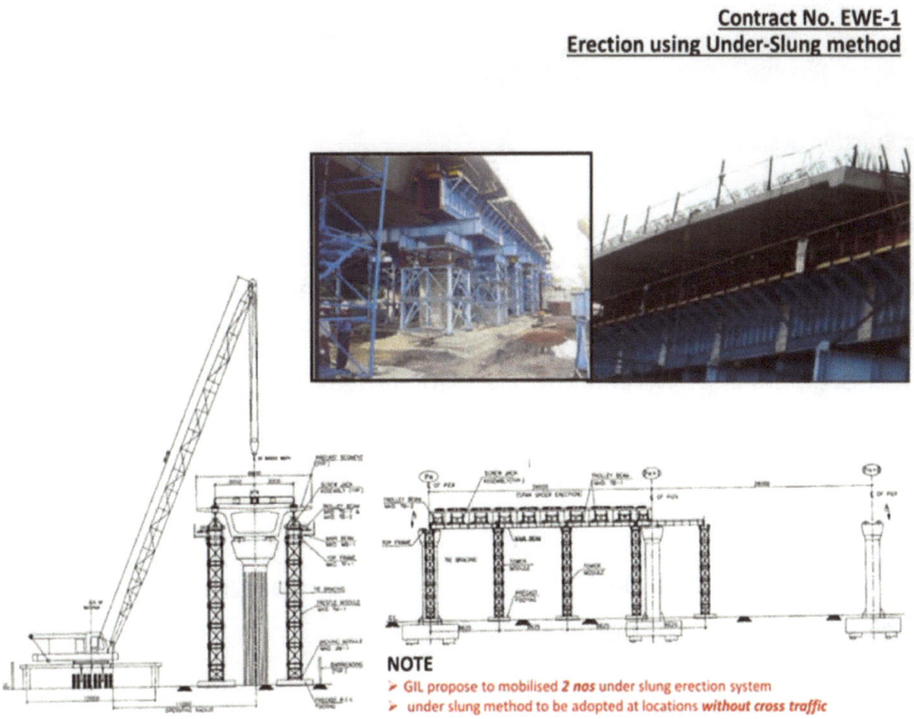

Fig. 6.74: Erection of Segments Using the Underslung Approach

Fig. 6.75: Underslung Erection System Work Under Progress

6.13.1 Precast Parapet

Precast parapets were made of concrete grade M-40, cast at the site and shifted to the stacking yard. The shape and size of the parapet were planned by a Noise Expert to minimise noise pollution, being a thickly populated area through which the corridor is traversing. The contractor was taken into confidence for the erection with minimum damage to the precast parapet using appropriate tools and machinery for correct and fast erection. The casting of the parapets started after the Client's approval

Precast parapets were around 1800 nos. plus wastage. A special casting mould had been prepared to facilitate repetitive work and a smooth finish.

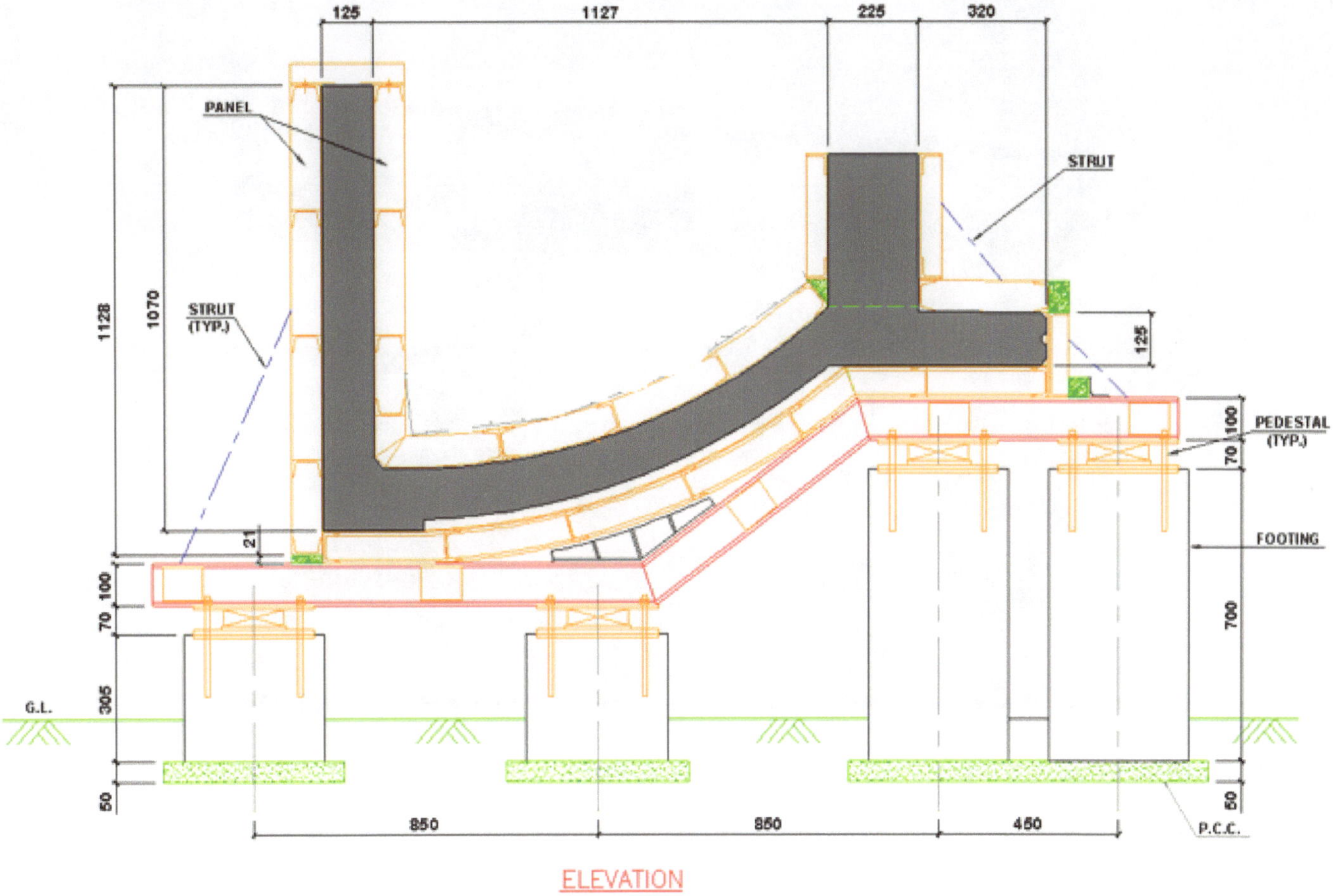

Fig. 6.76: Sketch Showing Parapet Mould and its Supporting Arrangement

Fig. 6.77: Precast Parapet in Formwork during Concreting

Fig. 6.78: Lifting of Precast Parapet

Fig. 6.79: Overhead Parapet Gantry for Erection

Figures 6.76 to Fig. 6.79 includes different activities to erect precast parapets in position.

6.14 Bearings

Elastomer Bearings: Bearings used for the elevated section were Elastomer Bearings. The design was as per the guidelines laid out in UIC 772R. Chloroprene was used in the manufacture of bearings. Grades of raw elastomer of proven use in elastomer bearings with low crystallisation rates and adequate self life used.

Fig. 6.80: Elastomer Bearing

The design, drawings and detailed method of statement for installation and the replaceability of the bearings were checked and certified by an independent agency and submitted to the GC for concurrence. The installation procedure conformed to the guidelines of the IRICEN publication and as per MORTH specifications.

Pot Bearings

Pot Bearings were used in the Depot area. As per contract the materials such as PTFE lubrication, confined elastomer, stainless steel and internal seal conformed to IRC:83 part-III. The pot base, saddle and top plate conformed to Cast Steel as per IS 1030 Gr.280-520 W and anchor bolts as per IS 1364, welding per IS 816 & 9595 and electrode per IS 814. Guides of sliding pot bearing were monolithic to parent component.

Permissible stresses in steel components and bearing stresses in concrete was as per IS 83, Part III. The bearings were visually examined both during the test and assembly. It was ensured that the steel bearing plate and steel piston maintain continuous and uniform contact during the test. Manufacturing tolerances were per IS 83, Part III.

Fig. 6.81: Pot Bearing

6.15 Construction of Ramp: Subhas Sarovar Area

The section between Stadium Station and Phool Bagan, is elevted to cross EM – Bypass and follows with a ramp down to connect Underground corridor abutting Subhas Sarovar, leading to Phool Bagan metro station. The elevated section from Stadium Station, descends according to the approved design, utilising a precast segmental box until it works to maintain vertical clearance, followed by a filled-in ramp till it reaches the ground. The ramp then continues below ground level. The initial section of this underground portion was done using the cut-and-cover method. When the depth of earth cutting was at a level where an earth retaining wall was required for stability, DW s are constructed on both sides of the ramp.

Fig. 6.82: DW at Subhas Sarovar

Construction of Diaphragm Walls was of Variable Depth in the Ramp Area, till it overlapped by 2/3m with two Tunnel corridors continuing from the Underground Station at Phool Bagan, constructed using Tunnel Boring Machine (TBM). The depth of DW s in the underground station area was the same. All the RCC Diaphragm Walls were of high-grade concrete and steel bars, except at the Entry and Exit of the TBM. The Diaphragm Walls, wherein the TBM entered or exited from the tunnel section after boring in the Underground station, were of GFRP (Glass Fibres Reinforced Polymer) bars & steel rebars. DW with hundred per cent GFRP bars, TBM could cut through the walls and continue through the earth. Wherein the Cutter Head punched through the DW, from the centre of the TBM cutter head, all the DW rebars were of GFRP bars with a tolerance of 500-600 mm area from the outer face of the cutter head. & the balance area was reinforced with Steel Rebars. **Fig. 6.82** includes placing of reinforcement cage for DW construction.

i) Top-down Construction of the Launching Shaft for TBM

The shaft plan size was commensurate with the final corridor dimensions, both transversely and longitudinally, to accommodate the TBM in length, plus Clearance on either side.

a. DW was cast on two sides of the tunnel section for a single track, for the other track to be constructed later, as per the activity schedule
b. To fix the rail, one set each on two sides (erected after the Launching Shaft was ready) to install the gantry to launch the TBM in parts.
c. Excavation up to the bottom of the roof level as per the location decided at the beginning of the covered corridor and levelling.
d. To plan, such as Chipping & taking out reinforcements from DW for the Roof Slab, fabrication & laying of reinforcements, cover blocks, formworks for Cut-Out (for launching of TBM parts) & both ends in the transverse direction, and others as required and concreting.
e. After the roof gained strength as per design, excavation below the roof slab was commenced, initially manual as required, followed by mechanical excavation till the Strut level or the bottom

level, including blinding concrete, waterproofing membranes, etc., as per approved drawing and removed the excavated spoils through top cutout in the roof.

f. Levelling the excavated surface, placing a waterproofing membrane, etc., per approved drawings.
g. Waterproofing Treatment was carried out on the PCC, based on approved drawings and as approved by the authorised agency, e.g., SIKA, FOSROC, etc.
h. Waterproofing membrane with adhesive on the Concrete surface wrapped below the Lower-level Rebar/coupler level of walls and terminated within the grooving in DW.
i. Once the waterproofing works were completed, the next sequence of activities followed. The bottom slab did not have any cutouts.
j. Next was the fabrication & placing of bottom slab reinforcements and concreting.
k. After the concrete in the bottom slab gained strength as per specification, the strut between the bottom slab and the roof slab, if any, was removed.
l. The top surface of the bottom slab was made good as required, and the platform was prepared to receive the TBM parts.
m. TBM parts launched one after the other in proper sequence and assembled on the platform.
n. The TBM assembly was checked before the commencement of TBM operation.

Note: All Construction Joints of Slabs at any level had waterproof Treatment.

II) Launching of the Tunnel Boring Machine

The following sequence of activities is followed for the launch of TBM.

a. Installation of Rail on top, both sides of DW.
b. Installation of Gantry Crane of Required capacity for Lifting of TBM,
c. Installation of Rail Path at Base Slab Level
d. Lifting of Cutter Head of TBM from Ground level and placing at base Slab level, Rail Track by Gantry Crane.
e. After the Cutter Head was placed at the base slab, the two remaining parts of the TBM were launched.
f. Once the TBM proceeded into the Tunnel, then step by step, balance Trailing Cabins of the TBM were launched progressively.

iii) Punching of Cross DW at the Entry point by TBM

The sequence of activities for running

a. Once the launch of TBM on the base Slab Rail was completed, the mechanical job was carried out to fix the exact lining position and the level of the TBM.
b. After installation, a dry run was done to check for any disorder or deficiency that may occur during actual work. If any deficiency/disorder was observed during the dry run of TBM, immediate action was taken to resolve the deficiency/disorder issues.

c. Marking on DW wherein TBM will punch the DW.

d. Ground Reality was checked, the chipping of Concrete on the side of DW was done till the GFRP rebars were exposed, ensuring that the TBM would not hit the rebars.

e. After the quality aspects and safety were ensured, the Green Signal was given to start the TBM for punching the DW.

f. Once the TBM punched the DW, it followed its course of action, boring into the soil.

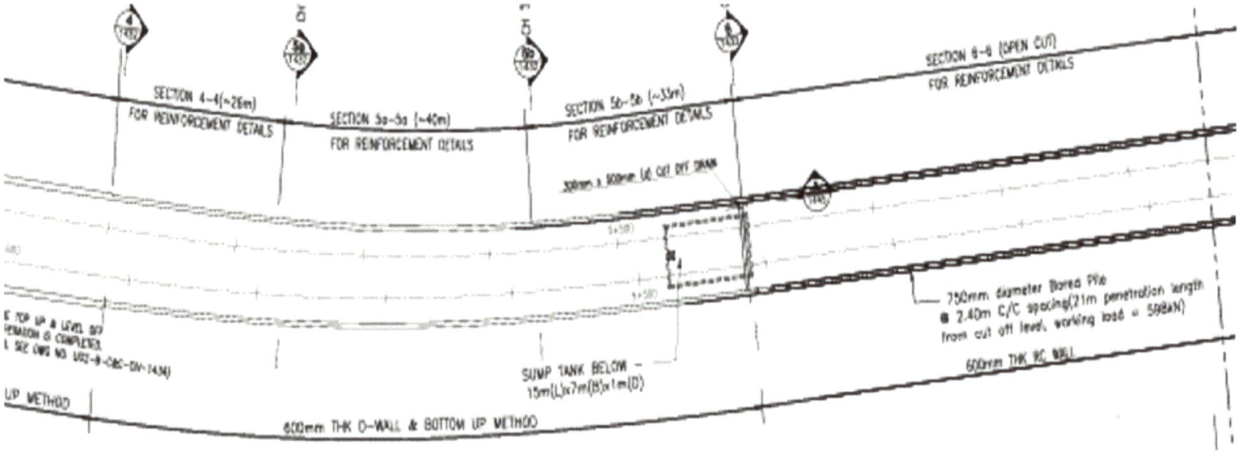

Fig. 6.83: Part Plan of Ramp showing Bored in situ Piles as Shoring Wall

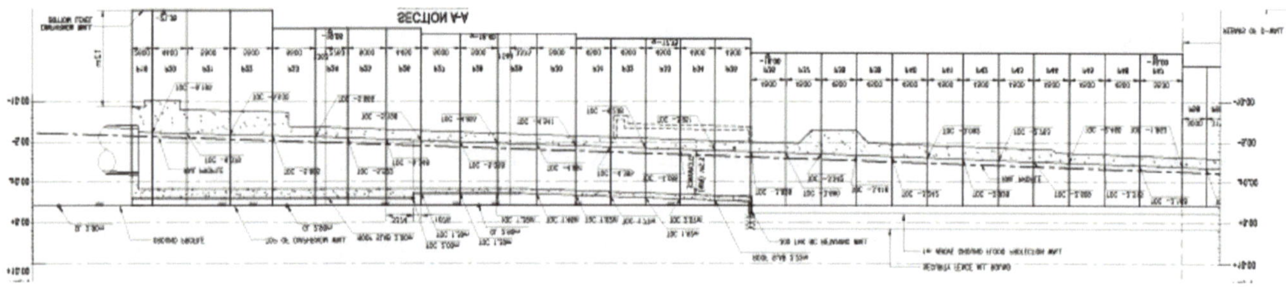

Fig. 6.84: Long Section with Top & Bottom Slab in Ramp Section

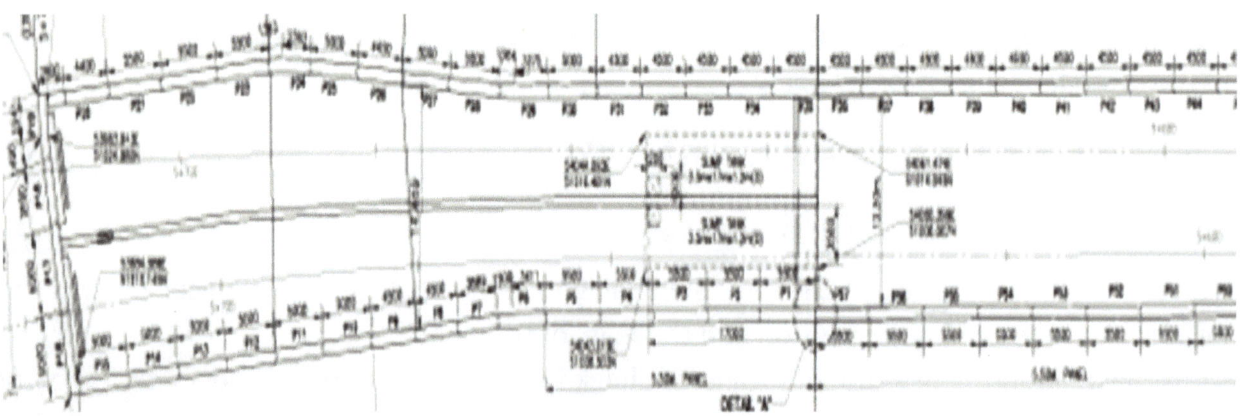

Fig. 6.85: Includes Plan View with DW Panels

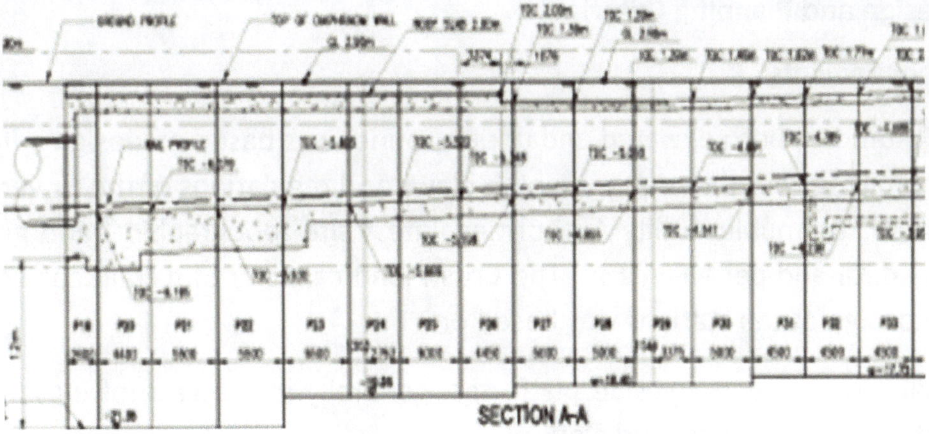

Fig. 6.86: Variable Depth of Diaphragm Wall – Underground Ramp Towards TBM Section

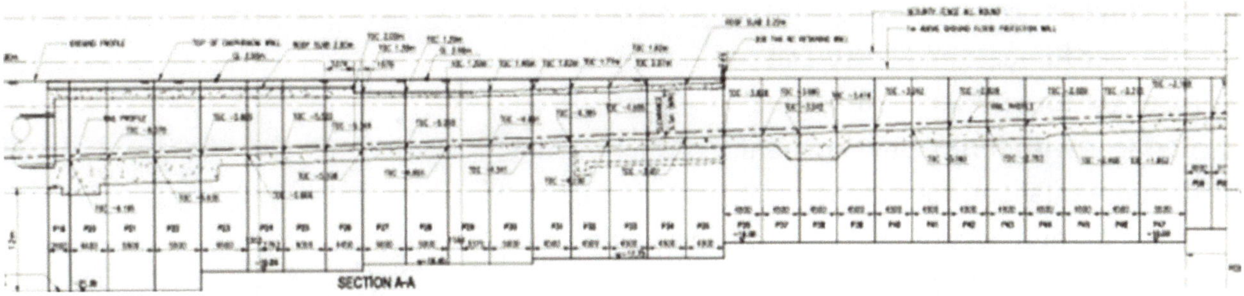

Fig. 6.87: Variable Depth of Diaphragm Wall – Underground Ramp Towards TBM Section

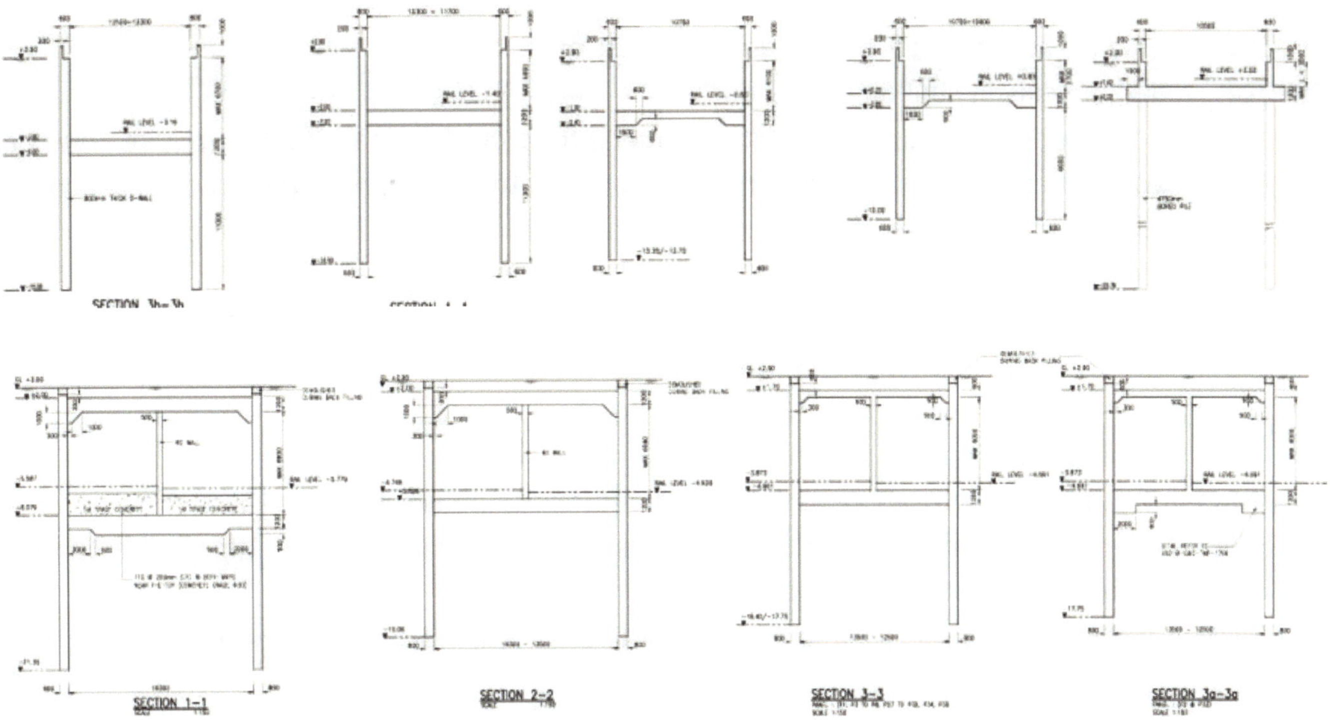

Fig. 6.88: Sections showing stages of UG Ramp Section, Including Top & Bottom Slab

6.16 Station Design and Planning Criteria

6.16.1 Design Requirements

The design of all stations, both elevated and underground, was based on design criteria included in the Contract, complying with the requirements of laws and regulations of the Government of India, requirements of Kolkata public Utility Authorities, latest site topographical and geophysical data, traffic/passenger data and per NFPA 130. The Crush load capacity of a train considered was 1800 passengers. The design of the stations was based on:

i. Operational requirements for side platforms, generally in the case of elevated stations and island platforms for underground stations.
ii. Traffic, Road and Pedestrian requirements
iii. Existing and planned Utilities
iv. Structural Provisions
v. Passenger forecast and Entry/Exit
vi. Passenger forecast and Entry/Exits
vii. Interfaces with future development
viii. Environmental considerations
ix. Disruption during construction and construction methodology
x. Flexibility to facilitate site requirements
xi. Provision of future expansion
xii. Parking requirement in discussions with the Client
xiii. Integration with Urban agglomerates, particularly for Heritage sites
xiv. Safeguards for adjoining properties and existing traffic
xv. Latest data on existing building conditions
xvi. In addition, the detailed plan and design included provisions for: Space for the required number & spaces for lifts for the handicapped persons, escalators to connect the ground, concourse and platforms for passenger forecast in the horizon year and as per E&M group requirement.
xvii. All works about MEP works
xviii. Detail design ensured coordination and interface with Mechanical contractors for fittings, openings, room sizes, installation sequence and space required.
xix. During finalisation of working drawings, the designer liaised with the SCADA contractor, the Employer, and the Kolkata Fire Services.
xx. Stipulation of all test procedures applicable to the various equipment and systems.
xxi. The detailed design included mechanical works like vibration isolation, equipment mounting, maintainability, removal and replacement.
xxii. Design included the general requirements of Mechanical works, Electrical requirements for mechanical systems.
xxiii. Uninterrupted Power Supply.

Three types of stations were used: intermediate, interchange and terminal stations. Interchange stations are those through which two or more corridors pass. The layout for each station was done considering the track geometry, operational requirements, predicted passenger flows, facilities required, and satisfying electrical and mechanical requirements. The station design was divided into two areas, namely paid and unpaid. Unpaid areas accommodated both travelling and non-travelling people. The paid area was designed for passengers and operational personnel as required.

The platforms were designed for required orientation, sufficient space to cater for boarding and alighting passengers, considering delay in train movement and to an appropriate level from the rail top.

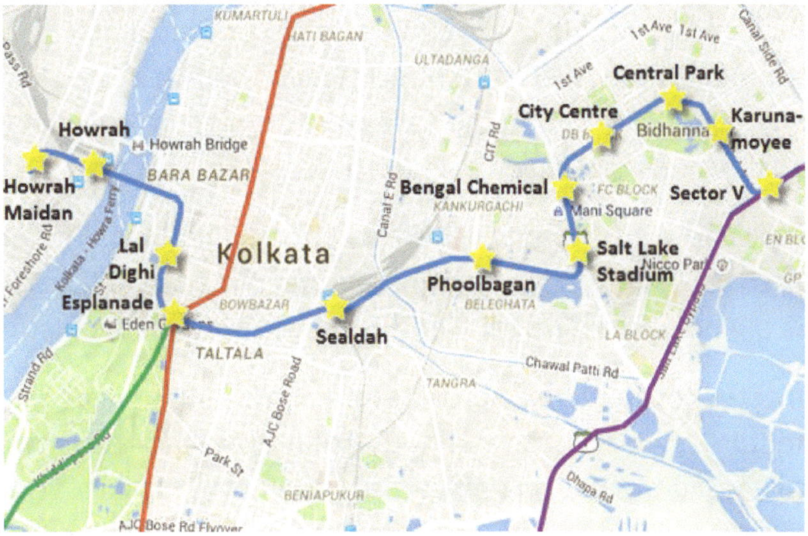

Fig. 6.89: Station Locations

Fig. 6.90: Front View of Sealdah Station

Island platforms have the advantage of requiring a minimum duplication of facilities, like stairs, escalators, lifts, lighting, and other services. The total platform width is generally less for Island platforms and designed to accommodate both the morning and evening peaks. In the case of side platforms, designed for elevated stations, each platform will be of sufficient width to accommodate peak traffic, stairs, escalators, lifts and other services. However, side platforms, on the other hand, are more adaptable to direct access from sidewalks. In the East-West Metro, all Underground stations had island platforms, and elevated stations had side platforms.

Two-level stations offer a means of controlling the number of passengers waiting on the platforms at any one time. By using the concourse area as a holding zone whenever the number of passengers on the platform becomes too large, scheduled train loading times can be maintained.

6.16.2 Concourse

A mass rapid transit system has its platforms only for the ingress and egress of passengers to and from trains. All other connected activities are dealt with at concourse, which is generally at a level between Entry level at ground/road level and platform level. There are cases where both, entry level and concourse are at ground. However, in the East-West Metro, the concourse, both for elevated and underground, is at the mezzanine level.

The concourse was designed with two distinct areas, unpaid and paid. The unpaid area is where passengers gain access to the system, travel information and purchase of tickets. On passing through the ticket gates enters the paid area. The paid area includes access to the platform level.

The principal function of the concourse is to provide for the movement of people between platforms and road access and for the activities associated with that movement. The main activity is paying the fare, so the concourse housed the booking offices, ticket vending machines and barriers to control entry into the paid area and exit from platforms. The concourse also provides several other facilities. Its size, shape and layout vary from station to station depending on the volume and nature of traffic and facilities provided. For concourse, a headroom of 3 meters had been used in general.

The other facilities provided were divided into two main categories: those required in connection with system operation in general and the stations in particular, and those provided for the convenience of and as amenities to the users. Of the first category, the most important was the Station Master's office. This housed the public address equipment for both concourse and platforms and means of communication with the Control Centre. CCTV monitors and AFC control were also located in that room.

At all stations, space had been provided not only for stairs, escalators and lifts proposed to be installed in the beginning, but also for those to be provided in the future. Space was also provided at all stations for air conditioning and ventilation equipment and electrical substations. Facilities required for the convenience of and as amenities to the passengers include illuminated direction signs, train indicators, information booths, counters for sale of newspapers and other selected commodities, restrooms (only for operational staff) and first-aid facilities. Although the primary design considerations of the concourse were efficient operation of the rapid transit system, all of them serve as a secondary but important function for free pedestrian movement. Therefore, the pedestrian facility has been designed wherever possible in such a manner that there would be free movement between all access points.

Typical floor plans, sections, columns and foundation details used in the corridor have been included below for reference.

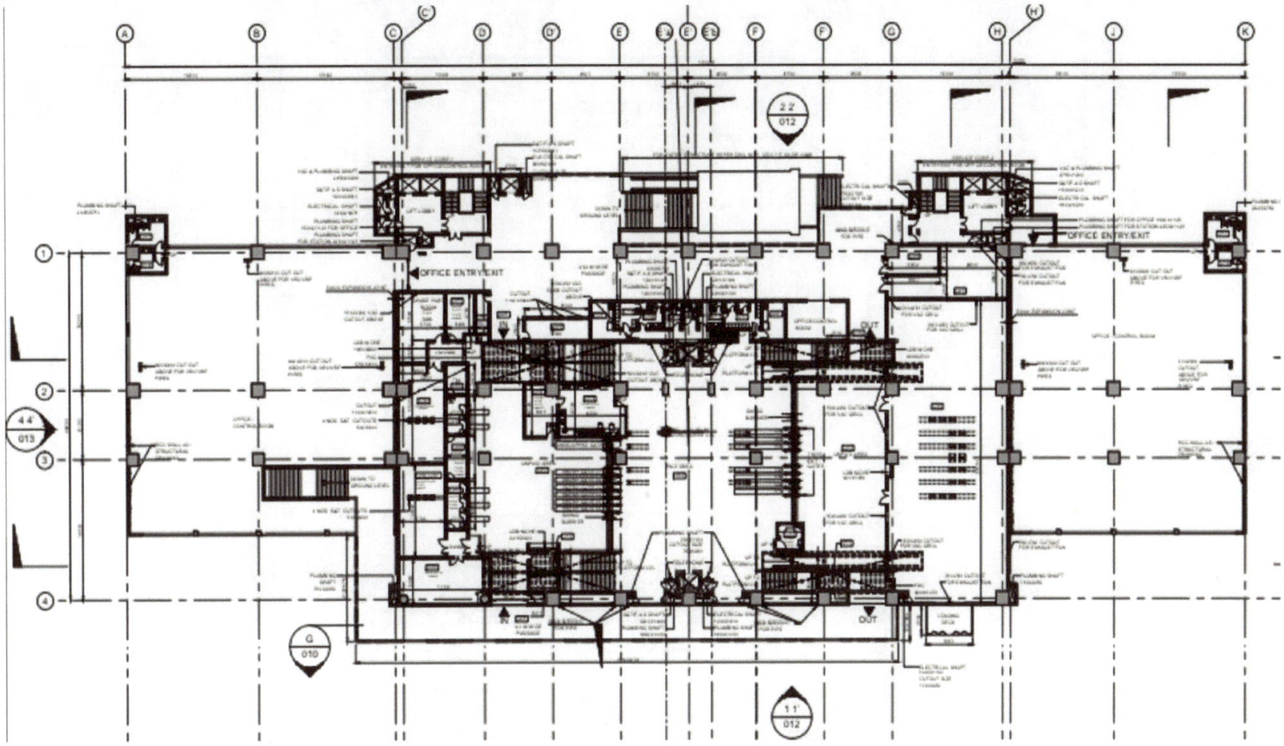

Fig. 6.91: Concourse Level Plan-Sector V

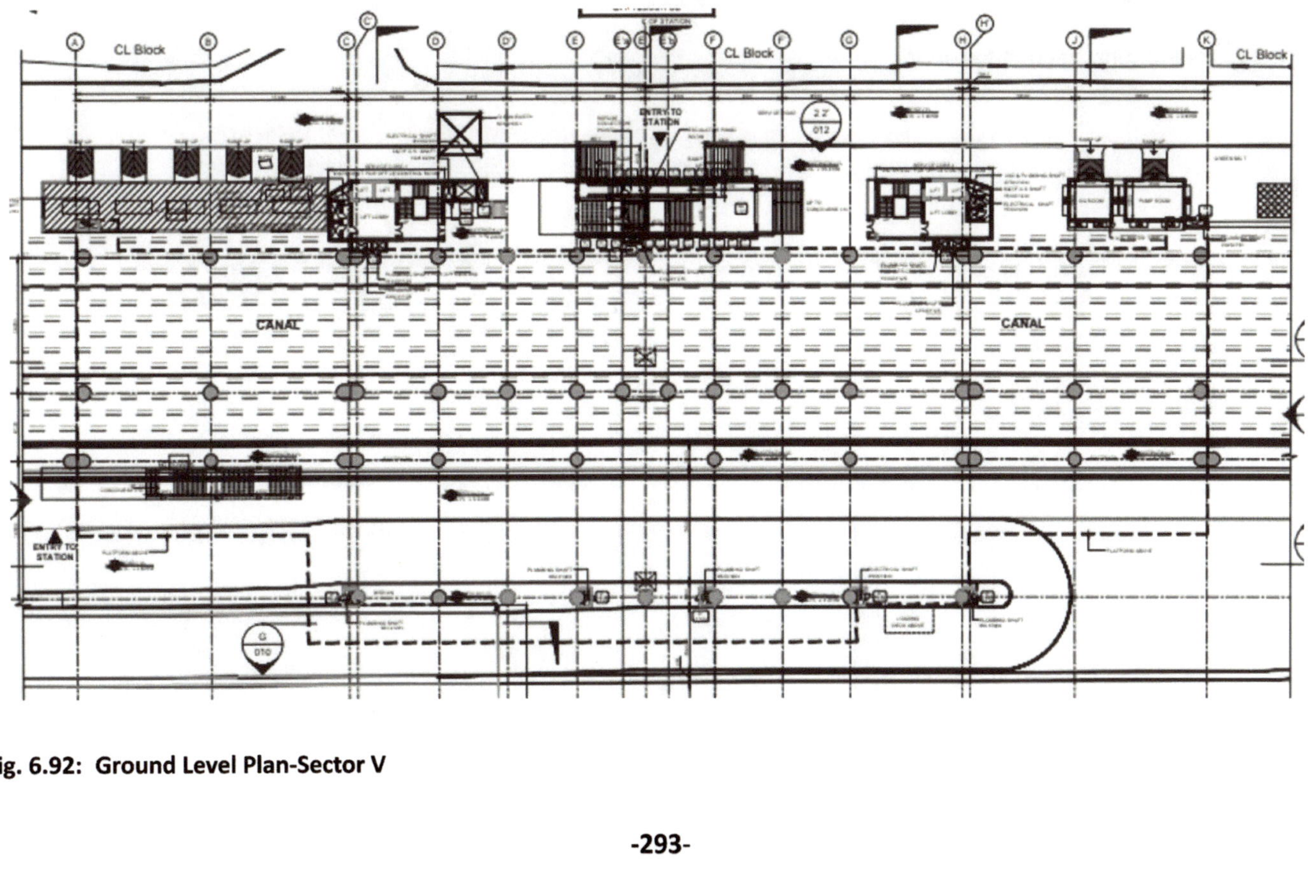

Fig. 6.92: Ground Level Plan-Sector V

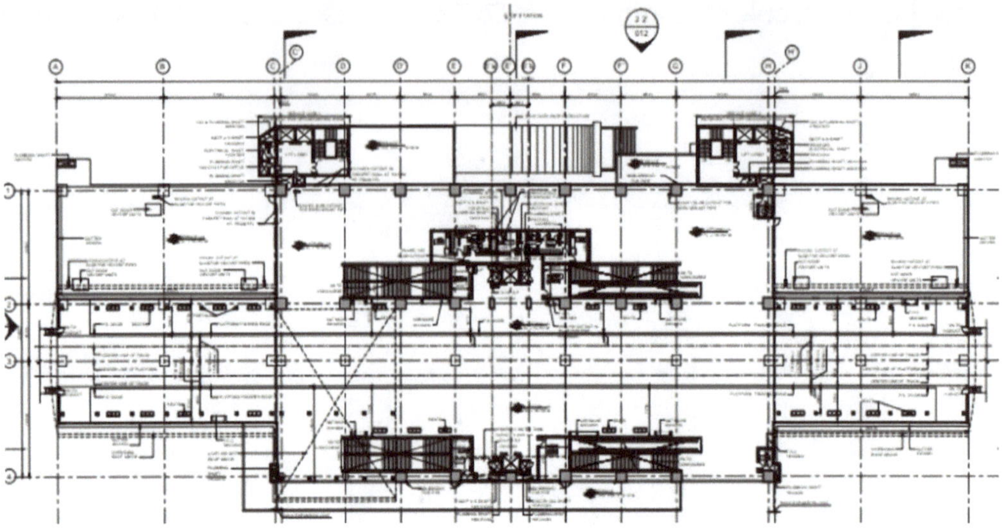

Fig. 6.93: Platform Level Plan-Sector V

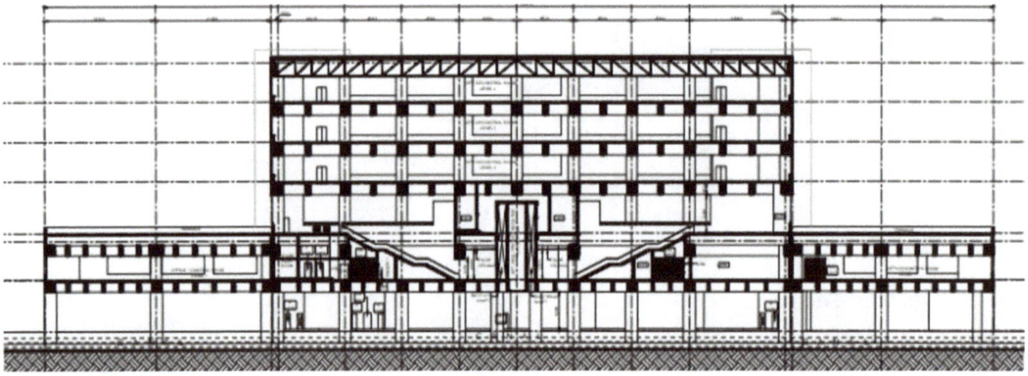

Fig. 6.94: Longitudinal Section-Sector V

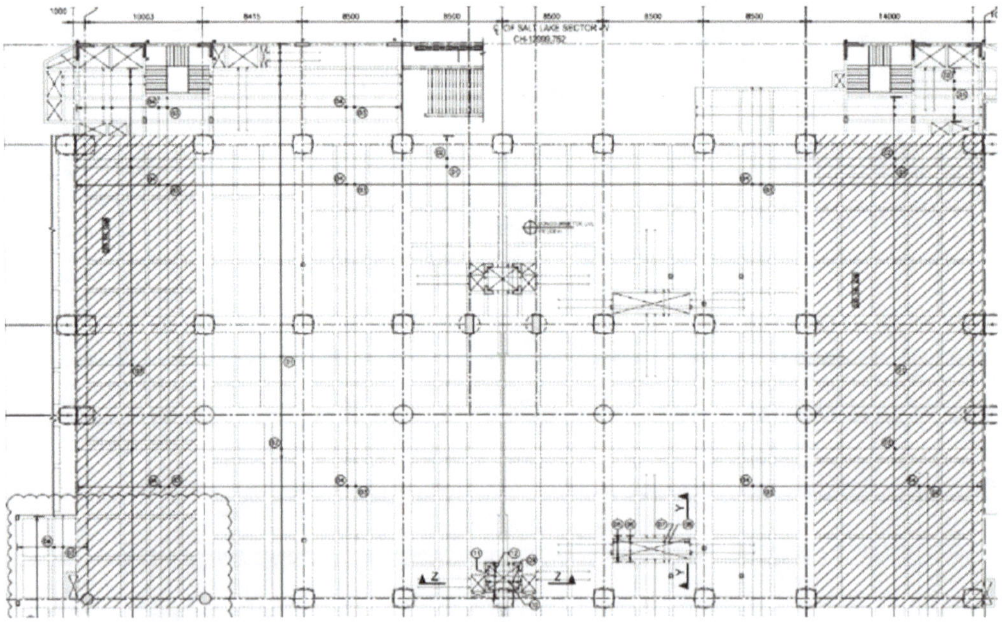

Fig. 6.95: RC Detail of Concourse Slab-Sector V

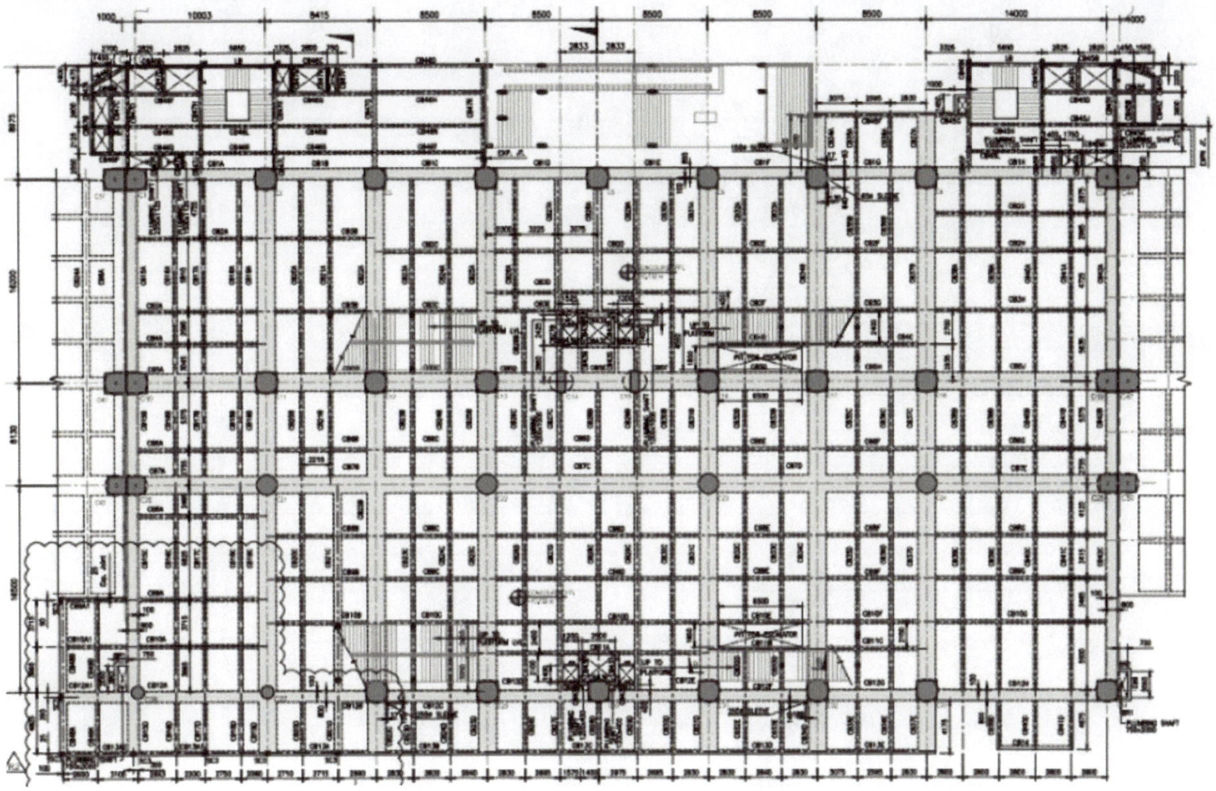

Fig. 6.96: Beam Layout at Concourse – Sector V

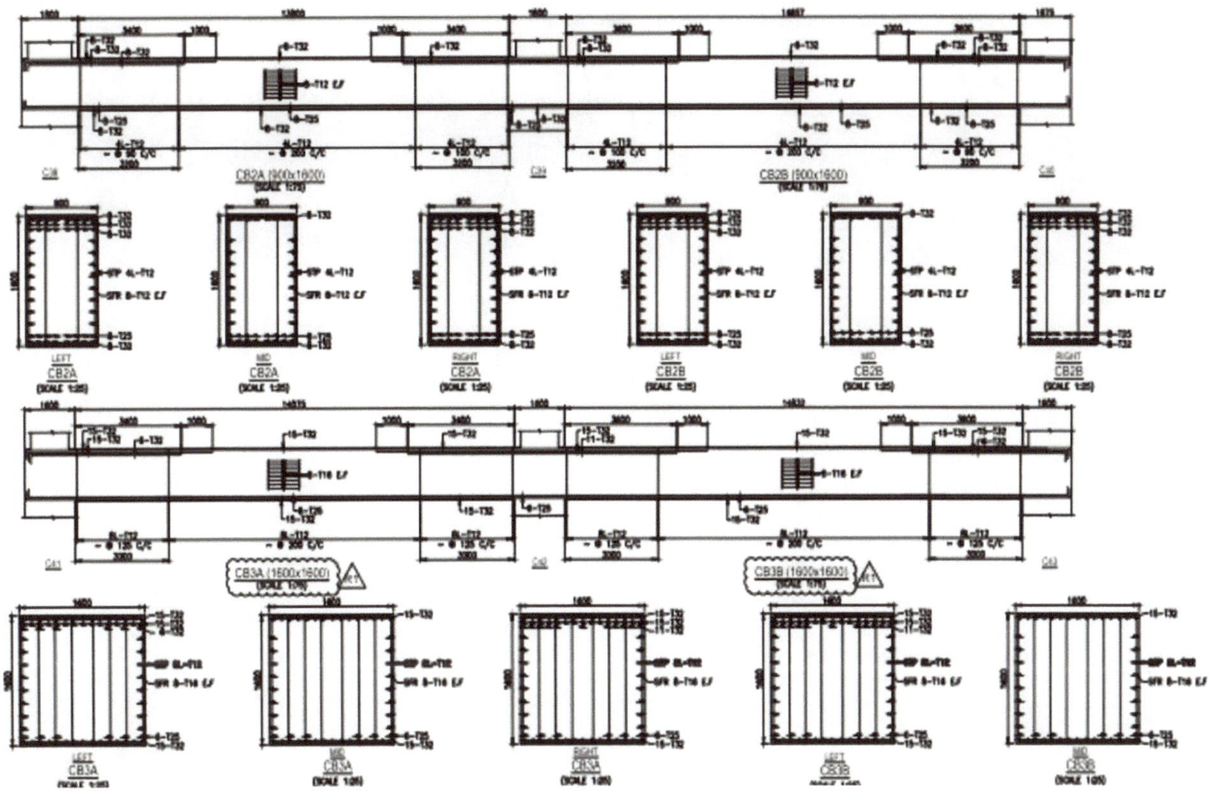

Fig. 6.97: Beam Details at Concourse Level – Sector V

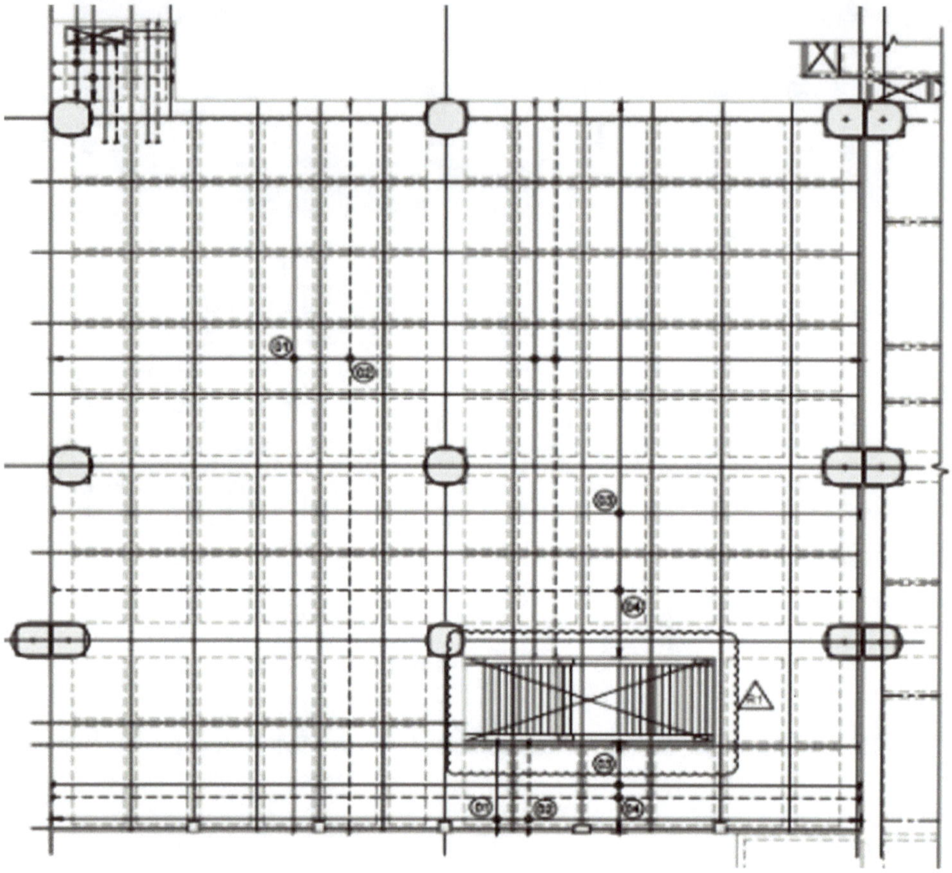

Fig. 6.98: Reinforcement Detail Concourse Level (Part Plan) – Sector V

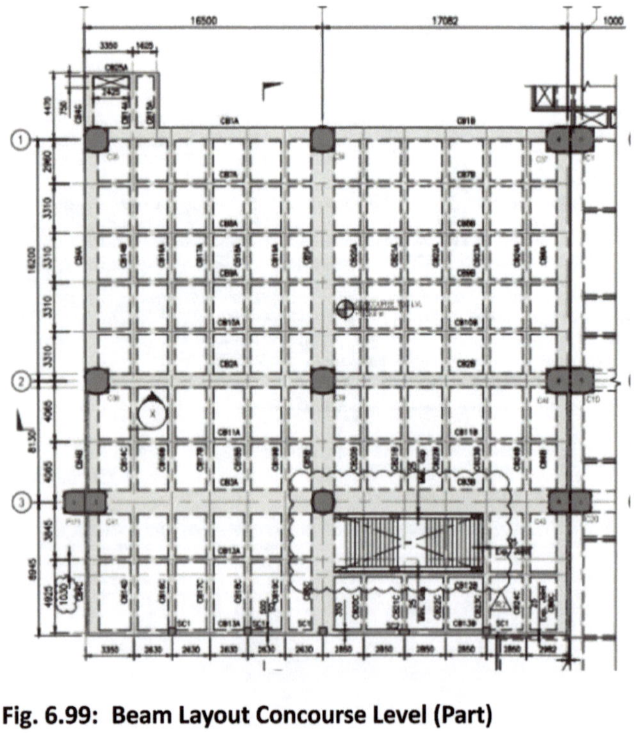

Fig. 6.99: Beam Layout Concourse Level (Part)

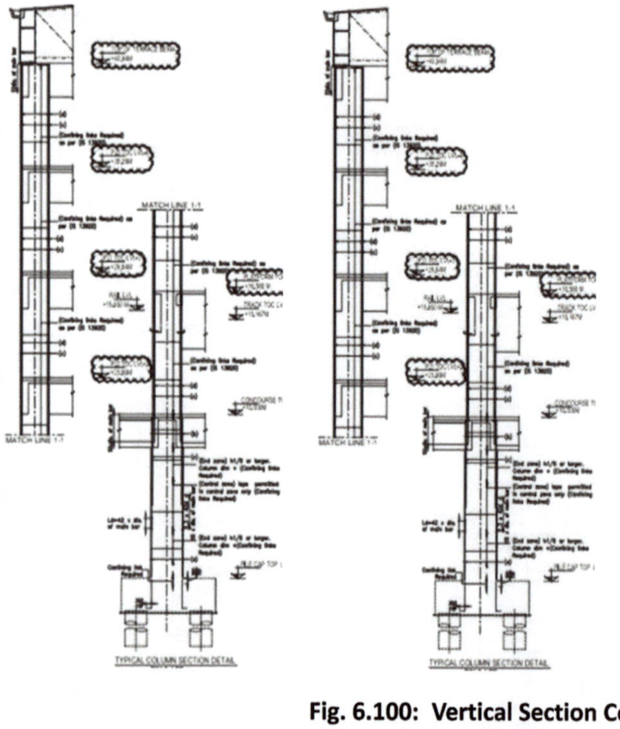

Fig. 6.100: Vertical Section Column, Beam & Pile Cap & Pile Cap

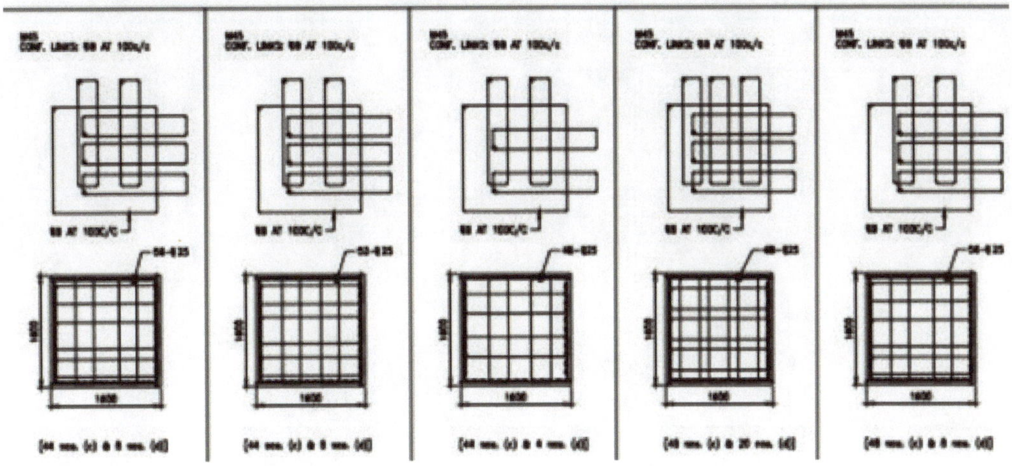

Fig. 6.101: Column Reinforcement – Sector V

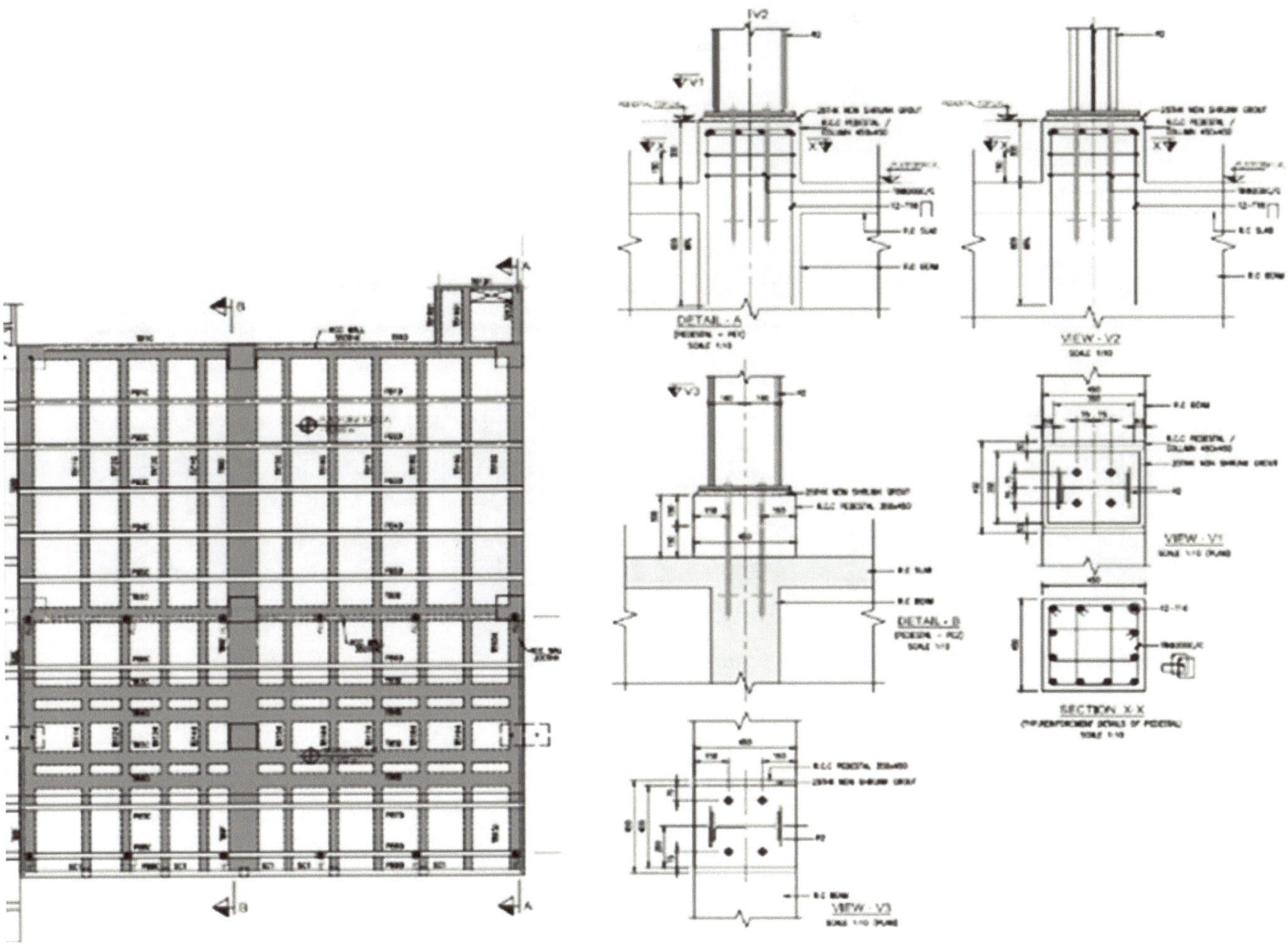

Fig. 6.102: Beam Plan (Part), Column-Beam Junction Details

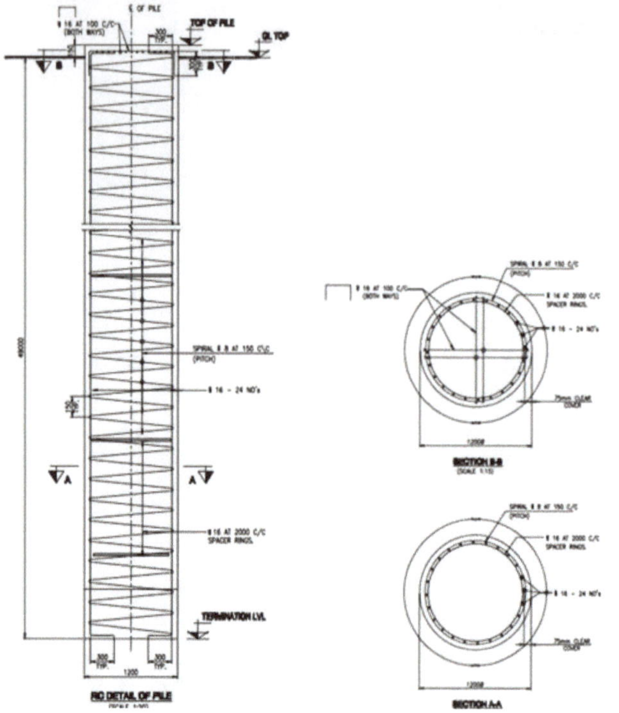

Fig. 6.103: Pile Renforcement

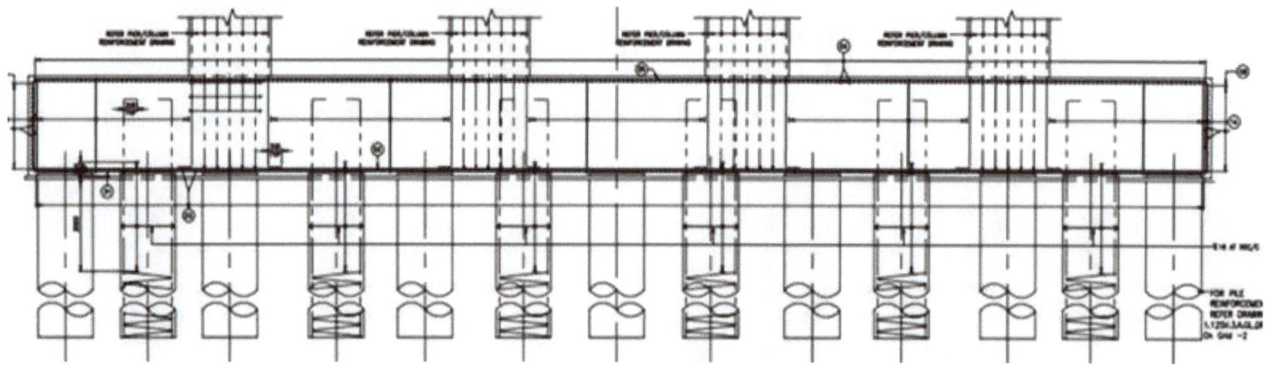

Fig. 6.104: Longitudinal Section through Pile Cap – Sector V

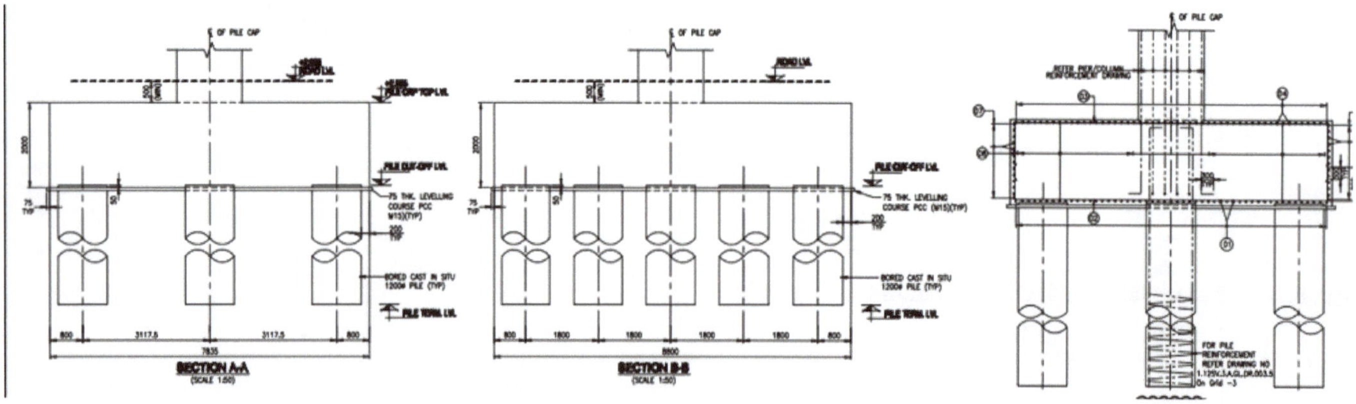

Fig. 6.105: Pilecap Cross Sections – Sector V

6.16.3 Station Sizing & Egress Requirements

To arrive at tentative sizes of station boxes for cut-and-cover stations, it is essential to determine the sizes of basic passenger facilities that Governance of the station's civil structure. These are essentially the widths of platform and combined widths of all vertical circulation elements (stairs and escalators) required for the emergency evacuation of passengers. The two are in turn calculated to satisfy the provisions of the National Fire Protection Act (NFPA) 130, 2010, a standard was followed the world over for the design of rail-based mass transit in the event of an emergency in the station requiring the evacuation of passengers from the station.

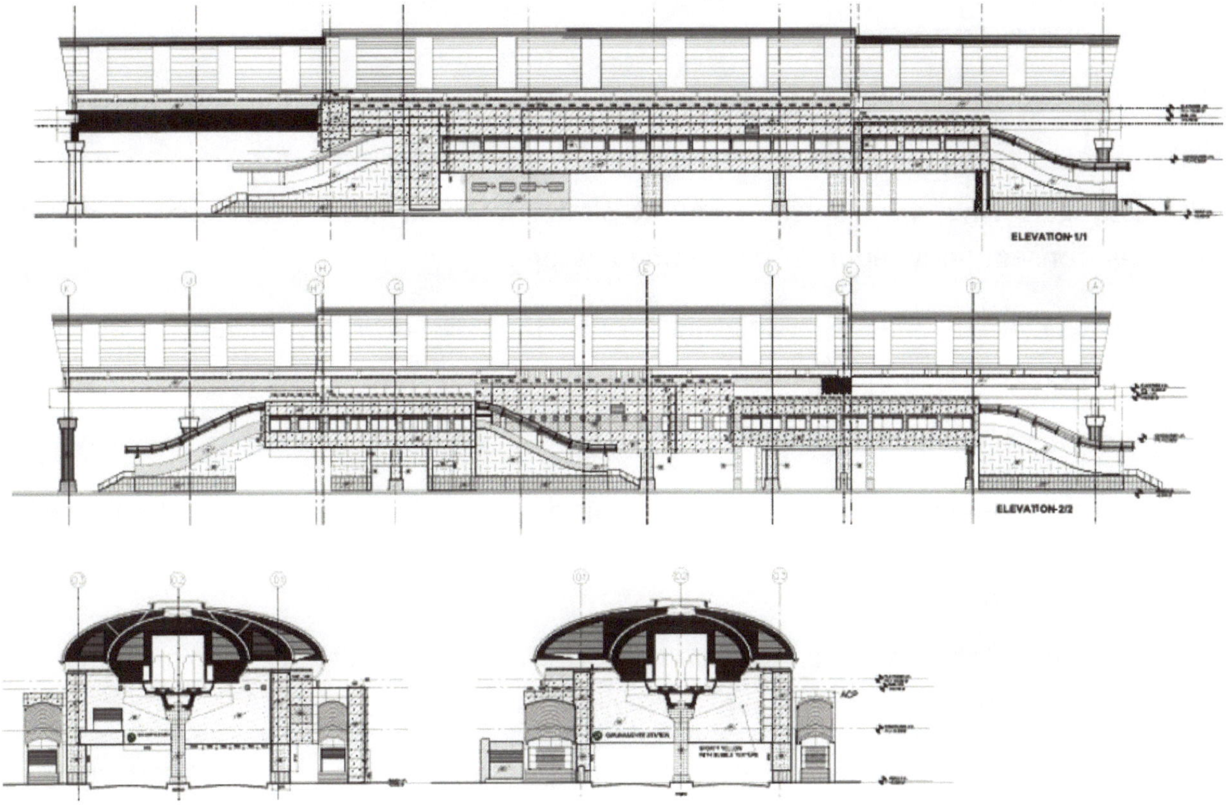

Figs. 6.106: Views of Completed UG Stations

Fig. 6.107: Typical Long and Cross Section – Elevated Station

NFPA 130 lays down conditions to be assumed for the 'worst case scenario', which are broadly as follows:

1. All passengers who were on the incident platform would need to be evacuated within 4 minutes, consequently requiring a combined escape.

   The capacity of the stairs & escalators serving that platform is sufficient to evacuate the platform occupant load in 4 minutes.

2. The platform occupant load will include-

   Waiting for passengers at the platform, factored to include a service disruption (between 5 to 10 minutes for train services of short intervals, which is assumed to be 6 minutes for the proposed MML3).

   Train loads on all the tracks serving that platform, which will mean the section loads of trains arriving from both directions.

3. Escape capacity will include the combined capacity of all stairs and escalators serving that platform.

Fig. 6.108: Kolkata Metro in Operation

4. Escalators moving in the direction of escape can be considered as moving and as a means of egress, subject to operational conditions such that those moving in the opposite direction can be stopped and restarted in the direction of escape.
5. One escalator will not be considered available at an egress capacity calculation on account of maintenance.
6. The maximum distance between any two means of escape will not be more than 100 m and the maximum distance of an escape stair from a dead end will not exceed 25 m.

   Two scenarios have been studied for the calculation of escape requirements. Scenario 1 is the 2031 ridership figures provided in the DPR, and Scenario 2 is the ultimate capacity scenario, which will arrive later.

6.16.4 Station Area Planning

The section includes station-specific structural form, station planning, entrance locations, and the sizing of station facilities for the East-West Metro corridor. Planning was carried out for both elevated and underground stations. Underground station planning norms are based on the following:

- The adequacy of road width where cut & cover construction is proposed
- Land for cut & cover access shafts is available where NATM construction is proposed.
- The availability, adequacy, and suitability of land are identified in the Project Report for street-level facilities such as entrance structures, ventilation shafts, and ancillary buildings.
- Worksites at station box ends need to be available to allow the launching/retrieval of TBMs and the removal of muck from the tunnels.

The construction of each underground metro station has been adopted based on site conditions and alignment.

6.16.5 Station Typology & Configuration

Various factors, such as land availability, construction methodology, and road widths, could divide the stations into the following categories.

Fig. 6.109: Elevated Corridor During Construction Fig. 6.110: City Center Station During Construction

6.16.6 Cut-and-Cover Station (UG)

A typical cut-and-cover station box **(Figs.6.108)** is about 285 m long, 24.5 m wide and about 17 m deep. This typology is applied to stations where a road width of more than 28 m is available for the construction of the station. This prototype is a two-level structure with a concourse at the upper level and an island platform, 185 m long and 12 m wide at the lower level. Space beyond the platform on either side is used for housing the auxiliary substations, one at each end and two tunnel ventilation fans at each end, along with the plenums for draft relief and tunnel ventilation system shafts. The concourse level has a centrally paid area, long enough to accommodate vertical circulation elements (lifts, escalators, and stairs). This paid concourse is flanked by unpaid concourses on either side, separated by fare collection barriers. These two unpaid concourse areas have ticket vending windows and space for the future.

For scenario 1, for the station, the traffic direction with the highest boarding (and therefore the highest waiting passengers) is identified and considered for the 'worst case scenario'. The other direction is treated as a non-incident direction. Accordingly, the platform occupant load is calculated. The required occupant load and platform widths are checked. Also, the requirement of the escape stair width is calculated. Station facilities for the East-West Metro corridor. Planning was carried out for both elevated and underground stations. Underground station planning norms are based on the following:

- The adequacy of road width where cut & cover construction is proposed
- The availability, adequacy, and suitability of land for street-level facilities such as entrance structures, ventilation shafts, and ancillary buildings.
- Worksites at station box ends need to be available to allow for excavation equipment and removal of muck from the tunnels.

The construction of each underground metro station has been adopted based on site conditions and alignment.

Fig. 6.111: Perspective of Howrah Station over Ground

Fig. 6.111 includes the concept of Howrah Station during design stage.

6.16.7 Access to Mezzanines

The underground stations have mezzanines under major roads at the level of the subway roof. Access to the mezzanines has been provided from both sides of the road. In the case of the important road junction above, efforts have been made to provide access to the mezzanine from all sides. Access from surface streets to the underground mezzanines is by stairways, except in a few stations where escalators have been provided.

In many areas of Kolkata, the pavements are inadequate to accommodate access to the mezzanines. In such cases, it is necessary to acquire adjoining areas/buildings and to provide access to the Mezzanine.

Fig. 6.112: Station at Platform Level

Fig. 6.113: Side end of Salt Lake Stadium Station

6.16.8 Access to Platforms

Access from the mezzanines to the station platforms is by stairway and/or by escalators, depending on the vertical separation of the two levels. For all upward movements of more than 5 meters, escalators are provided. At stations with exceptionally large passenger loads, escalators have been proposed even where the ascent is less than 5 meters. **Fig. 6.112** shows the Station Platform with Screen Door included.

Fig. 6.113 is the end View of Salt Lake Stadium station.

6.16.9 Platforms and Passageways

Platforms: The arrangement of entrances and exits to station platforms considerably affects the time required to load and unload trains. For a given train length and number of passengers to be loaded, loading time depends upon the distribution of passengers along the platform, the ratio of total door length to car length and the relative size of movements into and out of the train. The loading time is minimised when the number of passengers using each door is equal. Obtaining a uniform distribution of passengers along the platform is, however, difficult, and passengers tend to locate themselves at points which provide them with a favourable exit upon arriving at their destination, or at points which they feel will be less Crowded than other points.

The location of platform entrances affects the relative loading of each car in the train. If entry to the platform is at its center, the number of passengers in the end cars tends to be less. Similarly, passenger loading in the central area tends to fall off if there is access to the platforms only at the end. If two stairways are provided to access the station platform, they should ideally be located at the quarter points. This arrangement has been largely adopted in all corridors of the Kolkata Metro.

Platform widths depend largely on the maximum number of passengers expected to be on the platform at any one time. Given this number, acceptable loading densities (passenger concentration per sq. meter of platform) can be applied to determine the required width of the platform. The selection of an appropriate loading density for the platforms must be done carefully, keeping in view the nature of movement that takes place on the platform and the distance the passengers walk on the platforms. **Fig. 6.114** includes entry from unpaid area to paid area at concourse and from paid area one must move up to platform using separate stairs/escalators.

I) Length and width of platforms

The length of the platform depends on the length of the train. The length of a 6-coach train is 162.4 meters in the East-West Metro, and allowing a margin for inaccuracy of train stopping, a length of 170 meters has been provided. The width of the platforms, depending on the intensity of traffic, has been provided as follows:

Island-type platforms	-	10 meters standard
	-	8 meters absolute minimum
Side type platform	-	4 meters minimum to be increased to suit the traffic.

The platform levels have been kept equal to the level of the train floor under loaded conditions at 1080mm above the top of the rails. The headroom on the platforms has been kept between 4 to 4.5 meters, so that clearance below the platform signages is 3m.

In calculating the width of the platform, a minimum safety zone of 65 c'mon the train side and 25 cm on the offside has been kept. A column located on a platform is at least 150 cm away from the edge of the platform. In calculating the width of the platform, a density of 1.5 persons per square meter has been considered.

ii) Passageways

This term refers to all means of access to and from station platforms and includes corridors, stairways, and escalators. In determining the desirable capacity and arrangement of passageways, the following figures have been used in the Kolkata Metro to determine widths.

Fig. 6.114: Entry from Unpaid Area to Paid Area

iii) Capacity of Passageways

Table 6:1: Capacity of Passageways

Type of passageway	Capacity per meter of passageway per hour	
Level corridors and platforms	For movement in one direction	4000
	For movement in both directions	3500
Staircases	For the downward journey	3500
	For the upward journey	3200
	For both upward and downward journeys	3000

iv) Entries and exits at stations

The location and position of entries and exits, all of which start and terminate at the road level, and through staircases and corridors lead to the circulating hall (concourse/mezzanine) suit local conditions and needs. All entries and exits are provided with rolling shutters to prevent entry of outsiders into the subway system when the trains do not run during the night.

v) Staircases

In Kolkata Metro, to facilitate the handling of large volumes of passengers with ease and comfort, most stations are provided with escalators. Where escalators are provided, a conventional staircase is also provided. Where the heights of ascents and descents are fewer, only staircases are considered adequate. In some stations, only staircases have been provided for the present, but space has been left in the designs for the installation of escalators in future. In Kolkata Metro, generally, the staircases have been provided with steps of 32 cm tread and 14 cm rise.

VI) Facilities for the guidance of passengers

For the guidance of the passengers, platforms are provided with Train Indication Boards, giving only the destination or the name of the terminal station of the trains.

All platforms, waiting places and mezzanines are provided with loudspeakers so that continuous communication may be maintained between the Train Dispatchers at stations and the passengers. The system allows the Train Despatcher at Central Control to make announcements directly when required. Stand-by battery-operated UPSs have been provided so that in the event of any failure of the power supply, the system is still available, avoiding the confusion that would otherwise be created just at a critical time.

For the guidance of detraining passengers, indicators have been provided showing the direction of exit.

Fig. 6.115: Metro Train at Elevated Metro Station Platform

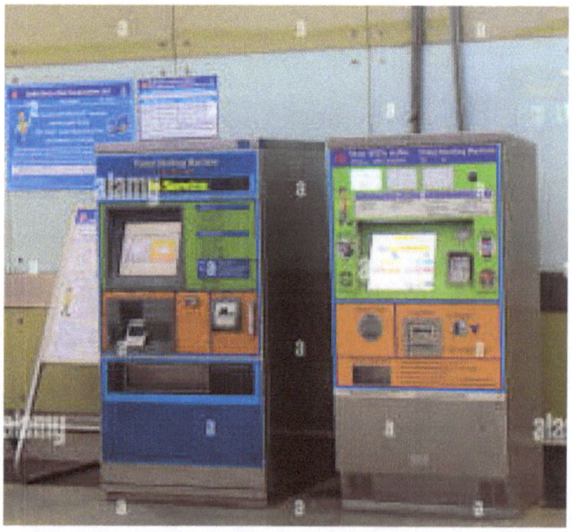

Fig. 6 116: Automatic Ticket Vending Machines

ECS (Environmental Controlled Systems) plant rooms, trackway exhaust, and under-platform supply fans are in the zone towards the end of the station box at the concourse level. Other station facilities, such as S&T (Signaling & Telecom) rooms, SCR (Station Control Room), and staff areas, are located at the concourse level beyond passenger areas.

At least four exits/entries are provided from the station box to the ground, two from each unpaid concourse.

vii) Provision for Truncated Operation

Kolkata Metro has a unique feature; it feeds the CBD from both the Northern & Southern sides of the city. To ensure that during any disruption to metro train services on the northern or southern side, the whole service is not affected, Y-siding has been planned at Maidan and Girish Park Stations. In case of any disruption on the south side beyond Maidan, trains can run between Dum Dum and Maidan, and similarly, in case of a disruption on the north side, trains can run between Tollygunge and Girish Park.

Analysis of Underground Stations and Software used is included in 6.27.

6.16.10 Design and Construction Methodology: Structures Other Than Tunnel

The construction methodology adopted was tunnelling through TBM and stations by cut-and-cover for the underground part, and an elevated viaduct consisting of a prestressed concrete segmental box girder on a single pier with pile foundations.

Metro Corridors, where feasible, for cost considerations, in thickly built areas with narrow roads and accommodating ground EPB (Earth Pressure Balancing) TBM or Slurry TBMS (Tunnel Boring Machine), are being used. Where land and existing structures/facilities create problems, a metro station is being built using the NATM (New Austrian Tunnel Method).

Underground metro corridors are commonly built using shoring walls like Diaphragm Walls (DW), for retaining soil and water, using either top-down or bottom-up approach. For stations, use of same approach is common. Where sufficient land is available, a pre-installed temporary retaining system for water and soil is used, with a cast-in-situ permanent corridor and station.

Construction techniques in Underground Corridors and station boxes are adopted considering parameters like geology, space availability, and excavation depth. Construction of the Box can be done using either bottom-up (as in Kolkata N-S Metro) or top-down (or a combination of both), as in Kolkata East-West (EW) Metro stations, the first Indian Metro under a major river.

6.17 Shoring Wall

Shoring walls have been used in cut-and-cover construction of UG stations. Alternative methods of design & construction of such walls have been discussed in detail in chapter 5. Briefs on these methods have been included for ready reference.

6.17.1 Diaphragm Wall (DW)

The continuous diaphragm wall is formed and cast in a slurry trench. The trench excavation is initially supported by either bentonite or polymer-based slurries that prevent soil incursions into the excavated trench. The term “diaphragm walls” refers to the final condition when the slurry is replaced by Tremie concrete, which acts as a structural system either for temporary excavation support or as part of the permanent structure.

Fig. 6.117: Operations at Surface Level to Start

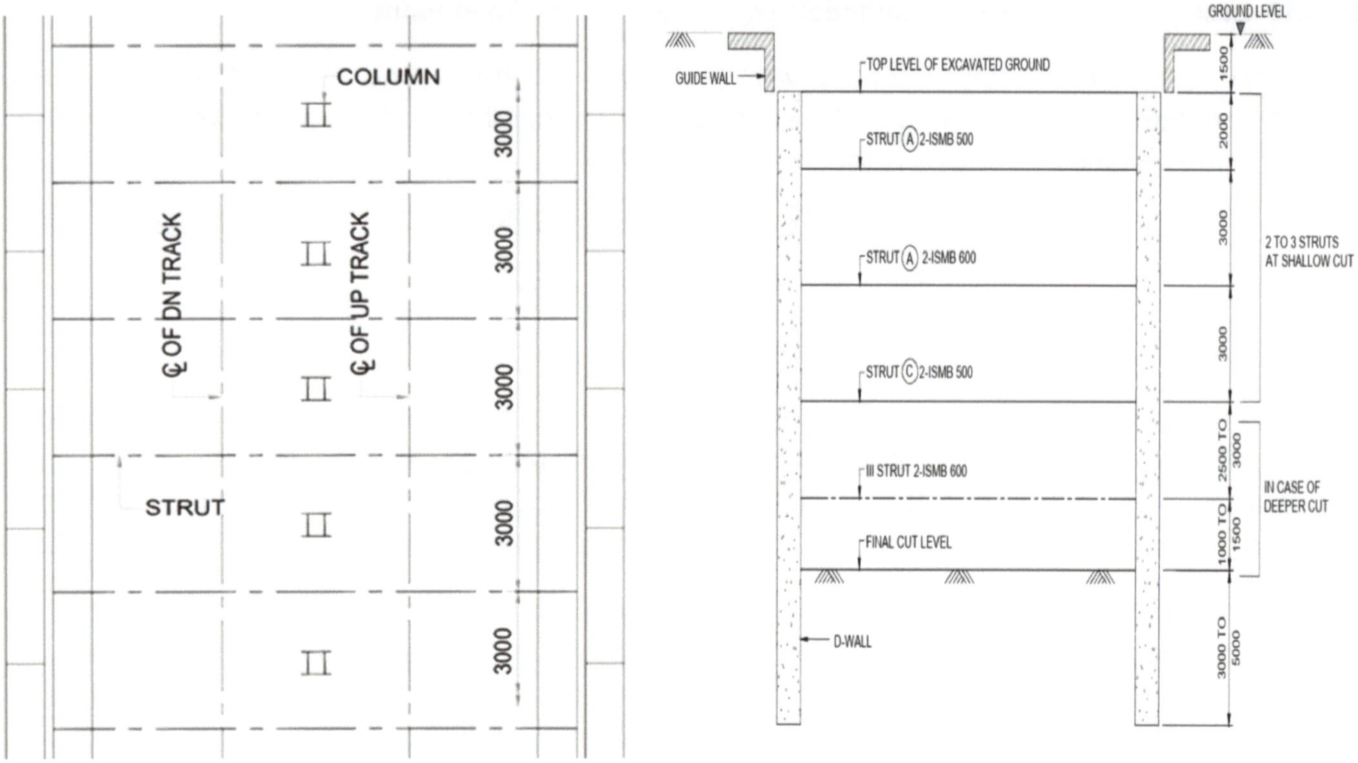

Fig. 6.118 & 6.119: Plan and Section of Strutted Excavation with Diaphragm Wall

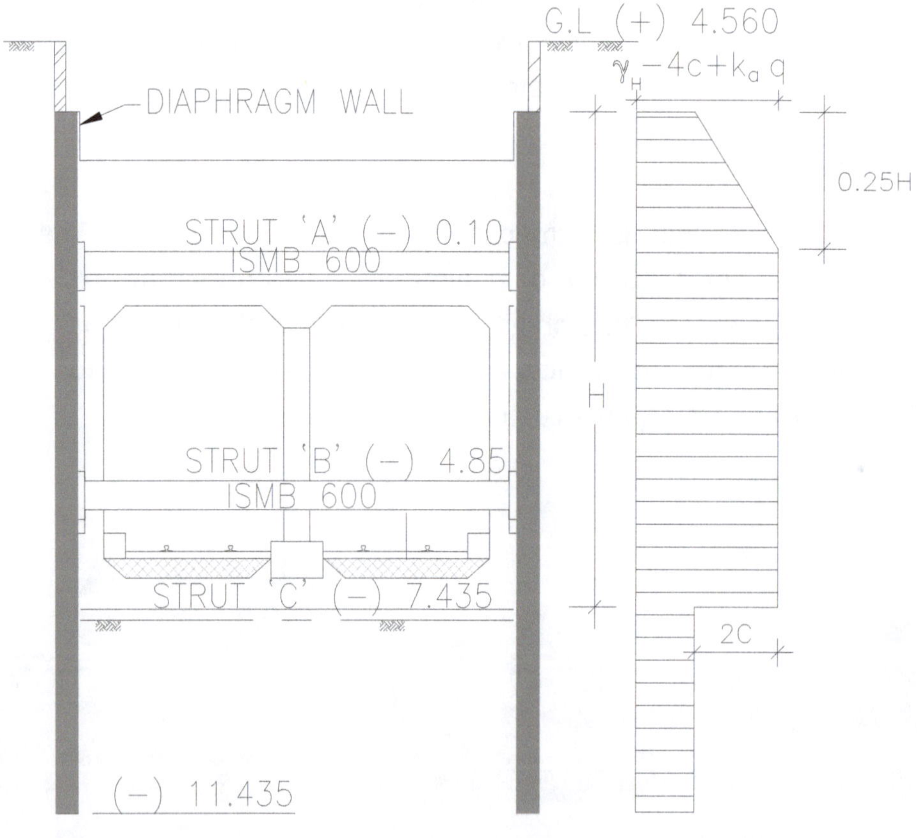

Fig. 6.120: Strutted Excavation with DW and Soil Pressure including Surcharge

Details of activities used in line 2 are included in **Figs 6.117** and **6.120.**

The main advantages of DWs are:

1. Increased wall stiffness compared to sheet piles.
2. Less noisy construction using a hydraulically operated Kelly is less loud than a Mechanical Clamp shell bucket. Both were used in the first Indian metro NS.
3. It can be used as a temporary and permanent wall. As a permanent structure, it can also withstand Vertical loads and earth and water pressure.
4. Provision for future Structures can also be kept, as was done in the First Metro, like the cross tunnel below, to be treated as the foundation of future flyovers in line with the City Master Plan.
5. Watertight, lately, with an improved design approach, leaking through joints has reduced considerably. In the seventies, steel tubes used at the junction of two contiguous DWs were now specially made watertight joints with neoprene membranes between two specially made male/female forms of steel sections are being used).

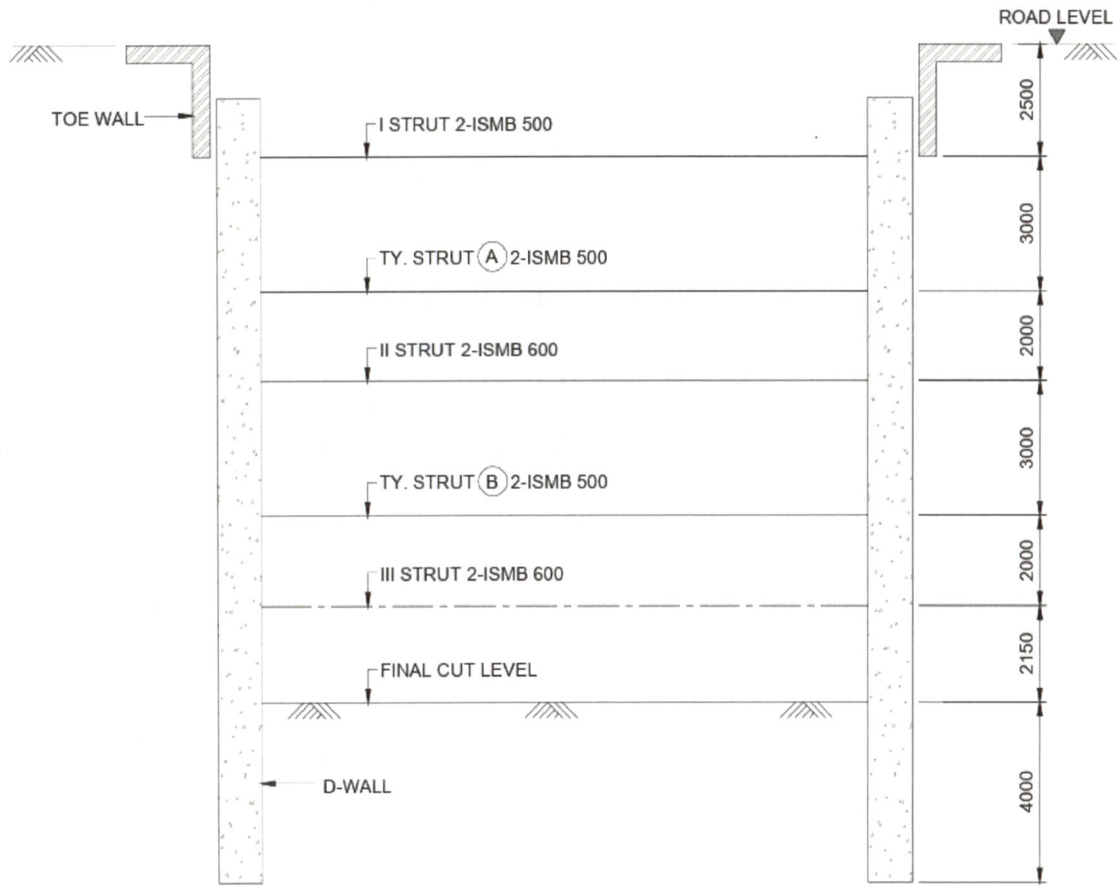

Fig. 6.121: Strutted Excavation for Deep Cut

The main disadvantages of DW walls are:

1. Requires special equipment.
2. Complete water tightness at the junction of the contiguous panels is difficult to achieve.

For deep excavations number of struts have increased, **(Fig. 6.121)** not to exceed 3.5 m maximum for safety and cost consideration.

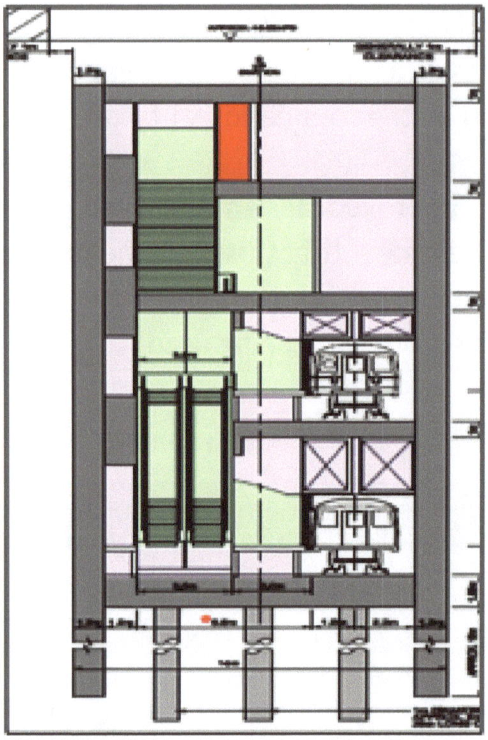

Fig. 6.14: Stacked Station-Proposed for Limited ROW (Repeat)

At the Central Station, the East-West metro original alignment would have intersected the existing North-South metro corridor and provided an interchange between the two underground metro corridors. The space constraints at the Central Station and mined tunnelling (NATM) below the N-S line were a challenge. The alternate proposal evolved to work on a possible interchange using each track at a different level to eliminate Land Acquisition. Implementing the design with two tracks warranted the temporary acquisition of a few structures on one side; the acquisition was in the advanced stage. However, finally the corridor had to be rerouted via Esplanade. This decision on alternate alignment substantially delayed the completion and increased the cost.

6.17.2 Secant Pile

A second pile wall can be a temporary or permanent structure in underground works. Secant pile walls are formed by constructing intersecting reinforced concrete piles. The piles are reinforced with either steel rebar or steel beams and are built by drilling through strata (can be a combination of RCC & PCC piles in case of temporary shoring walls). Primary piles are installed first, and secondary piles are constructed between primary piles once the latter have gained the required strength. Pile overlaps typically depend on the verticality tolerance rig and the pile depth.

The main advantages of secant pile walls are:

1. Increased alignment flexibility.
2. Increased wall stiffness compared to sheet piles.
3. Can be installed on difficult ground (cobbles/boulders).
4. Less noisy construction.
5. No need to install timber lagging.
6. Reduced water leakage.

The main disadvantages of secant pile walls are:

7. Verticality tolerances are hard to achieve for deep piles.
8. Total waterproofing is very difficult to obtain.
9. Increased cost compared to sheet pile walls.

6.17.3 Contiguous Pile

Contiguous pile walls are formed by constructing a line of circular reinforced concrete piles with a typical gap per design requirements. They may be built using segmental temporary casing, conventional heavy-duty rotary drilling tools, or continuous flight auger techniques. If required, a reinforced concrete guide wall can be constructed at the piling platform level to aid positional and vertical tolerances.

Concrete piles are constructed with full-length reinforcement cages.

In the tank area, the shoring wall was formed on the embankment side with a 16-m-long contiguous timber pile on one side. Local material was used with a joint in the middle (N-S Metro Kolkata) included in chapter 5.

The main advantages of Contiguous bored pile walls are:

1. Plan footprint flexibility.
2. Load-bearing capacity.
3. It can be installed on difficult ground.
4. Economy compared to Diaphragm and Secant pile walls.

The main disadvantages of bore pile walls are:

5. Verticality tolerances may be hard to achieve for deep piles.
6. Waterproofing is challenging to obtain.
7. Increased costs compared to the sheet pile wall.

6.17.4 Soldier Pile (Berlin Wall)

Soldier piles and lagging walls are some of the oldest retaining systems used primarily in shallow excavations. Soldier pile walls have successfully been used since the late 18th century in metropolitan cities like New York, Berlin, and London. The method is also known as the "Berlin Wall", when steel piles and timber lagging are used. In N-S Metro, the top 3m, soldier piles and timber lagging are generally used. Alternatively, caissons, circular pipes, or concrete piles can be used as soldier piles at an increased cost. Timber lagging is typically used, although reinforced concrete panels can be used.

Soldier pile walls are formed by constructing soldier piles (rolled I Beam used in N-S Metro) at regular intervals, excavating in small stages, installing lagging, and backfilling and compacting the void space behind the lagging. The soldier piles provide moment resistance in soldier piles and lagging walls. Passive soil resistance is obtained by embedding the soldier piles beneath the excavation grade. The lagging behind retains soil across piles and transfers the lateral load to the soldier pile system. Soldier pile and lagging walls are the most inexpensive systems compared to other retaining walls. It is also elementary and fast to construct.

The significant advantages of soldier pile walls are:

1. Fast to construct.
2. Cheaper when compared to other systems.
3. Installation is versatile, and adjustments can easily be made in the field to accommodate changes.
4. Lagging construction can be rapid.
5. Construction of walls does not require very advanced construction techniques.
6. Common lagging materials include timber, shotcrete, precast concrete panels, or steel plating.

Permanent transportation walls usually utilise precast concrete panels, while temporary soldier pile walls utilise timber lagging.

The primary significant disadvantages of soldier pile and lagging systems are:

1. They are primarily limited to temporary construction.
2. Cannot be used in high water table conditions without extensive dewatering.
3. Control base soil movements.
4. Poor backfilling and associated retaining systems. Ground losses can result in significant surface settlements.
5. They are not as stiff as others.

6.17.5 Sheet Piles

Interlocked sheet piles form a wall for permanent or temporary lateral earth support with reduced groundwater inflow, which is appropriate; anchors can be introduced to provide additional lateral support if required to withstand high pressure. Sheet piling is generally used in wet ground, but driving noise restricts its use in urban areas. The concept is to design narrow, interlocking sheets that can be connected and driven into the ground to form a wall. The shape and materials of the sheets define stability and strength. Steel is supposed to be the most appropriate material if the requirement is to withstand significant bending moments and pressure.

Major Advantages are:

a. Light to easy lifting and handling.
b. Sheeting is reusable and recyclable.
c. Long life for both under and above water.
d. Length is adaptable.
e. Stronger joints to withstand pressure.

The disadvantages are:

a. Most challenging to install in rock or boulder beds.
b. Driving causes noise and vibration problems in neighbouring areas.

For N-S Metro, the concept was prepared by Russian Engineers and sheet piles were proposed for shoring. During that period in India, the available length of sheet piles was 7 to 8 meters, and the section modulus was not good enough to use in medium to deep excavation. The appropriate length was proposed to be imported with the required section modulus. However, Indian engineers (Dr. BC Roy, CPPL for submitting the tender in the Maidan section) replaced the proposed imported sheet piles with DW to save on costs and foreign exchange.

6.17.6 Ground Anchors

Ground anchors are used to arrest the high stresses due to ground and water pressure. Ground anchors are usually considered as temporary system in place of Struts within the excavated trench. Temporary ground anchors are applied carrying high tensile forces onto the load bearing soil stratum for a limited period (say up to 2 years). A temporary ground anchor comprises root length, rod (free anchor length) and anchorage. Temporary anchors are primarily applied for:

*Support various temporary retaining walls, including sheet pile walls, secant piles, and diaphragm walls, designed as support systems for deep excavations.

*Supporting various types of permanent excavation support systems/construction. Elements are used for anchoring of bottom plates/slabs against uplift, stay anchoring of walls, test loading systems, etc. Ground anchors' primary advantage is that it provides unobstructed excavation within the area to be excavated like Station. However, in many cases, the theoretical bond between ground and soil mass is much greater than the actual bond value, which depends on the workmanship quality.

Fig. 6.122: Excavation with DW & Struts

Fig. 6.123: Holding of Existing Utility During Top-Down Excavation

6.17.7 Top-down construction: DW (Diaphragm Wall)

As of today, designers are commonly using FEM analysis and for which recently developed programs are available. For different analytical tools, the word "tools" must be emphasised here, they are widely used worldwide. The programs used are highly effective and efficient and Softwares are like STAAD, Etab, Plaxis, Midas, Lusas.

As a result, project execution is carried out in a highly condensed timeframe.

To ensure the integrity of the design, a detailed Instrumentation program was adopted; forces developed at strut ends were measured and tallied with calculated values.

The design considers provisions adopted in various codes in different countries, such as British Standards, Indian Standards, AASHTO, and Euro Codes.

The recent East-West (EW) Metro, the first metro in India to cross a significant water body, Hooghly River, started work in 2009. Underground metro stations adopted both Top-Down method and Bottom-Up method for crossover boxes and access shafts. A permanent diaphragm wall earth retaining system was part of the total design (Integrated). The EW metro corridor was designed and built using an earth pressure balancing Tunnel Boring Machine (TBM).

For EW Metro, the design is considered a rigid box subject to earth pressure at rest and other loadings. The underground stations on each riverside are deeper than the other stations.

A top-down construction technique is adopted where the geology is sand, soil, and soft material available up to substantial depth. To optimise construction costs and increase the execution rate, the permanent wall of the station box is also designed as temporary earth retaining. DW is usually designed for active pressure during the construction stages and at-rest pressure for the permanent stage, including water pressure for both construction and permanent stages.

DW is designed for soil pressure in a few Metro projects, i.e., active and at rest, with weep holes to avoid pore water pressure on DW. To retain the water pressure, separate the skin wall with a wall built with bottom-up construction. This construction methodology can be adopted where the water table is significantly below the ground level, and the station is proposed in a green field area. No significant water ingress is expected during the construction stage in the station excavation. Otherwise, it may lead to settlement in the nearby areas.

Construction stages for top-down execution are as given below:

- Diversion of traffic from the construction site.
- Installation of DW on two sides of the station box.
- Excavation up to the roof slab may be performed using strutted excavation (like soldier pile and timber lagging), which is dependent on the level of the roof slab.
- Divert the traffic within the station box area to normalise traffic movement.
- Excavate till the concourse and base slab and cast as per design.

Significant advantages of top-down construction are that one can avoid completely temporary work, i.e., a shoring system (except where the topsoil is of poor quality), that traffic can be placed before excavation after the casting of the roof slab, and that the road can be reinstated.

6.18 Tunnel Section

6.18.1 Track Alignment

Track alignment involves selecting the optimal three-dimensional path for the tunnels, considering factors such as the type of rolling stock, passenger comfort, maximum gradients, and geological conditions. The maximum allowable gradients and curves are constrained by the performance capabilities of the rolling stock and the need to provide a comfortable ride for passengers. The track alignment is designed to connect between stations while avoiding physical clashes with underground structures and utilities. Optimising track alignments to balance costs with operational and user requirements throughout the system's life cycle is crucial for a sustainable metro line. The critical aspects of metro design are interdependent. Track alignment is the primary driver, influencing the other factors. However, track alignment decisions are closely linked to station typology.

6.18.2 Below River

The tunnel alignment for the E-W Metro was finalised based on the following guiding factors below the river Hooghly:

a. Thickness of the overburden layer—It is often desirable to optimise a tunnel alignment by minimising the thickness of the overburden above the tunnel to reduce the length of tunnel approaches and ground loads on the tunnel.
b. For the tunnels in soft ground, a minimum clearance between the channel bottom and the top of the tunnel of 1.0 to 1.5 times the tunnel's diameter is commonly used.

c. Geologic conditions beneath the channel – It is critical to consider the risk carefully of encountering an adverse or unanticipated condition, such as an undetected fault zone, water-bearing feature, or geologic weakness zone.

   Clearance from existing shoreline structures-- The alignment is influenced by the proximity and arrangement of the shoreline structures, the type of foundations, and the potential for adverse impacts from tunnel excavation.

d. Other technical and non-technical factors include:

   * Subsurface conditions,

   * Schedule of dimensions of the project

   * Maximum grade requirements for the TBM,

   * Floatation and heaving

   * Scour depth of channel

   * Future channel improvements like dredging.

6.18.3 Challenges in Tunnelling Under Open Water

Tunnelling below open water is a significant engineering challenge as Colonel Henry Goslee Prout, a legendary figure in American history of Civil Engineering, made the following remark way back in 1898,

"Of all engineering work, that which is least certain is driving tunnels under rivers or other bodies of water."

Planning and executing open-water tunnels require coordinating multiple interrelated engineering disciplines, as the challenges increase manifold crossing such stretches. The most significant uncertainty for open-water tunnel projects is in the nature and characterisation of the ground where the tunnel is to be constructed. The success of an open-water tunnel project is dependent on the following fundamental principles:

i. Consideration of the character of the ground and its behaviour in the selection of the tunnel alignment and the excavation technique
ii. Optimisation of the alignment to pass through the impermeable ground to minimise difficulties from groundwater and potentially flowing ground changes in ground behaviour, are likely to occur.
iii. Planning for difficult and infrequent access to the TBM cutter head as tunnels under open water, opportunities for access from above the tunnel are complex and limited.

The great success of transportation tunnelling under open water bodies, which have connected cities and countries across water bodies, has been witnessed in many parts of the world, starting from the famous tunnel under the River Thames in 1843. Still, such a precedent was not available in India before. Therefore, the planning and execution of the Metro twin tunnel under the River Ganges at Kolkata has undergone sustained endeavour and demanded a high degree of precaution and care.

6.18.4 Horizontal Alignment of Below River

The horizontal alignment of the crossing was selected with the following considerations:

a. Jetty structure - It was ensured that the alignment does not infringe on any jetty structure or shoreline.
b. Location of Howrah Station in West Bank—The acquisition of land and the constructability of Howrah Station were significant factors, as the proposed location is inside an existing train terminal. A comparatively vacant area between platforms 15 and 16 was selected for the station construction by shifting some non-core activity centres of the existing station.
c. Location of Ventilation Shaft in East Bank - given the requirement of an intermediate emergency evacuation cum ventilation shaft at the east bank, locating a comparatively free acquirable area was required. A sizable area of Kolkata Port Trust was chosen, which contained some dilapidated warehouses, and this location also guided the alignment.
d. Regulation for the protection of Rabindra Setu (Bridge)—The proposed tunnels are about 320 m away from Rabindra Setu (Howrah Bridge) to avoid affecting the bridge ramps.
e. Horizontal Curve – No horizontal curve is provided for easy manoeuvring of TBM across the river.

6.18.5 TBM Assembly

East-west metro ug corridor was divided into two sections UG1 and UG2. UG1 was implemented by Afcons and UG2 by ITD Cementation. TBM assembly included in Fig. 6.124 is from AFCON's Presentation.

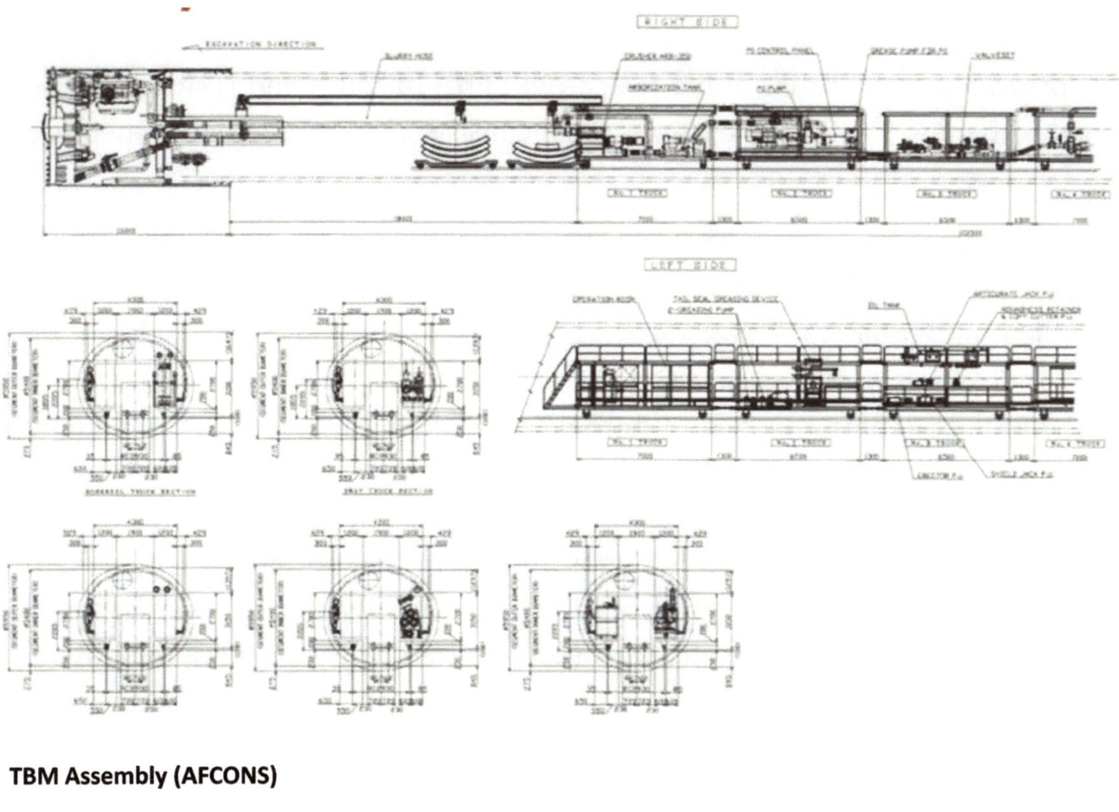

Fig. 6.124: TBM Assembly (AFCONS)

6.18.6 Vertical Alignment

The vertical profile of the tunnel under the river Ganges was a matter of paramount importance, and a study of hydrological parameters reveals the following:

a. Hooghly is a tidal river with a progressive type of tidal wave. From the sea mouth, the mean range of tide increases from 10 ft to 12 ft, and it falls again to 10 ft at Garden Reach 42 miles up and decreases rapidly on the upstream side
b. Bores occur in the Hooghly River, manifested as a standing wave about 5 feet high. The bores are strongest during Perigee Spring tides in the dry season when the river level is lowest and the depths are shallow.
c. As per reports available from Kolkata Port Trust, the highest scouring of the river at the location of the tunnel from 1988 to 2003 was in the order of 4.5m, while the highest silting was in the order of 6.5m, and the maximum velocity was in the order of 2.5m/second. In this stretch encompassing 500m upstream and downstream of the proposed river crossing, the flow hugs the right bank, i.e., the Howrah side, with stable deep contours is within a short distance from the high bank.
d. According to a survey conducted in February 2004, a maximum water depth of 17m was observed 90 meters from the right bank (Howrah side). The river's gradient is very flat. The deepest scoured riverbed at the tunnel's alignment is 23.865 m below M.S.L, per Kolkata Port Trust's report.
e. Before preparation of the Detailed Project Report, an initial feasibility study for this tunnelling was done by the Japan External Trade Organisation (JETRO), and the lowest rail level suggested was 28.74 m below MSL

The city is spread along the banks of the river from north to south and is situated predominantly on reclaimed wetlands. The Gangetic alluvial plain consists of several hundred meters of alluvium deposits.

A survey conducted in February 2004 showed a maximum water depth of 17m 90 meters from the right bank (Howrah side). The river's very flat gradient may be considered nil. The deepest Changes in Riverbed Profiles occurred, riverbed at the alignment of the tunnel, as per the report of Kolkata Port Trust, is stated to be -23.865m below M.S.L, as per the study in 2004. Before preparing the Project Report, the Japan External Trade Organisation (JETRO) conducted an initial feasibility study for this tunnelling. The lowest rail level suggested was 28.74 m below MSL. Further bathymetric survey was conducted in the DPR Stage in 2006 and after the award of the construction tender in 2010, and some changes in the riverbed profile were observed, as shown in **Fig. 6.125.** As per the latest survey, there is a tendency for silting on the right bank, and the deep channel is shifting towards the left bank (Kolkata Side) with a maximum depth of 14.434m at 280 meters from the left bank.

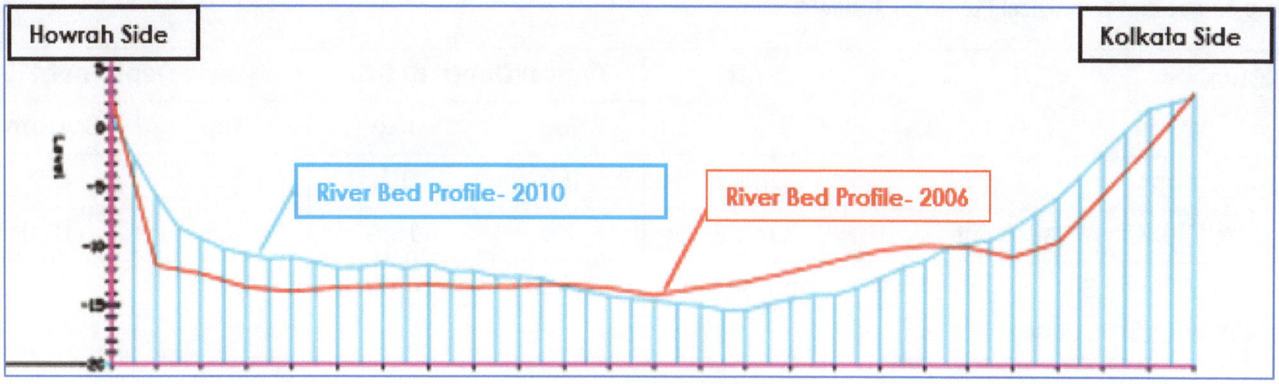

Fig. 6.125: Changes in Riverbed of Hooghly – Bathymetric Survey (Between 2006 and 2010)

6.19 Geotechnical Investigation

Kolkata soil primarily has alluvial silts, clays, sands and peat deposits. A meticulous geotechnical investigation was required for an underground construction. The geotechnical investigation was done along the tunnel alignment, mostly in 50m intervals, using land cable percussion in the land portion and marine cable percussion in the riverine section. Geotechnical investigations were done before the construction contract award, and confirmatory boreholes were made after the contract award. Geotechnical profile determined through investigation. And the engineering characteristics of subsoil in the tunnel zone are presented in Table 6.2.

The geotechnical investigation for the Kolkata East-West Metro was done with boring using a floating raft/boat for the river section. Given the waterproof capping layer of clayey subsoil in the upper range, it was essential to plug the boreholes to avoid water infiltration in the lower layers, which may be detrimental during tunnelling. This plugging was done concurrently with cement slurry when casing pipes were withdrawn for geotechnical boring. Details of boreholes inside the river are presented in Table 6.2.

Table 6.2: Boreholes for Geotechnical investigation in River Hooghly

DPR Stage				Post Tender Stage			
Total Boreholes	**Markings**	**Depth [MSL]**	**Depth below the riverbed**	**Total Boreholes**	**Markings**	**Depth [MSL]**	**Depth below the riverbed**
2	BH14/R14	-30m	20.45m	4	TTAMBH-1	-40m	30.45m
	BH15/R15	-30m	20.45m		TTAMBH-2	-41m	30.45m
					TTAMBH-3	-42m	30.45m
					TTAMBH-4	-39m	30.45m

The General Geological profile, as revealed through geotechnical investigation, is presented in Table 6.3:

Table.6.3: General Geological Profile for Tunnel Sections

Stratum	Strata	Typical Depth m BGL		Typical Depth m MSL	
		Top	Bottom	Top	Bottom
Made Ground	Unit 1	0.0	1.0	5.5	4.5
Organic Silty Clay/Clayey Silt	Unit 2	1.0	15.5	4.5	-10.0
Silty Clay/Clayey Silt (Locally with little sand)	Unit 3a/3b	15.5	45.5	-10.0	-40.0

Geology is relatively consistent across the site; the nature of alluvium deposited by windering river courses and the formation of channel fills, scrolls and levees has resulted in laterally discontinuous layers below 15m depth. The upper soft alluvium forms a consistent layer across the site, presumably made up of overbank silts and clays from the river Hooghly. This has recently been covered in a made-ground layer during reclamation activities. Variations can occur in Unit 3, where laterally discontinuous/continuous sandy silt (Unit 3b) was encountered. These sand silt deposits were probably related to the old river channel fill or meander scrolls. Elaborate laboratory tests are conducted on soil samples obtained through geotechnical investigations. The Engineering Characteristics of Soil in the River Section are shown in Table 6.4.

Table.6.4: Engineering Characteristics of Soil in River Section

Soil Stratum	Nature of Soil	Depth	Tunnel Alignment Section	Unit Weight (y)	N Value	Poisson's Ratio (ʋ)	Angle of Friction (ϕ')	Cohesion (c')	Undrained Shear Strength (Cu)	Drained Young's Modulus (E')	Undrained Youngs Modulus (Eu)	Earth Pressure at Rest (Ko)
				kN/m³			Deg	kPa	kPa	MPa	MPa	
Unit 1	Made Ground	5.5m mSL to 4.5m mSL	All	19		0.3	28°	0	n/a	12	n/a	0.53
Unit 3a	Firm to stiff clay Silt	-10 mSL to -40 mSL	Section 2 Hooghly River	19	25-31	0.2	30° 32° (below – 24m MSL)	0	(z+7.8814)/ -0.0847	(z+6.875)/ - 0.3125	1.25*E'	0.56
Unit 3b	Sandy Silt / Silty Sand		Section 2 Hooghly River	19	10-42	0.2	30° 32° (below – 24m MSL)	0	n/a	(z+6.875)/ -0.3125	n/a	0.56

6.20 Tunnel Boring Machine (TBM)

6.20.1 Selection

The tunnelling for Phase-II of Kolkata East-West Metro had been considered a continuous drive from Howrah Maidan to Esplanade (Distance of 3.8 Km), out of which the river crossing was a small stretch of 520m. Therefore, the selection of the Tunnel Boring Machine encompasses the requirement of the entire tunnelling scope and is not only specific to the river crossing. The soil profile encountered in the tunnelling horizon for the whole of the tunnelling scope is shown in **Fig. 6.126.**

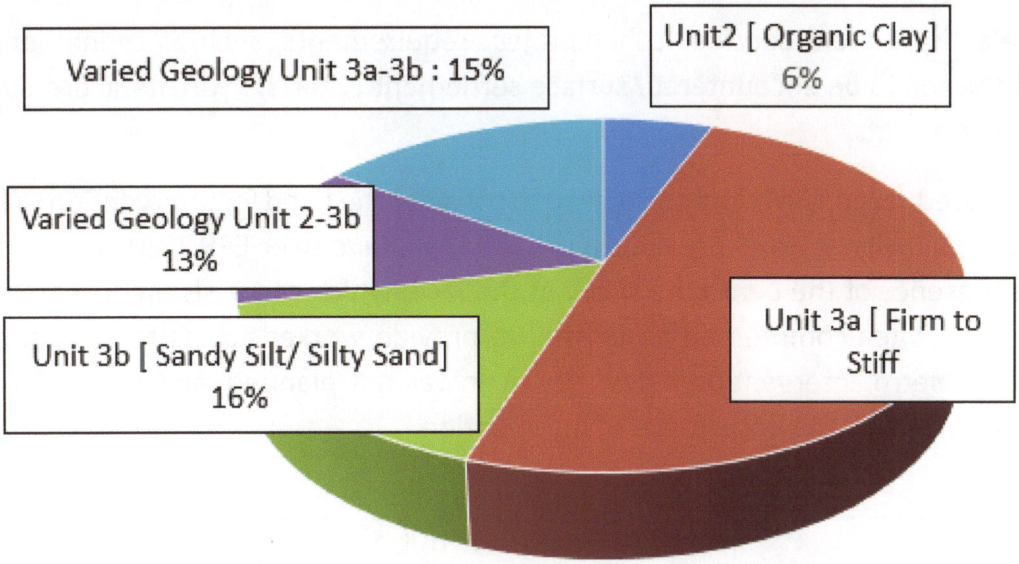

Fig. 6.126: Distribution of Tunnel Geology.

It was observed that 50% of the stretch of the tunnel passes through firm to stiff clayey silt. Therefore, TBM was chosen as a soft ground machine. Closed-face TBMs are effectively sealed from the whole ground & hydrostatic pressures. Those are usually demanded for soft ground tunnels beneath open water due to their ability to maintain control of the ground, particularly at the face of the excavation and their effectiveness at minimising ground loss and reducing the risk of developing a connection to the overlying water body. These TBMs provide positive support to the ground and minimise the potential for problematic groundwater inflows into the tunnel excavation. Closed-faced TBMs also offer a stable and safe underground work environment suitable for varying ground conditions. Two main types of closed-face TBMs exist: Earth Pressure Balance [EPB] and Slurry (mix shield) TBMS.

EPB Machines provide positive face pressure by allowing the excavation chamber to fill with soil behind the bulkhead, minimising the risk of ground loss at the excavation face. EPB Machines are essentially designed to work in predominantly cohesive soils where excavated material can readily form a plug in the system's extrusion screw conveyor. This design feature might be crucial for tunnels beneath open water with high groundwater pressures or risk of heterogeneous geology, since the compressed air buffer can immediately react to pressure changes.

Typical soil gradations for slurry and EPB TBMs are represented in **Fig. 6.127** (*R. L. Moncrieff*) with guideline values or zones that are only generic. The final selection of an appropriate closed-face TBM also depends upon several factors, including grain size distribution of the soil, presence of gas, presence of boulders, hydrostatic pressures, the availability of a mining site, spoil disposal and power requirements.

The initial choice of the closed face TBM should be guided by the degree of heterogeneity of the soils and their respective grain size distributions (**Fig. 6.127)** and provides a simplified flow chart of the selection process for closed-face TBMs based on anticipated soil types. (Pennington *Thomas W*)

The final TBM selection should consider other project requirements, such as tunnel length, the heterogeneity of the soil to be encountered, surface settlement criteria, hydrostatic pressures, and schedule/budget constraints.

The selection of a closed-faced TBM would also need to consider the need for intervention to facilitate maintenance and repair. Slurry TBMS provides a unique advantage over EPB TBMs in that they are enhanced by the presence of the bentonite slurry at the face, which can assist in developing a seal of the excavated face when compressed air is used to provide worker access to the cutter head. However, in Kolkata Metro, intervention below the river was not planned, and failproof protocols were established to minimise the occurrence of intervention.

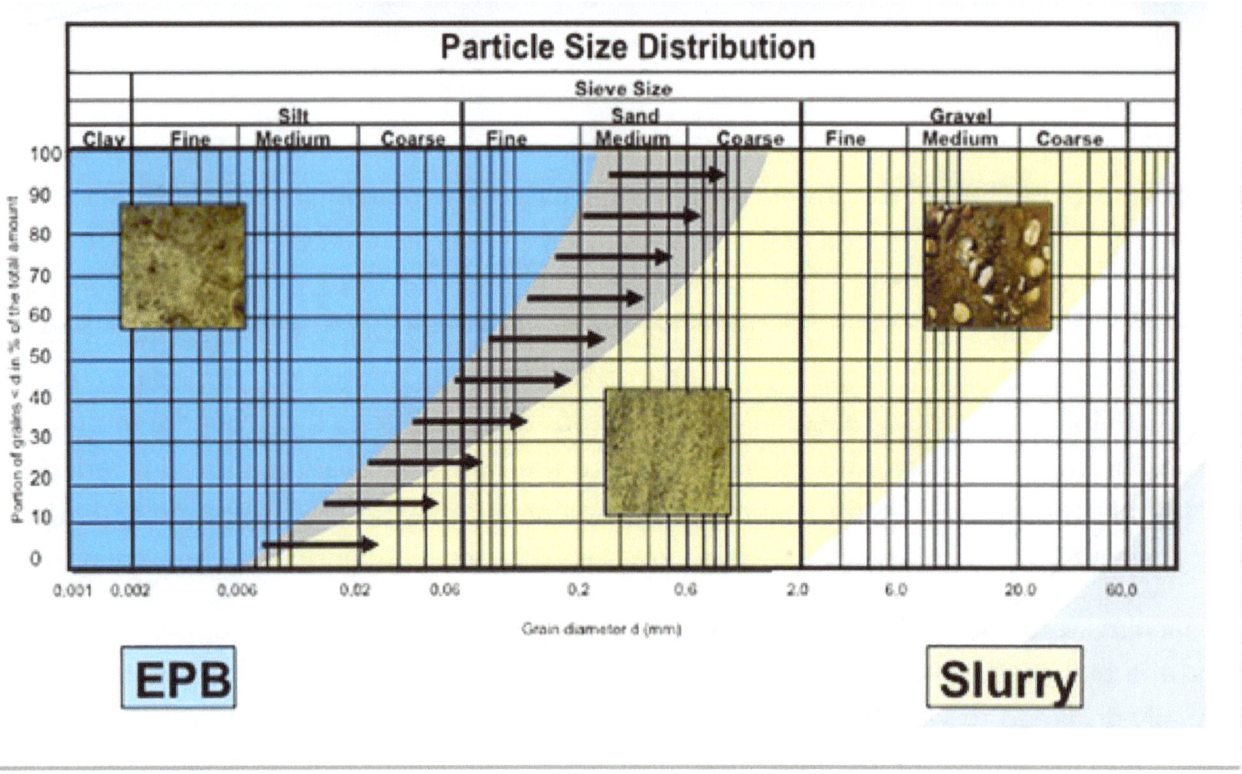

Fig. 6.127: Particle Size Distribution

The face pressure of about 3 Bar below the river played a pivotal role in selecting the machine's capacity. Earth pressure balancing Tunnel Boring Machine (Make: Herrenknecht of Germany) was chosen for this part of the project, which was used 1st time in eastern India for the Kolkata East-West Metro project. The boring tunnel machines used are designed for the requirement of the entire tunnel drive from Howrah Maidan to Esplanade, where the following geometric requirements persist as put up in Table 6.5.

Table 6.5: Geometric Requirements of Tunnel

Sl	Geometric/ Engineering Attribute	Requirement	Remarks
1	Single-Tunnel drive	3.8 Km	As cutter face intervention is a hazardous and non-recommended activity in a city environment, hard facing of the shield reduces the chance of cutter tools being worn.
2	Minimum Horizontal Curve	225 m	The machine is rated for 200m with articulation
3	Maximum vertical gradient	3%	

6.20.2 Tunnel Profile

- Length of Tunnel: 2 X 2962 m (Internal Dia. 5.4 m, Twin Bored Tunnel, except station lengths.

a. **TBM Data**

- Machine Type: Earth Pressure Balance Machine
- Power installed: 2100 kVA.
- Length TBM + Backup : approx. 95 m
- Working Pressure (at axis) : 4.5 bars
- Upward Gradient (Max) : 3%
- Downward Gradient (Max) : 3%
- Overburden (min/max) : 12m/38m.
- Inner ring diameter : 5550 mm
- Outer ring diameter : 6100 mm
- Segment Length : 1400 mm.
- Ring arrangement : 5+1.
- Segment weight (max) : 3.7 t.

b) TBM Progress:

- Segment production was 6 rings per day from January 2012 to April 2012 and 15 rings per day from May 2012 to March 2013.
- TBM was launched at HOM Cross over and pass-through Howrah Station using sacrificial segments.
- Considering the delay in design and land handover, the TBM passed through Mahakaran using sacrificial segments and the breakthrough Central Station.
- TBM progress rate was 6 meters (4 segments) per day per TBM in the beginning for the first one month and then 11 meters per day per TBM (8 segments) for the remaining period.

- There was a one-month lag between the operation of both TBMs so that there is at least a 100-meter distance between the two TBMs.

Fig. 6.128: Tunnel during Construction below River Hooghly

6.20.3 Flotation and Heaving

The important criteria for finalisation of the minimum cushion are the floating and heaving of the tunnel. Where tunnels are at a shallow depth, they should be checked for flotation due to differential water pressure as follows:

Overall safety factor = R/U ≥ 1.2 where:

R: Restraining force calculated by weight and shear resistance of soil

U: Net uplift force considering the weight of the tunnel in deriving the restraining and uplift force,

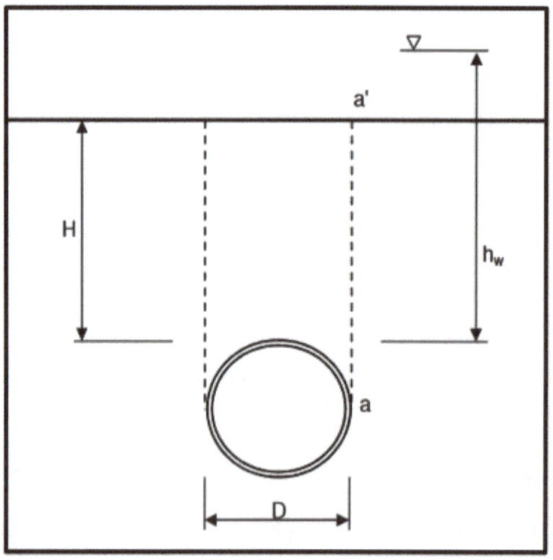

Fig. 6.129: Schematic Diagram for Floatation Check

The factors of safety for different soil covers are calculated below in Table.6.6.

Table.6.6: Comparative Factor of Safety with Different Values of Soil Cushion

Soil Cover [m]	Factor of Safety	Soil Cover [m]	Factor of Safety
14	3.93	6	1.787
12.4	3.50	5	1.518
10	2.86	4	1.25
8	2.32		

The tunnel's vertical alignment was finalised with a 3 % vertical grade (permissible limit) based on a bathymetrical survey done in 2010. The critical cushion kept is (15.855-4.5) = 11.355 m per maximum scour as reported by Kolkata Port Trust, as 4.5 m provides an adequate safety factor.

6.20.4 Sufficient Cushion Considering Shipping Route

Hooghly is a perennial river that is part of the Ganga-Brahmaputra River system. It has sufficient width and draught and connects numerous towns and cities in the northern and eastern parts of India. The river is vital for navigation and inland transportation. Tunnels beneath such open water courses are often subject to long-term operational risks due to activities from third parties. Future alterations to the river flow regime, such as dredging and channel navigation improvements, should be checked in the planning stage. In addition to alignment considerations, the tunnel lining design must consider the full range of loadings anticipated from these future developments if applicable. Impacts from the ship's anchors may need to be considered for design as they may impact tunnel construction (in the form of subsurface obstructions) or tunnel performance (resulting from increased impact load on tunnel linings. Modern ship anchors can weigh upwards of 20 tons and potentially penetrate significant depths below the bottom of the channel if soil conditions are soft.

Fortunately, the riverine port of Kolkata is located on the seaside of the tunnel alignment; therefore, there is hardly any effect on the significant ship route. There is inland transportation in the river for ferry passengers and goods in the form of steamers and barges. Information regarding the project was transferred to the Port Trust to confirm that there would be no effect on the shipping route, which had been ensured. A series of meetings had been organised with the Port Authority, which oversees the river from Farakka upstream to Sand Heads in the estuary. The Port Authority cleared up the tunnelling scheme.

6.20.5 Tunnel Alignment Below River Hooghly

The tunnel alignment below the river includes **Figure.6.130**.

Width of River Crossing: 520 meters, Internal diameter of the tunnels: 5.55 m

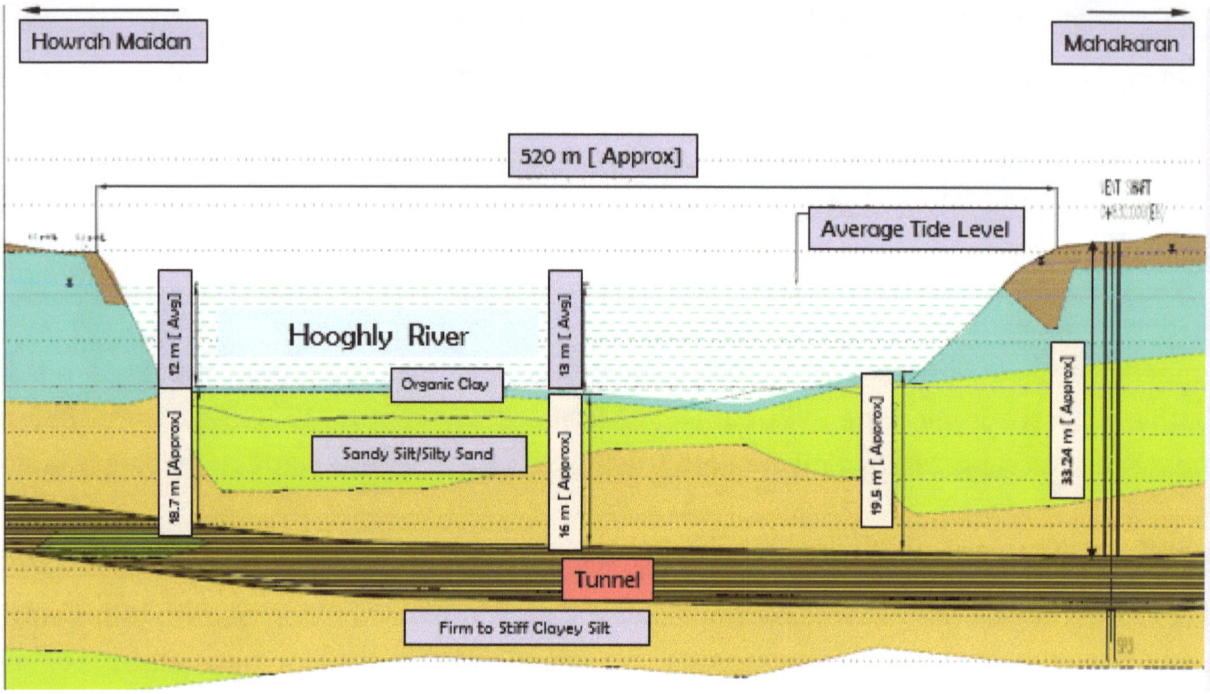

Fig. 6.130: Location of Twin Tunnels with C/C Distance of 16.1m Crossing the River

Spacing between tunnels: 16.1 meters center to center

Depth of Tunnel: 13 m water depth plus the depth of the tunnel crown extra dose of 13 m below the riverbed

Maximum Tunnel Gradient: 3% (Approx)

Nature of Soil: Predominantly Stiff Clayey Silt in tunnelling horizon.

The tunnel is set up using 275 mm-thick segmental precast reinforced concrete liners of M50 Grade, erected in position using the Tunnel Boring Machine. The complete circular tunnel has six segments, including one key segment.

The maximum face pressure required by the machine during the tunnel's execution below the river is approximately three times atmospheric pressure. The tunnel liners were designed adequately to withstand the forces beneath the river during construction and the O&M Stage.

6.20.6 Tunnelling Near Writer's Building

Writers' Building is the most prominent and iconic structure in Kolkata. Designed by Thomas Lyon in 1777, it was the first three-story building in eastern India. Subsequently, there were several additions. The building was constructed as a service residence of the 'writers' of the East India Company, headquarters of the British Raj. In 1821, adding a 128-foot-long veranda with 32-foot-high circular columns in the southern part imparted a Greco-Roman look with an entrance in the central bay and was further facelifted in 1883 with the terrace lined with spectacular statues. After independence, this building was the administrative headquarters of West Bengal. It is a brick masonry structure with load-bearing walls and a roof on each floor. It is constructed with lime concrete supporting a steel frame. A few years back, a significant structural modification was undertaken, and the offices were shifted to another building.

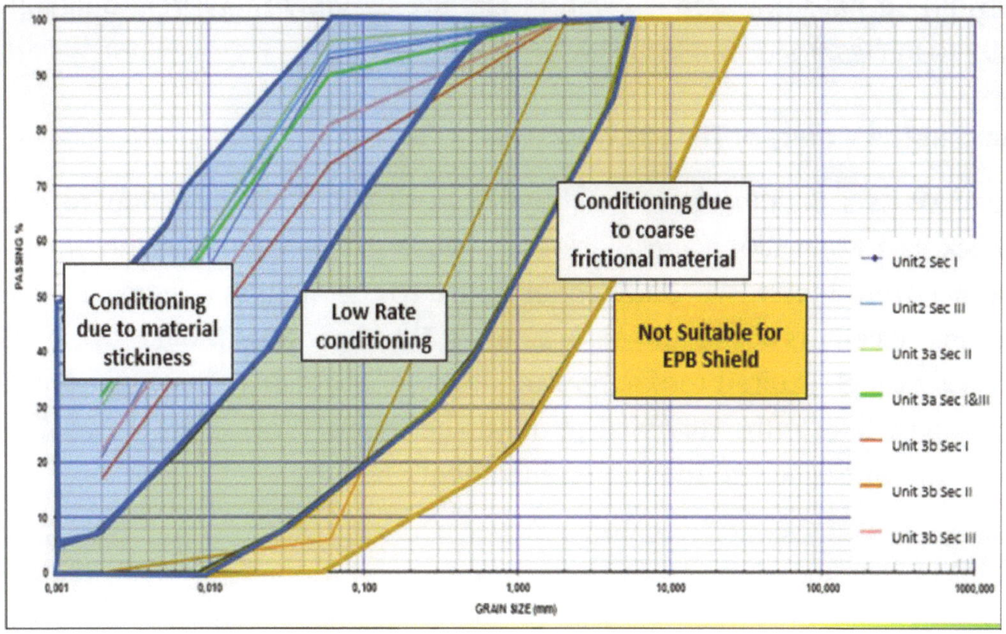

Fig. 6.131: Grain Size Distribution

Fig. 6.132: Writer's Building

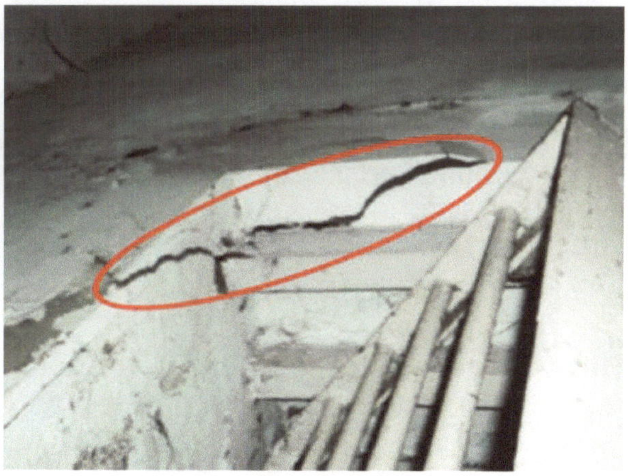

Fig. 6.133: Pre-Existing Cracks

6.20.7 Building Condition Close to Tunnel

Considering the age of the building as well as its immense importance, a thorough building condition survey was conducted. In absence of existing plan, a digital plan was prepared. Being centuries old, there was rigorous engineering surveillance and maintenance, and the general condition of structural health was found to be workable. However, the natural degradation of structural material due to ageing in corrosive, humid atmospheric conditions with heavy rainfall resulted in considerable damage to the building. The building's foundation details have been determined from drawings and field surveys. The foundation consists of stepped brick masonry work with a bottom width of 2.5m and a depth of 1.8m from ground level.

Grain size distribution near the area is given in **Fig. 6.131**. View of Writers building and present condition included in **Figs.6.132** and **6.133.**

6.20.8 Tunnelling in The Vicinity of The Structure

The present-day Writers Building covers a huge area at the Dalhousie Square (BBD Bag). Out of Thirteen Block, the Main Block and Block-1 are close to the tunnel alignment. The minimum horizontal distance from the Westbound Tunnel centre to the Writers Building is 5.87m, while the distance between the centers of the tunnels is 12.65m. The roof of the Main Block is a Georgian roof structure with a Jack arch vaulted roof supported by steel I-sections. The I-sections of the roof of the main block run parallel to the tunnel alignment. The roof of Block-1 is a flat roof supported by I sections, which run transversely to the tunnel alignment.

6.21 Structural Impact

6.21.1 Impact Assessment

The impact assessment on structures due to tunnelling begins with the study of induced ground movement. The tensile stress in structures due to induced ground movement is analysed to assess the resultant effect. Precautionary measures and ground improvement techniques are devised to avert any untoward effects.

6.21.2 Ground Surface Settlement Due to Tunnelling

The construction of a shield tunnel leads to a more significant amount of soil being excavated than being replaced by the tunnel volume. The volume loss quantifies is the amount of over-excavation.

This volume loss is the extent of the total ground disturbance. It causes the settlement trough at the surface and in undrained conditions; the volume of this settlement trough is the same as the volume loss. Attewell derived the volume loss into four categories for shield tunnelling: face loss, shield loss, tail loss and long-term radial loss after grouting. **Fig. 6.131** shows a schematic section of the advancing tunnel shield, which depicts the three components of volume loss due to soft sub-soil shield tunnelling. The first component is the 'face loss'. It is mainly due to the inward flow of the ground from a zone of influence ahead and to the side of the face (Zone 1) into the shield. The settlement component caused by the face loss is referred to as ∂f. The rotating cutters of the simultaneous confinement pressure are applied. Despite the confinement, the ground tends to protrude a shield to remove material from the tunnel face and out of the face from a zone of influence ahead and around the tunnel face. To minimise the volume loss at the face, the shield advancement parameter must be controlled through face pressure (pf).

To advance the shield to reduce the chance of the shield being stuck and to allow steering the shield in a curved alignment, it is necessary to excavate an oversized hole, as the cutter diameter is slightly more than the TBM body diameter. An annular void is created after the cutter has passed, and the surrounding soil tends to move inward radially to fill the void. This movement of the soil creates a loss of ground, namely 'shield loss', which leads to the settlement component ∂s in Zone 2. The entity of this convergence depends upon the soil's deformation rate concerning the rate of tunnel advance, as per Surarok.

Due to the slightly smaller size of the tunnel lining compared to the Shield/TBM Body, there is an annular space usually filled with grout. Thus, there is an opportunity for the soil to displace radially onto the lining until the grout has hardened and gained enough strength to resist the earth's pressure. This soil movement causes the last component of volume loss, 'tail loss', which occurs in Zone 3. Tail void loss is a function of the pressure, volume, and accuracy of the injection filling of grout into the tail void around the installed tunnel lining.

A further increase in the radial convergence can be observed over time. Such long-term loss can occur due to the lining deformations after the grouting, caused by the transfer of the overburden pressure (unless time-dependent soil behaviour is observed, this component is negligible). The ground movement has been calculated according to an established empirical formula, as the settlement trough transversely to the tunnel axis can be approximately expressed by an inverted Gauss distribution curve, where dimension and shape can be defined mainly through two parameters, as per Attewell et al. and Rankin.

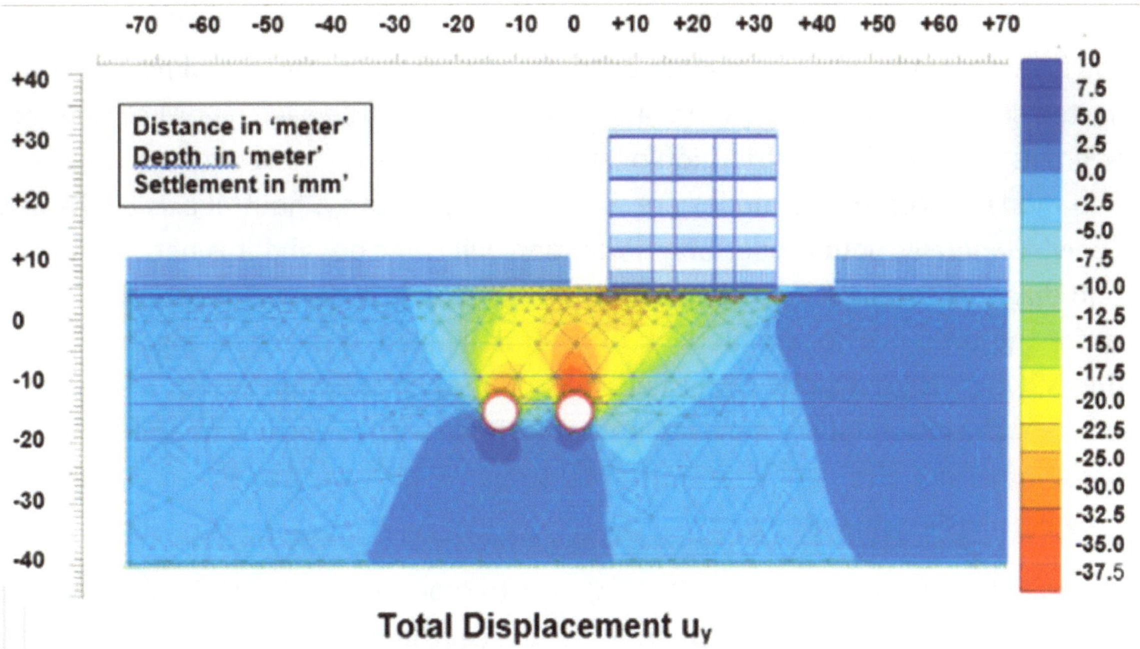

Fig. 6.134: PLAXIS Output for Ground Settlement after Passage of Both Tunnels

The ground movement was calculated by Numerical Method, with a more accurate FEM method using 2-D Plaxis analysis and done to assess the settlement. The building was modelled as plate elements, and a live load of 5 kPa is applied on each floor. For the analysis of the geotechnical model using Plaxis, the following sequence was considered:

- Stress condition and water pressure initialised
- Surcharge applied.
- Tunnelling and installation lining for 1st tunnel,
- Tunnelling and installation lining for the 2nd tunnel,
- Consolidation after construction is completed
- Change the lining properties to long-term modulus.
- Apply seismic acceleration
- Consolidation after seismic acceleration.

The analysis considers a 1.5% ground loss, and relaxation of ground stress is assumed before the lining is installed. **Fig. 6.134** shows a typical Plaxis output.

Figure 6.167 compares maximum ground settlement vasssslues derived from the empirical formula and Plaxis analysis. The latter shows a maximum settlement of 36.94 mm compared to 42.71mm by the empirical formula.

6.21.3 Stress Analysis in Building Due to Ground Settlement

As discussed above, ground settlement induces tensile stress in buildings/structures, which causes distress/cracks in the building. The settlement analysis is therefore used to determine the tensile strain in the building. The approach adopted by Burland and Wroth was used for analysis, where the building is represented by a rectangular beam to calculate the tensile strains in the beam for a given deflected shape of the building foundations and to obtain the sagging or hogging ratio Δ/L at which cracking is initiated. The distribution of strains within the beam is determined based on the deformation mode. Two extreme modes are 'bending only' about a neutral axis at the center and 'shearing only'. Generally, both modes of deformation will coincide, and it is necessary to calculate bending and diagonal tensile strains to ascertain which type is limiting. Timoshenko has given the expression for the total mid-span deflection of a centrally loaded beam having both bending and shear stiffness, which is used along with the derivation of the maximum extreme fibre strain in the replicated beam. The ground surface movements associated with tunnelling not only involve sagging and hogging profiles but also significant horizontal strains, which were analysed as per the method suggested by Bos Cardin and Cording.

The tensile strain induced in the Writers Building due to ground movement caused by tunnelling is derived using the above principles and summarises the calculated tensile Strains at critical sections for anticipated surface settlements due to tunnelling operation.

6.21.4 Pre-Emptive Measures Adopted

The following pre-emptive measures were adopted:

Control of Tunnelling Parameters:

- Maintaining a steady rate of tunnelling
- Maintaining proper face pressure
- Proper tail skin grout pressure and proper grout volume
- Rig Tertiary grout in case of requirement, secondary grouting at every third ring of the tunnel
- Tertiary grout in case of requirement

Fig. 6. 135: Tunnel During Advanced Stage of Construction

Supporting System and Propping in Vulnerable Locations

- Vertical propping at critical load transfer locations is provided as an alternative load path.
- Props should always have proper jacking facilities to ensure positive support.
- Staircases have been supported.
- To avoid new cracks and widening of already existing cracks, all the openings in walls, such as doors and windows, have been supported
- Differential settlement under the building is expected to be around 20-25mm. To avoid unusual structural instability, vertical and horizontal propping must be installed to control the differential settlement and restrict the propagation of cracks.
- The point of inflexion is a potential crack-developing zone where the building's dome is located. Hence, extensive supporting measures are provided in this zone from the ground floor to the top floor to ensure that no structural member fails.

Permeation Grouting: The surface grouting has been done with cement, sand and water as per IRC Bridge Code. The typical water-cement ratio of the grout was 0.4 to 0.5 with the addition of non-shrink admixtures. The grouting pressure adopted was 1 to 2.5 kg/sq. cm transverse direction using G.I. or PVC pipes of 12mm to 20mm diameter up to 2m depth from GL. The grouting pipes were drilled in such a way as to avoid other utilities, depending upon the situation.

6.21.5 Face Pressure and Grouting Control

The computation of tunnel face pressure during advance and stoppage was significant, and maintaining the parameters of face pressure during tunnelling was equally essential in open water to have effective control. Unlike normal circumstances, maintenance of machines under the river and interventions were expensive and involve higher risk factors. Face pressures were computed using values of overburden and groundwater parameters along the tunnel alignment at tunnel depth using the following principles:

- Face Pressure to be maintained during advancement - Mean Values determined at 20m intervals along the alignment by COB Method (Dutch Center Underground Bowen)
- Face Pressure during Stoppages - Mean of values determined at 20m intervals along the alignment by Max Ground Water Load plus 20 Kpa

For ease of EPB pressure regulation, the mean values are further stepped in a 40 m interval, and a mean graph is generated. An upper envelope with a maximum face pressure of 0.2 bar more and a lower envelope with a minimum face pressure of 0.2 bar less than the mean value was also plotted.

Though stoppages during river crossing are not recommended, the chamber must be pressed by injecting Bentonite if required for more than 24 hours during breakdowns and weekends. The face pressure graphs thus generated are cut off to a maximum value of 3.6 bar, considering the machine's maximum rated face pressure of 4.5 bar. A typical face pressure plot for advancement is shown in **Figure 6.136.**

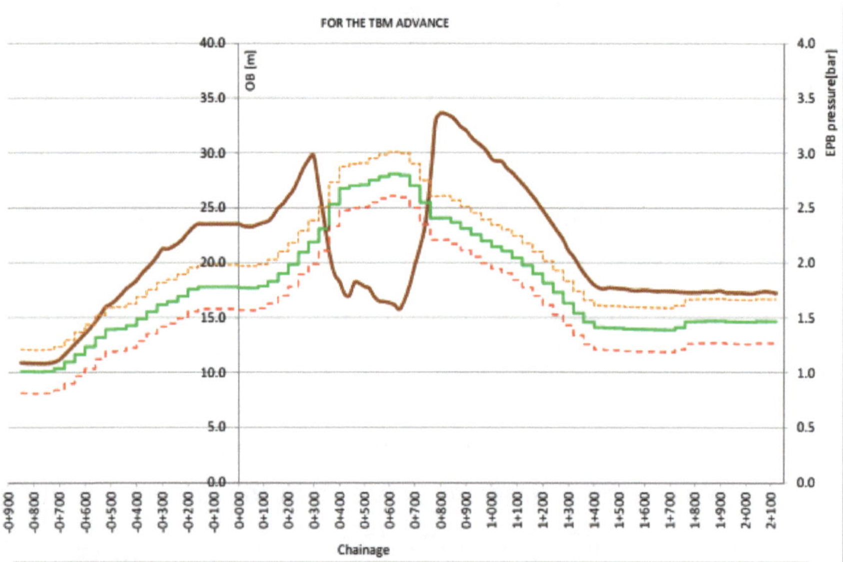

Fig. 6.136: Typical Face Pressure

The density in the chamber shall be kept as high as possible. The density of the material in the working chamber should never be lower than 12.0 KN/M³. It would be desirable to keep a value of 14 KN/M³. For ease of EPB pressure regulation, the mean values are further stepped in a 40 m interval, and a mean graph is generated. An upper envelope with a maximum face pressure of 0.2 bar more and a lower envelope with a minimum face pressure of 0.2 bar less than the mean value is also plotted.

Though stoppages during river crossing are not recommended, the chamber must be pressed by injecting Bentonite if required for more than 24 hours during breakdowns and weekends. The face pressure graphs thus generated are cut off to a maximum value of 3.6 bar, considering the machine's maximum rated face pressure of 4.5 bar.

The chamber's density shall be kept as high as possible. The density of the material in the working chamber should never be lower than 12.0 KN/M³. Keeping a value of 14KN/M³ or higher would be desirable to achieve the closest pressure gradient between the front soil and the chamber's conditioned material.

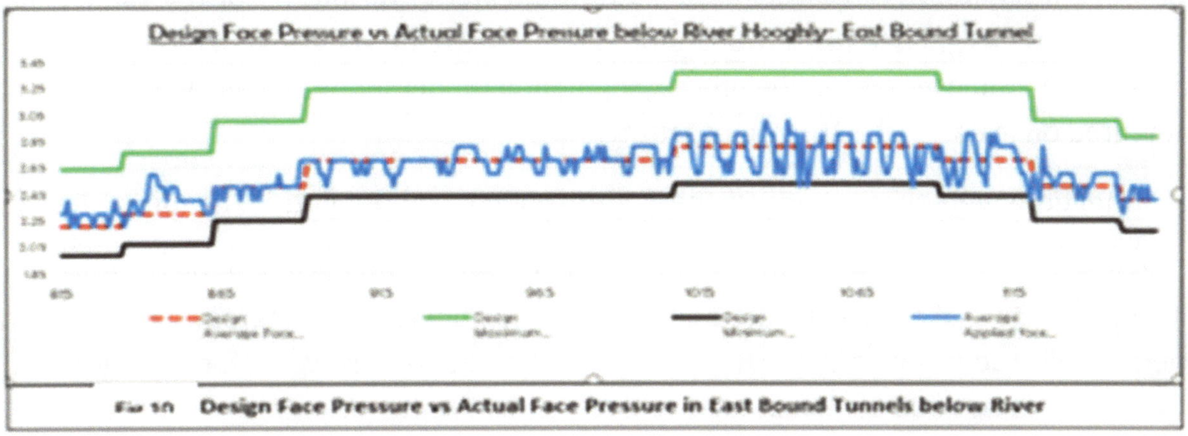

Fig. 6.137: Design Face Pressure vis-vis Actual Face Pressure

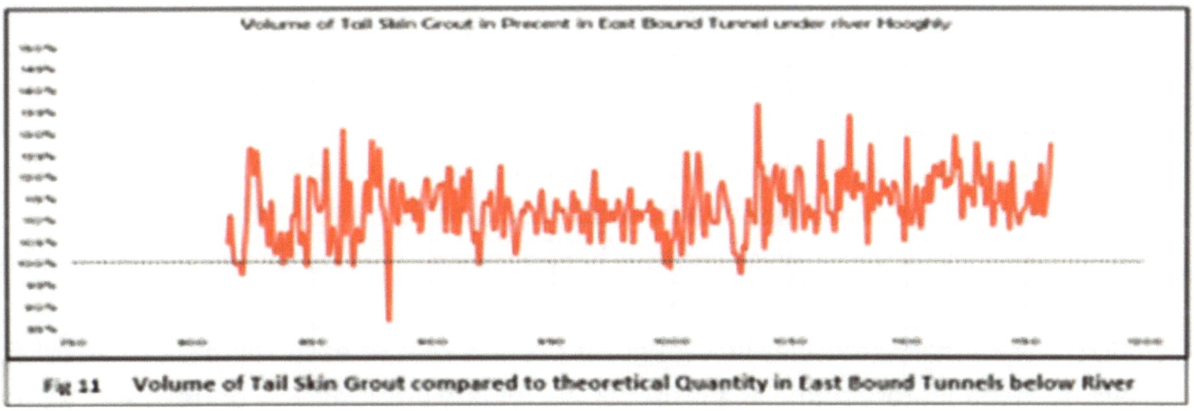

Fig. 6.138: Volume of Skin Grout Compared to Theoratical

Fig. 6.137 includes design face pressure vis-vis actual face pressure and **Fig. 6.139** includes volume of skin grout.

Tunnelling below the river, Hooghly was done with meticulous control over the parameters. A 24/7 team of supervision personnel was formed to monitor the matters, and a high-level management information system protocol was developed.

Tunnelling was smooth below the river with no significant surprises. All the protocols mentioned were strictly followed, and a uniform rate of progress was achieved. The tunnel below the river started on 14th April 2017, and both the tunnels crossed the river stretch in 66 days. depict the status of tunnelling parameters during the river crossing.

6.21.6 Waterproofing of Tunnel

Underground Structures and Tunnels, especially beneath open water, were designed to withstand the detrimental effects of groundwater, which were more aggravated in the case of the segmental tunnels with several joints in liners. Protection against groundwater can be in the form of waterproofing membranes or gaskets installed on the exterior or interior of the tunnel structure. The structure was designed to resist hydrostatic loading and to be watertight over its design life. The standard of permissible leakage in the precast segmental tunnel set by Kolkata Metro is 0.2 lit/Sqm/day, measured on average of a 10m tunnel stretch.

Several protections were taken to ensure the waterproofing requirements in the tunnel are as below:

- Defence 1: The concrete for tunnel liners was mixed with fly ash, chemical admixtures and micro-silica to make the concrete minimally permeable to water ingress. The ceiling limit of permeability determined by the Rapid Chlorine Penetration Test Value for concrete used in the tunnel liner was 1000 Coulombs.
- Defence 2: All the segmental joints had one neoprene and one hydrophilic gasket to prevent water leakage. The hydrophilic gasket material expands five times its dimension if in contact with water, making a water-resistant seal. It is essential to erect the tunnel liner segments in proper alignment to control step and leap so that the effective surface contact between gaskets from adjoining segments is not reduced. A very stringent value of stepping and lipping was allowed in this project.
- Defence 3: During the segmental liner's erection, the extra dose of the concrete liner was filled with high-pressure grouting mixed with sodium silicate for the quickest setting. The applied grout pressure and volume were constantly compared with the design grout pressure and theoretical grout volume.
- Defence 4: Secondary grouting through grout ports is kept at each segment to arrest further leakage. As a mandatory tunnelling protocol, secondary grouting was done in every third ring at a target pressure of 1 bar more than the hydrostatic pressure.

Even after all such elaborate defenses, some water leaked. An intricate arrangement of sumps and dewatering pumps was made.

6.22 Design Basis

i)Tunnel Section

Precast Concrete Lining: Characteristic Strength- 50 N/sq.mm., Ec (short term)-35355 N/sq.mm, Ec (long term)-16 836 N/sq.mm., Poisson's ratio-0.2, Moment of Inertia-0.000878 m4.

ii)General Load Cases

DL - In situ concrete & reinforcement (including wt.):as given in earlier sections.

Imposed Loads: General Surcharge- 50 kpa at ground, Live Loads from Road- 20 kpa, Wheel load from train- Railway loading in line with modern train for Standard Gauge 1435 mm, Maximum Axle Load (type and position of the train to produce adverse effect

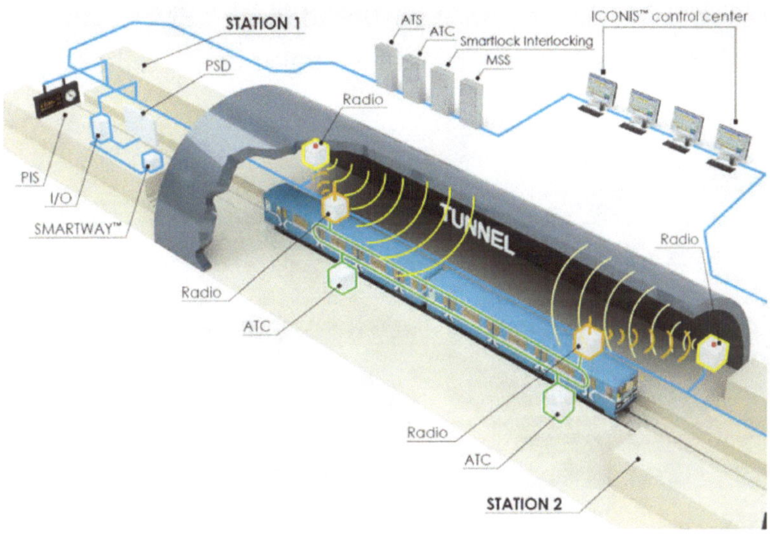

Fig. 6.139: Long Section of a Tunnel Boring Machine (TBM)

Fig. 6.140: Part Long Section of TBM

6.22.1 Tunnel Segment Forces

a) Static Lining Force

The continuum analysis for ground and groundwater forces uses a closed-form model developed by Duddeck & Erdmann (1982). This is based on a general plane frame analysis that will be carried out for varying soil stiffness and differential hydrostatic load. A full bond between the lining and the subsoil is assumed. Calculations have been carried out for the soil's average effective unit weight, considering the soil layers and groundwater table. The effect of differential hydrostatic pressure along the circumference of the tunnel has been considered in the analysis. The analysis is carried out at 20m intervals along the tunnel alignment. Critical sections are identified for detailed numerical analysis considering seismic loading.

6.22.2 Soil-Structure Interaction

A more appropriate numerical soil-structure interaction analysis is done through numerical models for specific critical locations identified through static analysis. Numerical analysis has been undertaken using PLAXIS version 9. A volume loss of 1.5% is assumed for the PLAXIS analysis. This is conservative compared to typical earth pressure balance tunnelling values with tail-skin grouting in soft clay.

Fig. 6.141: Stacking of Segment at Site

Fig. 6.142: Segment Stacking Yard at Casting Yard

Fig. 6.143: Segment Transportation

Fig. 6.144: Muck Car Tipping

Numerical analysis is undertaken with the following construction sequence:

- Initialise stress condition and water pressure.
- Apply surcharge.
- Excavate tunnel and install lining for 1st tunnel.
- Excavate tunnel and install lining for the 2nd tunnel.
- Consolidation after construction is completed.
- Change the properties of the lining to long-term modulus.
- Apply seismic acceleration; and
- Consolidation after seismic acceleration.

Relaxation of ground stress is assumed before installation of the lining.

Fig. 6.145: Segment Mould Cleaning

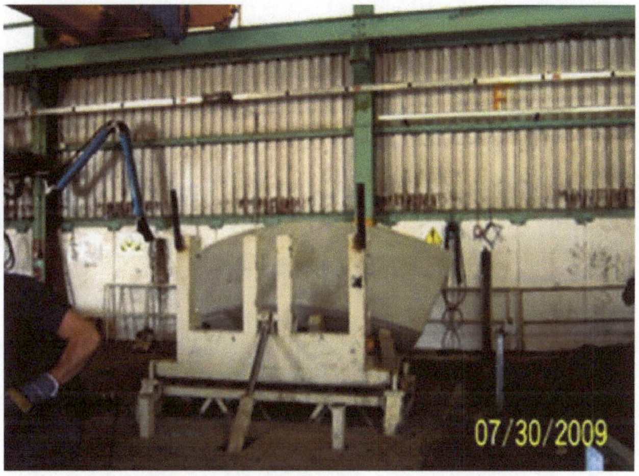

Fig. 6.146: Segment Turner

Fig. 6.147: Vacuum Lifter

Fig. 6.148: Segment Removal from the Mould

6.22.3 Effects of Imposed Distortion

The maximum allowable deflection of the tunnel lining is 25mm on the radius of the tunnel. The induced bending moment by applying this distortion is derived using the Morgan method for jointed lining cases with reduction of lining moment of inertia based on Muir Wood's recommendation. The predicted bending moment and minimum thrust force would be the input for reinforcement calculation.

Seismic Lining Forces

Kolkata is in seismic Zone 3 as per IS-1893: 2002, giving a Zone Factor, Z = 0.16g, corresponding to the Maximum Considered Earthquake (MCE). According to IS-1893: 2002, the design methodology adopted was to ensure that the structures possess at least a minimum strength to withstand a DBE event "without significant structural damage" and an MCE event "without collapse." The design of tunnels to resist seismic loads was based on internationally recognised design methods from published literature.

Factors Affecting the Design of Tunnel Liners.

In addition to the factors mentioned above, there are other factors.

- Analysis of the Effects of Poor Ring Build Design of Radial Joints
- Analysis of the Effects of Jacking for Propulsion at the Circumferential Joint
- Effects of Uneven Shield Shoving Loads
- Grouting Loads
- Analysis of the Handling and Stacking

Fig. 6.149: TBM Assembly at Howrah Maidan (1)

Fig. 6.150: TBM Assembly (2)

6.22.4 Face Pressure During Tunnelling Below the River

Calculating tunnel pressure in advance and during stoppage is extremely important, and maintaining the parameters of face pressure during tunnelling is equally essential in open water to ensure effective control.

Face pressures **(Fig. 6.137)** are computed using values of overburden and groundwater parameters along the tunnel alignment at tunnel depth using the following principles:

- Face Pressure to be maintained during advancement - Mean Values determined at 20m intervals along the alignment by COB Method (Dutch Center Underground Bowen)
- Face Pressure during Stoppages - Mean of values determined at 20m intervals along the alignment by Max Ground Water Load plus 20 Kpa

6.22.5 Blow Out Verification

Due to the application of face pressure, the riverine portion of the tunnel is most vulnerable to the blowout phenomenon. The resisting force R in **Fig6.151** should have an adequate factor of safety against uplift blowout pressure U provided by the face pressure. This aspect was appropriately verified before the finalisation of the tunnelling scheme.

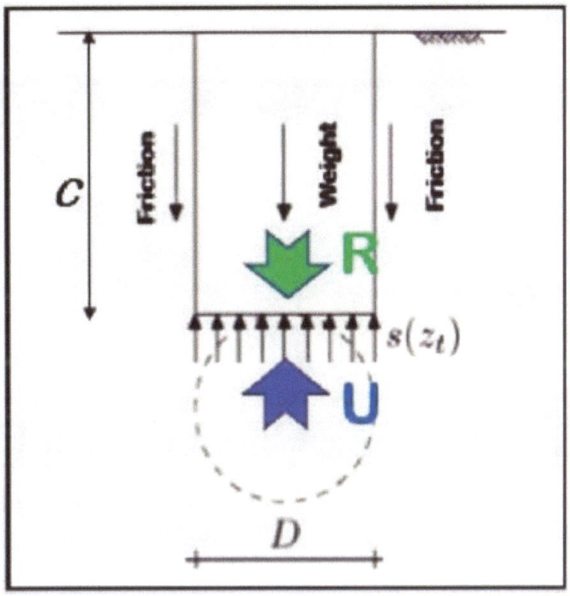

Fig. 6.151: Diagram for Blow Outs

6.22.6 Tunnel Safety

Safety during tunnelling operations is paramount. Control measures for tunnel safety were taken during implementation as below:

a. **Working in Compressed air**: Only medically fit skilled persons are allowed under the supervision of a doctor with a medical lock established.
b. **Ground Collapse during installation** – The proper instruments were used for continuous monitoring.

c. **Support Failure**

Fig. 6.152: View of UP Tunnel After Inauguration

Fig. 6.153: Down Tunnel During Operation

6.23 NATM

6.23.1 Cross Tunnel

According to the excavation procedure, two methods are used for tunnel construction. One is continuous tunnel excavation, which is excavation by a boring tunnel machine (TBM), used where continuous excavation is necessary. NATM with the cyclic advance of the tunnel face. The tunnel excavation can be subdivided into different sequences/stages when required. NATM is based on the concept that the ground around the tunnel acts as a load-bearing element. Typically, the excavation and support activities are continuously adjusted to suit the ground conditions, always considering the technical/design requirements. The ground reaction, in the form of lining displacements, is measured to check the stability of the opening and to optimise the excavation and support process.

6.23.2 Principles of Conventional NATM

The support is determined depending on the sub-soil conditions, such as shallow soft ground, deep rock strata and the results of the geotechnical parameters. Contracts should be flexible to ensure the appropriate and cost-effective support system is used. Shotcrete and rock dowels, which have been used since the mid-seventies, are common supports. Steel ribs or lattice girders provide limited early support before the shotcrete hardens and ensure correct profile geometry. If ground conditions require support at or before the excavation face, face dowels, shotcrete, spiles or pipe canopies are installed as needed.

The excavation cross-section is subdivided into a top heading, bench, and invert depending on ground conditions and logistical requirements (i.e., to facilitate the use of standard plant and machinery). Side drift galleries are provided to limit the size of large excavation faces and surface settlements.

Conventional tunnelling (NATM), in connection with various auxiliary construction methods, enables safe and economic tunnel construction even in situations with changing or unforeseen ground conditions. It allows reacting in both directions depending on whether the ground changes to a less favourable or more favourable site. This flexibility makes NATM the most advantageous method for tunnel construction.

Using the standard set of equipment, the following changes can easily be applied during construction if ground conditions change or if monitoring results require action:

- ✓ Increase or decrease of support, e.g. the thickness of shotcrete, number and/or lengths of rock bolts per linear meter of the tunnel, spacing and dimensions of steel arches
- ✓ Number and lengths of spiles, application of shotcrete at the tunnel face, bolting the face, etc.
- ✓ Variation of ring closure time - the time between excavating a tunnel section and applying partial or complete support, or variation of ring closure distance from the excavation face.
- ✓ Introduction of primary support ring closure.
- ✓ Variation of explosives charge per blasting round and variation of detonator sequences. Other variations in the design enable one to react to changes in the stand-up time of the ground encountered.
- ✓ Increased or decreased length of excavation round (typical round lengths vary from 0.5 m to 4.0 m)
- ✓ Partial excavation by splitting the excavation face into the crown, bench, and invert excavation steps or even further in pilot and sidewall galleries and staggered bench/invert excavations.

In case exceptional ground conditions are encountered, regardless of whether predicted or not, the NATM method can react with a variety of auxiliary construction technologies.

6.23.3 Grouting

Different types of grouting in use are consolidation grouting, fissure grouting, pressure grouting, compensation grouting.

Technologies to stabilise and improve the ground ahead of the actual tunnel face include foreboding, pipe umbrella, horizontal jet grouting, and ground freezing.

Conventional tunnelling (NATM) enables:

- ✓ A more significant variability of the shapes.
- ✓ Better ground knowledge by using systematic exploratory drillings at tunnel level ahead of the face.
- ✓ Greater variability in the choice of excavation methods according to the ground conditions.
- ✓ Greater variability in the choice of excavation sequences according to the ground conditions.
- ✓ Easier optimisation of the primary support using the observational method in special cases.
- ✓ A more significant variability in the choice of auxiliary construction methods according to the ground conditions.

Conventional tunnelling (NATM) is incredibly convenient for:

- ✓ Difficult ground with highly variable ground conditions.
- ✓ Projects with highly variable cross-section shapes.
- ✓ Projects with a higher risk of water inflow under high pressure.
- ✓ Projects with difficult access.

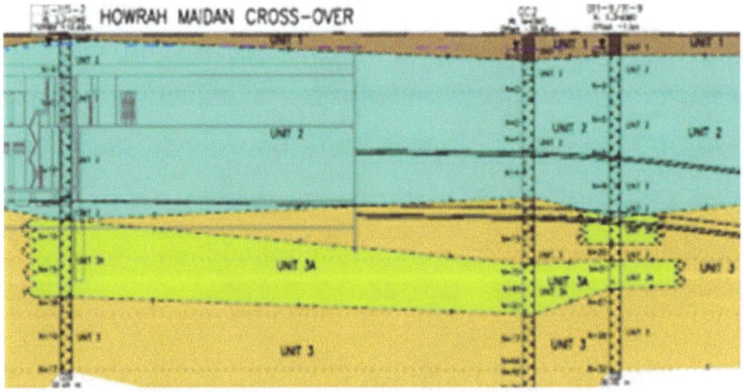

Fig. 6.154: Soil Profile at Howrah Maidan Cross Over

6.24 Approach for Dewatering During Construction

6.24.1 Dewatering must follow

- Drain moderate-high permeability soils to permit excavation
- Limit the risk of hydraulic heave, blowout, boiling or piping risks to excavating into higher permeability
- Avoid ponding of rainwater during excavations

6.24.2 Dewatering must not

- Cause excessive drawdown and consolidation of the very soft cohesive soils
- Pump silt/sand, causing ground loss and instability

6.24.3 Design of Dewatering System

- Based on site-specific stratigraphy and geotechnical conditions
- Possible pumping trial for deep excavation boxes in difficult sub-soil conditions
- Use of pumping wells within deep excavation boxes (slotted pipe, submersible pumps)

Fig. 6.155: During Dewatering Operation

6.25 Quality

(More details included in the later part)

Quality Plans, Method Statements & Job Procedures: Inspection and Test Plan, Testing Formats, Concrete Design Mix, etc., were developed, reviewed, and recommended to the Client

i) Quality Control:

* Testing equipment, including the calibration records, schedule, etc., was verified before use.

* Third-party testing laboratories were selected based on the facilities.

* Verified testing equipment, including calibration records and schedules, prior to use

* Selected third-party testing laboratories based on their facilities

ii) Material Testing and Approval:

- Conducted at the Contractor's Laboratory
- Conducted by a Third Party

iii) Quality **Certification and Quality Audits:** Conducted

6.26 Construction Facilities for Tunnel Section

Construction of Tunnel Section, activities required included in photographs

i) Precasting Yard: Layout plan in line with number of segments to be cast in a time frame of erection schedule, two types of mould fabricated in steel one for normal segment and the second was for key segment. Mould geometry as per approved drawing.

Fig. 6.156: Inside the Casting Shed

Locomotive used for segment transportation from casting yard to tunnel side. Batching plant erected at site to produce highstrength quality concrete satisfying the design requirements. Wet concrete transported by using transit mixer and poured to hopper for casting of segments in the casting shed.

Cast segments were stacked at a prelocated site in a manner as given in the approved drawings.

Fig. 6.157: Locomotive used during the Cast segment Operation

Fig. 6.158: Batching Plant at Casting Yard

Fig. 6.159: Lifting of Cast Segment in the Casting Shed

Fig. 6.160: Stacking of Pre-Cast Segments at Casting Yard

Fig. 6.161: Transport of Concrete for Casting of Precast Segment

6.27 Underground Stations

6.27.1 Type Construction Methodology

Construction techniques in Underground station boxes are adopted considering parameters like geology, availability of space, and depth of excavation. Construction of the Box is done using Cut-and-Cover method. Cut-and-Cover method can be adopted using open excavation, top-down & bottom-up approach, or a combination of both, as done in Kolkata East-West Metro.

Cut-and-Cover Method: Kolkata metro had to deploy different types of design and construction techniques to suit the availability of material (indigenous, & cost), site conditions and to keep future provisions. The cut-and-cover method can be open excavation where sufficient width is available for sloped excavation, like in General Tank's area. For top-down and bottom-up construction to build RTS box, it is required to build shoring wall on either side of the box. Shoring walls can be Diaphragm Walls (DW), heavy duty sheet piles, contiguous bore piles, soldier piles (H-pile) and timber lagging, as shoring wall. Few of these walls, like DW or contiguous bore piles or continuous secant piles, can be used both as temporary and permanent structures.

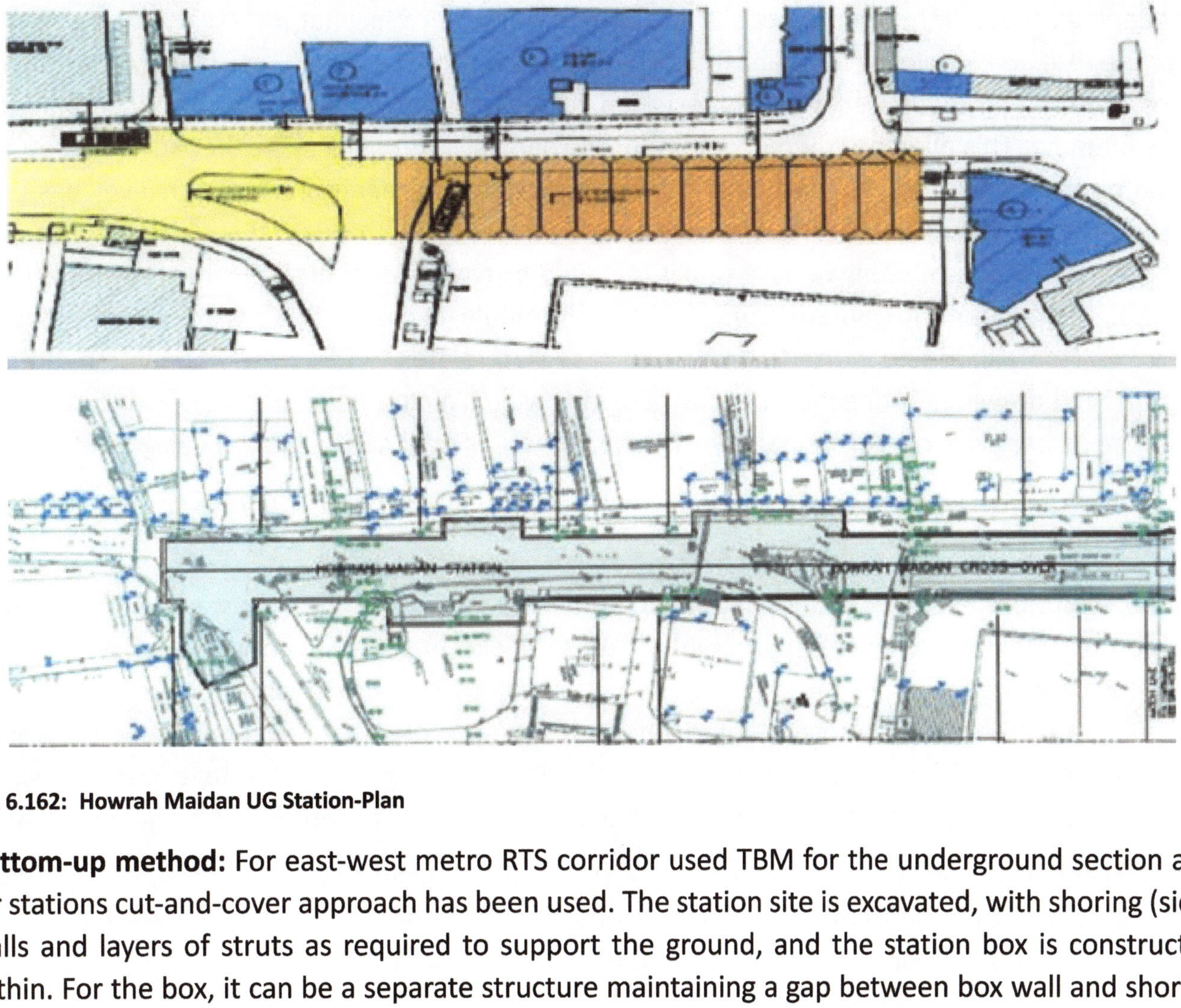

Fig. 6.162: Howrah Maidan UG Station-Plan

Bottom-up method: For east-west metro RTS corridor used TBM for the underground section and for stations cut-and-cover approach has been used. The station site is excavated, with shoring (side) walls and layers of struts as required to support the ground, and the station box is constructed within. For the box, it can be a separate structure maintaining a gap between box wall and shoring wall, or box-DW can be semi-integrated or integrated. For north-south metro all the three types of construction have been used where east-west metro used mostly integrated. However semi-integrated and integrated structures using other types of shoring structures have not been tried. Station box is constructed with cast-in-situ concrete. Concreting starts from bottom with layer of struts on. With the progress of box concreteing, base slab with side haunches & longitudinal beam at center cast first, over constructed lean concrete, including layer of granular layer below as per approved drawing. Rows of ceter columns at middle and walls on both sides are cast then. Struts have been removed sequencially as per design & construction sequence. After casting of roof slab and required strength achieved the trench is backfilled as per specifications, above the tunnel roof, and the surface is restored.

Top-down method: In case of top-down construction, shoring walls on two sides of the station box are cast (DW, Contiguous bored piles)/erected, first, for a length & at a level as per drawing. Then the top part of the trench for RTS box, is excavated with or without support depending upon depth of excavation. Depending up on top level of shoring wall, unsupported depth of excavation and soil properties, temporary wall is constructed from ground level, to retain topsoil. After assuring the stability of excavation, and profile of excavated surface, as per drawing, top (roof) slab of RTS is cast keeping an opening. Opening size is in commensurate with facilitating to carry out activities, under the roof, to complete the RTS box. Casting of station roof used in situ concrete. After the roof attains required strength, backfilling is resumed to restore the original surface, maintaining the access opening. This allows early reinstatement of roadways, services and other surface features. Excavation then takes place under the permanent station roof. After attaining design level, base slab is cast over lean concrete including granular base below to accommodate dewatering which adds safety against floating. Casting of base slab is followed by casting of central columns and side walls, in case of independent box, whereas only columns at middle, in case RTS box is integrated with DW. That is top (roof) slab is supported on DW, and base slab relates to DW & its' connection, both as per approved drawing. Connection of roof slab with wall/DW and its connection with columns at center are as per drawing. All connections can be hinged or rigid depending on design-detail.

In case of, strut is required between roof & base (rarely) slabs, this is done as per construction methodology given in approved drawing.

In a few Metro projects, DW is designed for soil pressure, i.e. active and at-rest soil pressure, with provision of weep holes (where applicable) in DW to avoid pore water pressure on DW. To retain the water pressure, separate the skin wall from the wall built with bottom-up construction. This construction methodology can be adopted where the water table is significantly below the ground level, and the station is proposed in a green field area. During the construction stage, no major water ingress is expected in the station excavation; otherwise, it may lead to settlement in the nearby area.

Major advantages of top-down construction are that one can avoid completely temporary works in most cases, i.e. shoring system (except where topsoil is of poor quality), also before excavation of the box part, traffic can be placed in after casting of the roof slab, and road can be reinstated with backfilling and placing road crust in position.

Design method: During the design of the North-South Metro, designers used the empirical method, prevailing worldwide at that time. Now, designers are using finite element-based software for more accurate and optimised designs. Because of the advancement of technologies, engineers can use more sophisticated tools to handle different complex scenarios. Good engineering design includes analysis of system performance as part of design development. Many tools are available to design details of underground structures—for example, the thickness of supports or the appropriate size and location of ventilation shafts. Empirical and analytical design methods, ideas and procedures based on experience, building codes, and analytical and computational software packages all contribute to the optimal design of underground structures.

Fig. 6.163: Station Neighbourhood at Howrah and Phool Bagan Construction Stage

Fig. 6.164: Mechanised Excavation in South Kolkata

However, their effectiveness directly correlates with the quality of data input and the knowledge base and experience of those applying the tools.

Empirical design methods rely on quasi-quantitative approaches to material characterisation based largely on experience-driven judgments. They are not necessarily based on fundamental mechanisms of ground and structure behaviour. For example, soil and rock classification systems are based on the comparison of observed material behaviours from the infrastructure site to similar observations made elsewhere. Behaviour is predicted based on those comparisons, but the underlying reasons for the behaviours may not be well understood. Even so, the soil and rock classifications based on empirical characterisation often inform support requirements and design specifications. In some cases, empirical characterisation may be based on many prior observations, and resulting conclusions may be robust. In other cases, few data exist to enable comparisons, and empirical characterisations may be more like "best guesses." Empirically based conclusions can be considered first-order estimations, and further observations may be necessary to confirm or refine those estimates. In other words, they can be used to interpolate but not to extrapolate. Improper use of empirical tools can result in safety hazards, poor performance, or unnecessary expenses. Educating professionals about the limitations of empirical methods is one way to improve their use. Improving the databases of observations and the method for their expansion is another.

6.27.2 Analysis of Braced Cut

Extensive analysis has been done for the Plaxis modelling of the station sections to determine the stability of the excavation and the ground/building settlement profile along the excavation. In the Howrah Maidan area, there are buildings in proximity to the excavation. Even before the present construction near the Howrah station commenced, most of the Metro alignment through Kolkata City had been done by braced excavation with diaphragm wall and multiple struts. Ground settlement evaluation and the condition survey of all buildings along the alignment had been done before construction, and the buildings had been monitored closely during construction.

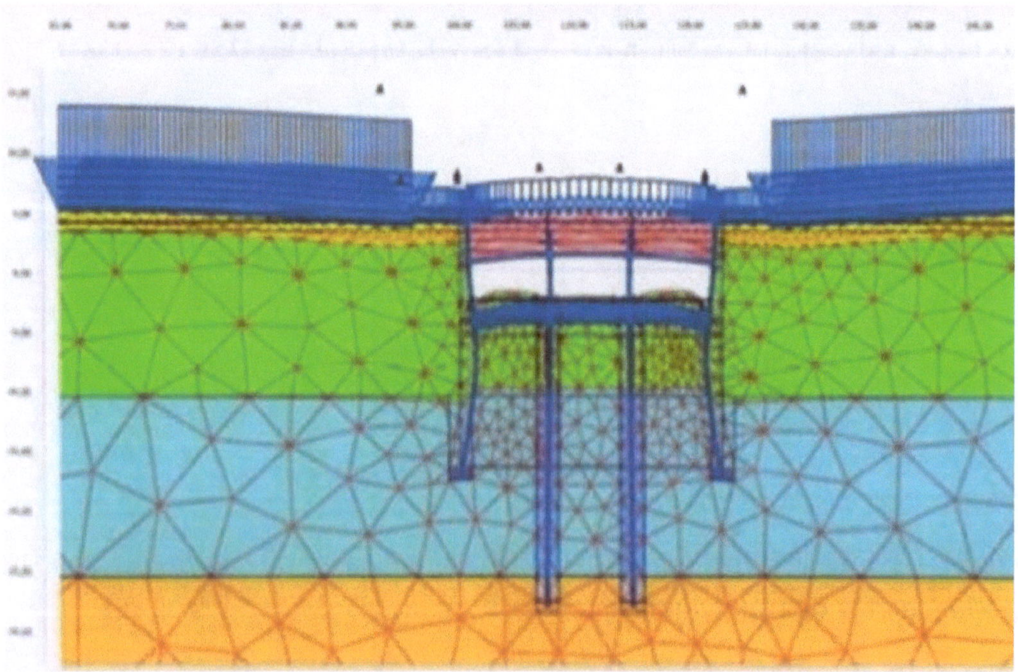

Fig. 6.165: Plaxis Modelling of Howrah Maidan Excavation

Problems were anticipated at Narkeldanga during the present phase of construction when the twin TBMs were required to go below two railway bridge abutments. The deformed shape of the ground for the twin tunnel was predicted from the Manual of Design and Construction of Road Tunnels (1965), according to which two parallel tunnels three or more diameters apart (center to center) would give surface settlement well predicted by adding the individual settlement of the two tunnels. The center-to-center distance between the two tunnels is 17.344 m from their combined. No distress was, however, noticed on the railway platform during the construction of the TBM tunnel. **Fig. 6.164** includes ground settlement profile.

6.27.3 Excavation near Howrah Maidan Station

Figure 6.163 gives the layout of buildings near Howrah Maidan Station/Howrah Maidan Crossover. The analysis of the excavation by Plaxis modelling gave the deformation pattern around the excavation with diaphragm wall, the concourse and base slabs, wherever required for preventing uplift of the bottom slab. The design objective was to keep the angular distortion within 1/300. The buildings were subjected to close monitoring during construction, and contingency measures were taken to ensure the safety of the city.

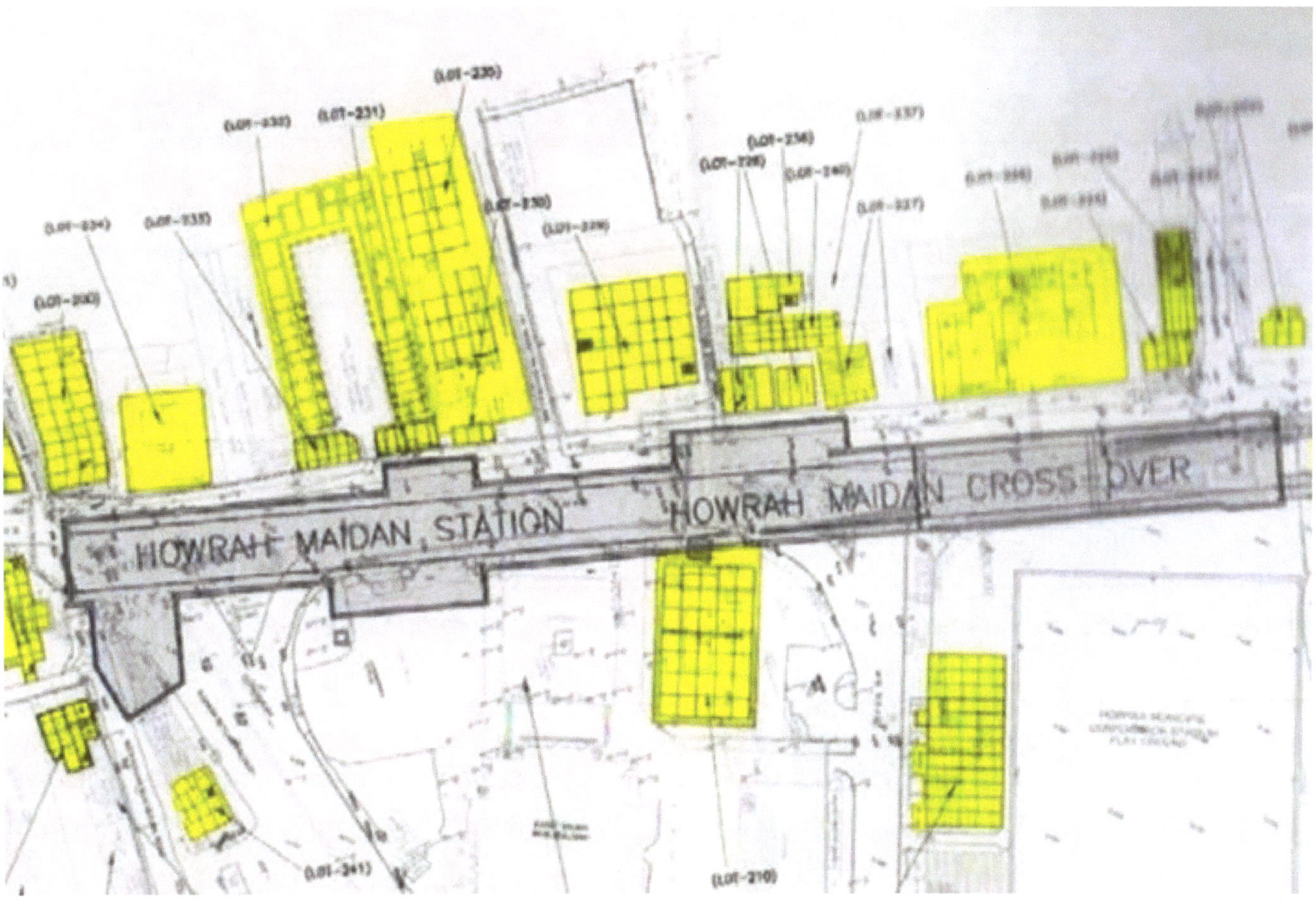

Fig. 6.166: Howrah Maidan Station and Neighbouring Buildings

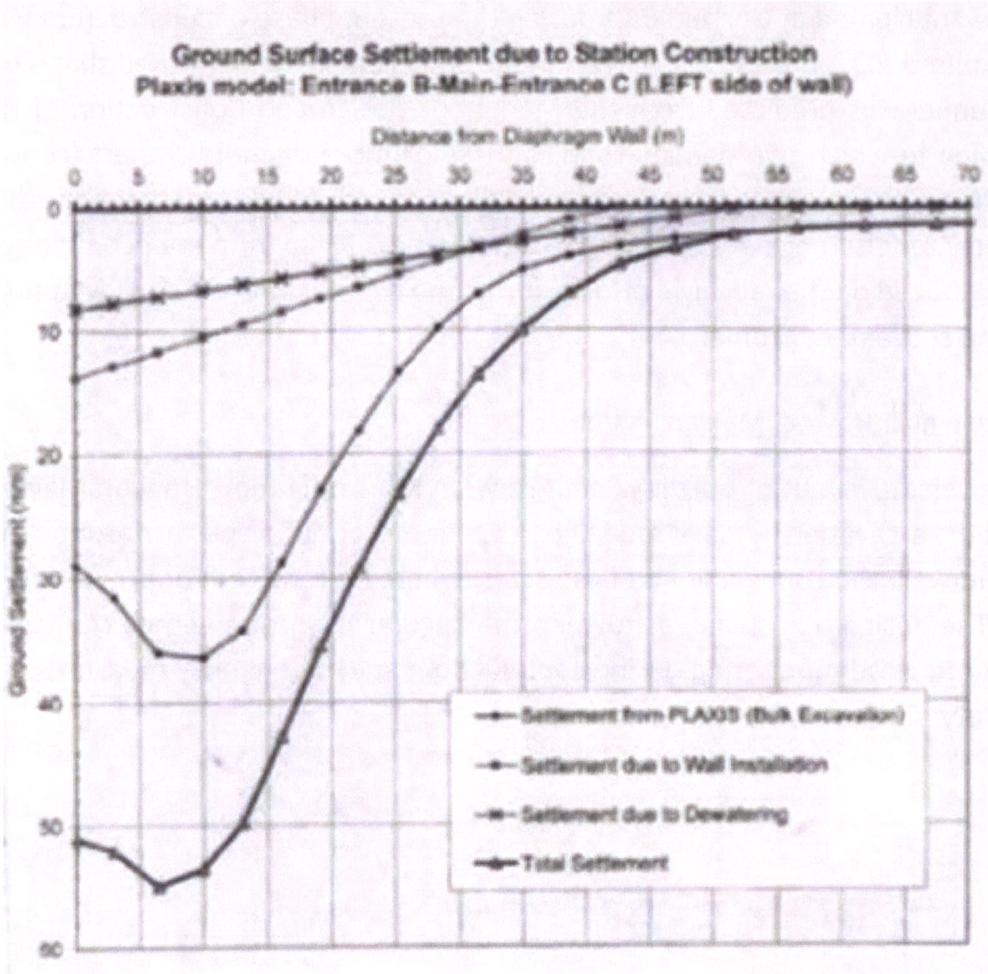

Fig. 6.167: Ground Settlement Profile

6.27.4 Soft Ground Tunnelling

Initially, in the first phase of construction near Belgachia station (North-Soth Metro), shield tunnel with compressed air had been done to cross a busy railway yard. Although progress was slow and several buildings in the vicinity had been distressed due to ground subsidence, the tunnelling was done without any major hazards. The observed ground settlement resembled a dish-shaped profile with maximum settlement of 200-250 mm, which could be predicted well by error function distribution based on the maximum settlement occurring at the midpoint of two tunnels.

6.28 Analysis of Underground Station Box

As mentioned earlier that the Design of the East-West Metro Station primarily used Cut-and-Cover construction with the top-down method of construction. And the design considered the Diaphragm wall as an integral part of the station structure.

The underground metro station analysis and design consist of two parts. The first one is the construction stage analysis, and the second one is the permanent analysis. The construction stage analysis is the retaining system analysis doing which uses WALLAP software. Using the WALLAP software calculates the maximum moments and shear force in soil and the excavated face of the diaphragm wall. For underground metro station design, two types of load combinations were considered. One was based on ultimate strength conditions, and the other one is on ultimate serviceability conditions. For ultimate strength conditions, consider the multiplication load factor is 1.5 based on IS: 456(2000). For ultimate serviceability conditions, consider the multiplication load factor is one based on IS: 456(2000). Permanent stage analysis is made by using STAAD Pro software.

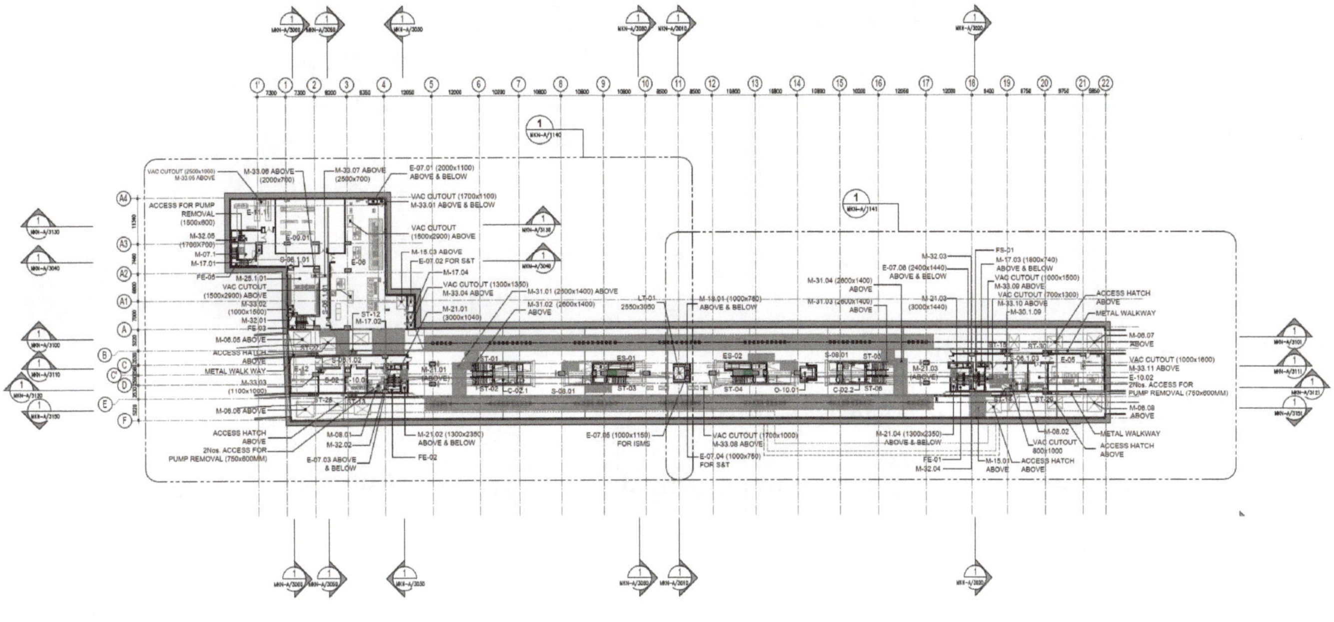

Fig. 6.168: Platform Level Plan Mahakaran

From the load combinations based on IS: 456(2000) the maximum bending moment and shear force based on ultimate strength and serviceability conditions were calculated. To analyse the diaphragm wall, divide the entire length of the diaphragm wall into small parts. Each part must be analysed by both the construction and permanent stages. Finally, compare the maximum moments and shear force from the above two results. The maximum bending moment in both soil and excavation sides were calculated. The reinforcement detail was prepared in the form of a drawing. For the analysis and design of the roof slab, concourse slab, and base slab, use STAAD Pro software. The reinforcement detail must be prepared in the form of drawings. In this analysis, the structure must retain the load of the construction stage and thus take into consideration all the geotechnical details and incorporate in the software called WALLAP to find the embedment depth and the envelop moments on the wall by giving the proper commands to the software. Thus, the bending moment behaviour of the elements is studied through this software, and then the moments for each member are noted down for comparison of moments in permanent analysis. The embedment depth for this project is found to be 5m below

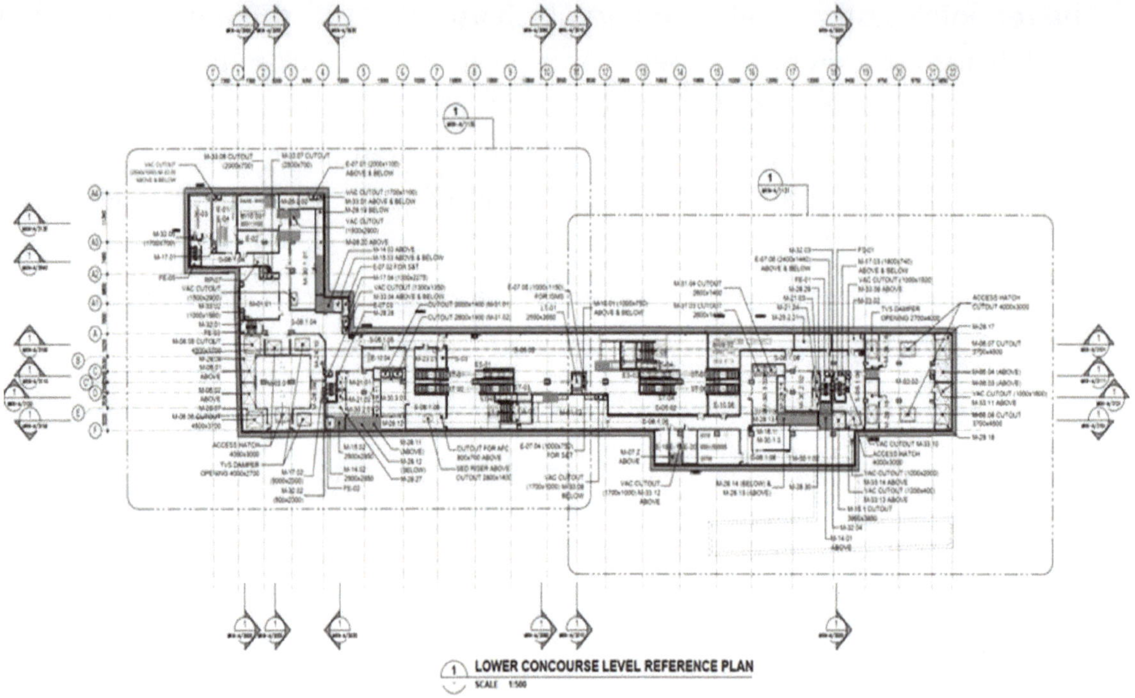

Fig. 6.169: Lower Concourse Plan Mahakaran

Components of an Underground Metro Station

The various components of the underground metro station box are as follows.

- Diaphragm wall
- Roof slab
- Concourse slab
- Base slab

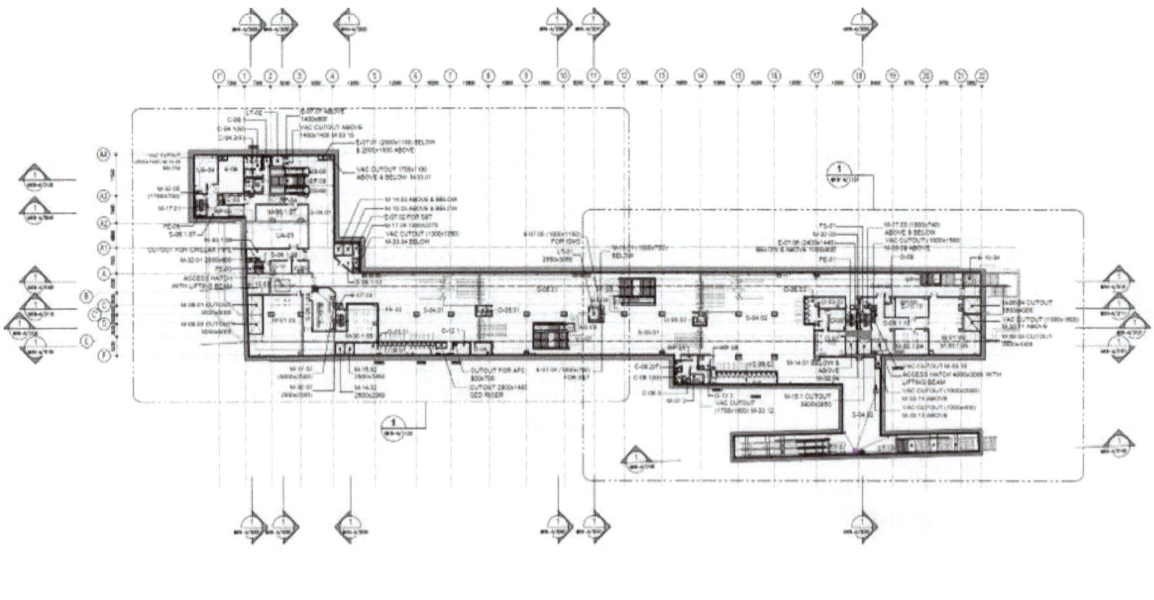

Fig. 6.170: Upper concourse Plan for Mahaaran

These structures Diaphragm wall (DW) is a reinforced concrete wall constructed in the ground using the under-slurry technique advantage that it has a smaller number of Joints compared to other types of shoring walls and can go for greater depths. It improves the water tightness. DW can be designed to take the high structural loads.

The roof slab is below the ground level and the top-most layer of the underground station box. Above the roof slab, the soil backfill is placed. The station is 200m long.

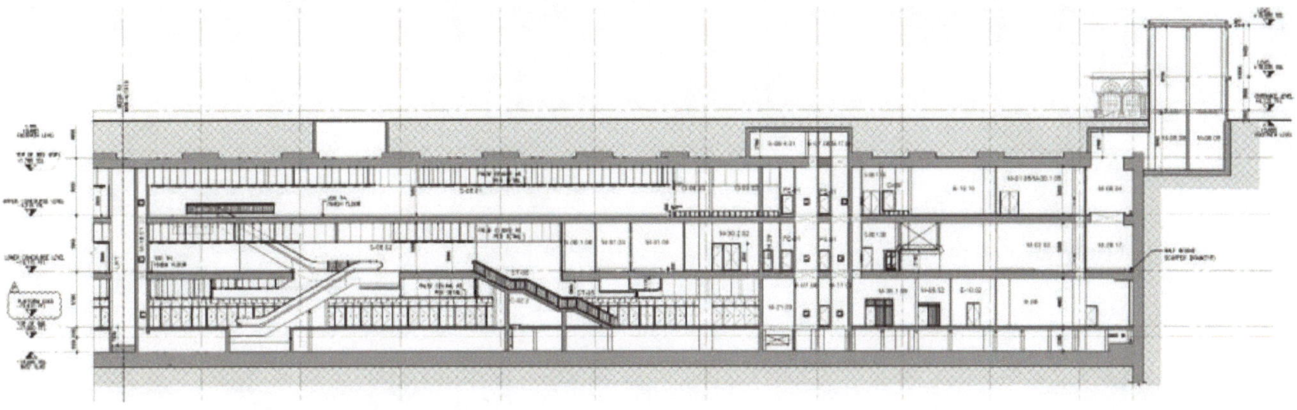

Fig. 6.171: Section of Mahakaran Station

The concourse slab is placed below the roof slab. At the concourse level, ticket counters and amenities like Stairs, escalators and lift facilities are available.

The base slab is the bottom-most layer of the underground metro station box. Base level is the platform level and includes tracks. Passengers are to board/alight trains. Stairs, escalators and lifts are provided.

6.28.1 Deepest Metro Station

The Howrah Metro station goes as deep as 30m below the surface and is located adjacent to the busiest railway hubs of India — the Howrah Station. Constructed using the top-down methodology, it proudly holds the record for the country's deepest Diaphragm Wall, reaching an impressive depth of 55m. The station features have five distinct slabs: the roof slab, upper concourse, mechanical slab, lower concourse, and base slab. Its thoughtful design facilitates seamless connections to the old and new complexes of the Howrah Station through four entrances at the surface level and two entrances through the subway at the concourse level. "Through a hydrographic survey followed by geotechnical investigation, it was identified that the most suitable pathway for the underwater tunnel is 37m depth from the ground level along the Hooghly River stretch. Careful geotechnical studies and design stipulations further suggested the crown of the tunnel should be, at least, 13m below the riverbed. To ensure seamless integration, the Howrah Metro Station had to be strategically adjusted to a level where the tunnels could effortlessly navigate the river stretch with minimal undulation. This adjustment required a deeper excavation, resulting in the station's increased depth of 30m.

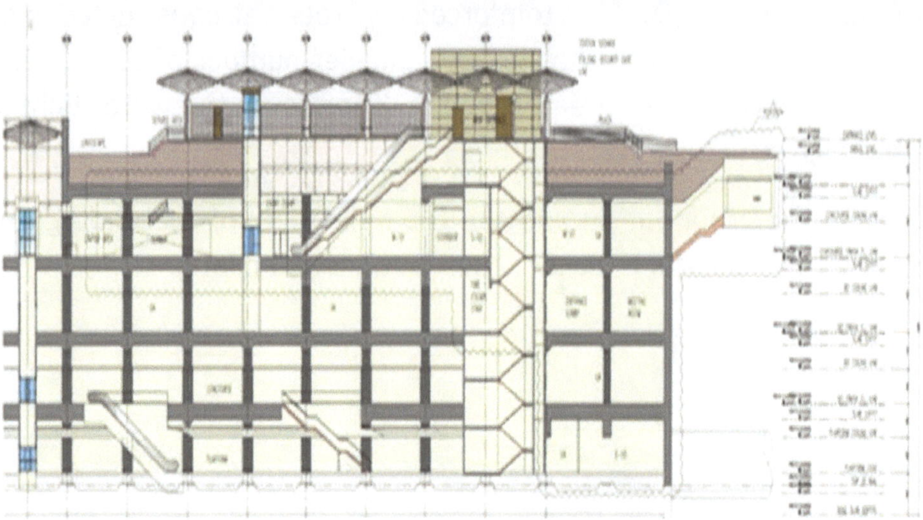

Fig. 6.172: Section of Howrah Station

Fig. 6.173: 3D View of Karunamoyee Station (Elevated)

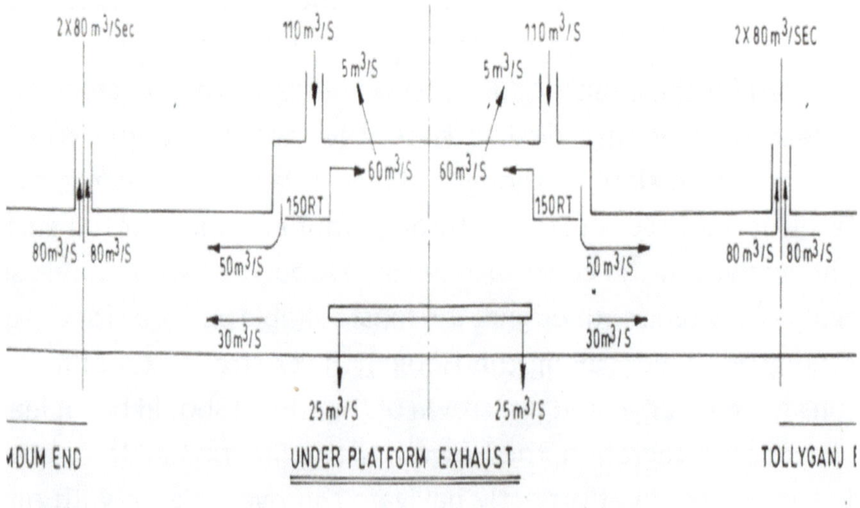

Fig. 6.174: Under Platform Exhaust

6.29 Central Park Depot

Depot Designs encompasses various activities, including dimensioning facilities, planning the buildings and track layout, specifying technical equipment and project management, and commissioning and handover. An essential part is to consider the most appropriate maintenance procedures and the logistics requirements for spare parts and consumables. The Central Park Depot is in Central Park of Salt Lake City, Kolkata. It was constructed on about 39 acres of land, and its primary function was stabling and providing operating and maintenance facilities for the rolling stock of East West Metro. The aim was to build a green and sustainable maintenance depot to become a benchmark for an environmentally responsible Mass Transit system depot comparable to leading global standards. The depot was designed, constructed, and commissioned per the Employer's requirements.

Fig. 6.175: Central Park Depot

Fig. 6.176: Rake at Maintenance Workshop

6.29.1 "Green" Depot

The depot was planned to have built-up and open-to-sky areas. The built-up area includes Workshop buildings, Maintenance Workshops, Administrative, facility Buildings and Stabling Areas, and open-to-sky regions for Outdoor Utilities, Roads, and greens. At the instance of KMRCL, the client, "Green," was customised for each. For the open areas, the focus was on capturing as much rainwater as possible and ponding it to replenish groundwater. The landscape is contoured appropriately. As an additional "Greening" effort, native species were used. The built-up area offered several features for "Greening". Daylight level was assessed, using daylight analysis and internal partitioning that considered the materials and their reflectivity. While glazing using double layers or spectrum-selective glasses increases the thermal efficiency of the building and reduces the power requirement, it is likely to add to the cost of construction and total embodied energy. As in the case of every "greening" effort, an attempt was made to be struck. Sun-path analysis was carried out to take advantage of solar lamination and shades as appropriate.

When a concrete structure is demolished, the remnants can be aggregated for subsequent concrete production. And the process recoups the embodied energy and counts as "green." Recent evaluations show that the crushed concrete aggregates, both coarse and fine, may require more cement-intensive binders. In case of a ship breaking, there is a way to reuse Steel. In earlier days, structural elements were brought into a molten state and reworked to satisfy current requirements. In the process, the energy intensity at the gate was significantly reduced. Recently, steel from ship breaking ought to be used after minimal processing. The design was to adhere to "Green" norms with water management, including sedimentation ponds, wetlands, retention ponds, mounds, and swamps, as many as possible.

The team planned to design to achieve the highest standards. The outline design and construction specifications include green guidelines.

6.29.2 Recent Trends

Recent vehicles incorporate new materials and production approaches that can minimise maintenance. They use electronics more in train control, condition monitoring, and fault diagnostics. Operational and business-related issues should not be underestimated. As such, service and basic repairs should be undertaken as closely as possible so that vehicles are operated, reducing unproductive positioning. A few factors must be considered when planning new or remodel maintenance facilities. Primarily, the depot is designed for a stand-alone line or part of a network.

The Planner/Designer may consider the essential servicing routine, including supplying fresh water and disposing of waste, internal and external cleaning, and litter removal. In dimensioning the depot, the planner must decide whether to make provisions for accident repairs or periodic overhauls. This, in turn, would help to make provisions for the storage of spare parts, the overhaul of workshop space, or the manufacture of specific components.

Local factors were considered, including climatic conditions, such as cold areas requiring snow thawing and heat and areas with rainfall.

In most of the Metropolis, there are many metro lines. The lines may be of similar or different types of gauges, coaches, spare parts, parts to be manufactured and so on. For metros with similar specifications, the maintenance depot may accommodate all the services required, provided the capacity is sufficient to tackle all repairs or manufacturing as needed and without affecting the operations. If different technologies are used in a metropolis, then different maintenance shops are required based on technology requirements. In such cases, the stable yard can be for each line or clubbed for metro lines with similar specifications. Then, the maintenance depot may be separate from the stabling yard, and the number will be technology-driven.

Fig. 6.177: Aerial View of the Depot During Construction

6.29.3 Using BIM Methodology for Depot Layout Design

Given the number of metro lines and the different technologies warranting separate maintenance depots and tables, Building Information Modelling (BIM) is essential for designing depots. BIM allows for creating detailed 3D models encompassing every depot aspect, from infrastructure and equipment to workflows and utilities. By using BIM, designers can visualise the entire depot in a virtual space, making it easier to identify potential design issues, optimise space, and ensure that the layout supports smooth operations.

BIM also helps simulate real-time operations, showing how trains, personnel, and equipment interact within the depot. This methodology enables better decision-making by allowing stakeholders to collaborate and adjust the design before construction begins. Furthermore, BIM ensures that the depot is designed with scalability, making future expansions or modifications easier to implement without significant disruptions to current operations. The use of 'BIM' is gaining popularity in India for any big, multidisciplinary infrastructure projects.

6.29.4 Essential Features for Depot Design

The essential feature of infrastructure design was Sustainability, and the Goal was Climate neutrality. Issues. The parameters considered for the depot design for the East-West Metro depot were

- Water Conservation
- Energy Efficiency
- Material and Resources Conservation
- Indoor Environment Quality
- Operations and Maintenance

Though the depot's planning and design considered several elements of a 'Green' design, the construction could not adhere to them. This was primarily due to the unavailability of clear land until the scheduled completion date.

6.29.5 Sustainable Approaches

While planning and design considered the sustainable approaches, given below, in a graphical format for easy understanding:

Building Orientation

The building complex is oriented on the East-West axis, reducing low altitude solar exposure to the building skin.

Solar Shading

Self-shading building overhangs at south facades and fritted glazing and interior blinds at east, west and north facades reduce solar heat gain and mitigate glare, thereby reducing cooling loads and increasing comfort.

Operable Windows

The use of operable windows facilitates natural ventilation, reducing operational costs within the work environment. With access to fresh air, productivity rises and absenteeism decreases.

Green Roof

Green roofs serve as accessible building amenities and provide a vegetated cover atop of the roof membrane. This reduces degradation due to solar exposure while also mitigating the heat island effect.

Pervious Paving and Xeriscaping

Landscaping with vegetation that requires little to no irrigation reduces dependency on water sources. Pervious paving prevents stormwater runoff and the flow of pollutants off site.

Stacked Ventilation and Sky Gardens

Sky garden loggias located central to the buildings create opportunities for daylight penetration and view connections to the outside. Sky gardens can be fully open, partially open, or enclosed interiors.

Fig. 6.178: Sustainable Approach

6.29.6 Watershed Management

Watershed management, mainly water conservation, was the prime design issue. A few sketches have been included, as follows. However, due to the long delay in handing over the entire site, the watershed management could not be implemented in its entirety. **Nevertheless, the design aspects included here may help in future design. During the design stage, the designers should be able to identify and contain the elements that can be implemented.**

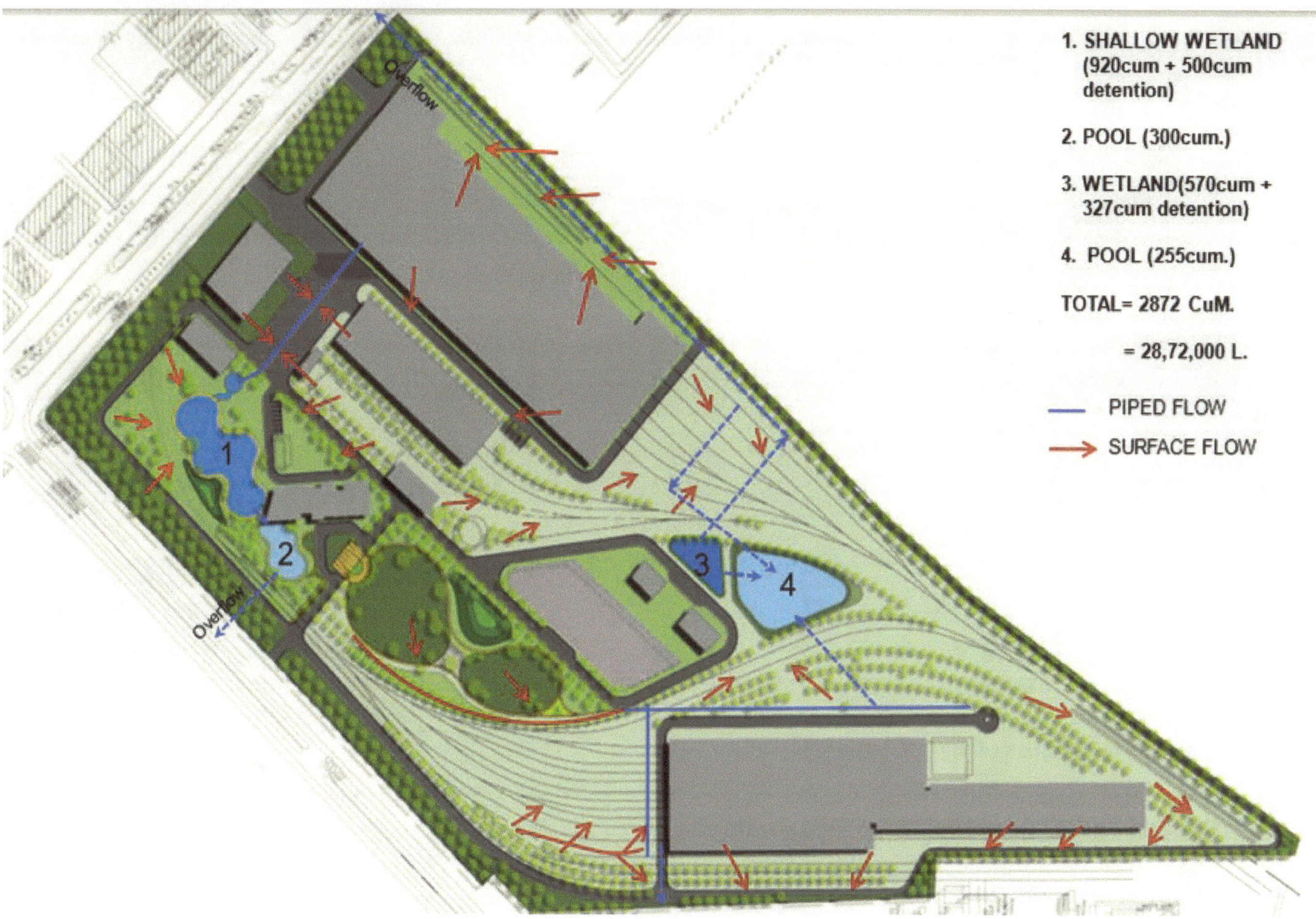

Fig. 6.179: Watershed Management Plan

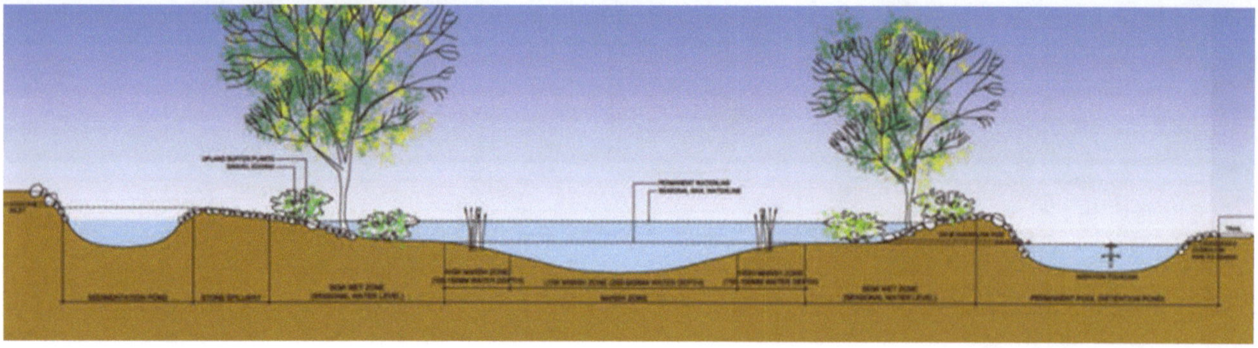

Fig. 6.180: **Wetland Proposed Longitudinal Sectional View**

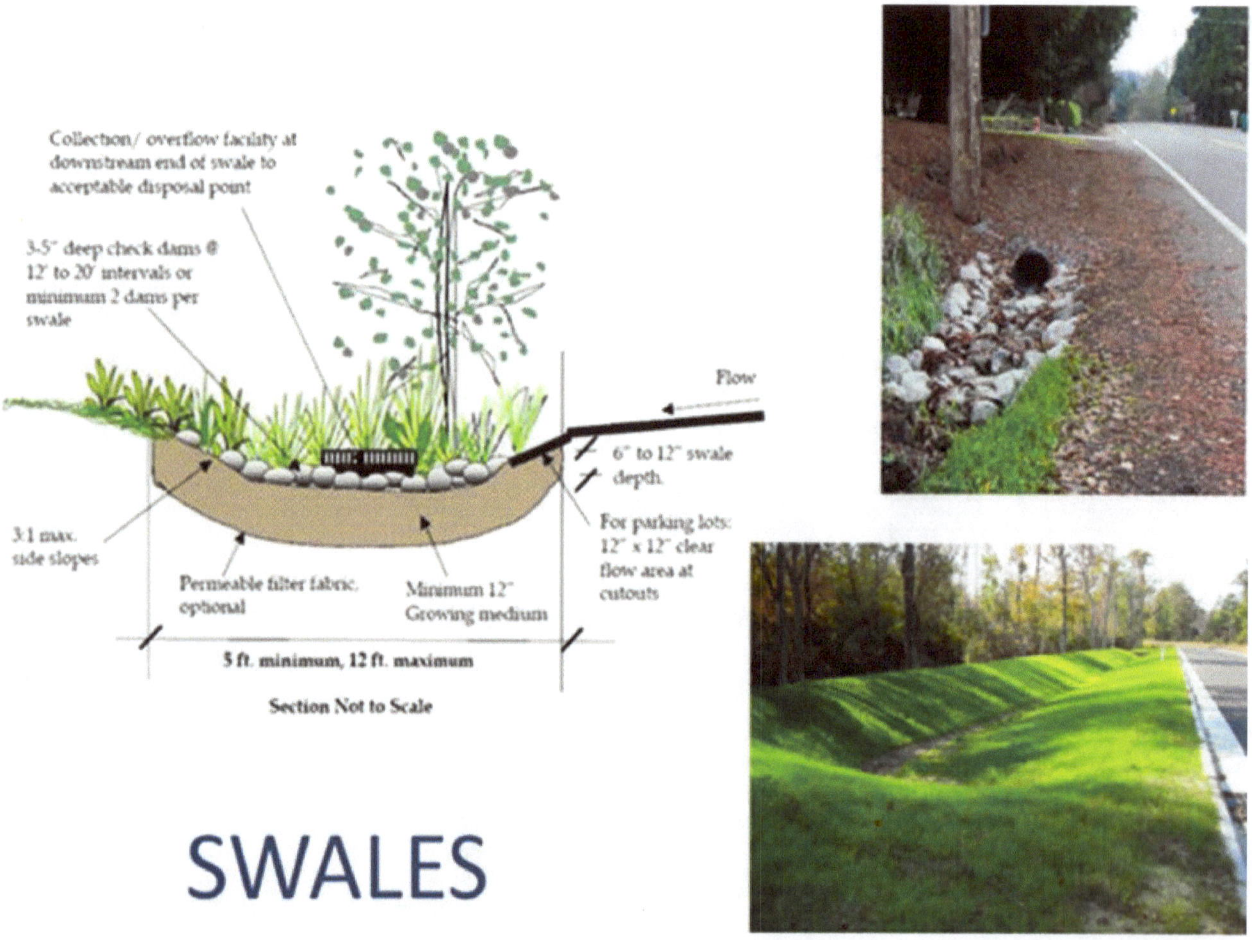

Fig. 6.181: Proposed Details at Different Locations

6.29.7 Central Park Depot - Pillars of Sustainability

- Water Conservation
- Energy Efficiency
- Material And Resources Conservation
- Indoor Environmental Quality
- Operations And Maintenance

6.29.8 Goal: Climate Neutral

The slide also includes a rating system for points scored:

- 50-60 points: One star
- 61-70 points: Two stars
- 71-80 points: Three stars
- 81-90 points: Four stars
- 91-100 points: Five stars

6.29.9 Daylighting Approach

- The model was created per the given drawing. Zones were created per the data furnished. Internal partitions have been created per the section, and the materials considered.
- The visible light transmission (VLT) for Exterior glazing is 30%.
- Parameters Considered for Daylighting Analysis
    - Floor Reflectivity - 20%
    - Wall Reflectivity - 30%
    - Ceiling Reflectivity - 50%

6.29.10 Lighting for Admin Building

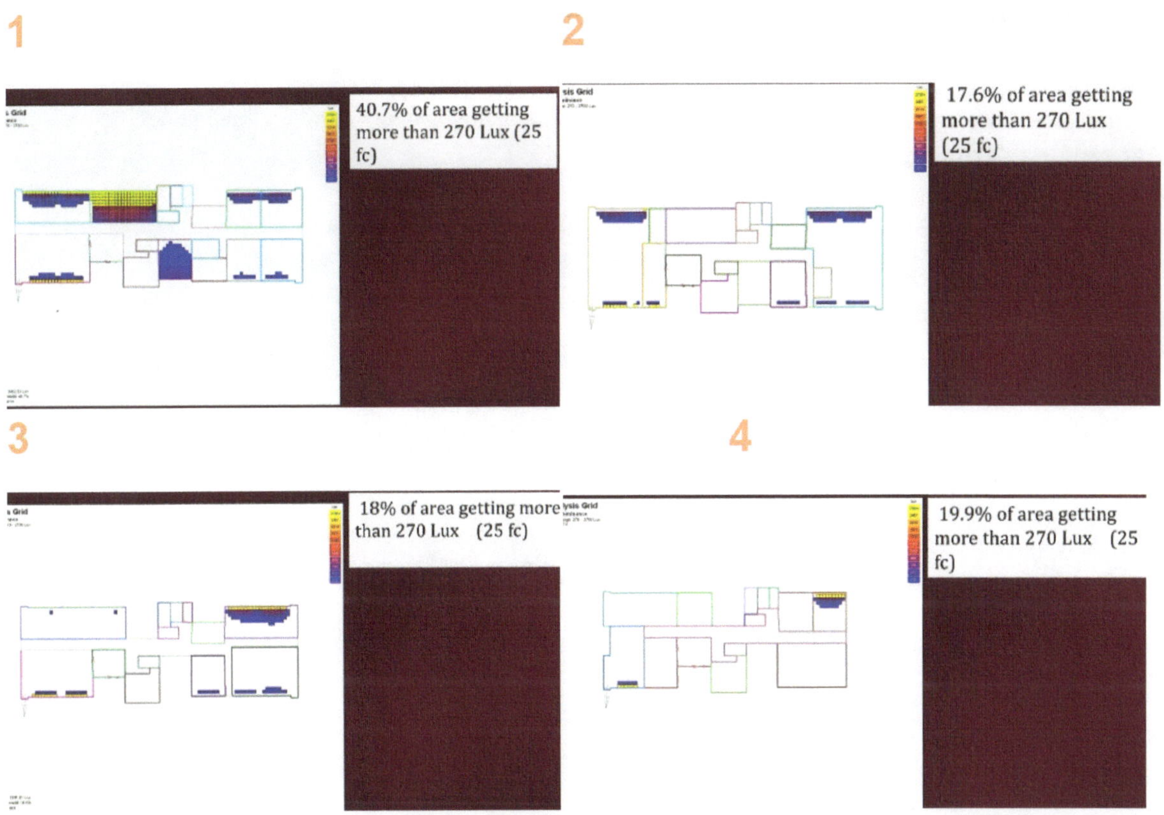

Fig. 6.182: Proposed Illuminations

6.29.11 Calculation

Table 6.7: Calculation EWMP

EWMP - KOLKATA			
Floor name	Total regularly occupied area Sq.ft	Total Daylighting area Sq.ft	Percentage
G	6099	2482	40.70%
1	5436	957	17.60%
2	5769	1050	18.20%
3	5661	1019	18.00%
4	1484	295	19.90%
5	3576	190	5.30%
Percentage of Day light	28025	5992.811	21.38%

6.29.12 Central Park Depot - Sun Path Analysis

The sun path analysis is used to understand the movement of the sun and the fall of shadows and shades on the proposed building by its own and by the neighboring structures.

We took sun path analysis for different periods as shown below:

1. Mar,ch 21st (Equinox)
2. June 21st (Summer Solstice)
3. July 21st (Local Equinox)

a) Sun Path Analysis in Pictorial Form

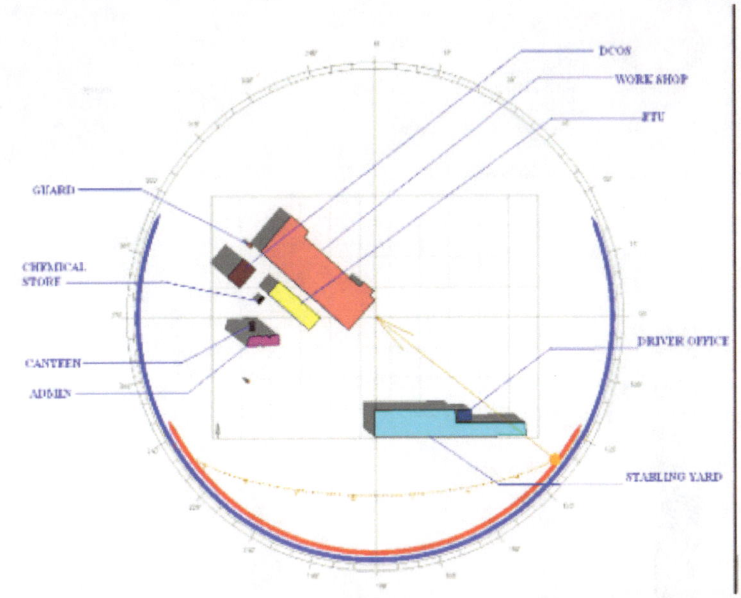

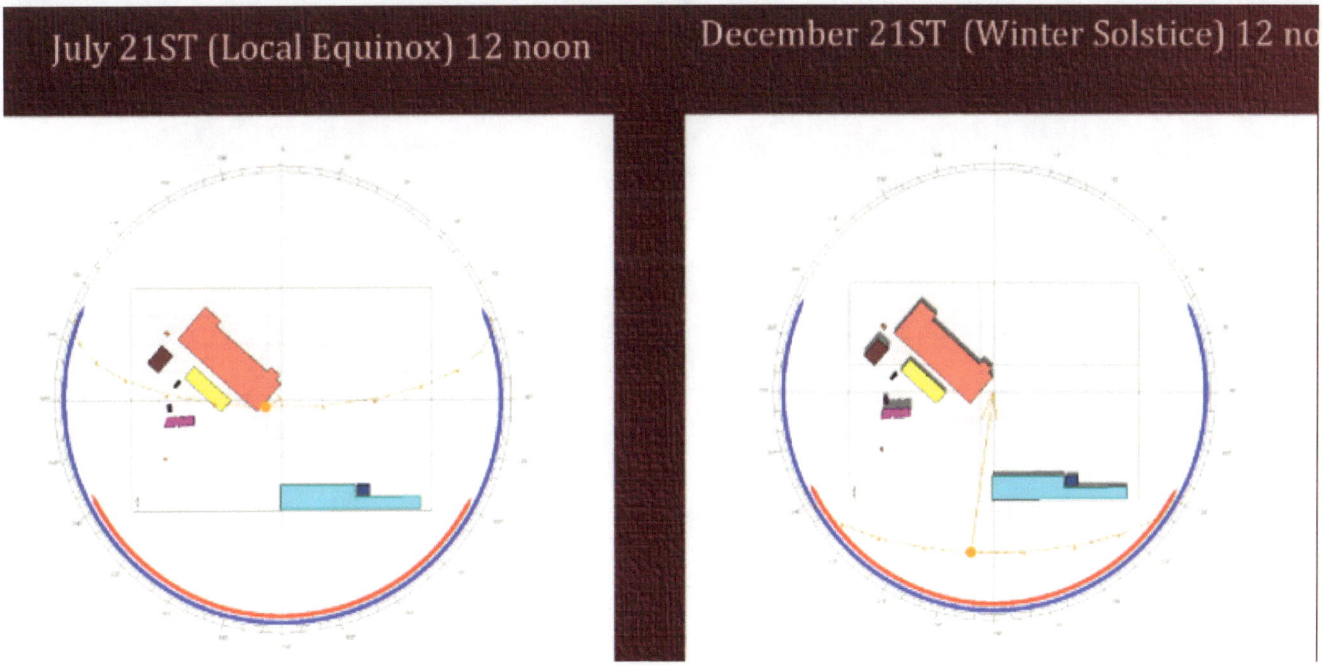

Fig. 6.183: Sun Position at Different Periods of Time

b) Sunlight and Sun Position at Different Times for Workshop Buildings

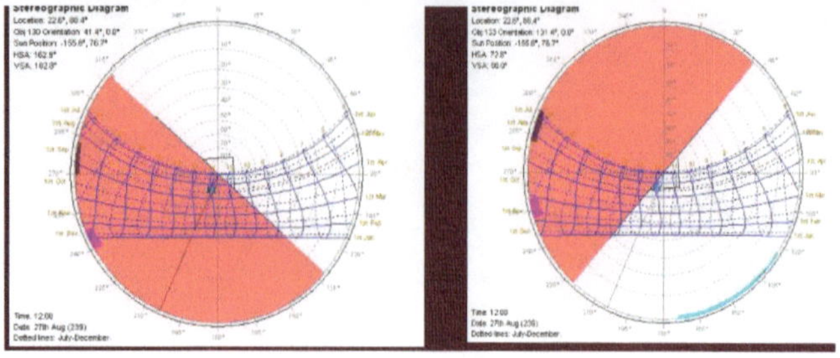

Northeast Southeast

Stereographic Design

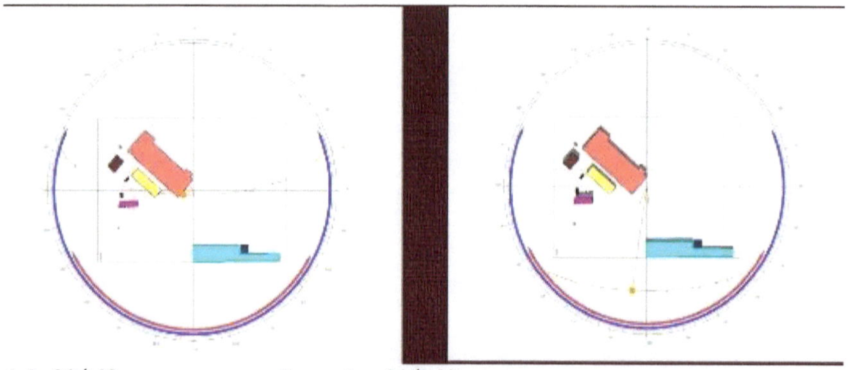

July 21st-12 noon December 21st -12 noon

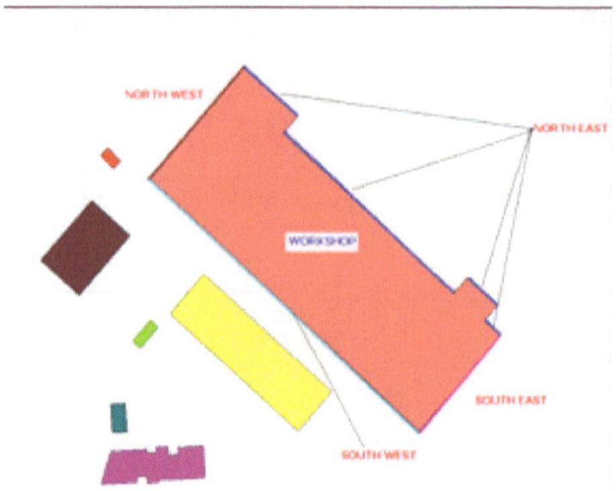

Fig. 6.184: Sun Position at Different Times

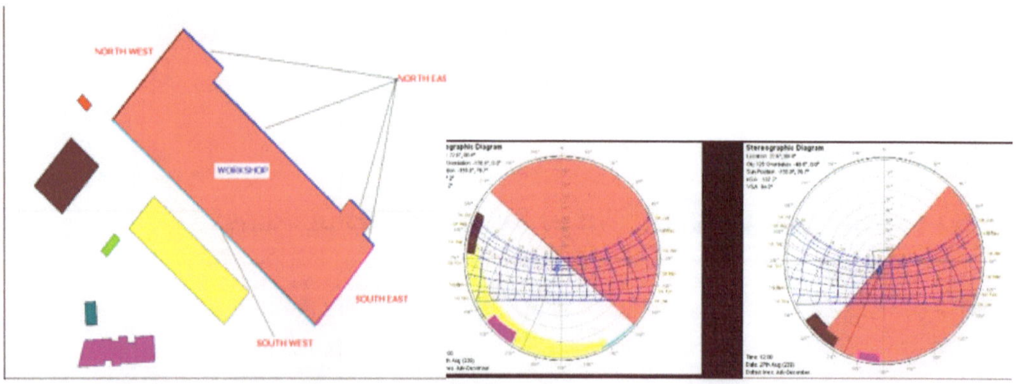

Northeast Northwest

Fig. 6.185: Wall Nomenclature at Each Direction Stereographic Design

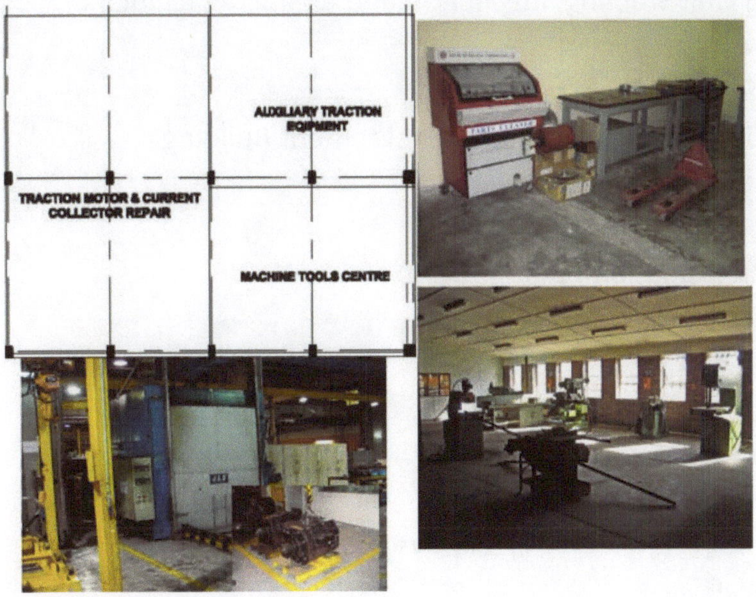

Fig. 6.186: Traction Equipment Maintenance

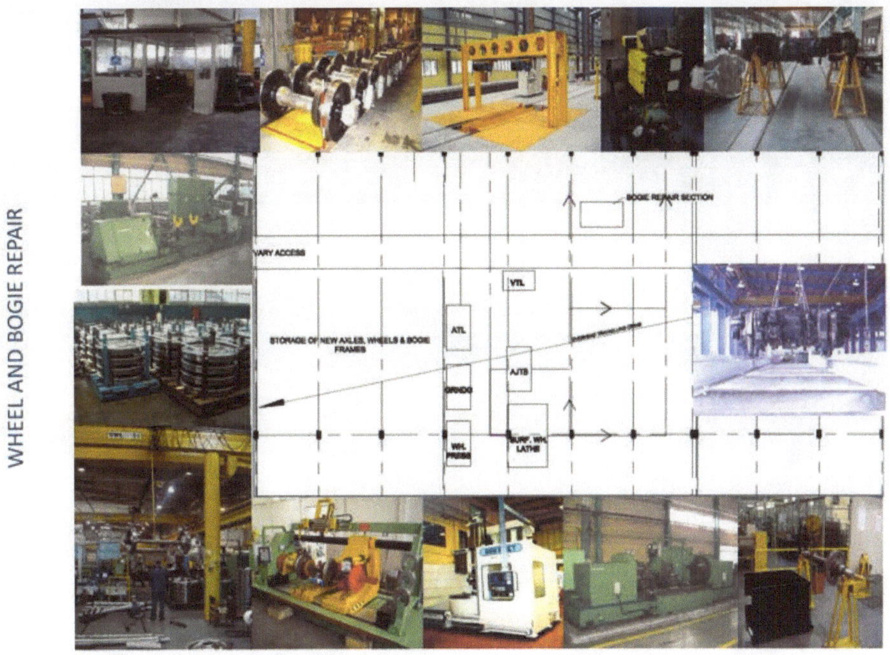

Fig. 6.187: Wheel and Bogie Repair

6.29.13 Workshops and Depot Buildings

The Central Park Depot includes the Following Buildings and Services.

i. Administrative Building, including the Operations Control Centre (OCC)
ii. Depot Control Centre (DCC)
iii. Stabling shed for Stabling of Metro trains.

iv. Engineering Train Unit Stabling, Inspection and Cleaning facilities.
v. Driver's Booking Office
vi. Time Office and Security Building.
vii. Depot Stores (DCOS) and Commercial Development Building
viii. Storage facility for inflammables.
ix. Canteen building
x. Underground water tanks and pump room.
xi. Overhead water tank.
xii. S&T Cable Store
xiii. Waste Store
xiv. Effluent treatment plant and room.
xv. Firefighting facilities for the entire depot
xvi. Roadways and Parking (including clearways for Emergency Vehicles
xvii. Site access and Walkway
xviii. Boundary walls and Fencing

Fig. 6.188: Workshop Building-Internal View

Fig. 6.189: Internal View of a Workshop

In addition, depot design includes

Services, Electrical fittings (light, fans, etc.), including laying of LV cables from the sub-station to all buildings, workshop, ETU shop, inspection shed, stabling shed, yard, footpath, security gates, etc.

Plumbing (water supply, soft water, treated water, stormwater, sewage and effluent and compressed air).

Communications (Cable galleries/conduits for Integrated Access control system, Public address system, telephone, LAN and CCTV)

Compressors and compressed air lines

Centralised AC system for the OCC, Administration and Property Development building. AC system using window cooler, Split Type for specified areas in the workshop, ETU, canteen building and the driver' booking office (xxiii) Sustainable Landscaping (xxiv) Power, trackwork, communications, signalling, rolling stock, E & M, M & P and others as advised by the Employer and Engineer.

Water Supply and storage tanks. Installation and testing of Pumps, including sump pumps and pipes for Drainage/sewerage.

Architectural works at the Central Park Depot.

A water supply system consisting of a deep tube well, overhead water tank, and associated pumps, valves, fittings, and pipes for the CPD Depot will be provided.

STABLING YARD

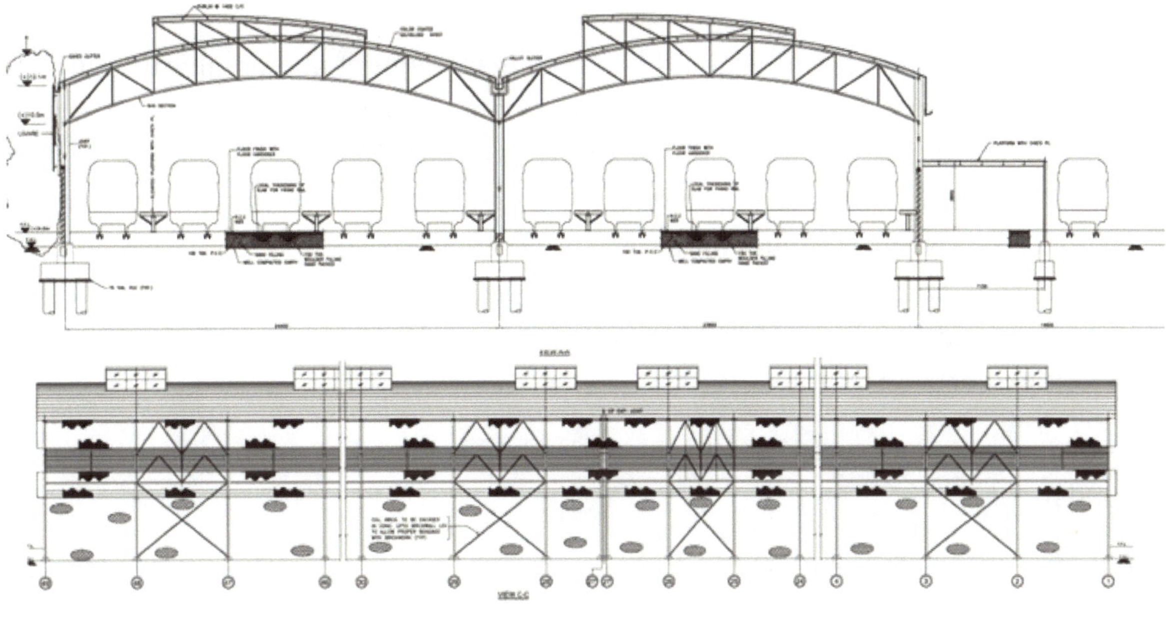

Fig. 6.190: Stabling yard

Fig. 6.191: View of Workshop Buildings including ETU (CCCL)

Fig. 6.192: External View of Stabling Yard (CCCL)

Fig. 6.193: Workshop Building

Fig. 6.194: Internal View of ETU Building

i) Administrative Building, Including the Operation Control Center (OCC) and Depot Control Centre (DCC), Excluding the System

Administrative office facilities are provided for various management and administration functions, including general managers, department managers, system support, IT services, human resources, finance and budgeting, document management, and others. The OCC and DCC were accommodated in the same building, comprising a central control room and associated offices and support facilities, equipment rooms for signaling, communications, SCADA, fare collection, and building facilities—electrical and mechanical systems, power supply, including UPS. The building had a central air-conditioning system and lifts. The air conditioner system had three nos., each with a chiller plant of 140 TR (Ton Refrigerant).

The admin building was constructed as a (G+5) storied building, an RCC-framed structure resting on RCC pile foundations.

ii) Workshop for inspection and repair—The main workshop buildings were designed as a prefabricated steel simple erection type structure with galvalume pre-painted roof sheeting. The Workshop Building included inspection bays, measuring tracks, periodical overhaul bays, mechanical repair, wheel and bogie repair, bogie wash, new car delivery, air compressors, and other specialised Workshops. Two-level side annexes supported the workshop, providing offices, amenities, local tools, consumables stores, and selected specialised component maintenance and testing.

The workshop building had brick cladding resting on RCC pile foundations, RCC columns, beams, etc. Also, 80 self-propelled turbo ventilation fans and intermittent transparent sheets have been provided on the rooftop to keep the sound entry of natural lighting and good ventilation within the shed.

6.29.14 ETU Building:

As per the contract provision, the building was designed and constructed similarly to the Workshop and was equipped with one X 5T capacity crane. The ETU building tracks were embedded with two center pits of 1.5m depth for water, power, compressed air outlets, pit lighting, and 25VDS and 3-phase access power for welding and other such plant and equipment. The design provided adequate natural lighting and ventilation.

A conceptual copy of the approved GAD of ETU is enclosed for reference. Like Workshop Buildings, ETU Buildings are also made from fabricated steel structures resting on RCC pile foundations, RCC columns, beams, etc., with pre-painted galvalume roof sheeting on top. The firefighting arrangement with sprinklers (inside), fire alarms (like smoke detectors, heat detectors, manual call points inside the building), Hydrant points (around the building) and extinguishers has been made available for the building.

6.29.15 DCOS Building

DCOS Building, during finalisation, was constructed as a B+G+4-storied RCC-framed structure resting on RCC piles. The basement was built for car parking, providing 1st, 2nd & 3rd floors for commercial development. Four lifts of 16-person capacity were installed from the basement to operate between all floors. The basement's whole area and the 1st floor's specified area's firefighting arrangement were included in addition to the basement portion for well ventilation, fresh air and jet fans with an automatic CO2 sensor were installed. One 3T capacity Monorail had been installed in the loading and unloading bays on the ground floor.

6.29.16 Stabling Yard

The Stowing shed building was built with a prefabricated steel structure with a third rail to allow rakes to move under power. The stowing area was closed from the side to protect the vehicles from direct sun, storms, and rain. The adopted building structure was to accommodate future extensions. The car access platform was provided with transverse access, enabling drivers and cleaners to enter the vehicles.

Like the Workshop Building and ETU Building, the Stabling Yard was made of fabricated steel structures resting on RCC pile foundations with pre-painted galvalume roof sheeting. Five nos. inspection bays, including pit lighting arrangements, were made. An overhead system was provided instead of the Third Rail system for safety purposes.

6.29.17 Canteen

According to local uses, a Canteen for the staff operating and maintenance and KMRC personnel was built in the Depot for fifty diners. The canteen was a single-storied RCC-framed structure with brick cladding. An indoor dining space of (11.730 x 7.740) sq.m. with seven split AC of 2TR capacity each was installed. The firefighting arrangements with sprinklers (inside), fire alarms (like smoke detectors, heat detectors, and manual call points inside the building), Hydrant points (around the building) and extinguishers were made available for the building.

6.29.18 Driver's Booking Office

The Driver's Booking Office building was dedicated to driver booking and included a management office, rostering room, waiting and rest room, lockers, showers, and toilets for the drivers. Due to the early morning beginning of operation, five small sleeping rooms and a dining room were provided. This building was constructed as a double-storied RCC-framed structure resting on RCC foundations.

6.29.19 Site Security Posts

This building was a light, prefabricated, small structure, and its aesthetics were related to the overall depot site. Two Guard Houses, i.e., the primary guard house and the secondary guard house, were constructed at the site. It is an RCC framed with brick cladding, glazed windows, water lighting, and toilet facilities. The firefighting arrangements with fire alarms, Hydrant points, and extinguishers have been made available for the building.

6.29.20 Small buildings like the Chemical Store, Waste Store and S&T Cable Store

Three separate small buildings were constructed: one for the chemical store, one for the waste store, and another for the S&T cable store.

6.29.21 Water Supply and Drainage

As per the contract, the depot was provided with a sufficient water supply for various requirements. The water was Potable for the Canteen, amenities, water fountains, locker and shower rooms, and sinks in toilet rooms. Water was also processed for maintenance, servicing, and firefighting.

In addition, there are **Underground Reservoirs and Pumps, an Effluent Treatment Plant Firefighting system, a VAC System, Electrical Works including UPS and DG Sets, an Internal Drainage System, an**

Internal Sewerage System, an External Drainage and Sewerage Disposal System, Horticulture work, Sustainability and Green Depot Concept, Fencing, Roads, Carparking, walkways, and Fencing.

A brief about Sustainability and Green Design has been included earlier in this section. To mitigate the above requirement, the agency (M/s CCCL), the Depot Contractor, appointed one consulting agency named "LEED Facilitator, Energy Modelling and Lighting Simulation Consultant: M/s Conserve Consultants Pvt. Ltd.", to submit and to receive the Certificate to IFS Business (Green built certificate of CPD) issuing authority named "USGBC Breen Business Certification Institute Pvt. Ltd., Noida, Uttar Pradesh-201301. The appointed agency on behalf of M/s CL to get the Depot Green Building concept certificate has already applied and deposited money for registration and the required certificate.

6.30 Track Structure

The East-West Metro in Kolkata utilises both elevated and underground sections, with a track structure designed for a rapid transit system. The project includes elevated and underground stations, and the track sections are engineered for reliable and efficient train operation.

In the Indian Metro System, both ballasted and ballastless tracks are used. Ballastless tracks are being used for elevated and underground metro corridors. In the Depot area and tracks at grade, ballasted tracks are used. Ballasted tracks in the general Indian Railway specifications are being followed with appropriate modifications to suit the alignment and gauge adopted. Kolkata Metro was the first to bring ballastless track in India, and the first to be used in the North-South Metro, i.e. Line 1 of the Kolkata Metro network. Extensive pulsometer tests were carried out on the assemblies in the RDSO at Lucknow.

In 1964, the ORE of UIC formed a committee to study various ballastless tracks to evolve a standard measuring method for noise, vibration and performance of fitting. The composition of the ballastless track is better suited for a hard surface, like a concrete base, replacing ballast with a bed of concrete. The rails rest on rubber pads placed over concrete sleepers, fixed on the concrete bed. The ballastless track eliminates dust and noise pollution and can achieve high speeds. It is, with minimum joints and better load distribution, more stable. Ballastless track can be i) Fixed base, such as Continuous RCC Plinth or Continuous slab ii) Floating Slab with Continuous Resilient Medium or Resilient Pads at intervals and iii) Sleepers inserted in Concrete.

Basic advantages of a ballastless track are Stability, Resilience, Durability, Fastening System and ease of replacement of sleeper or fastener. In addition, ballastless track has lower dead load in compared to ballasted track and less overall thickness, which helps in reducing the Tunnel size for underground Metros.

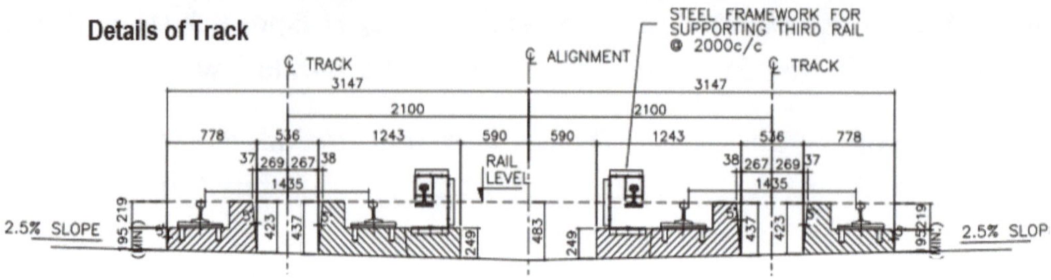

Fig. 6.195: Track Structure used

Fig. 6.196: Track Structure for East-West Metro (Ref. Construction World)

The Recommendation of MOHUA

Rails: All the metros are using 60 kg/m rails, and to adopting such rails for metro systems. Head-hardened rails of 60 kg/m are used in metros at sharp curves (500m or sharper). Gauge and SOD. Both the Broad Gauge and Standard Gauge systems may be considered by cities. Curves and Gradients: Minimum curve radius of 120m for SG system. The gradient is not more than 3% (desirable) compensated for curvature, and max 4% in exceptional circumstances, with compensation for curvature.

A design speed range of 85-95 kmph and an Operating Speed range of 80-90 kmph for standardisation for the MRTS. Check Rails on sharp curves, as assumed to be necessary on sharp curves, following the legacy. Research has been conducted to check whether check rails are at all necessary. Briefs on the East-West metro are as follows.

Elevated Sections: A portion of the project features elevated sections, with 6 elevated stations and corresponding track structures.

- **Underground Sections:** A significant portion of the line, particularly in areas like the Howrah Maidan-Salt Lake stretch, is underground, requiring specialised tunnel construction and track layouts.

- **Track Gauge:** The East-West Metro utilises a 1,435 mm (standard gauge) track gauge.
- **Power Supply:** The metro line uses a 750V DC third rail electrification system.
- **Signaling and Train Control:** Modern signaling and train control systems are integrated into the track structure for safe and efficient train operation.

6.31 Supervision

Supervision is crucial in Indian metro projects to ensure construction quality, safety, and adherence to timelines and budgets. Effective supervision involves monitoring construction activities, ensuring adherence to standards, resolving technical issues, and coordinating with stakeholders. It also plays a key role in operational monitoring and maintenance of the system to ensure its efficiency, safety, and longevity.

The important elements of supervision:

a) Ensuring Construction Quality and Safety

- **Adherence to Standards:**

  Supervisors ensure that all construction work adheres to the relevant safety standards, quality control measures, and project specifications.

- **Monitoring Progress:**

  Supervisors track the progress of construction, identifying potential delays or problems early on.

- **Resolving Technical Issues:**

  They are equipped to handle technical issues that arise during construction, ensuring they are resolved efficiently and effectively.

- **Safety:**

  Supervision is crucial for maintaining a safe work environment for all personnel involved in the construction.

b) Efficient Project Management

- **Adherence to Timelines and Budget:**

  Qualified supervisors help to ensure the project stays on schedule and within the allocated budget.

- **Stakeholder Coordination:**

  To facilitate communication and coordination between stakeholders, including contractors, suppliers, and government agencies, to ensure smooth project execution.

- **Quality Control:**

  Dedicated supervision helps maintain high-quality standards throughout the construction, leading to a reliable and durable infrastructure.

c). **Operational Efficiency and Maintenance**

- **Monitoring System Performance:**

  Supervision is essential for monitoring the performance of the metro rail system once it is operational.

- **Maintenance and Repair:**

  To ensure the metro system is properly maintained to minimise downtime, ensure passenger safety, and optimise operational costs.

- **Emergency Response:**

  Qualified supervisors are crucial in managing emergencies that may arise both during implementation and operation.

d). **Indigenisation and Technology**

- **Developing Indigenous Technologies:**

  India's development of its Automatic Train Supervision (ATS) system, like the i-ATS, is a significant step towards reducing reliance on foreign vendors and enhancing the capabilities of Indian metro systems.

- **Integrating New Technologies:**

  Supervision is important in integrating new technologies and innovations into metro rail systems, ensuring they are implemented effectively and safely.

e) Independent Testing of Materials

* Physical & Chemical testing of reinforcement steel is done at an independent Government Laboratory.

* Materials like cement, aggregates, bentonite, water, and admixtures are also tested at an independent Laboratory.

* Pile Integrity Testing by a Third Party have been carried out.

* Testing of pre-stressing materials and bearing is done by GC.

Overall, supervision is an integral part of metro rail projects in India, contributing to the successful completion of construction, ensuring the safety and efficiency of the system, and fostering its long-term sustainability.

6.32 Building Condition Survey

6.32.1 Introduction

Building Condition Survey (BCS) is a critical and important activity to construct elevated and underground transportation structures, particularly in an Urban Context. As such, this activity is a must in a thickly built-up area. For East-West Metro GC of the project MYCEL, comprising AECOM, Yachio, CES, EGIS & Lea Pomeroy, conducted the BCS for the original alignment of the metro corridor. A copy of the report, prepared in two volumes, was distributed to the bidders. In addition, as per the contract, the design-build contractors also carried out the BCS before the construction was resumed. **Annexe 3** includes the excerpts of the BCS carried out by GC includes the basic approach of the Survey. As such, a few slides were included from the survey conducted for Howrah Maidan to Sealdah and presented by Afcons Infrastructure. This will help readers to get a background of the conditions of the existing structures on both sides of the corridor.

The building condition survey has been done within the influence zone, 25m to 40m from the center line of the Tunnel/Cut-and-Cover section on either side, to take proper precautionary measures before taking up the construction to prevent/minimise the extent of damages during construction of the underground structures.

Fig. 6.197: Buildings Surveyed

Fig. 6.198: Extent of Damage

Table 6.8: Category of Building Conditions

TYPE	STATUS					TOTAL
	GOOD	AVERAGE	BAD	VULNERABLE	U/C	
G+0	24	32	6	1	0	63
G+1	18	55	47	10	0	130
G+2	21	41	34	6	1	103
G+3	30	16	32	2	2	82
G+4	26	8	10	0	1	45
G+5	21	8	2	0	1	32
G+6	6	3	0	0	0	9
G+7	4	2	1	0	0	7
G+8	1	1	1	0	0	3
G+9	3	1	1	0	0	5
G+10	2	1	2	0	0	5
G+11	1	1	0	0	0	2
TOTAL	**157**	**169**	**136**	**19**	**5**	**486**

Table 6.8, includes the Category of 486 buildings surveyed.

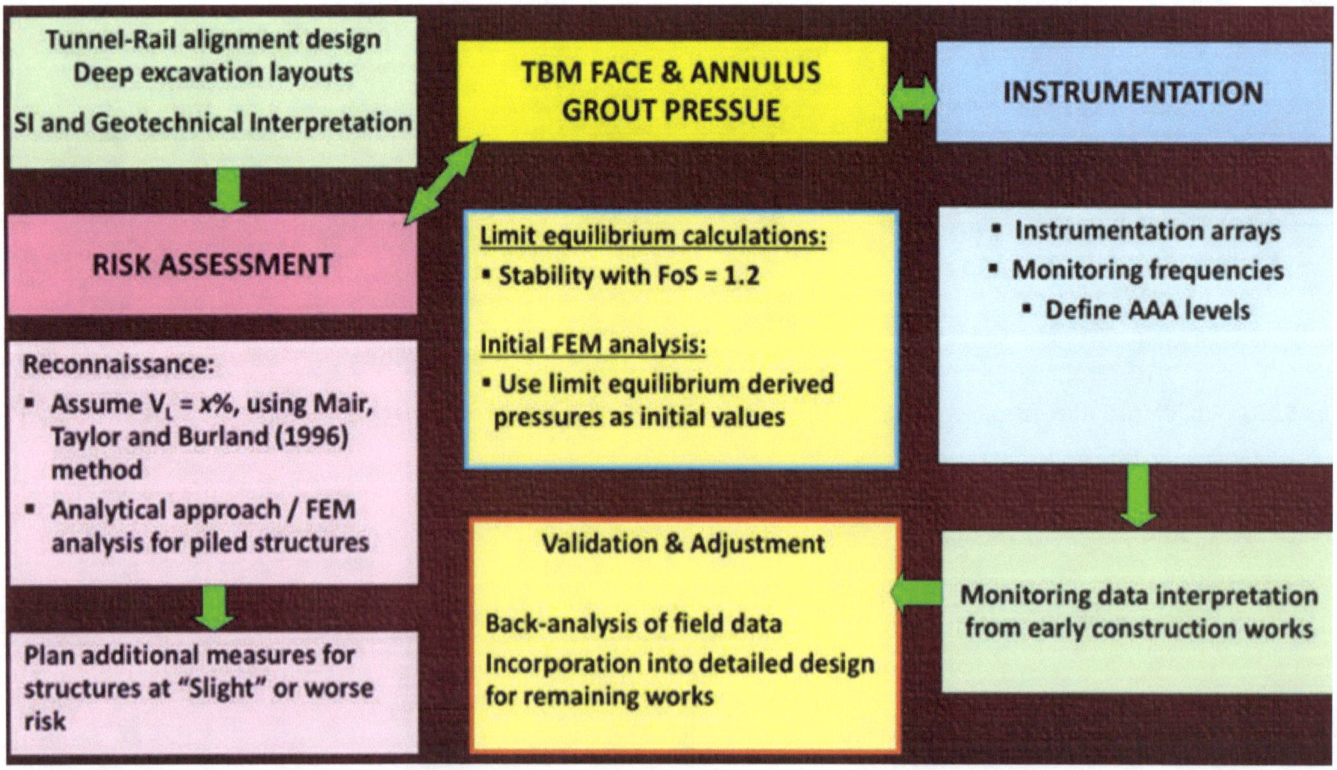

Fig. 6.199: Assessment of Building Damage (Slide 1)

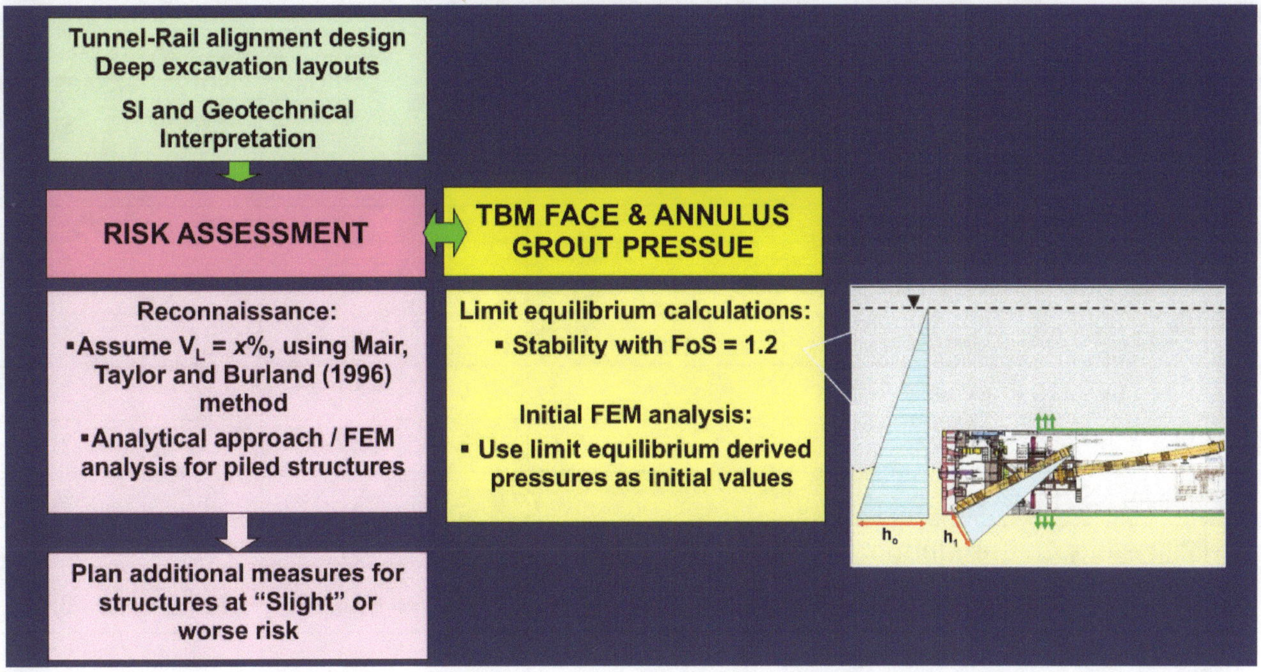

Fig. 6.200: Assessment of Building Damage (Slide 2)

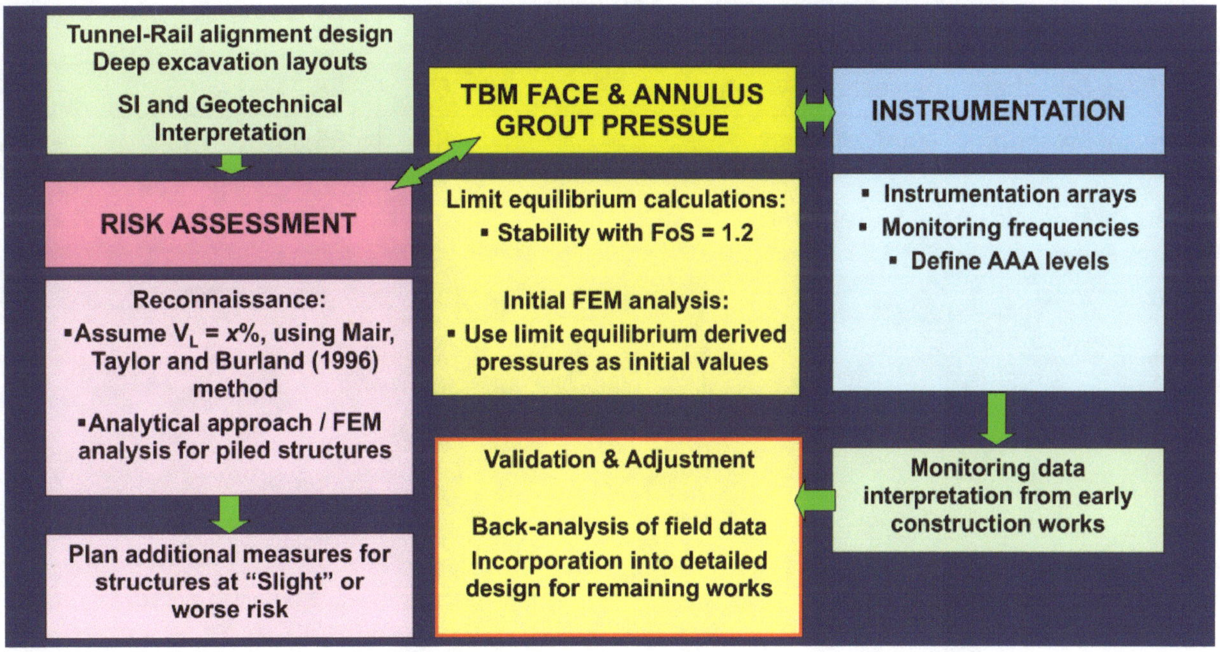

Fig. 6.201: Assessment of Building Damage (Slide 3)

Figures 6.199, 6.200, & 6.201: Includes the BCS in tabular format.

6.33 Noise Level Monitoring

The noise level assessment survey has been carried out at different locations along the alignment of the Elevated corridor to initiate the mitigation measures.

Table 6.9: Noise Levels Measured at Different Locations

Sl no.	Noise monitoring location	Area Code	Leq Day	Noise Std.	Excess over Std.
1	Place: Wipro-near Main Gate, Sector-V, Kol-91	C	69.52	65	+4.52
2	Place: Near Balurghat Bhawan	R	51.30	55	-3.70
3	Place: Anand Lok Hospital-Main Gate	S	67.34	50	+17.34
4	Place: Hindustan Sweets	C	63.56	65	-1.44
5	Place: Vikash Bhawan	C	61.90	65	-3.10
6	Place: Indira Bhawan	R	57.82	55	-2.82
7	Place: ILS Hospital	S	56.14	50	-6.14
8	Place: City Centre – Infront of HSBC Bank	R+C	57.34	65	-7.66
9	Place: Vidyasagar More-Vidyasagar Univ. Main Gate	S	64.99	65	-0.01
10	Place: Shyamali Apparetment	R	54.13	55	-0.87
11	Place: 115 Duttabad Road-TMC Party Office	R	71.15	55	+16.15
12	Place: Apollo Gleangles Hospital-Main Gate	R	70.08	50	+20.08
13	Place: Nandankanan Club, 262 Anandpur Road	R	69.13	55	+4.13
14	Place: Beleghata Intersection-Swabhumi	C	68.74	65	+3.74
15	Place: Swabhumi-Near Main Gate	R	56.55	55	+1.55

C-Commercial **R**-Residential **S**- Sensitive

Fig. 6.202: Noise Monitoring-Instruments & Software Used

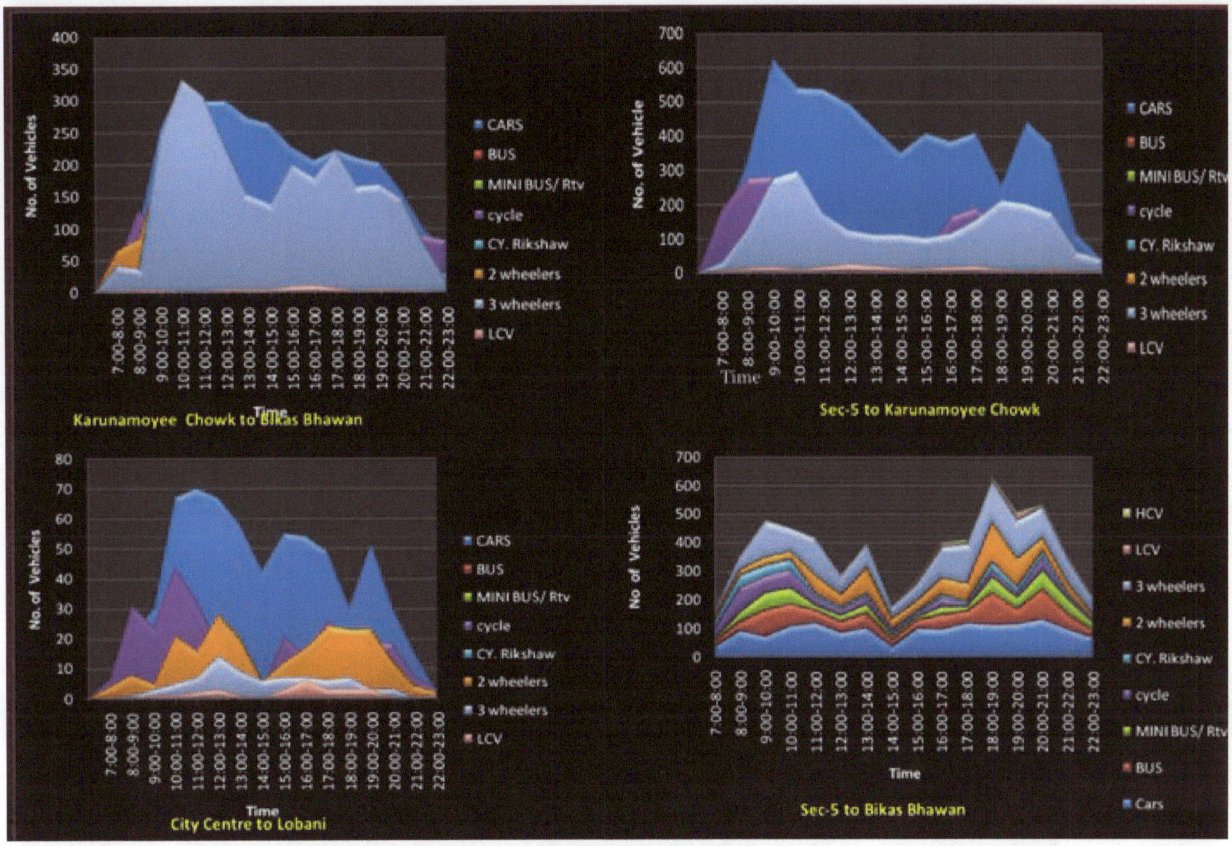

Fig. 6.203: Noise Monitoring- as per Number of Cars at Different Locations

6.34 Quality Control and Monitoring

(In continuation of the earlier Paragraph)

Contractors submitted a detailed Quality Assurance Manual demonstrating the proposed method of achieving the required quality standards of the Employer. The quality manual addressed the quality system as per ISO 9001-2008. The quality manual lists out the procedures, activities, checks, standards and any such thing, such as a schedule of tests, which he proposes to carry out during the progress of the works and are necessary to comply with the quality objectives of the Employer. The manual also lists the corporate policy adopted by the contractor.

Important items to adhere to the Quality brief below;

a) Quality Plans, Method Statements & Job Procedures: Inspection and Test Plan, Testing Formats, Concrete Design Mix, etc., were developed, reviewed, and recommended to the Client.

b) Quality Control:

* Testing equipment, including calibration, schedule, etc., was verified before use.

* Third-party testing laboratories were selected based on the facilities.

- Verified testing equipment, including calibration records and schedules, before use.
- Selected third-party testing laboratories based on their facilitie

c) Material Testing and Approval:

- Conducted at the Contractor's Laboratory
- Conducted by a Third Party

d) Quality Certification and Quality Audits: Conducted

The following photographs include the materials tests at the site laboratory and the production of concrete per design strength and specification. Three Batching plants were used to produce concrete separately for the elevated and two for the underground sections, one on either side of the river.

Fig. 6.204: Testing of Materials at Site Laboratory

Fig. 6.205: Batching Plant at Casting Yard for Box Segment, Elevated Corridor

6.35 Safety, Health and Environment

The compliance of Safety, Health and Environmental Protection safeguards as given in the combined Safety, Health, and Environmental Manual (SHE) provided by KMRCL. below. The Contractors provided necessary access, assistance and facilities to enable the Engineer and Employer to carry out surveillance to verify that the site safety and safety assurance plans were being fully implemented per Contract. Contractors had to maintain housekeeping at sites and ensure safe surroundings. Like mishandling of bentonite used for DW construction and elsewhere, splashing of bentonite outside the barricading or non-cleaning of tyres of dumpers & transit mixers. All issues to cover safety, health & environment are included in the following bullet points, which are self-explanatory.

a) Safety in Design & Safety in Execution

- Health & Environment
- Enabling/Temporary Structures
- Fire/Smoke
- Evacuation
- Building Condition
- Noise Monitoring

b) Risk of injury is not just confined to difficult operations, but also to activities relatively simpler in nature Accidents happen due to

- Lack of foresight
- Poor or insufficient planning
- Carelessness
- Deficiency in the assessment of risk and mitigation measures

There are inherent hazards--Safety methods are not properly in place, and are often left to an individual worker or operator

c) Safety, Health & Environment (SHE)

C1) Project Goals and Objectives

- Zero Dangerous Occurrences
- Zero Fatalities
- Minimised Reportable Accidents
- Ensure adequate measures to avoid accidents (safety)
- Occupational illness (health)
- Harmful effects on the environment

Safety can impact on budget and schedule

  - Production should not circumvent safety requirements

- To take a proactive and prescriptive approach to safety management
- To identify, manage and mitigate all risks considering the safety challenges

C2) Types of accidents

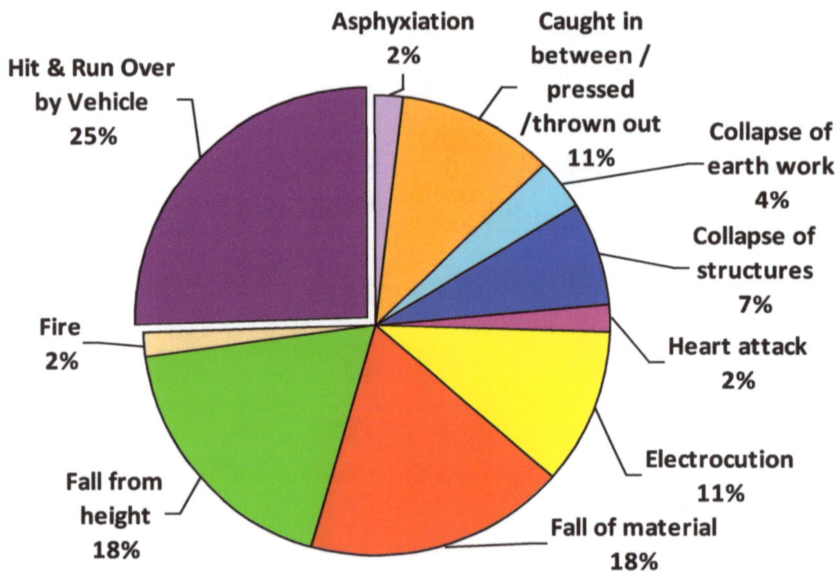

Fig. 6.206: Pie Chart Includes Types of Accidents

d) Safety as Cost Control Measures

d1) The importance of Safety as a cost-controlling measure

- Defects or failures at construction sites result in huge costs and delays.
- Dedicated commitment ensures project success within cost & time

d2) Safety is a critical item on all construction projects

- Welfare of employees - Safe equipment and work procedures, basic amenities like drinking water, toilets, canteens, rest places
- Safe work environment - Client, consultants and contractors deploy experts on safety
- Controlling construction costs - Controlling/eliminating all possible hazards, including Noise, Dust, Heat and Cold
- Statutory benefits applicable to construction workmen and additional compensation through the Labour Welfare Fund

e) Occupational Health

- Health issues in construction
- Health and the law & surveillance
- Occupational health policy statement

f) Security & Resilience

- Prudent overreaction, Preparedness & Response plans
- International obligations and national laws
- Threat levels, Inspections, validation, and verification

6.36 GC Recognition and Intervention

- Child Labour during work deployments
- Unplanned utility shifting operations
- Unplanned and unsafe geotechnical surveys
- Unsafe lifting (methods and devices)
- Unsafe traffic planning
- No PPE issuance to employees
- No Safety Planning (lifting, excavation, etc)
- No Access and Control

6.37 Prevention and Control of Pollution

Prevention and Control of pollution are among the most important issues for construction, as mandated by the Air (Prevention & Control of Pollution) Act, 1971, and the Environment (Protection) Act, 1986. In this regard, all Contractors were supposed to follow the activities of the Project Site; a few are mentioned here.

a. Wrap the construction area with geotextile fabric, installing duct barriers, or as appropriate.
b. To apply water and maintain soils in a visible damp, and apply water before levelling or any other earth-moving activities.
c. Vehicle speeds were limited to 15 mph on the work site.
d. Cleaning of wheels and undercarriage of haul trucks before leaving the site.
e. To apply a cover or screen to stockpiles and stabilise stockpiles.
f. To stabilise surface soils where loaders, support equipment and vehicles operate.
g. To stabilise adjacent disturbed soils.
h. Stabilise sloping surfaces.
i. Disposal of debris in consultation with the local authorities
j. Ambient noise level not to exceed 65 dB (A)s,

Typically, underground stations are spaced at 1 km, interconnected through a tunnel

*Heat is generated by Traction and other equipment

*Today's state-of-the-art modern rolling stock, provided with air-conditioners, dissipates substantial heat inside the subway, resulting in hot air in the tunnel and pulsating heat load on the platform

Consists of

- Ventilation of the Tunnel
- Mechanical Ventilation for Plant Rooms
- Smoke Control and Extraction

Components

*Tunnel Ventilation Fans

*Attenuators, Dampers & Nozzles

*Air Compressors

*Tunnel Booster Fans

*Trackway Exhaust Fans

Tunnel Ventilation System

*Provide required ventilation for different operations inside the tunnel, including

- Normal operating conditions; to provide relief from the piston effect.
- Congested operation; to provide air flow inside the tunnel to control temperature rise.
- Emergency scenario of fire.

  * Extract smoke from the tunnel and

  * Provide a safe evacuation path for the passengers

  * Provide a safe path for the firefighting personnel

- During the non-revenue period, to provide ventilation inside the tunnel for maintenance staff

Tunnel Ventilation Fans

Emergency scenario of fire.

  * Extract smoke from the tunnel and

  * Provide safe evacuation paths for the passengers

  * Provide a safe path for the firefighting personnel

- During the non-revenue period, to provide ventilation inside the tunnel for maintenance staff

Emergency operation

Resorted to when smoke is generated in the tunnel or station trackway from any variety of occurrences.

The tunnel ventilation system is set to operate to control the movement of smoke and provide a safe and smoke-free path for evacuation of the passengers and entry for the fire-fighting personnel.

Prevent back-layering of the smoke layer and discharge the smoke to the exhaust without passing through occupied areas.

As per NFPA-130 section 7.2.1, the emergency ventilation system shall be designed to do the following

- Provide a tenable environment along the path of egress from a fire incident in enclosed train ways
- Produce airflow rates sufficient to prevent back layering of smoke in the egress within enclosed train ways.
- Be capable of reaching full operational mode within 180 seconds.
- Accommodate a maximum number of trains between ventilation shafts during an emergency.

* Exhaust Fans

6.38 Fire Safety Design

Fire Prevention Measures

*Fire Detection Measures

*Fire Control Measures

*Means of Escape

*Access for Firemen

*Means of Fire Fighting

*Housekeeping and Staff Training

Fire Prevention Measures

Potential sources of fire can be reduced by

*Layouts to permit ease of maintenance and cleaning of stations & accessibility.

*Use of non-combustible & smoke retardant materials.

*Appropriate location and installation of equipment

*Special storage for combustible materials

*Prohibition of smoking in a "no-smoking zone"

*Always have good earth connections

Fire Detection Measures

Automatic Smoke/Heat Detectors and Manual Call Points connected to the Fire.

The Control Panel is in the Station Control room, along with

*Mimic panel indicating fire status on a Zone basis

*Fire panel detection of all designated areas.

*Fire Audible/Visual alarms raised in staff areas, linked to telecommunications, sprinkler discharge controls, fire dampers, ventilation equipment & suppression systems

Fire Control Measures

Spread of fire and smoke controlled by

*Compartmentalisation of Fire Risk Areas

*Smoke Extraction

*Smoke Compartmentalisation

*A compartment is a portion of the building, separated from adjoining portions by walls/ floors/doors

*Every opening is protected to contain fire and the spread of smoke

- All openings sealed
- Ducts opening sealed with fire dampers
- Fire-rated doors

*Depending upon the degree of fire load and the degree of protection required for life safety, Fire Resistance Periods (FRP) are prescribed for different areas of the station

Smoke Management

To ensure that the escape routes are not full of smoke

*To provide a safe passage for firefighters to reach the fire location

*Smoke extraction system

*Smoke containment

Pressurisation- firemen and emergency staircase

Smoke barrier– At 3 m height every 60m, around openings around staircases, escalators

Mechanical ventilation – tunnels

Means of Escape

Evacuate passengers from the platform and train to a point of safety without obstruction

*Calculations made for each station, at least 2 means of egress, remote from each other, planned

*NFPA 130 guidelines followed

- The total Evacuation time (NFPA- 5.5.6)
- From platform to a point of relative safety 4 mins
- To a point of ultimate safety - 6 mins
- (NFPA-5.5.6.1.1); Max. Travel distance on the platform to the point of egress shall not exceed 100 m.

Smoke Control and Extraction System

To prevent back-layering of the smoke layer and discharge the smoke to the exhaust without passing through occupied areas.

*To provide a safe and non-contaminated path for the evacuating passengers and the fire personnel.

*To avoid the propagation of smoke into other non-affected areas by providing fire-rated ducts and dampers

Vibration Level

- State-of-art modern technologies to control vibration level during operation of the system including:
  - Use of modern rolling stock so that ground vibration level during operation is imperceptible
  - Provision of suitable track components including rubber pads between rail and bearing plate and also bearing plate and the concrete to avoid ground borne transmission of noise and vibration
  - Based on the finally chosen rolling stock characteristic, the provision of floating track slab on mass springs may also be examined at the detailed design stage

6.39 Systems

Systems in Metro Operation

* Track

* Power Supply/Substations

*Traction/Auxillary-33 kV cabling

*Rolling Stock

*Signal and Telecommunication

*Automatic Fare collection

6.39.1 Power Supply

Siemens supplied the complete railway electrification system for the East-West metro line. The company received a corresponding order from the Kolkata Metro Rail Corporation (KMRC), which included project management, design, installation, and commissioning work, as well as personnel training and maintenance throughout the three-year warranty period. Construction of the metro line was done in two phases, with the first section, an elevated corridor and six stations, being finished by the end of 2013. The second section, which is entirely underground, was scheduled to open in two phases. The power supply activities were completed at the end of 2024 for the full corridor, including six underground stations. It also includes a tunnel underneath the Hooghly River, the first underwater metro line in India.

The line is completely electrified by Siemens Mobility with a 750 V DC traction power supply collected from a third rail system. The power for the nine substations and the depot was fed into the system via a 33 kV AC medium-voltage ring. The Calcutta Electric Supply Corporation Ltd was selected to supply the power. A SCADA system was adopted to monitor and control the entire power supply system. A Sitras SMS type stray-current monitoring system was adopted to eliminate any damage due to stray current corrosion, thereby lowering the cost of maintenance.

Detailed breakdown

- **Traction Power Supply:** Siemens supplies a 750 V DC traction power supply for the trains.
- **Power Distribution:** 33 kV AC medium-voltage ring powers the substations and the depot, with power provided by the Calcutta Electric Supply Corp.
- **Project Scope:** Siemens' contract encompasses project management, design, installation, commissioning, staff training, and three-year maintenance.
- **Line Details:** The East-West Metro, also known as the Green Line, is an 8.9 km line With six stations, including a tunnel under the Hooghly River for the UG section
- **Operational Status:** While the line is complete, the line's sections have been operational in phases, with the Salt Lake Sector-V to Sealdah section opened in early 2020.
- Siemens energised in 2011, providing the 750 V DC traction power supply, with nine substations and the depot to be fed from a 33 kV AC...

6.39.2 Traction System

Traction Distribution Two types of traction supplies are used in Metro Railways: A. Conventional 25KV AC Overhead supply. The traction supply is made through a conventional overhead catenary system, though some design features are different from the Indian Railways system. B. Third Rail system with DC supply. Traction power supply is made through a 750V DC system through a conductor rail laid parallel to the track (called third rail)

Generally, three options for power supply systems are considered in Indian Metros, which are

- 25 kV & 2x25 kV AC Overhead Catenary system
- 750 V DC third rail system
- 1500 V DC Overhead Catenary system

Fig. 6. 207: Overhead Catenary System

Supporting structure: Poles, beams and other structures used to install the various lines and wires over the permanent way.

Fig. 6.208: Third Rail Traction for Bangalore Metro

Fig. 6.209: Kolkata Metro to Replace Steel by Aluminium Third Rail

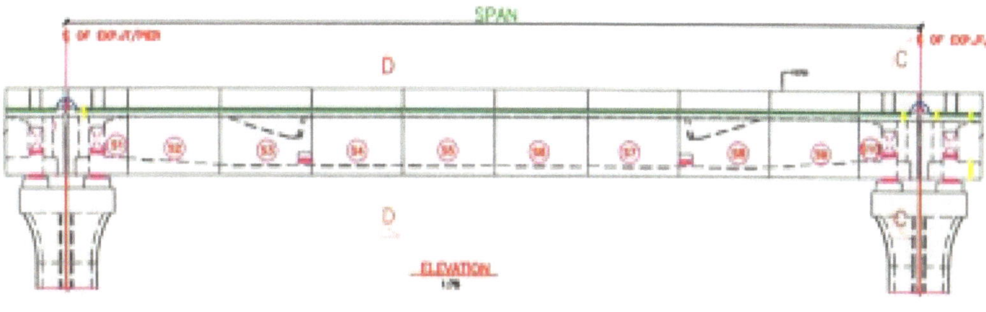

Fig. 6.210: Typical Earthing Arrangement for Elevated Track

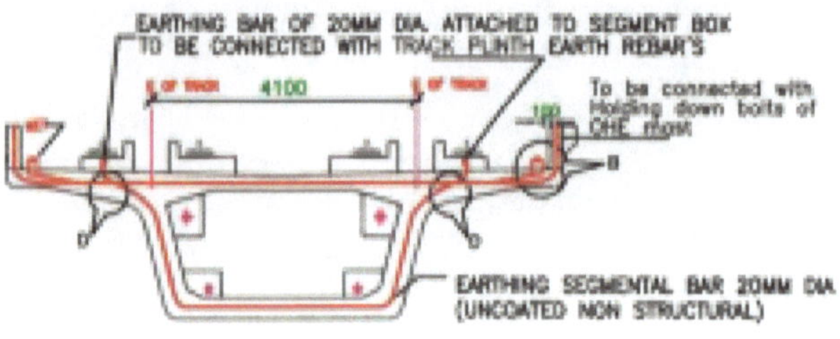

Fig. 6.211: Earthing Arrangement for Viaduct Box Section

Merits and Demerits of various traction systems

a) 25 kv AC with OCS (Flexible/rigid): -Merits

- **Reduced cost** – Unlike DC traction, this system does not require substations at frequent intervals due to high voltage, reduced current levels and lower voltage drops. As a result, there is a substantial cost reduction. The cost of 25 kV AC traction systems is about 15% less compared to the 750v DC 3rd rail traction system for the estimated level of traffic.
- **Energy regeneration & line losses-** Energy regeneration is more than 30% in the 25 kV AC traction system compared to 18% in the 750V DC 3rd rail traction system. In the 25 kV AC traction system, line losses are 12% less compared of the 750V DC 3rd rail traction system
- **Cost of rolling stock—**The cost of rolling stock and the maintenance cost of the traction system are comparable.
- **Capacity—**The system can accommodate traffic needs of more than 75000 PHPDT, but other constraints restrict this.
- **Easy capacity enhancement—**Capacity enhancement can be easily achieved by simply improving the transformer and its associated equipment at the receiving substation.

Typical Earthing at Tunnels

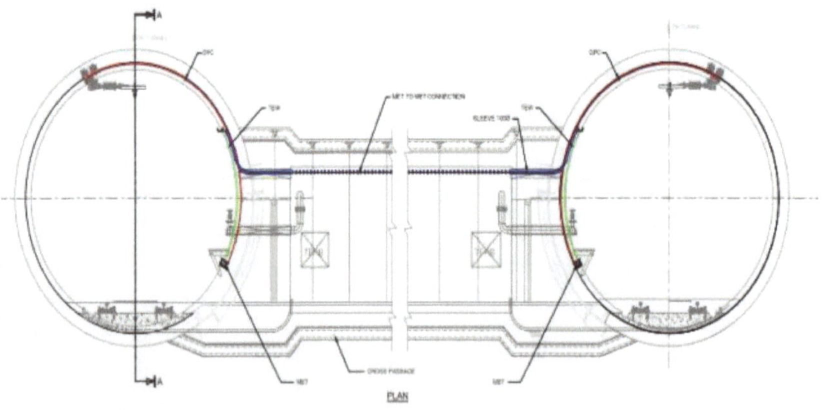

Fig. 6.212: Typical Earthing Arrangement at TBM Tunnel Section

- **Higher efficiency of operation—**The efficiency of regeneration is substantially higher than that of DC systems, and line losses are very low, on the order of 5%. 100% recovery of regenerated energy is possible in the case of 25 kV AC traction, compared to 75% in the case of 1500 V DC systems and 60% in the case of 750 V DC systems.
- **Less Fire hazards- The AC** system poses fewer fire hazards as current levels are much lower than the DC system.
- **Stray current** - There are no problems with stray currents, and hence, nearby metallic structures are not affected by corrosion. However, there are problems with EMC/EMI, which can be controlled by using return conductors & screened cables in signalling applications & fibre optic cables in telecommunication systems without booster transformers as per recent developments. This also helps avoid booster transformers, which cause 2%-line loss and excessive voltage drops, besides involving maintenance & reliability issues.
- Traction equipment in 25 kV AC systems is standardised & mostly indigenously available.
- In the underground section, the higher side tunnel diameter is an issue, but this is not the case here.

b) 750-850 V DC third rail traction system

- High operating currents and High voltage drops necessitating reduced spacing of substations- This leads to larger voltage drops along the Third Rail distribution system, which necessitates closer spacing of substations at an interval of almost every 2 Km, leading to higher costs.
- **Low levels of regeneration—60% of regenerated energy in a 750 V DC system must** be retrieved.
- **Safety hazards with the use of high voltage at ground level—**The "live" third rail exists at ground level, which can make this system hazardous to the safety of commuters and maintenance personnel if they fail to adopt safety precautions.
- **Line losses—**Line losses are higher due to higher current. Transmission line losses on a 750 V DC traction system are around 21% as against 5% of a 25 kV AC traction system.
- **Phenomenon of stray current-** In a third rail system, where the running rails are used as a return path, a part of the return current leaks into the track structure. This current is called a stray current. It is necessary to manage the stray current to ensure minimal corrosion effect and consequent damage to metallic components in the track structure, metallic reinforcement, and metal pipes of the building metro and public areas adjacent to the Metro alignment.

c) 1500 V DC system with Overhead Catenary System: -Demerits

- Higher maintenance requirements and costs compared to 750V DC third rail system.
- Theoretical traffic capacity with a 1500 V traction system is less as compared to a 25 kV AC system.
- Line losses are more due to higher current as compared to 25 kV AC. It may be in the range of 10 to 12% as against 5% of the 25 kV AC system.

d) 2x25 kV AC single-phase AC traction system

The following are the benefits of the 2x25 kV AC traction system.

The electromagnetic interference in 2X25 kV systems is significantly less; hence, no BT/RC system is required, which has its own maintenance and reliability issues. However, additional space for auto transformers should be required.

- The line losses in the 2X25 kV system are almost half as compared to the 25kV conventional system.
- In the event of 2X25kV traction systems being adopted, the number of neutral sections shall be reduced.
- Load balancing will be better on all three phases by using a connected transformer.
- Since the OHE current is much lower in the 2x25 kV systems hence the OHE voltage profile will be better than a conventional system. The voltage drop at the terminal end for a similar load will be almost half in the 2X25 kV system.
- Harmonics: Since the fault currents are much higher in 2x25 kV and 220/132 kV sides, the system has fewer harmonics. Fifth harmonics on the system are in the range of 1% or less, compared to around 3% or higher in conventional systems. Due to fewer harmonics, the reliability of the Traction installation equipment and electric rolling stock is expected to be better.
- Substation spacing: The inter-spacing between substations is almost double in the 2X25 kV system as compared to a conventional system. Therefore, the number of substations is almost reduced to half in the 2X25 kV AT system (with proper planning) as compared to the conventional system. This brings in substantial savings of costly space of around 2000 sq.m. in the Mumbai metro area, besides savings due to the use of lower voltage cable.
- The 2X25 kV system is highly suitable for high-density loads, as the voltage is fed to the system at a 25kV AC system due to better voltages, which improves the rolling stock's efficiency.
- Arcing problems are encountered at BT overlap in the 25kV system, and there have been melting cases of contact wire at the BT overlap location when bridged by stationary pantograph. No such problem will be encountered in 2X25kV systems.

A 2x25 kV system will replace the return conductor with a feeder wire, and the design will be finalised accordingly.

In view of the above techno-economic considerations, a 2x25 kV AC traction system is suggested for this corridor. Otherwise, a 25 kV AC system is preferred.

Merits and Demerits of various traction systems

e) 25 kV AC with OCS (Flexible/rigid): -Merits

- **Reduced cost** – Unlike DC traction, this system does not require substations at frequent intervals due to high voltage, reduced current levels and lower voltage drops. As a result, there is a substantial cost reduction. The cost of 25 kV AC traction systems is about 15% less compared to the 750V DC 3rd rail traction system for the estimated level of traffic.

- **Energy regeneration & line losses-** Energy regeneration is more than 30% in the 25 kV AC traction system compared to 18% in the 750V DC 3rd rail traction system. In the 25 kV AC traction system, line losses are 12% less as compared to the 750v DC 3rd rail traction system
- **Cost of rolling stock—**The cost of rolling stock and the maintenance cost of the traction system are comparable.
- **Capacity—**The system can accommodate traffic needs of more than 75000 PHPDT, but other constraints restrict this.
- **Easy capacity enhancement**—Capacity enhancement can be easily achieved by simply improving the transformer and its associated equipment at the receiving substation.
- **Higher efficiency of operation—**The efficiency of regeneration is substantially higher than that of DC systems, and line losses are very low, on the order of 5%. 100% recovery of regenerated energy is possible in the case of 25 kV AC traction, compared to 75% in the case of 1500 V DC systems and 60% in the case of 750 V DC systems.
- **Less fire hazards—**AC systems pose fewer fire hazards as their current levels are much lower than those of DC systems.
- **Stray current** - There are no problems with stray currents; hence, nearby metallic structures are not affected by corrosion. However, there are problems with EMC/EMI, which can be controlled by using return conductors & screened cables in signalling applications & fibre optic cables in telecommunication systems without using booster transformers as per recent developments. This also helps avoid using a booster transformer, which causes 2%-line loss and excessive voltage drops, besides involving maintenance & reliability issues.

The Green Line is the first corridor to implement the CBTC signalling system, which enables safe and automatic train operations. In addition to automatic train protection (ATP) to prevent accidents, the system allows trains to operate in Automatic Train Operation (ATO) mode, ensuring train composition.

6.39.3 Rolling Stock

The Design Parameters and Specifications of Rolling Stock and General Conditions of Contract for East-West Metro, as included in the contract, are as follows:

Operational Environment and Performance

Vehicle Driving Modes

Vehicle Design and Architecture

Vehicle Car Body

Train Driver Cab

Passenger Saloon Interior and Doors

Bogies, Pneumatics, Air Supply and Brake System

High Voltage and Electrical Propulsion Equipment

Auxiliary Supply Electrical Equipment

Ventilation and Air Conditioning

Electrical and Control Equipment

ATP, Automatic Train Protection

Train Management System (TMS)

Communication System, including emergency communication from the Driver's Cabin & CCTV for video communication.

Material and Workmanship

RAMS: Reliability, Availability, Maintainability, Safety, including Fire safety.

The manufacturer had to furnish details for inspection, testing and commissioning plans and programs and to facilitate Inspections, Tests, & Commissioning and Car Body Mock-ups. They had to interface with other contractors as required. During the finalisation of the Architecture & design, if needed, the Manufacturer had to refer to International Standards, in discussions with the Client.

After multiple rounds of tendering, the Kolkata Metro Rail Corporation (KMRC) awarded the Rolling Stock contract for the East-West line to Bangalore-based BEML, which emerged in September 2015. As per the contract, they supplied 14 train sets of 6 coaches each for a total of 84 coaches. Before this, in 2009, the Kolkata Metro Rail Corporation (KMRC) had awarded the Rolling Stock contract to Spain's Construcciones Auxiliar de Ferroparticulas (CAF), but due to cost escalation issues, the company pulled out of the project in 2014.

Fig. 6.213: Typical Rolling Stock of Indian Metro

Fig. 6.214: Internal View of E-W Metro Coach

BEML (Bharat Earth Movers Limited) supplied the rolling stock for Kolkata's East-West Metro (Line 2 or Green Line). BEML provided 14 train sets, each with 6 coaches, total of 84 coaches. Key details about BEML's East-West Metro rolling stock:

- **Contract:** BEML secured a contract for 14 train sets, each with 6 coaches.
- **Train Sets:** The train sets are designed, manufactured, and delivered by BEML as per design parameters & specifications in the contract
- **Technology:** The train sets utilise advanced features like a bogie for 750V DC Third Rail application, 3-phase AC motor control, and regenerative braking.
- **Design:** The car body is made of stainless steel with customised interiors and air conditioning.
- **Operation:** The BEML-built train sets have been operational since February 2020 and are operating successfully.
- **Additional Orders:** KMRC has placed additional orders for more rakes, including 3 rakes in 2022, increasing the total to 17 rakes.
- **Indigenous Design:** BEML's design, development, and manufacturing of cars is a significant step towards self-reliance in the rolling stock industry.
- **Technology Transfer:** BEML has a technology transfer agreement with South Korea's Hyundai Rotem.
- **Future:** KMRC plans to add more rakes, potentially increasing the fleet to 25.
- **Rake Designation:** The rakes are numbered MR-60n, where 'n' is the rake number
- **Indigenous Manufacturing:** Kolkata Metro plans to increase its fleet of 131 rakes by 2026, with 117 rakes being indigenously manufactured by Integral Coach Factory (ICF) and BEML.

6.39.4 Signaling & Communication

Ansaldo STS Spa was awarded a contract to supply the signalling and train control system for the Kolkata East-West Metro (Green Line), also known as the East-West Metro. This consortium comprised Ansaldo STS S.p.a. and Ansaldo STS, Australia, as members. The system includes elements like MICROLOK II, MICROCAB, AF904, UM71, point machines, and Power Frequency Track Circuit. The East-West Metro uses a Communication-Based Train Control (CBTC) system, which allows for high-frequency train operations.

Fig. 6.215: Signal at the entry of the Platform (indianbureacracy.com)

The Green Line utilises a Communication-Based Train Control (CBTC) signaling system for safe and automatic train operations. This system, equipped with anti-collision software, allows for shorter train headways, enabling operations with 90-second frequencies. The CBTC system is integrated with a fibre optic cable, SCADA, radios, and a PA system. In addition to the Automatic Train Protection (ATP) to prevent collisions, speeding and red light passing, the system allows the Metro rake to operate in Automatic Train Operation (ATO), where all rakes can communicate their position, speed, acceleration and deceleration.

The system is designed in a way that the train driver acts as an observer with the primary objective to close doors and the train start button. In the case of an emergency, a single button press at the control center will stop all trains simultaneously.

At present, the east-west corridor operates in two parts, and the missing link is between Esplanade and Sealdah. Once the missing link is installed, the system is integrated as one complete unit. KMRCL desired to conduct a trial for the full corridor.

Fig. 6.216: Kolkata Metro Gears Up for Driverless Operation

Key features and functionalities of the East-West Metro's CBTC signaling system:

- **ATO:** Trains can operate in automatic mode, with the driver acting as an observer.
- **Continuous Automatic Train Control (ATC) with Automatic Train Protection (ATP):** Ensures safe train operation and prevents accidents.
- **Position, Speed, and Acceleration Communication:** Trains communicate their position, speed, acceleration, and deceleration to the control center.
- **Emergency Stop:** In case of emergencies, a single button press at the control center can stop all trains simultaneously.
- **Balise Technology:** Tracks are fitted with balise, electronic beacons that locate the position of trains, with balise fitted every 250 meters.
- **Integrated System:** Includes a fibre optic cable, SCADA, radios, and a PA system.
- **Train Information System:** Provides information to passengers about train schedules and destinations.
- **Control Telephones and Centralised Clock System:** Facilitates communication and coordination within the metro system.

6.39.5 Automatic Fare Collection System (AFC)

The Kolkata Metro's East-West corridor utilises an Automatic Fare Collection (AFC) system for efficient and contactless fare collection. This system allows passengers to use contactless smart cards and tokens to access stations. The AFC gates at Howrah Maidan station are QR-code enabled and designed to handle a high volume of passengers.

Hitachi is involved in providing Automatic Fare Collection (AFC) systems for the Kolkata East-West Metro, which includes state-of-the-art gate technologies and integration with multi-modal transportation networks. The AFC system optimises ticketing operations, regulates passenger movement, and can integrate with other transportation modes. Hitachi also supplies train control systems for the metro, enabling Grade of Automation (GOA) Level 2 operation, which is a high level of automation.

Fig. 6.217: Automatic Vending Machine

Fig. 6.218: East-West Metro System at Salt Lake (www.dreamstime.com)

- **AFC Systems:**

  Hitachi Rail GTS provides advanced AFC systems that improve the efficiency of fare collection and enhance the passenger experience. These systems can use contactless smart cards for multiple journeys and contactless tokens for single journeys.

- **Train Control Systems:**

  Hitachi's Communication-Based Train Control (CBTC) solution allows the metro to operate at GOA Level 2, enabling features like Automatic Train Protection (ATP) and Automatic Train Operation (ATO). This means the trains can be controlled automatically, with the system ensuring safety and efficient operation.

- **Integration with Multi-modal Transportation:**

  The AFC systems can be integrated with other transportation modes, allowing passengers to use the same smart card or ticketing methods for various transportation services.

- **Power Automation:**

  Hitachi Energy's role in the East-West Metro project by provide power automation and control solutions, including Micro SCADA X, to manage power distribution and train operations.

Salient features of the East-West Metro's AFC system are.

- **Contactless Technology:** The system relies on contactless smart cards and tokens for fare payment.
- **RFID Technology:** Radio Frequency Identification (RFID) technology is used for faster and more efficient communication between the tokens/cards and the AFC gates.
- **QR Code Integration:** QR code-enabled AFC gates, like those at Howrah Maidan station, allow for quick and easy fare payment, particularly for passengers using mobile ticketing.
- **High-Capacity Gates:** The AFC gates at stations like Howrah Maidan are designed to handle a high volume of passengers, with some gates even allowing for bi-directional movement to manage peak hours.
- **Accessibility:** A few gates are designed to be accessible for wheelchair users.
- **Ticket Vending Machines (TVM) and Recharge Terminals:** Passengers can purchase tickets and recharge their smart cards at TVM and Recharge Card Terminal Machines (RCTM).
- **Centralised Server:** The system utilises a centralised server to track and manage all fare transactions and token/card information.

6.40 Benefits of East-West Metro

The list of benefits that the East-West line would confer extends across the gamut. First and foremost, it should be mentioned that this is a cross-river link. Even considering that in recent times the connectivity between the two banks of the river has been enhanced through the Second Vivekananda

Bridge, it is still only the Rabindra Setu and ferry service in the vicinity of the Howrah Station. The new metro line will add to the river-crossing significantly, particularly because the metro station at this location and Howrah Station will be integrated for passenger convenience. The main line passengers' access to the various strategic points of the city will undergo dramatic improvements. A similar advantage accrues to people using Sealdah station.

It is expected that the interchange at Esplanade will go a long way in improving the passenger mobility between east, west, north and south, time-effectively. The lines will be more fruitfully utilised if feeder bus services to and within neighbourhoods of stations are introduced – an aspect of multi-modal travel. There is a proposal for the East-West Metro to have passenger integration with the existing North-South Metro line at Central Station. Therefore, either from Howrah Station or from Sealdah Station, the North-South stretch of the city between Dum Dum and Garia becomes easily accessible.

In Mahakaran Station, the CBD Area of Kolkata, many commuters will benefit; it is proposed that an SPV would run a shuttle bus operation along the corridors of B.B.D. Bag, from this station at regular intervals on a free-of-cost basis. Similar bus facilities will be given at the three elevated stations at Salt Lake City Centre, Central Park and Karunamoyee stations.

6.41 Comparison Between North-South (1970) and East-West Metro (2000)

The approach of design and construction in the North-South metro was based on a contemporary method in line with the approach used globally. After a decade, the implementation of the East-West metro was again based on a contemporary global approach. However, in between, a vast improvement in design and construction was achieved. Software was developed, with soil-structure integration, and the use of TBM was taken up mostly, replacing the cut-and-cover method for underground corridors, though cut-and-cover is being continued in stations and short underground corridors. The author was involved in both corridors. A comparative table of design & construction, as used in both corridors, is included in the following table.

Table 6.10: Comparison of North-South and East-West Metro

East-West Line	North South Line
Train Composition *Start of opera tion:4Coach (2 MC+2 TC) Future: 6 Coach (2 MC+4 TC)	Train Compositio,wn 8 car train (6 M + 2 T)
Standard Gauge: 1,435 mm	Broad Gauge: 1,676 mm
Carrying Capacity of Coaches *Motor Coach 252 (42 seat+210 standing) *Trailer Coach 260 (50 seat+210 standing)	Capacity: 2500 passengers per train (approx.)
Headway (Initial): 5 min. Headway (Intermediate):2.5min. Headway (Final):1.5 min	CATP system planned for final headway of 116 sec.

Design Aspects: Cut & Cover U/G Box

East West Line	North South Line
Top-Down method for stations and Bottom-Up for the Crossover box and access shaft. Permanent diaphragm wall earth retaining system for Bott-Up box. Rigid box subject to earth pressure at rest and other loadings.	The Bottom-Up method was adopted, but the top-down method was also studied. Vertical retaining D-wall ground supports and inside concrete boxes. In some portions, the retaining D-wall was semi-integrated with the inside concrete station box. Rigid box. Retaining D-wall is designed for full earth pressure and other loadings. In-situ concrete box wall designed for hydrostatic pressure

Construction Aspects: Cut & Cover U/G Box

East West Line	North South Line
Temporary strut-waling system reduced by the Top-Down technique Limited use of temporary decking during construction. Less C&C box footprint in the Top-Down technique	Strut-waling support system for Bottom-Up technique Temporary traffic decking is used at most locations. More width is required for constructing a temporary earth retaining D-wall
Hydraulic rig with grab for d-wall. Special equipment for monitoring the trench profile. Stop end and water bar used at the D-wall panel joint. D-wall panels up to 1.5m thick and ~ 60m depth. Membrane waterproofing to the base and roof of the box. Grade of concrete M40	A conventional system used with the Kelly-Grab attached to the crane, and a mechanical grab in the Maidan Area Tube form articulation at panel joint. Thickness of D-wall 0.6m-0.8m. All-round PVC water bar at box-box joints. Grade of concrete M20/M25

Design Aspects: Tunnels

East West Line	North South Line
Bored Tunnels linking the u/g stations for the entire corridor. Segmental precast concrete tunnel lining, joined by high tensile bolts. Analytical + Numerical analysis for segmental lining subject to earth pressure and other loads.	Cut & Cover box tunnels linking stations. Bored tunnelling for short critical stretches using conventional TBM and blade shield tunnelling. Precast cylindrical concrete/cast iron ribbed lining jointed by high tensile bolt. Short portion using precast segmental lining. Analytical methods using an empirical formula for lining.

Soil Structure Interaction analysis using the Plaxis program Seismic analysis performed Radial segment joints	Earth pressure calculations using Terzaghi-Peck empirical formula
Numerical analysis for construction and permanent stages Soil-structure interaction models employed using Plaxis program Seismic modeling performed for u/g structures	Analytical methods of analysis Earth pressure calculations using Terzaghi-Peck empirical formula

Construction Aspects: Tunnels

East West Line	**North South Line**
EPB Machine for bored tunnels. The sensor is fitted with face pressure monitoring. Hydraulic segment erector. Two-level gaskets for joints. VMS guidance and Safety features. M40/M50 grade concrete	Link tunnels are generally C&C boxes, bored tunnel portions using the Blade Shield tunnel machine and a double 'D' tunnel by TBM is also used. Conventional TBM with air pressure 0.5-1 atmosphere. Conventional system for bored tunnels Conventional TBM with air pressure regulation. M40 grade concrete

CHAPTER 7

Kolkata Metro Line 3

(Purple Line - Joka-Mominpur Corridor)

7.1 Introduction

The Joka-BBD Bagh Broad Gauge Metro Railway line (Line 3) was sanctioned in the year 2010 as an elevated corridor. Subsequently, to avoid several constraints and to maintain the skyline of the Maidan Area, the Mominpur-Esplanade Section of the corridor was converted to underground. The corridor length is 16.72 km. Initially, the Authority decided to implement it in two sections. First is the Elevated Section between Joka and Mominpur; the remaining is underground. In between, the route was shortened and planned to terminate at the Esplanade. The execution of Line 3 has been entrusted to RVNL.

Fig. 7.1: Line Diagram of Line 3

	Chainage	Location	ROW	LHS C.W.	RHS C.W.	Median Width	Planned Width
JOKA	0	Start	38.18	6.61	6.94	7.88	2X7.5
		Center	38.88	6.24	6.45	9.92	
		End	36.69	5.41	6.74	11.90	
MAJERHAT	7607	Center	15.45	40.06		Off the Road	Flyover
MOMINPUR	8625	Start	25.99	13.32	5.83	1.06	2X7.5
		Center	26.33	12.53	5.89	1.07	
		End	25.87	13.36	8.56	1.06	

Fig. 7.2: Details of Diamond Harbour Road – Basic Route of Line 3

7.2 Project Brief

To start with RVNL engaged CES for the detailed design of elevated corridor, from Joka to Mominpur and the Depot. RVNL decided to build a separate Depot for Line 3, in Joka. While CES carried carried out the detailed design RVNL, deployed IIT Delhi, to carry out the proof checking and engaged Simplex Infrastructures for construction. The remaining part, the underground section, has been awarded to L&T. The extension of this line to IIM and Diamond Park for 2 km has been approved in the Budget of 2012–13. The Ministry has also approved extension from Esplanade to Eden Gardens. Line 3 of the Kolkata Metro System, known as the Purple Line, has the first eight stations as elevated and the last four as underground.

Fig. 7.3: Kolkata Metro Line 3 (Tenneco)

Line 3, the purple corridor, was planned to be implemented in four phases, which are.

- Joka to Mominpur (now Operational)
- Mominpur to Esplanade (Under construction)
- Diamond Park to Joka (planned)
- Esplanade to Eden Garden (Under Planning)

Important Data of the corridor are.

* MRTS Corridor Length is around 9.7 Km, (Excluding Depot and UG Section) including Eight Elevated Stations

*Depot is located around 1.8 Km away from Joka Station

* The alignment is along the existing Diamond Harbour Road (NH 117)

* Eight Elevated Stations are Joka, Thakurpukur, Sakher Bazar, Behala Chowrasta, Behala Bazar, Taratola, Majerhat & Mominpur

* Average Inter-station distance is 1.225 Km, with minimum distance of 0.878 Km & maximum is of 1.445 Km

Figs.7.4: Elevated Corridors during Construction

Hon'ble Prime Minister Shri Narendra Modi inaugurated the Joka-Taratala stretch, including the Joka metro station of Kolkata Metro's Purple Line, on 30 December 2022

The analysis and design of Kolkata Metro line 3 was prepared based on the given design parameters, loadings, load combinations, and simulations of the construction stage, durability aspect, and other structural requirements.

7.3 Transport System

Kolkata, a metropolis with a population of 22.4 million in 2020, faces the challenges of managing urban transport for a dense population with limited road infrastructure. Despite having limited road infrastructure, Kolkata boasts a well-established rapid transit system. As of 2023, the city had 96 kilometers of metro lines, contributing to an average of 5 kilometers of rapid transit per million population, exceeding the national average of 2. This extensive metro network plays a crucial role in serving Kolkata's population, with 33% having convenient access to public transport services, surpassing the South Asia average of 19%. However, challenges persist in Kolkata's urban transport landscape. While the metro system provides a backbone for public transport, a significant portion of the population still relies on private modes of transport, contributing to congestion and air pollution. Data from Google Environmental Explorer (2023) reveals that 66% of trips are made using private modes, leading to concerns about emissions and air quality. As Kolkata continues to grow, integrating its existing public transport infrastructure with sustainable solutions and improved urban access will be vital in creating a more efficient and livable city. (ATO)

7.3.1 Kolkata Transport System

Kolkata's transport system includes different modes such as Tram Services, Water Transport System, Road-based System, Suburban Rail System, Circular Railway, and Existing Metro Corridors. Out of all these services, at present, the road-based system carries the maximum passengers and covers all parts of the city. More than 2 million people were distributed over 1851.41 per sq. km with an overall population density of 7890 such, bus services, which Include both public and private operators, can not handle, particularly when most congested area in KMC, population density as high as 23149 persons per sq.km. Total ridership of the State-owned Bus system in the city was as low as 2441,82 lakhs in 2005, and a yearly increase in capacity was also insignificant, like 2433.54 lakhs in 2008. To solve the problem, high high-capacity transport system is essential. And in 2008 rail rail-based urban transport started operation. After the implementation of Line 1 and Line 2, the Government approved the implementation of Line 3, besides extensions of Line 1 and Line 2.

Fig. 7.5: Existing Traffic Scenario

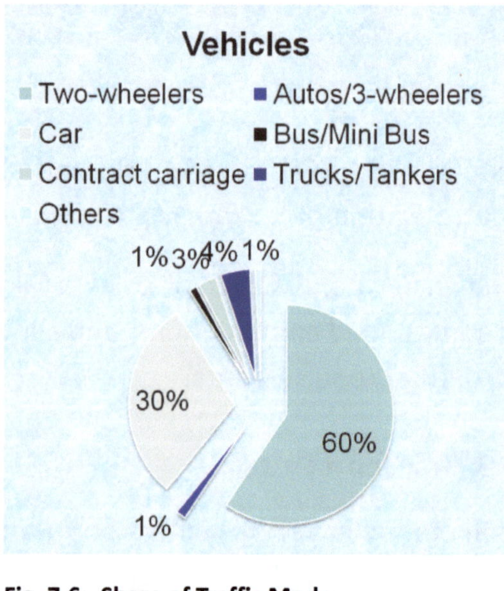

Fig. 7.6: Share of Traffic Mode

7.3.2 Ridership Assessment of Line 3

The detailed Ridership assessment was conducted by M/s RITES, and the projections are as per the follows:

	Year	2015	2025	2035
1.	**Joka Thakurpukur**	2900	3800	4800
2.	**Thakurpuk ur Sakher Bazar**	3500	4500	5800
3.	**Sakher Bazar Behala**	4500	5800	7400
4.	**Behala Behala BaZar**	7700	9800	10700
5.	**Behala Bazar Taratala**	7700	9800	12600
6.	**Taratala Majerhat**	8800	10700	13000
7.	**Majerhat Mominpur**	11100	13500	16500
8.	**Mominpur Khidderpur**	11200	13700	16700
9.	**Khidderpur Victoria**	12400	15100	18500
10.	**Victoria Park Street**	12400	15100	18500
11.	**Park Street Esplanade**	7300	8900	10900

7.4 Salient features

For Line 3, Track Geometry, Track Structure and Rolling Stock for the design of the Viaduct and Stations are listed below. SOD of Metro Railway and correspondence with Kolkata Metro Railway with RVNL and vice-versa:

i. Gauge: Broad Gauge 1676 mm.
ii. Track: Ballast less track with minimum plinth thickness of 185mm.
iii. Partly Elevated and Partly Underground
iv. Design, Proof Check, System Design, Supervision and Testing – Carried out by Indian Engineers

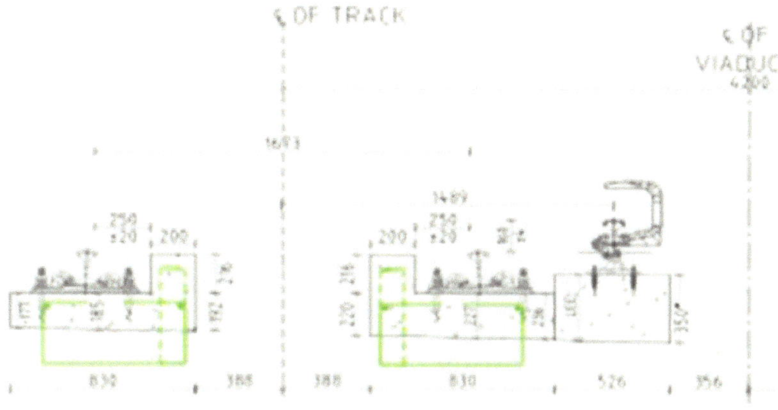

Fig. 7.7: Ballastless Track

Fig. 7.8: Ballasted Track used in Depot

The radius of curvature was considered as: a) Generally not less than 300 m; and b) at difficult locations, not less than 200 m.

i. Rolling Stock: RTS Coach (as per loading diagram of Metro SOD)
ii. Track: Ballast less with a plinth of 195 mm minimum thickness on all structures, like Viaducts, Embankments and stations. The track consisted of 60 UIC Rails with bearing Plates, fastenings, and other accessories. The design load of the track plinth was the load calculation based on the existing track plinth of the Kolkata Metro with a maximum super-elevation of 150mm.

7.5 Alignment

The alignment has been designed based on the following approved planning parameters by Metro Railway letter No. MRTS/W-90/49 (RVNL) and MRTS/W-30/32 (HQ) (20) and prevailing practice for similar works. Between Joka and Mominpur, all stations viz. Joka, Thakurpukur, Sakher Bazar, Behala Chowrasta, Behala Bazar, Taratala were planned at the central verge of the Diamond Harbour Road, whereas the station at Majerhat was planned over the existing railway yard. Details of the final alignment, being executed as follows,

- **Track centers** C/C = 4.2 m for BG.
- **Platform** width for side platforms = 5.5 m and for island platforms = 12.1 m.
- **Horizontal distance** between C/L of track to platform = 1.465m to 1.475m. Extra clearance for curves to be adopted.
- **Length of platforms** = 180 m.
- **Minimum horizontal** curve radius = 200 m
- **Gradient at stations** to be level or 0.5% max. At mid-section, desirable 2% and in critical locations 3% max.
- **Points and crossings** are 1 in 12, except in yards where up to 1 in 8.5 has been used.
- **Vertical curve** radius has been kept 3000m and minimum 2500m as per Metro letter No. MRTS/W-30/32 (HQ)/20.

- **Minimum overhead** clearance for road/highways to be 5.5m and preferably 6.0 m and for Railway lines 6.25 m on OH electrified railway track.
- **Horizontal clearance** for piers from C/L of existing Railway track is 1.720 m
- **General span length** is kept 28m and maximum up to 34m.
- **Width of viaduct** has been kept 10.50m. Additional width as required on curves.
- **Rail top level** has been generally kept 10.0 m above the rail top level has been generally kept 10.0 m above the existing ground/road level for mid-section and 13.2 m at stations accommodating concourse for elevated section. To provide adequate clearance of 5.5 m minimum and 6.0 m preferably.
- The **proposed alignment** is envisaged elevated from Joka to Mominpur and underground from Khidirpur to BBD Bagh (later reduced to Esplanade).
- **Curves** - Maximum allowable cant of 150mm and Cant deficiency of 75 mm on Metro tracks is considered for safe speed.
- **Horizontal curves**: Curve radius in mid-section –

  Minimum: 300 m

  Absolute minimum: 200 m

  Minimum radius of curve at stations: 1000m

  Maximum permissible cant (Ca): 150mm

  Maximum cant deficiency (Cd): 75mm

- **Gradients:** Normally the stations shall be on level stretch. In limiting cases station may be on a grade of 0.5% maximum. Between stations, generally the gradients may not be steeper than 1.0%. However, where steeper than 2%, gradients up to 3% (compensated) can be provided in short stretches on the main line on case-by-case basis.

  * **Stations:** Maximum gradient at stations: 0.5% Desirable gradient at stations at level

  ***Elevated section**: Maximum gradient in mid-section: Desirable: 2.0% Maximum: 3.0%.

  ***System adopted:**

  Third rail current collection system: The system has been designed for system operation and headway up to 150 seconds. Railway Board's directives issued vide letter No. 2002/ Elect(G)/181/1 to GM/Metro Railway to adopt existing 750-volt DC traction with 3rd line current collection for new corridors. 25KV system's OHE in Metro Railway system shall change the skyline. In 25KV system all motor coaches must be provided with transformer. Cost of transformer carrying in train is substantial with 150 sec. frequency. During rainy season city depends on Metro Railway, OHE is vulnerable system in front of storms, so DC Metro system is reliable and efficient compared to 25KV system with 150 sec. frequency.

  ***Signalling system**: The proposed signalling system will comprise of – Continuous Automatic Train Control (CATC), which shall consist of following subsystems: o Automatic Train Projection

(ATP) o Automatic Train Supervision (ATS) o Automatic Train Operation (ATO) o Computer Based Interlocking (CBI)

***Transport System Capacity**: The proposed elevated corridor shall have a design capacity of about 42,000 passengers per hour per direction for Joka-Majerhat corridor.

* **BG coach carrying** capacity would be as under:

Gauge- BG, Train Length- 8 coaches, Coach Width-2.74, Coach Length-19.5 m, Rake Cap Train length (coaches) Coach width (m) Coach length (m) Rake capacity (pax) 1728 Pax, designed peak hour Freq. (Sec)-150 secs, Designed peak hour Cap.(pax)- 41,472 Pax

* **Rolling Stock**: Gauge-BG, Length-19.5, Max.Width-2.74, Axle load- 17 T, Tare wt. of MC-: 42 T Tare wt. of TTC:28T Capacity of 8 car-siting 420 Standing 1696 Speed: 80 Kmph. (max) E.B.D.: 180m Traction Power supply 3rd rail: 750 V

* **Passenger Interface** with other modes of suburban section as under: (i) Ballygunge – Budge Budge and Circular Railway at Majerhat. (ii) That the existing N-W corridor at Park Street/ Esplanade, (iii) Bus stand and tram depot at B.B.D Bag.

Ridership assessment: - The detailed Ridership assessment was conducted by M/s RITES

7.6 Design Philosophy

All the structures follow the limit state design philosophy, including Water Tanks with storong water, where crack widths are limited to 0.1mm. For section without water crack width of 0.2 mm is maintained.

In the limit state of design, each member was designed to satisfy two limits: the limit state of collapse and the limit state of serviceability. The IRS concrete bridge code was followed throughout the design. The steel roof structure was designed following the latest revision of IS:800 and IS:801. In addition, ductile detailing was followed at appropriate locations as per IS:13920& IS 1893 Part 3 to ensure additional safety during an earthquake. Appropriate Codal provisions as per IRS CBC, IS 456 & guidelines, for anchorage length, lap length, etc., were provided to reinforcements. Concrete composite structure design conformed to IRS-CBC requirements, while steel-concrete composite components were designed per the IRS-Steel Bridge Code and another relevant standard.

7.6.1 Limit State of Collapse

At this state, all the structural elements were designed for axial force, flexure, shear, punching shear and torsion. In the case of slender columns, the effect of induced deformation was incorporated while estimating reinforcements at a critical section. In the case of concrete sections, the complete rectangular-parabolic stress-strain curve for concrete, as mentioned in all the design standards (IRS CBC, IS:456 and IRC:112), was used in conjunction with simplified elastoplastic stress-strain curves for reinforcements to determine the flexural capacity at ultimate state, following the strain compatibility and equilibrium of internal forces. Critical sections were designed for shear and torsion as prescribed in the case of steel members. Flexural capacities were computed depending upon the section type, such as plastic, compact, semi-compact, slender, etc. as per IS:800. Intermediate and bearing stiffeners were provided at appropriate locations to reduce web thickness and to enhance section capacity.

7.6.2 Limit state of Serviceability

The serviceability limit states were checked to ensure the structure remains functional during its service life. Three limit states were checked, which were i) the limit state of cracking, ii) the limit state of deflection, and iii) the limit state of floor vibration. These limit states predominantly governed the design of flexure members. These are, however, not critical when a section is subjected to significant axial compression, e.g., column members, and thus are not checked. In addition, flexural stress levels under SLS combinations were kept within elastic limits as prescribed by IRS CBC and IS:456 to ensure the longitudinal cracks, micro-cracks and concrete creep, which were within acceptable limits. In reinforced concrete flexural members, the crack width was estimated using the formula in IS:456 Appendix F, as it provides more conservative values than the formula in the IRS CBC. Considering a moderate environmental exposure condition, the estimated crack width was limited to 0.20 mm (as per IRS CBC). In addition, adequate reinforcements were provided at every exposed surface at a maximum spacing of 150 mm to prevent shrinkage cracking (early-age thermal cracking). Excessive deflection of a structural member in service conditions may adversely affect its proper functioning or appearance and/or cause discomfort to facility users.

7.6.3 Expansion gaps

The expansion gap was calculated based on movements caused by temperature, creep, shrinkage, and out-of-phase movement caused by earthquakes and structural deformations. Gaps provided in the viaduct superstructure = 100, Gap in the Building = 50 mm.

7.7 Design Life

The life of the structures considered in the design were

i. **Main structural systems** - 120 years.
ii. **Bearings & Movement joints** - 20 years.
iii. **Service life of minor components**- at least 20 years.
iv. **Paint system for steelwork** - 15 years.
v. **Road works** - Minimum 15 years.

7.8 General Arrangement of Viaduct (GAD)

The **GAD** includes foundations, substructures, superstructures, and clearances. The clearances for Road Traffic is 6.0 m at 0.50 m from the pier shaft outer line and pier shaft line (0.45 m width of the 1.25 m-high Jersey type crash barrier + 0.050 m clearance between crash barrier and pier shaft).

Clearances for Rolling Stock and Vertical Clearance were not applicable. However, the high voltage overhead transmission line near Joka station was planned to be removed in the portion (infringing on this alignment)—horizontal **Clearance** beyond the fixed structure line as per the Schedule of Dimension (SOD). Viaducts Emergency Walkway width of 1045 mm, including railing, was proposed.

7.9 Materials

7.9.1 Concrete

The minimum grade of concrete was according to the durability provisions for structures and considered moderate conditions of the environment by IRS CBC: 1997. The minimum grades of concrete for various structural elements, satisfying durability and strength requirements, are being proposed as follows.

Piles, Pile cap and open foundation: M35

Cantilever Pier cap & Pier shaft: M50/M60

Hammerhead Pier cap & Pier shaft: Generally, M40, M50/M60

in special cases Portal Pier & Inverted L-type Pier: M50/M60

Portal beam integral to Pier: M50/M60 Pedestals M60

Precast segmental superstructure: M50

Cast-in-place deck slab & diaphragm for I-girder: M40 Precast Crash barrier

Parapet over deck slab: M40

C**rash barrier around the pier**: M40

Other miscellaneous structures: M35

Instantaneous modulus by IRS-CBC-1997: For fck = 60 MPa E = 36,000 MPa
For fck = 50 MPa E = 34,000 MPa
For fck = 45 MPa E = 32,500 MPa
For fck = 35 MPa E = 29,500 MPa

Modular Ratio

Modular Ratio including long-term effects such as creep, taken (IRS-CBC 1997 and A&C slip no 12, dated 02.06.2009). For tensile reinforcement -**mt** = 280/fck **compressive** reinforcement -**mc** = 420/fckc

7.9.2 Other miscellaneous data

Coefficient of thermal expansion for steel and reinforced concrete 11.7 x 10^{-6} (IRS-BR-2008). Coefficient of thermal expansion for plain concrete: 10.8 x 10^{-6} (IRS-BR-2008)

Unit weight for load calculation: As per IRC6-2010

Creep & Shrinkage coefficient for long-term losses, as per IRS-CBC

Poisson's ratio - 0.15

7.9.3 Reinforcement steel

Thermo-mechanically treated (TMT) reinforcement bars of grade Fe 500 (min) conforming to IS: 1786, Modulus of elasticity - 200 GPa (IRS CBC)

Minimum yield strength - 500 MPa (IS: 1786)

7.9.4 Pre-Stressing Steel

Low relaxation, uncoated stress-relieved strands were considered per IRS – CBC 1997, and IRC: 18 -2000

a. **Pre-stressing units** - 12 O 6 and 19 O 6 per IS: 14268
b. **Breaking strength of strand** - 260.7 KN per IS: 14268

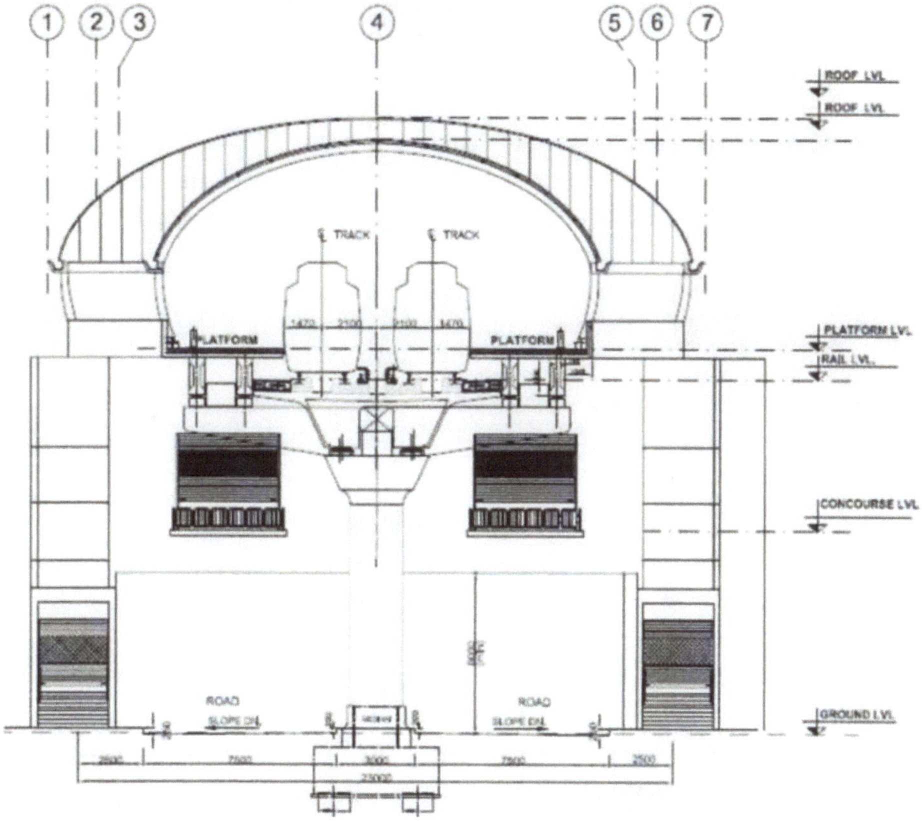

Fig. 7.9: Station Section-Without Concourse

c. **Nominal area of strand -** 140 sq.mm per IS: 14268
d. **Minimum breaking stress -** 1860 MPa

e. Design was based on a **Maximum anchoring force** equal to 70% of the characteristic strength of Tendons immediately after anchoring. An increase of up to 80% as per IRS – CBC 1997 was planned if required

f. **Modulus of elasticity** - 195.0 GPa IRS – CBC 1997 for the Strands conforming to IS: 14268.

g. **Corrugated HDPE ducts** conforming - IRS – CBC 1997

h. Duct characteristics for **friction loss** as per IRS – CBC 1997.

i. **Coefficient of friction** (-) 0.17/radian by IRS – CBC 1997.

j. **Wobble coefficient** (-) 0.002 as per IRS – CBC 1997.

k. **Anchorage slips** during lock off - 6.0 mm

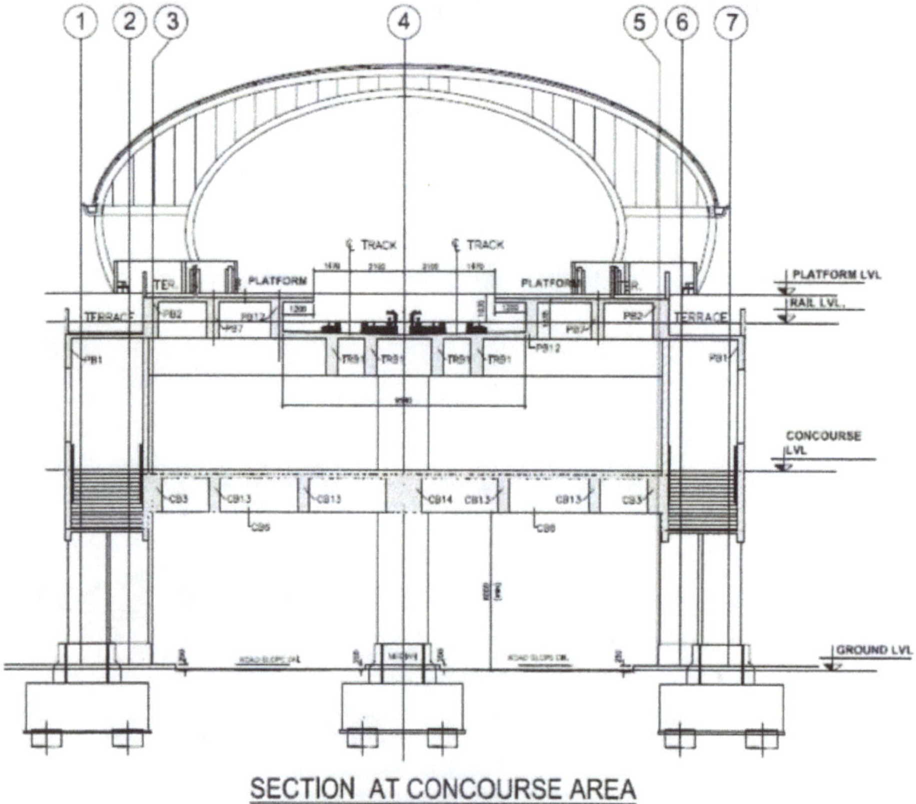

SECTION AT CONCOURSE AREA

DIMENSIONS ARE INDICATIVE ONLY

Fig. 7.10: Station Section with Concourse

l) **Long-term relaxation** - 6.375% of initial stress by IRS-CBC-1997 (Long-term relaxation) loss for low-relaxation loss as per IRS concrete bridge code: For initial prestress equals to 70% of characteristic strength (1860 MPa), 1000 hr relaxation loss = 2.5% of initial prestress, which reduces to linearly to 0% loss for initial prestress of 50%. of characteristic strength. Interpolating for initial prestress of 67% of characteristic strength (average), 1000 hr relaxation loss is 2.125%. Long term loss = 3 * 1000 hrs loss = 6.375%

7.9.5 Structural Steel

Minimum grade of structural steel

a. Grade of high strength steel - Fe 540 (E410) as per IS: 2062 – 2006
b. **Grade of mild steel** - Fe 410WC (E250C) as per IS: 2062 – 2006
c. **Grade of hollow steel sections** - Yst 310 as per IS: 4923
d. **Modulus of elasticity** - 200 GPa
e. **Coefficient of thermal expansion** -12 x 10-6 per (IS-800 – 2007)
f. **Unit weight** - 78.5 kN/m3
g. **Poisson's ratio** - 0.3
h. **Modulus of rigidity** - 77 GPa

Structural steel for composite spans and other miscellaneous uses was as par steel-concrete composite bridges.

Live loads are considered as per the IRS & IRC as applicable

7.10 Other Important Parameters

7.10.1 Dynamic Augmentation

The augmentation in vertical train loads due to dynamic effects, considered as per IRS Bridge Rules 2,4.1 & 2.4.2, is given by the CDA=0.15+8/(6+L), where the coefficient of dynamic augmentation (CDA) is subject to a maximum of 1.0, and L is the loaded length in m.

7.10.2 Forces Due to Curvature

Considered as per Indian railway bridge rules. Design Speed for various radii of curvature, as per; C= Wv2/127R, where W is the axle Load, v is the maximum design speed in Km/h, and R is the Curvature in meter. The force was considered at a height of 1.83 m from the rail top.

7.10.3 Temperature Load

i) Permanent Structures

a. The temperature effect was in line with Indian railway bridge rules. A temperature difference of ± 35º Celsius was used to calculate axial forces (if any).
b. In the absence of any provisions in IRS for the temperature gradient effect, this was considered as per IRC 6 – 2010 to calculate the stresses due to moment.

ii) Due to track structures (Long Welded Rail Force)

Horizontal load was considered in the longitudinal direction per the Rail Structure Interaction analysis, depending on the fastening system adopted. The load was considered with live loads but without seismic loads. The methodology for computation of loads due to LWR was as per UIC 774-

3R, 2001. The longitudinal force due to the case of one rail fracture was also considered in the design of the structures, with an increase in allowable stresses by 33.33%.

iii) Resistance to Movement of Elastomeric Bearings

Resistance to the movement of the elastomeric bearings due to changes in temperature and deformation of the superstructure used as (SL L – SR R), where SL and SR are the shear ratings of the elastomeric bearings supporting the left and right span, respectively; L and R are the deck movement above them respectively. In the case of identical spans and bearings, 10% of SL L was considered when designing the substructure and foundation.

7.11 Longitudinal Force

Longitudinal forces arise for the reasons considered while designing the substructure and foundation. The braking force was proposed to be 18% instead of 15%, which aligns with the design of other Metro Rail Projects.

a. Tractive effort of the driving wheels - 20% of live load per track.
b. Braking force from all braked wheels - 18% of live load per track. When trains are on both tracks, longitudinal force was considered the sum of the tractive force from one track and the braking force from the other. The horizontal lateral loading due to racking was considered a uniformly distributed load of 4.0 kN/m per track acting at the top of the rail head without any impact effect.

Parapet The parapets were to resist lateral horizontal force5.9, and a vertical force of 1.50 kN/m was applied simultaneously at the top of the parapet, as per IRS-BR 2008's 2.10.

7.12 Wind and Earthquake

The effect of wind pressure was calculated as per § 2.11 of IRS-BR-2008 and IS: 875 (Part 3) and the design wind speed of IS: 875 (Part 3). Seismic Design Provisions, as stated in § 2.12 of IRS-BR & A&C Slip No- 42 dated, updated by IS:1893 Part-I 2002 and further updated by IRC 6:2010, resulting in more severe effects on structures. The seismic analysis of viaducts was conducted as per recommendations of IRC 6:2010. The peak ground acceleration, denoted as zone factor in IRC: 6-2010, was considered 0.18 instead of 0.16 for Kolkata to cover the design life of 120 years as per General Planning Criteria.

The horizontal design coefficient is given by Ah = (Z/2) * (I/R) * (Sa/g) Where, Ah = horizontal seismic coefficient. Z = peak ground acceleration or zone factor = 0.18 I = importance factor = 1.5 R = response modification factor as per IRC: 6-2010 Sa/g = normalised pseudo-spectral acceleration for 5% damping of load resisting elements depending upon the fundamental period of vibration T as recommended in IRC: 6-2010.

Vertical Seismic Coefficient The vertical seismic was considered as two-thirds of the horizontal seismic.

7.13 Erection Force

The weight of all permanent and temporary materials and all other forces and effects that can operate on any part of the structure during erection was considered in the design. The load arising from the most onerous condition of the Construction Scheme/Launching Girder Scheme was as par with the approved method submitted by the Contractor. However, initially, the force had been conservatively taken as 530 tons applied at 0.85 m from the centre line of the pier, which must be verified before the launching of the girders. The contractor must take special care during construction.

7.14 Derailment Force

Derailment forces were considered per Appendices (xxviii) and (xxix) of Bridge Rules with due proportioning of axle loads. The structure was checked for derailment conditions of one or more wheels, assuming that one set of derailed wheels would be confined within the guard rails. No increase in the impact effect needs to be considered, provided permissible stresses are not increased. Additionally, "Sacramento derailment criteria" must be checked for derailment loads. This criterion corresponds to applying 40% of one coach's weight horizontally as a 3m long uniform impact load on the box girder top flange. This derailment load corresponds to a ULS load. For SLS, derailment forces will be taken as per Appendices (xxviii) and (xxix).

7.15 Geotechnical Detail

Lithological log indicates the presence of a clay bed at the top of the geological succession with thickness varying from 10-40 mts. There is a further alternate clay and sand bed. A group of granular aquifers is found between 250-650 mts. below ground level.

Table.7.1: Zone wise Locations

Zone Number	Approximate Chainage (km)	Location
Zone-I	0.000 to +1455	Joka Station to Thakurpukur Station
Zone-II	+1455 to +2675	Thakurpukur Station to Sakher Bazar Station
Zone-III	+2675 to +4128	Sakher Bazar Station to Behala Chowrasta Station
Zone-IV	+4128 to +5465	Behala Chowrasta Station to Behala Bazar Station
Zone-V A	+5465 to +6260	Behala Bazar Station to Taratala Station
Zone-V B	+6260 to +6600	
Zone-V C	+6600 to +6900	
Zone-VI	+6900 to +7744	Taratala Station to Mazerhat Station
Zone-VII	+7744 to +8755	Mazerhat Station to Mominpur Station
Zone-VIII	0.000 to -1250	Joka Station to Depot

Typical Subsoil Profile

Table.7.2: Subsoil Profile

Stratum No.	Layer Details: Description	Depth below EGL (In R.L) From	Depth below EGL (In R.L) To	Average Field N-Value	Bulk Density (t/m3)	Liquid Limit (%)	Plasticity Index (%)	Shear strength Parameters
I	Top surface filled up by loose soil mixed brick bats, sand etc.	0.0	-1.5	-	-	-	-	
II	Firm brownish grey clayey silt with traces of sand & mica.	-1.5	-5.8	5	1.84	43	18	C= 3.5 t/m^2, ϕ =0 deg
III	Soft greyish silty clay / clayey silt with few percentage of decomposed wood.	-5.8	-17.8	2 to 3	1.67	50	22	C= 2.7 t/m^2, ϕ =0 deg
IV	Stiff bluish / brownish grey silty clay with traces of brown spots and calcareous nodules. Sand at bottom.	-17.8	-22.6	9 to 12	1.88	43	21	C= 7.0 t/m^2, ϕ =0 deg
VI	Medium dense to dense brownish / yellowish grey micaceous silty fine to medium sand.	-22.6	-32.8	18 to 26	1.90*	Non-Plastic		C= 0.0 t/m^2, ϕ =30* deg
VII	Very stiff brownish grey silty clay with traces of sand and brown spots.	-32.8	-36.9	26	1.92	48	22	C= 11.0 t/m^2, ϕ =0 deg
VIII	Dense yellowish grey micaceous silty fine to medium sand.	-36.9	-50.6	34 to 49	1.94*	Non-Plastic		C= 0.0 t/m^2, ϕ =33* deg

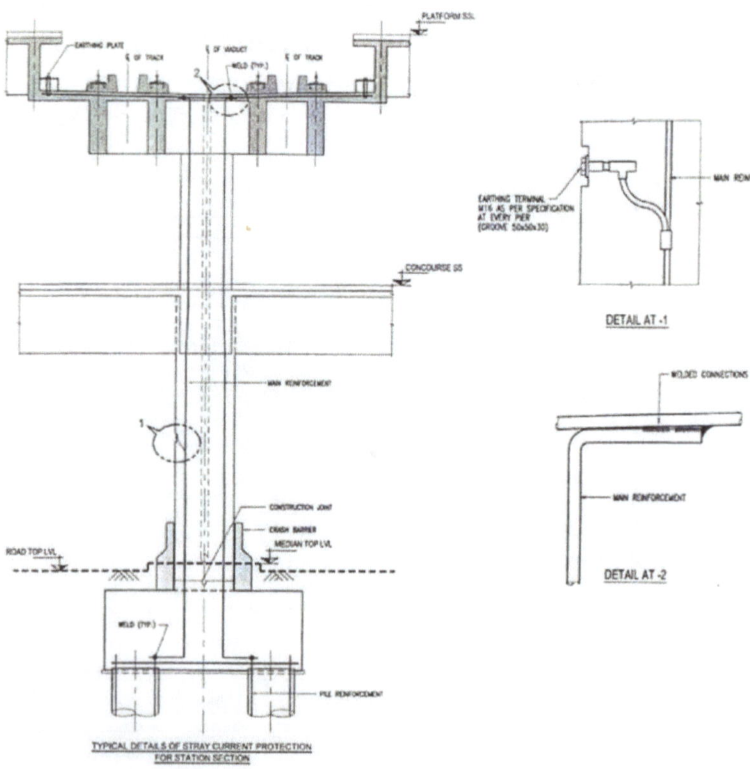

Fig. 7.11: Details of Stray Current for Protection of Station Structure

Fig. 7.12: Photos during Piling

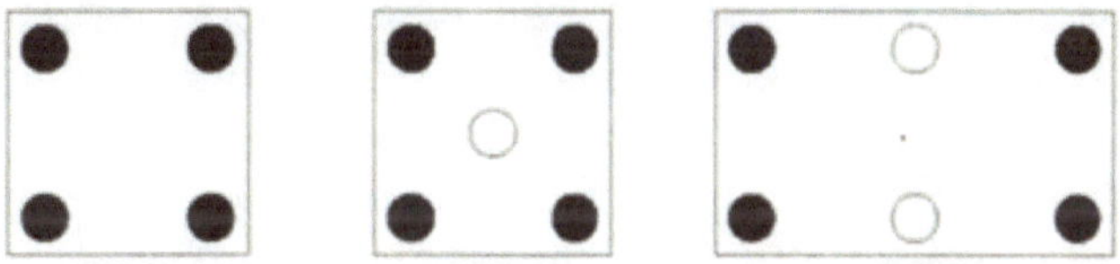

FIG.1 FIG.2 FIG.3

Fig. 7.13: Plans of Pile Foundations

Fig. 7.14: Pile Cap and Column Reinforcement

Fig. 7.15: Pier Form

Fig. 7.16: Pier Staging

Fig. 7.17: Pier after Cast

In Summary Structural Briefs are as follows

Superstructure

- Precast Post tensioned segmental single cell box girders varying from 19m to 31m with increments of 3m for standard spans.
- 50m Cast-in-place post tensioned box girder.
- Precast post tensioned I-girder made continuous for 4 spans at cross-over location.
- 2-span continuous post tensioned precast segmental box girder.

Bearings

- Elastomeric bearings for transferring normal loads.
- Reaction block with/without cable restrainer for seismic loads.
- POT-PTFE bearings for long spans.

Substructure

- Concentric piers of 1.95m circular section with heights varying from 11m to 18m (difference between Rail Top Level to Top of Pilecap).
- Inverted L-piers up to 2.0m maximum eccentricity between centre line of track and pier section.
- Portal piers at obligatory locations.

Foundations

- **1200mm dia bored cast-in-situ reinforced concrete pile.**
- **Pile shaft length varying from 25m to 40m depending upon various types of soil strata.**

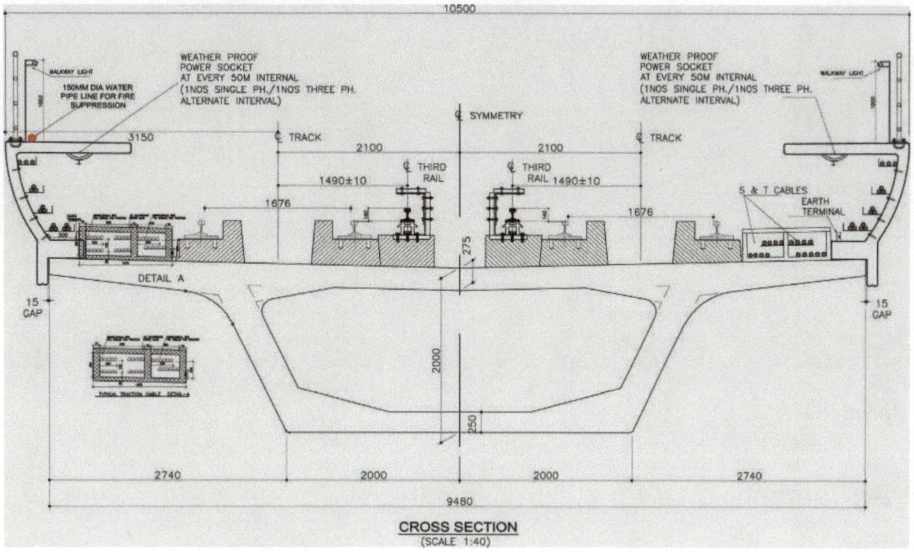

Fig. 7.18: Cross Section of Segmental Box Used

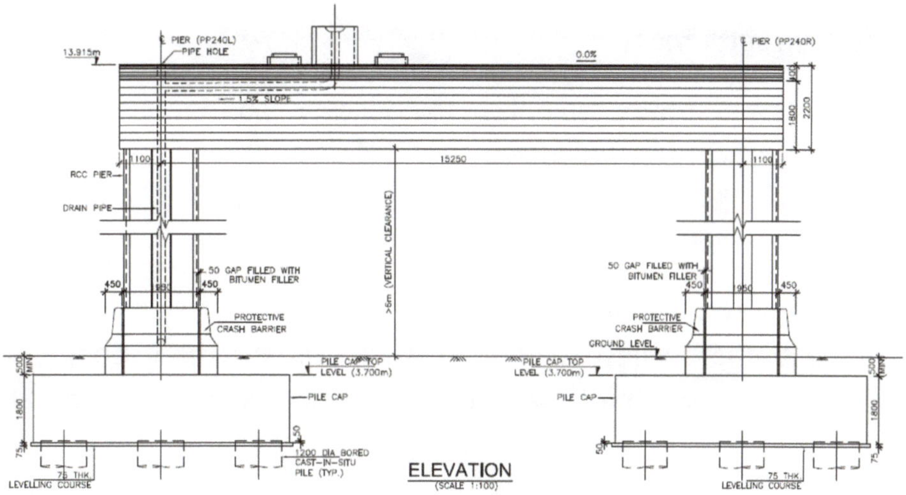

Fig. 7.19: Portal Pier and Horizontal & Vertical Clearance

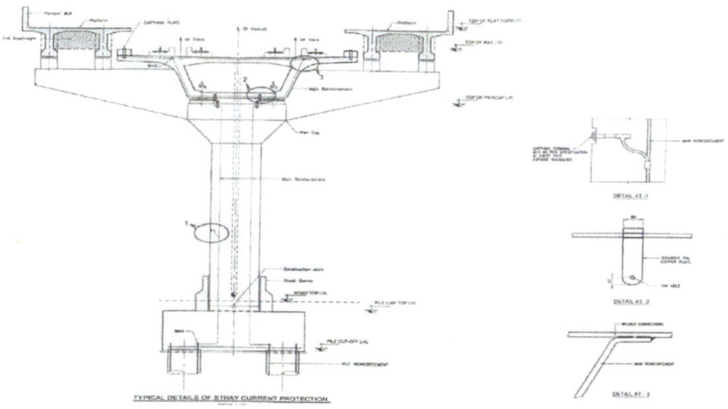

Fig. 7.20: Stray Current Protection for Sections Beyond Concourse

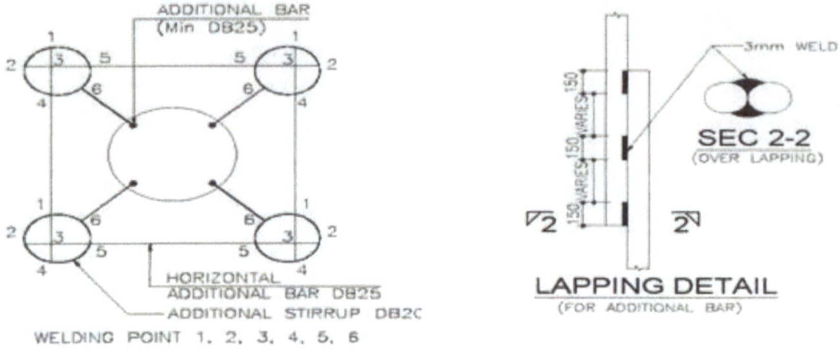

Fig. 7.21: Earthing in Bored Pile

7.16 Differential Settlement

Differential Settlement between two adjacent viaduct piers was notionally taken as based on experience gained on previously designed metro viaduct structures:

i. 10 mm for Long Term Settlement.
ii. 6 mm for Short-Term Settlement

The differential settlement was considered only in the design of continuous structures, if any, which may be required to be constructed at crossover.

7.17 Traffic Diversion

Traffic Congestion during Metro Construction is the primary motivation for this project. Several areas in city are facing traffic congestion due to metro construction. Congestion triggered by continuing metro work impacts operation, particularly for numberer of hospitals along the corridor and influenced area.

Fig. 7.22: Diversion of Traffic in Stretches

Road diversions for the construction of the **Joka Metro (Purple Line)** extension, specifically in the Mominpore to Esplanade stretch along **Diamond Harbour Road (DH Road)** and surrounding areas, are in effect and subject to ongoing changes as construction progresses. Along the DH Road till mominpur traffic diversions were made Section by Section as shown in **Fig. 7.22.**

7.18 Ultimate Limit State for Superstructure

Ultimate strength checks for flexure were made as required in the IRS Concrete Bridge Code, 1997, cl. 16.4.3. Appropriate formulae or software might have been used. Shear and torsion were checked under IRS CBC 1997, no 16.4.4, while maximum shear stress was calculated as per 16.4.4.5. 12.

7.19 Durability

Crack Width: The width of reinforced concrete members for SLS combination LC 1 was checked—crack width allowed as per 15.9.8.2 of IRS CBC. Crack width was not to exceed the admissible value as per Table 10 with A&C slip No. 12 of IRS CBC based on the exposure conditions. For crack control in columns, cl. 15.6.7 IRS CBC was referred.

7.19.1 Other Considerations for Ensuring

The following specifications are intended to be adhered to deemed satisfied and meet the durability requirements:

- Complete and adequate drainage;
- Sufficient concrete cover;
- Limiting crack width
- Appropriate concrete mixture design and good pouring, acceptable permeability and surface finishing (IRS-CBC-1997 § 5.4 with correction slip No 12 dated 02-06-2009)

7.20 Design Codes and Standards

The design and loading requirements for the structures were based on Indian Standards and Codes of Practice, together with all applicable amendments, addenda, corrigendum slips, errata slips, etc. The preference of codes, standards, specifications, technical reports, special publications, special committee recommendations, and manuals was used in the order as per RVNL guidelines.

Indian Railways/Research Design and Standards Organisation, Lucknow RDSO report no. BS-14 was to be used to comply with durability requirements. If it is found to be inferior, the provision of the document listed lower in order of preference will be adopted, and exclusive approval will be taken from the RVNL/KMR to this effect.

i. Indian Railways/Research Design and Standards Organisation, Lucknow/Kolkata Metro Railway
ii. Indian Road Congress/MORTH
iii. Bureau of Indian Standards
iv. British Standards/European Standards

v. AASHTO Standards
vi. UIC recommendations and publication
vii. ICE, International Electromechanical Commission recommendations and
viii. NFPA 130 Standard for Fixed Guideway Systems: to be used for design relating to fire safety.

7.21 Stations Area

The following figures show vertical and horizontal clearances for road traffic both in the concourse area and outside. Vertical clearance after median kerb (which comprises centre column/pier shaft p

7.21.1 Analysis Requirements

Structural analyses are generally performed using in-house developed spreadsheets for simple problems and the general-purpose finite element package STAAD Pro for performing linear static and dynamic analyses of framed structures. The following sections include the underlying assumptions for the study of the structures protected by crash barriers. It is kept at 6.00m above the highest road at the top level.

7.21.2 Station Viaduct

Fig. 7.23: Erection of Segments on DH Road

The station viaduct comprises a cast-in-situ box girder placed over a concentric pier cap and a circular pier column over the pile foundation. The superstructure was analysed in the longitudinal direction as a simple beam subjected to uniform loading due to self-weight, superimposed dead load and moving load due to trains with an appropriate coefficient of dynamic augmentation, following the principles of statics. In the transverse direction, the box section of unit width is analysed as a frame section supported over two bottom supports at the soffit level. The necessary reinforcement is calculated for various working and erection stages in permanent and temporary conditions. The pier cap was analysed as a corbel or a cantilever beam depending upon the depth to shear span ratio. The shear span is taken as the distance from the load point projecting from the equivalent square inscribed within the circular column section, as per IS:456. The pier column was analysed as a cantilever column, fixed at the pier base and free at bearing level, for the various loads coming from the superstructure. The pile cap is analysed for the loads coming from the part of the structure above its soffit level using both the strut-and-tie model and simple beam model, as well as the more onerous case used for reinforcement designs. While computing loads on the pile top, it is assumed that the pile cap is infinitely rigid compared to individual piles. All the analysis work is done using in-house developed spreadsheets and validated by employing general purpose FEM package STAAD Pro.

Fig. 7.24: Elevated Corridor of Line 3

7.21.3 Station Building

Fig. 7.25: Entry to Platform (link.com)

Fig. 7.26: Joka Station Platform (Timesnownews.com)

Station structures with the concourse Main Station building portion comprise a special moment resisting RCC framed building supported over pile foundations. Miscellaneous structures such as corridors, staircases, E&M structures, entry/exit structures, etc., were designed as RCC-framed buildings on pile/open foundations. All the building frames were modelled in the general-purpose FEM package STAAD Pro using centreline dimensions of respective beam-column elements. The effect of the RCC element is designed appropriately. Each component was assigned to its respective cross-section dimensions, as shown in the drawings, along with material properties. The floor slab was modelled by imposing kinematic constraints (master-slave node in STAAD Pro) to simulate its rigidity under in-plane loadings, such as wind and seismic forces, and flexibility under out-of-plane loadings, such as gravity loads. All the columns were assumed to be fixed at the foundation level with only a vertical spring to account for differential settlement; foundation flexibility was ignored to get a stiffer structure, increasing seismic force demand. The design forces for the underground water tank were taken from the design tables published by the design forces for underground water tanks was as per BIS- IS:3370 Part IV, for appropriate end restraints.

Fig. 7.27: Top View of a Station

7.21.4 Foot overbridges

Foot-over bridges were designed as per the provisions of IS:800-2007 without fatigue consideration (IRS Steel bridge code). Loading will be defined in Section Steel-Concrete Composite Structures and the DBR. Deflections were limited to the girder (L) length, divided by 325 as per cl. 4.17 IRS Steel Bridge Code.

7.21.5 Steel-Concrete Composite Structure

Since the IRS steel bridge code does not address design requirements for steel-concrete composite structures, such structures were designed as per IRC:22-2008/IRC:24-2010/IS:8002007 with permissible steel stresses defined for 10 million cycles as per Clause 3.6.5 of the Steel Bridge code.

7.21.6 Concrete- Concrete Composite Structures

The platform supporting precast beams was planned to be composite with the floor slab using shear connectors as per the IRS CBS/IS-456/IRC -112. To facilitate the design, the characteristic strength of the precast members and the slab had been kept the same.

7.21.7 Wall Type Pier

Wall-type piers were designed per IRS-CBC (like viaduct piers), with a response reduction factor of 3.0 (Table 3 IS:1893 part-3) for ductile wall-type piers.

7.21.8 Foundations

All the major structural systems, including the main station building and viaduct portions, were supported on pile foundations. Minor systems, such as single-storey buildings for pump houses, generator rooms, water tanks, etc., were supported over open foundations. In general, 1200 mm diameter bored cast-in-place RC circular piles were mainly employed, whereas 750 mm diameter piles were used for minor structural elements, as required. The vertical geotechnical capacity of a bored cast-in-place circular pile was determined using two approaches, and the lesser of the two was used to assess its capacity for a given pile length. First was the static formula in IS:2911 (Part 1/Section 2) using the following parameters.

a. Layer-wise geotechnical data, such as angle of internal friction, cohesion, bulk density, etc, down the pile length.
b. The angle of wall friction between the pile and soil was taken to be the same as the angle of internal friction.
c. The coefficient of earth pressure K is taken as 1.5 per IS:2911 (Part 1/Section 2) and other literature.
d. The Maximum overburden pressure used to compute shaft and tip resistance was 20 times the pile diameter from the existing ground level.
e. It was assumed that the groundwater table could reach the existing ground level, and thus, submerged density was considered for computing overburden pressure. o
f. The factor of safety was taken as 2.5 when pile capacity was computed using a static formula.

The pile's lateral load capacity was evaluated using the empirical formulae in IS:2911 (Part 1/Section 2), limiting lateral deflection to 1% of the pile diameter at the scour level, as per The permissible lateral deflection at the top of the pile for the computation of lateral capacity was limited to 10 mm for a 1200 mm diameter pile.

Second, the initial load test was done as per IS:2911 (Part 4) for a load amounting to 2.5 times the safe vertical load computed using a static formula. The estimated vertical capacity was then enhanced for seismic loading by 25% and for temperature loading by 15% of standard capacity (IRC:78 and IS:1893 Part 1). The lateral geotechnical capacity of the pile was also determined from both empirical formulas as given in IS:2911 (Part 1/Section 2) and from the initial load test as per IS:2911 (Part 4), and the least is taken as capacity under normal loading. The lateral capacity is estimated for a lateral deflection of 10 mm for 1200 mm dia. piles and 7.5 mm for 750 mm dia. piles, considering fixed pile head condition when the empirical formula is used. The permissible displacements are taken as per amendment no 6 to IRC:78, where permissible lateral capacity is defined for a lateral displacement of 1.0% of pile diameter at the pile cut-off level. Due to the end user's limitation, the allowable deflection is limited to 10 mm instead of 12 mm for 1200 mm dia. piles.

7.22 Station Planning

Length of the proposed Joka-Mominpur Corridor is about 9 km, and 8 nos. of stations out of 12 total stations, have been planned to take into consideration of the traffic catchment area, constraints in land acquisition and congestion issues. All these stations have been planned as an elevated station and average inter station distance is 1.2 km varying from 0.98 to 1.25 km depending upon the site and operational and traffic constraints.

Majerhat is an elevated metro station on the Joka-Esplanade Metro corridor between Taratala and Mominpur. It is the seventh station out of 12 on the line. Majerhat station was planned, designed, and executed under the RVNL, which also delivers other stations and Viaduct portions for the Kolkata Metro. This metro station is located over the Suburban Railway station, which comes under Eastern Railway's jurisdiction. The station has been planned with extensive property development and a weight like the operation.

The station, quite substantially big compared to other stations, is built over one of the busy railway yards and crosses a canal. Majerhat station was constructed with two buildings: a backup control center and a property development building. It has entry structures and office, equipment, and property development spaces on the two ends across the Railway yard and the canal. The central portion connecting the two portions on both sides is built with unique piers with pier caps at three levels for supporting girders at three levels—concourse, Track, and Platform.

The station has a total length of 180m and covers a total area of 12,281 square meters. Based on the total covered area and multilevel buildings, the requirement for ancillary structures like underground water tanks and electrical substations is also much more significant for this station than for other stations. The two parts on either side of the central part have a nearly circular shape and are nomenclature as Property Development Building (PD Building) and Back-up Control Centre Building (BCC Building). These two buildings also have basements, which increased the complexity of construction and design because of the high-water level and proximity to a canal, large water pipelines, a railway platform, signaling, and power cables. Being in a very densely populated city, finding adequate areas for adjusting all the amenities was always a challenge.

7.23 Design Philosophy

The station structure supports both the dynamic loading of the trains and a general static live load of people & and equipment, so it needs to be analysed and designed based on IS and the IRS code. The station structure has been analysed with the FEM Package of STAAD Pro using centerline dimensions of beam and column elements. The floors are modelled with kinematic constraints to simulate rigidity under in-plane loadings, such as wind and seismic loading. The structure has been designed for all the anticipated gravity loading, such as self-weight, superimposed dead load, static live loads, and dynamic loading due to the train's movement. It also takes care of lateral loading due to trains' raking, braking, traction loads in addition to general wind and seismic loads.

The RCC building structure has been mainly designed with the IS 456 code, considering the seismic detailing provided in IS 13920. However, the parts that also hold dynamic loading due to train movement are also checked under the provisions of the IRS CBC guidelines. The steel structure and steel roofs are generally designed with provisions of IS 800, and similarly, portions where the steel structure is supporting the tracks are designed using the IRS-Steel Bridge code. The structure has been designed against an essential wind speed of 50m/s (i.e. 180 km/hour) with other factors as per IS 875 Part-3. IS 1893 has been used for seismic loading, considering zone III for the structure, as it is in Kolkata. The peak ground acceleration (PGA) for this zone is 0.16, but considering the structure's life to be 120 years, the probability of exceedance is more than 50%, which is for a 100-year lifetime. So, the PGA has been increased from 0.16 to 0.18 to account for a design life of 120 years. The important factor considered is 1.5 for the structure, considering its pre- and post-earthquake needs. The station frame structure uses the multi-modal response spectrum method to compute earthquake forces. Like the viaduct portion, the station track portion is also designed to account for "Sacramento Derailment Criteria" in an accidental situation. The central steel structure holding the tracks has been designed for fatigue loading as per the provisions of the IRS- Steel Bridge code. The durability aspect has also been considered in the design of the structure. It is checked that each building component has a flexural crack width limited to 0.2mm as per the IRS CBC, considering Kolkata to have moderate environmental exposure. However, the code has been revised, and the permitted value has become 0.25mm.

7.23.1 Design Standards

Elevated Stations (General Priority)

Indian Railway/RDSO, Kolkata Metro Railway

Bureau of Indian Standard, IRC/MORTH, AASTHO Standard

UIC recommendation and publication

ICE, International Electrotechnical Commission recommendation

NFPA 130 standard for fixed guide way system for fire safety design

Viaduct Portion (General Priority)

Indian Railway/RDSO, Kolkata Metro Railway, IRC and MORTH, Bureau of Indian Standards

AASTHO Standard

UIC recommendation and publication,

ICE, International Electrotechnical Commission recommendation

NFPA 130 standard for fixed guide way system for fire safety design

7.23.2 Durability Aspects

To ensure good structural performance of structures under this moderate environmental exposure, the following aspects are considered.

1. By limiting the crack width of flexural members under service loads to 0.20 mm, as recommended by the IRS/CBC.
2. A minimum cover for any reinforcement is provided as per the IRS/CBC and IS:456 recommendations.

   Location and Minimum Clear Cover

   Pile 75 mm

   Pile Cap 75 mm

   Pier, L-Pier & Pier Cap 50 mm

   Stitch Beam 40 mm

   Portal Beam, parapet 40 mm

   Pedestal 50 mm

   Segment 40 mm

3. Minimum cement content and other aspects, such as acceptable permeability, workmanship during concreting work, etc., are maintained whenever design mix concrete is used as per IRS/CBC and IS:456 guidelines.
4. For the permeability test, the depth of penetration of moisture shall not exceed 25 mm as per clause 5.4.2 of the IRS/CBC.
5. The durability requirements are as per Cl 5.4 of IRS/CBC.

   Type of structure Type of cement used Min. Cement content (kg/m3)

   Pile, Pile Cap PSC 400

   Pier, Pier cap OPC 43 Grade 430

   Segmental Box Girder, Pedestal OPC 53 Grade 430

7.23.3 Soil Conditions

i. Number of Bore holes done for Ground Investigation: 52 Nos
ii. Category of Zone as per soil condition: Zone I and Zone II
iii. Zone-I (Stabling line, Shed1 & 2, Admin Building, Electric Sub Station, Sewage Treatment iv) Plant, Track portion after Bridge J2 towards Depot End) – OGL at RL (+)1.00 m

   a. From Top RL (+)1.00 m to (-)2.00 m CI-CH (Clayey Silt/Silty Clay)
   b. (-)2.00 m to (-)15.00 m OH (Organic Clay)
   c. (-)15.00 m to (-)21.00 m CI/CH-MI (Clayey Silt/Silty Clay)
   d. (-)21.00 m to (-)27.00 m SM (Silty fine Sand)
   e. (-)27.00 m to (-)39.00 m CI/CH (Clayey Silt/Silty Clay)
   f. Zone-II (Store ward, Pit Wheel Lathe, ACWP, Security Building before Bridge J2 towards Depot Start) – OGL at RL (+)1.00 m
   g. From Top RL (+)1.00 m to (-)8.00 m CI-CH (Clayey Silt/Silty Clay)
   h. (-)8.00 m to (-)29.00 m SM (Silty fine Sand)

7.23.4 Design Approach

i) Foundation System

The major structural systems, such as the main station building and viaduct portions, are supported over a pile foundation with a 1200 mm diameter bored cast-in-place RC circular piles. The Vertical Capacity of bored cast-in-place circular piles is determined using two approaches, and the lesser of the two is used to determine its capacity for a given pile length. First, it is calculated by a Static formula as given in IS:2911 (Part 1/Section 2) and then checked by the initial pile load test. The lateral geotechnical capacity of a pile is also determined from both the empirical formulas given in IS:2911 (Part 1/Section 2) and from the initial load test as per IS:2911 (Part 4) and at least is taken as capacity under normal loading. The pile spacing is kept at a minimum of 3 times the pile diameter.

ii) Pile Cap

The pile caps are designed as flexural members, and thus both ULS and SLS checks are performed and checked for crack width under service loads. (As per clause 35.3.2)

According to IS 456 -2000, the upper limit of 0.2 mm is for moderate exposure.

iii) Pier & Pier Cap

The pier analysis and design are based on clause 15.6 of the IRS Concrete Bridge Code: 1997. The effective length of the pier column for this purpose is taken as 2.3 times the actual height measured from the pier base to the bearing top level. The pier cap is designed as a corbel (if the span/depth ratio is less than 1.0) or as a Flexural member (if the ratio is more than 1.0).

iv) Bearing System

Elastomeric Bearings are designed as IRC:83 (Part II) 1987 for normal vertical loads only, whilst the concrete shear key is designed as a corbel under the action of lateral forces, such as braking and tractive forces and seismic forces.

v)Post-tensioned Segmental Box Girder

The box girder is designed as a prestressed member in the longitudinal direction and as a reinforced member in the transverse direction, following the guidelines of IRS/CBC and IRC: 112. At the ultimate limit state, critical sections along the longitudinal directions are checked for flexure and shear, considering Prestressing forces after all losses. At service, the sections are checked for residual compression at the soffit and deflection at mid-span. There is no limiting residual compression limit mentioned in the code. Decompression is generally checked at service for the segmental span.

vi) Building Column & Beam

Beams and columns of the station buildings are designed for ultimate loads. The flexural and shear capacities at critical sections are determined using the formula given in IS:456.

Concentric piers are of 2000 mm diameter and have been cast with M40 grade RCC.

L-piers have been cast of size (1950 x 3550 mm) with M50 Grade of RCC. Portal piers have been cast in size 1950x1950 mm and generally cast with M60 Grade of RCC.

Floor Slab:

The design moments for floor slabs under ultimate loads are computed as per Annexes D of IS:456 for various end restraints.

Roof Structure:

The main structural elements are made from structural steel plates of Mild steel conforming to IS: 2062. Bolts and nuts are of property class 8.8, conforming to IS:1367 and holding down bolts are of property class conforming to IS:1367 and used under IS:5624.

7.23.5 Segment Casting Long Line

(M50 grade Box Segments have been cast of 2000 mm overall depth.)

The length and width of the bed for straight and curved spans are taken as per the drawing. In a typical casting sequence, two pier segments are cast on pier segment beds separately and placed at their final locations on either side of the long line bed. One intermediate segment is a match cast starting from one end. In case of casting of a non-standard span (length), the sequence of casting is the same as that of the standard span, except during casting of mid-segment/adjacent segment, two neighbouring segments are lifted and re-positioned on the bed, maintaining the correct length required for the middle segment/adjacent segment to be cast.

Fig. 7.28: Formwork (Complete Set)

Fig. 7.29: Placing of Reinforcement Cage

i) Fixtures in Segment

Embedded items such as HDPE pipes, anchorages, etc., are fixed as per the drawing. All other fixtures, such as drainage spout, etc., are provided at specified locations as detailed in the drawings.

Fig. 7.30: Fixing of rail for inner trolley

Fig. 7.31: Fixing of HDPE Sheathing Pipe

ii) Lifting of Segment

The match cast segments have been fitted with a lifting frame through the lifting holes provided as per the drawing at the time of casting of the segment. The segment has been lifted from the soffit shutter by the 65T capacity Goliath Crane and placed at the stacking Yard. The lifting of the segment has been done after it attains a minimum cube strength of 25 Mpa.

Fig. 7.32: Setting of Form work with Cast Segment

Fig. 7.33: Transfer of Segment to Site In front of Matching Face

Fig. 7.34: Segment Erection at Site

Fig. 7.35: View from Bottom after Erection in One Span

Fig. 7.36: Erection of Segment Span by Span

7.24 Basement Design and Construction

The PD and BCC buildings are planned and designed to be eight-story structures accommodating various facilities for the station and a basement. The requirements generated high loads that had to be transferred to the ground. Due to the geotechnical conditions, long piles were suggested. A model of the raft with soil springs was modelled to simulate integrated behaviour. The piles were modelled at their respective position, with linear springs only, as piles to carry both tension and compression. The introduction of compression-only springs increased the analysis time, as each load combination was required to be formed as a separate load case because the superimposition method was no longer valid for a nonlinear analysis. Due to a water table at a very shallow depth and a water canal nearby, the basement was also checked for flotation due to the buoyancy effect during construction.

Many uncharted services were identified during basement construction, which were then adequately diverted and rerouted. During the excavation of the BCC basement, it was found that there was an existing masonry bridge foundation. The old city map (innovative plan) shows that the earlier alignment of the Majherhat Bridge differed from that of the present alignment. The ruins of the old, abandoned Bridge structures became a significant bottleneck towards pile installation and construction of the Raft foundation. It was initially decided to demolish the existing structure, but it was later figured out that it was huge, and the strength of the masonry was robust.

Fig. 7.37: Aerial view of Majerhat Metro Station across the Railway Tracks

It wasn't easy to break. After excavating the soil surrounding the Bridge foundation, it could be demolished, piles driven, and rafts constructed over it.

The challenge faced during the basement construction was building it near the canal. It was noted that if excavation were carried out very close to the canal, there would be a big possibility of damaging the side slopes of the canal and disrupting the water flow. So, it was checked with the canal authority whether it could temporarily be made nonfunctional till the excavation of the basement and construction of a wall at that location was complete. In that way, whatever damage occurred during the excavation and construction of walls could be made good before the canal was opened again.

Fig. 7.38: Old brick Foundation of Bridge Structure Found at the BCC Side of Majerhat Metro Station

As the timing and period offered by the canal authority were unsuitable, alternatives like using diaphragm walls, sheet piles, contiguous piles, etc., were investigated, and 750mm diameter contiguous bored piles around 3m from the canal's edge were suggested. These contiguous bored piles, propped at the top with the existing building structure, are designed to withstand a fill height of 5m approx. and the submerged soil pressure on the wall for around 40m in length. The critical stage was during construction when they had the highest utilisation ratio.

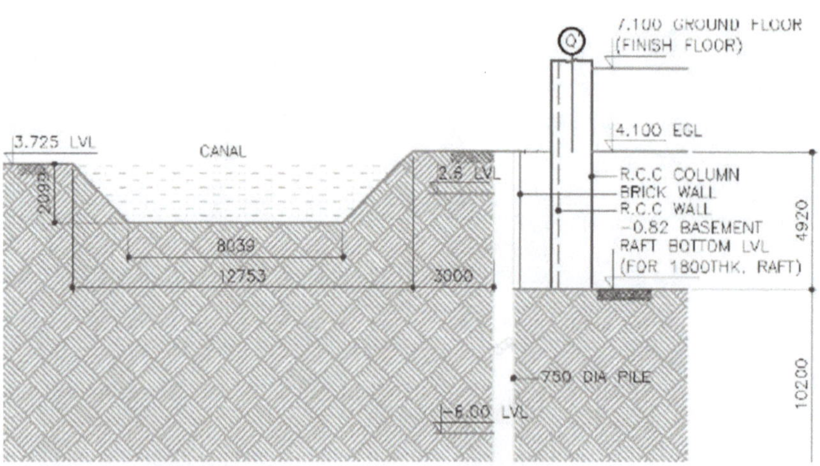

Fig. 7.39: Contiguous Bored Piles Between the Canal and the BCC Building

19 m pile lengths were determined based on clayey strata in line with soil parameters at the location. These piles are also constructed in a sequence such that boring is adjacent to any pile with a sufficient time gap so that the already built pile concrete achieves enough strength to resist lateral load during the new boring. By introducing this solution, the building's construction could be done independently of any interference from canal issues.

Many utilities were safely maintained or diverted at the PD building site during construction. A defunct 1800-diameter pipeline of PWD was removed before piling; at some locations, pipes were around 6m to 7m below ground level, and OFC cable lines were crossing the PD building area. They span across the yard during the Launch of Steel Girders for the Station. During the construction of the PD building, the water supply main pipe valve was leaking, flooding the already excavated portion of the basement. It required rapid restoration of the pipe and the site and protection of the platform of the existing Majerhat Railway station. Many sheet piles were installed with strutting and round-the-clock monitoring of sway in sheet piles and the water level. When the excavated area for the PD building part of Majerhat station got flooded, and the construction was held up, there were many other requirements, such as protection of the adjacent Railway platform, etc.

Fig. 7.40: Leakage of Water During Construction of the PD Building Basement

After restoration, construction work resumed, and the building was safely completed. Some sheet piles were required to be left as sacrificial to ensure no damage to the water main. Some could not be accessed at all. Many railways signaling and telecom cables were in the construction area and were shifted carefully.

Fig. 7.41: Cable Shifting at PD Building Area

7.25 The Tragic Accident

All construction work on the Metro Station was stopped from September 2018 to September 2020 because the adjacent Road-Over Bridge (ROB) collapsed. During this period, the PWD West Bengal built a cable-stay bridge that was wider and longer than the bridge, making the worksite for the Metro station and viaduct narrower and making the work even more challenging.

Fig. 7.42: ROB Collapsed Adjacent to Metro Station Under Construction

7.26 Central Portion of the Station

The design and execution of the central portion, which spans over seven railway tracks and a canal, was challenging. Due to the presence of railway tracks with electrified lines, any structure over it must be constructed with minimum hindrance to railway traffic. Due to the span length and the operational requirements of the electrified railway lines, the lightweight steel structure was preferred over the other precast systems. In addition, it wasn't easy to mobilise heavy machinery at that location. A minimum clearance of 6.5m was required over the railway tracks, resulting in a height constraint within which the station concourse, platform and track levels were accommodated. This dictated that designers work with structural form and sections that provide a minimum depth of the floor system so that enough clearance can be maintained between each floor level. The substructure was designed with RCC elements to take care of collision loads; the foundation system was designed to suit all the site constraints and a stable foundation with minimum width; a 6-pile group of 1200mm diameter piles with staggered arrangement was used, keeping the width of the pile cap restricted to 3400mm which could be the minimum possible.

The topsoil strata were loose, so it was decided to use a steel liner extended to a much lower depth to avoid undue circumstances. This reduced soil friction, and so pile lengths were increased. The metro line and at-grade railway lines are skewed, so it was decided to go for a steel liner extended to a much lower depth to avoid undue circumstances. The metro line and at-grade railway lines are in a skewed position to each other. Thus, the substructure and its foundation system were planned at a skew to the superstructure, placing the pile caps within the gap between the railway tracks. A power block of adjacent railway lines was taken during erection.

Fig. 7.43: Concrete Pier with Pier-caps and Bearing Location

Behala Chowrastha

Sakherbazar

Thakurpukur

Joka

Fig. 7.44: Views of a few Stations of Line 3

7.27 Floor System of Concourse and Platform

The floor system of the concourse and platform has been designed with a steel composite concept with continuous girders. Keeping continuity helped reduce the depth of steel girders to 1350mm deep, which is used to span 75m at the concourse level with a 200mm deck slab thickness. At the platform level, 1500mm deep girders with 200mm thick platform slabs were used to span similar distances. As pier caps in this location extend at track level to support the platform level girders, track level girders could not be made continuous, and a supported system with 1700mm deep girders with deck slab thickness varying from 225mm to 340mm was used to span around 22m with railway loading. Track-level girders are supported over elastomeric bearings, but on other levels, metallic plate over-plate connections were provided to reduce maintenance requirements. In this arrangement, slotted holes are provided to eliminate the temperature-induced stresses in the system. The condition has been checked, as during service, there is a chance that the movement of slotted holes may get blocked, and temperature stresses start developing. The girders are placed below the rails at track level to induce minimum stress in the deck slab. This helped control the dead weight of the superstructure.

Proper planning and a foolproof erection scheme were required so that all the risks could be mitigated. The contractor had designed the erection scheme, which was then checked and reviewed by the Detailed Design Consultant, Consulting Engineering Services-JACOBs and proof-checked by Jadavpur University. The design and construction were more complex due to the multilevel structural erection over the railway line. To have minimum interference to the railway's section, a launching scheme was developed with trestles and temporary supports placed outside the railway yard on the PD building side. There were around seven pairs of trestles that supported the rolling beams. Rolling beams are the units over which permanent girders are placed so that the same can be easily rolled over them.

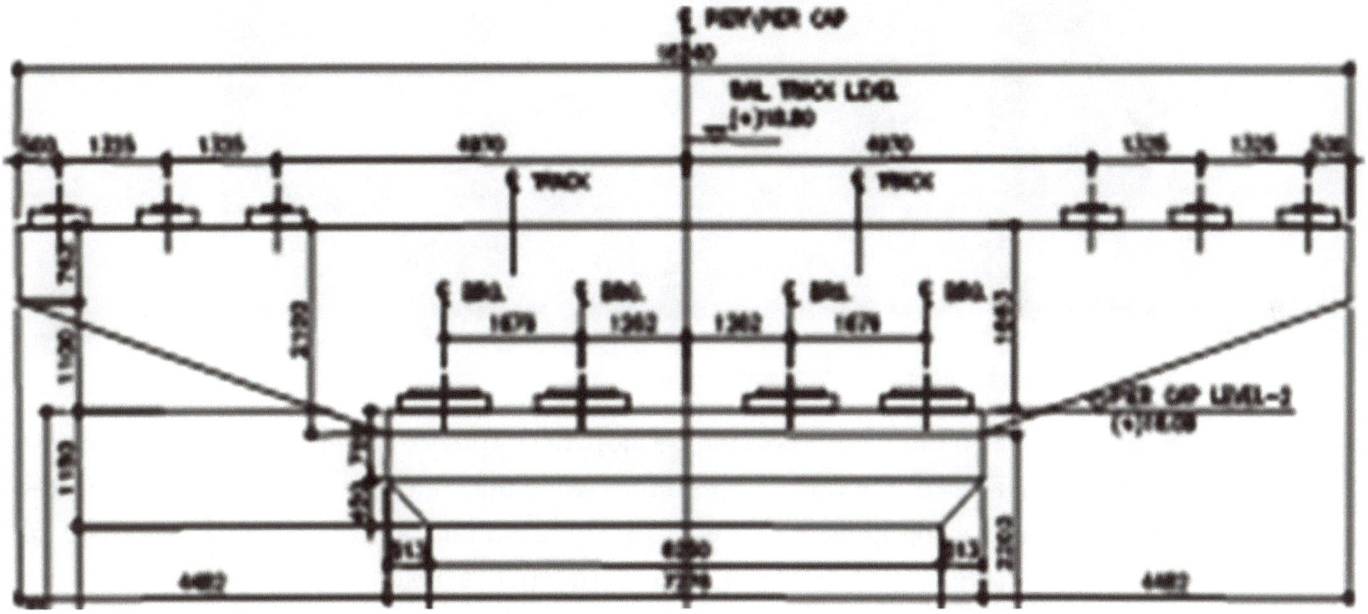

Fig. 7.45: Pier Cap including Bearing Locations at Upper Level

One of the trestles was placed over the railway platform (Platform number 1), up to which rolling beams were laid. The maximum span of ISMB 600 used for rolling permanent girders was 8.45m. All the girders were launched with the help of a nosing girder, a box trestle of 12m length and varying depth of 1500mm to 750mm.

Simultaneously, a pair of permanent girders could be launched. The nosing girders were spliced with permanent girders to behave as a continuous girder system while pulling over various supports located over the railway portion. Noising girders related to the permanent girders were designed to be placed nearest to the OCC side, and as the whole system rolled, the next girders were spliced behind the previous ones. The last girder to be connected was the one closest to the PD building side, where the launch was initiated. Once all the girders were launched and the nosing girders got into the BCC building side area, splices of the nosing girders with permanent girders were removed. During the election process, the permanent girders were erected in pairs with top and bottom flanges braced with the horizontal cross-bracing system to keep the unbraced length of the girder to a minimum.

One of the trestles was placed over the railway platform (Platform number 1), up to which rolling beams were laid. The maximum span of ISMB 600 used for rolling permanent girders was 8.45m. All the girders were launched with the help of a nosing girder, a box trestle of 12m length and varying depth of 1500mm to 750mm. Using a box trestle helped provide additional stability.

Fig.7.46: Nosing Girder launching the Permanent Girder at Concourse

Simultaneously, a pair of permanent girders could be launched. The nosing girders were spliced with permanent girders, acting as a continuous girder system while pulling over various supports located over the railway portion. Noising girders related to the permanent girders were designed to be placed nearest to the OCC side, and as the whole system rolled, the next girders were spliced behind the previous ones. The last girder to be connected was the one closest to the PD building side, where the launch was initiated. Once all the girders were launched and the nosing girders got into the BCC building side area, splices of the nosing girders with permanent girders were removed. During the erection process, the permanent girders were erected in pairs with top and bottom flanges braced with the horizontal cross-bracing system to keep the unbraced length of the girder to a minimum.

Each of the temporary erection steps was analysed, and all the members were checked to ascertain adequate strength and stability during these temporary conditions. It was seen that while launching, when the nosing girder travels between two supports acting like an overhang or cantilever, it generates uplift over the rolling beam. A proper Kentledge was provided over the rolling beams/ trestles to counter the uplift, which was removed after the erection.

The erection process was similar for all the floor levels except at the track level, as all spans for the track girders were simply supported. Hence, a separate splice portion was created between the two spans of track-level girders, connecting the back end of the previous girder with the front end of the subsequent girder. The whole track-level girders were erected over temporary stubs at each pier location. Once all the girders were launched, the central splice portions were removed, and the girders were lowered to a permanent bearing level. During the erection, precautions were taken:

1. Pulling girders were carried out once the power and railway traffic was blocked.
2. All machinery and labour were kept ready before the traffic and power block so the block period could be utilised to the maximum extent possible.
3. Before the launch work starts, all the winches and jacks are checked to ensure they function correctly during the traffic and power block period.
4. To avoid undue impact, the girders were hauled at a very low speed, not exceeding 100 mm/min, and correct alignment was ensured at each step.
5. Wind speeds were noted before and during the erection so that work could be stopped if the wind speed exceeded 25km/h.
6. The nose's deflection was monitored at every step to ensure that the erection was as per the design.

7.28 Casting of Deck Slab at Each Floor Level

A corrugated steel deck sheet was used as a sacrificial shutter for casting the deck slab at each floor level. This helped cast the in-situ deck without additional support from the base and was a very light member. The total weight of the system also remained unaltered. These reduced the blocks required for slab casting and avoided any chances of accidents. Sacrificial shuttering sheets were made of Zinc-Coated steel.

Fig. 7.47: Sacrificial Shuttering for Spans across Railway Yard

This whole process of launching girders and casting decks for all levels was executed in 124 hours of block work over 35 days. The work was executed according to plan with minimum deviation.

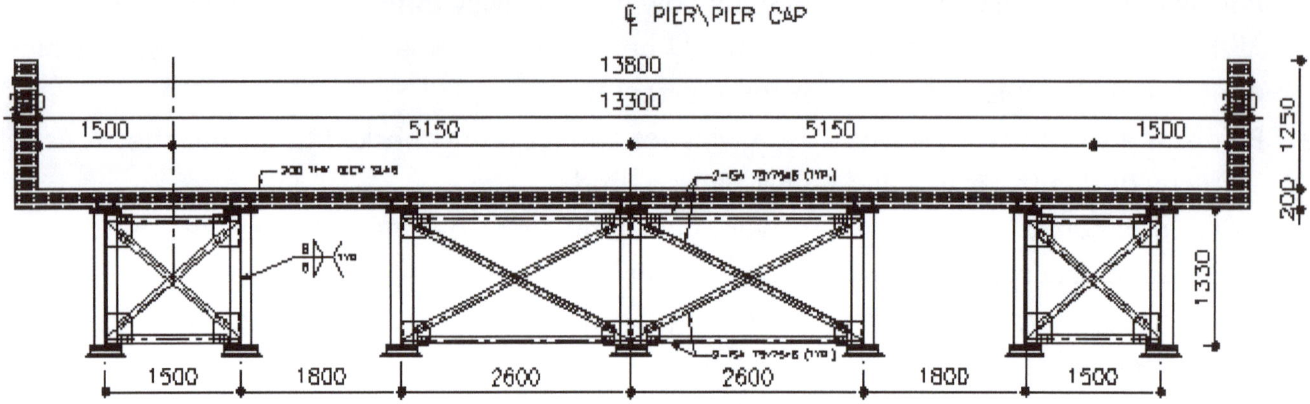

Fig. 7.48: Cross Section at Concourse Level

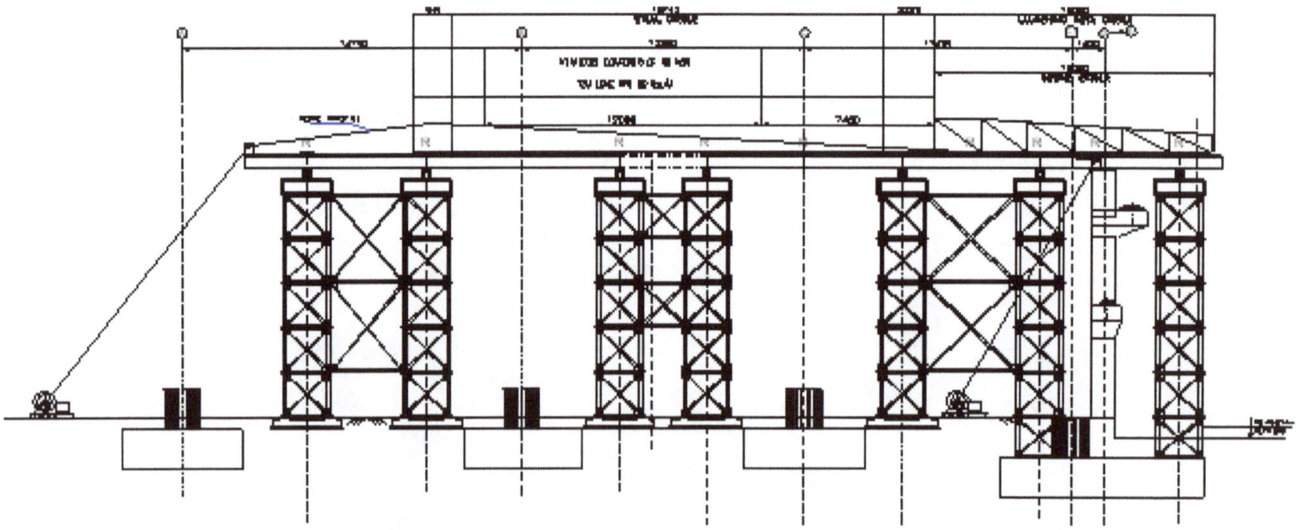

Fig. 7.49: Sketch showing the launch Nose Girder and Kentledge over Temporary Trestles

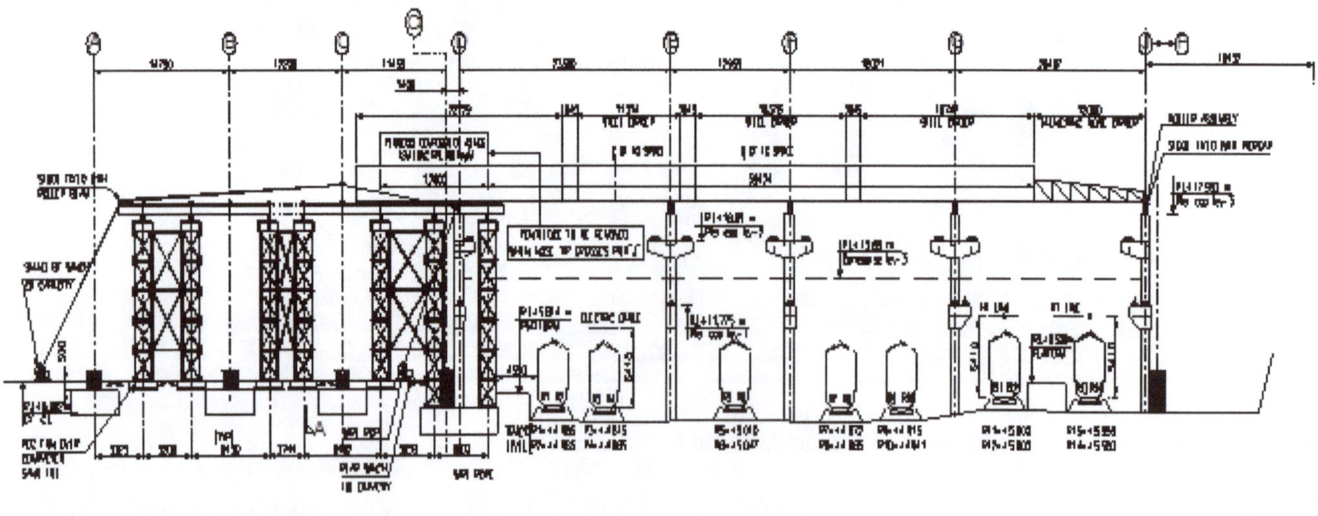

Fig. 7.50: Pictorial Representation of the Erection Scheme

7.29 Roof Structure

The roof structure consists of 40 nos of curved steel I-beams connected to steel columns of I-section at either end. These relate to beams (ISMB 250 section) and purlins. There is a provision for a Monitor Truss arrangement at a middle-top location, which is covered with a polycarbonate sheet and allows wind to pass through. These sheets have a thickness of 2mm and rib 20mm to 30mm with a pitch of 200 mm. The rest of the area is covered with troughed Zincalume steel sheets. These Zincalume steel sheets are of rib height of 28 to 30 mm with a pitch of 200 mm. The thickness of the sheets is 2 mm, with a bare metal thickness of 0.53mm

Fig. 7.51: View of Metro Station over Railway Yard Area

Fig. 7.52: Viaduct Across a Railway Track

7.30 Casting Yard

7.30.1 Operation

There are 2 nos of casting yards developed near IIM/Joka, having all necessary facilities of a batching plant, gantry cranes, casting arrangement, stacking arrangement etc. In the yard, Segments (weighing 40 to 50 T) are cast with Metro Site. The segments are taken to the site of erection with a trailer.

Fig. 7.53: Casting of Segments in Casting Yard and a Dedicated Batching Plant

7.30.2 Launching Girder

The launching scheme for segmental construction is the same as for any metro construction. The only thing that varies is the busyness of the road and other local constraints. Launching the girder main box structure rests on the 4 supports, i.e. front erection is checked and executed in a very safe manner. The typical diagram of the shifting of segments is as under support, middle support, rear support and rear trolley. The erection of girders is done by taking suitable blocks on the road and erecting them with the road Cranes. The method of statements for each of the activities are prepared, and checklists are prepared before

Fig. 7.54: Erection of Launching Girder in DH Road

Fig. 7.55: Erection of Precast Box Sections

F ig.7.56: Erection of Launching Girder at Diamond Harbour Road

7.31 Joka Depot

Joka maintenance depot is a stand-alone system, and the location has been finalised near includes a stabling yard and all the facilities required in a maintenance yard. The facilities included are given below.

- Facilities for Scheduled Maintenance, including Periodical Overhaul (POH
- Facilities for non-scheduled maintenance of Metro Rakes
- Facilities for the Stabling of Rake Unloading line
- Facilities for the maintenance of all components like Wheel, Bogie, Traction Motor, RMPU, etc.
- Bakrafat Road on 25 hectares of land, and the facility is designed for 17 rakes.
- The depot Test track for testing of rakes before putting them into passenger services Store ward
- Machinery and Plants
- Electrical equipment
- Depot Control Centre & Administrative Office
- Security Office
- Self-propelled relief vehicle with rescue and restoration
- ART Parking Bay: Including one ART parking Bay.

Fig. 7.57: Administrative Building

- Shed 1 (200 m x 82 m covered shed) for routine and unscheduled maintenance consisting of (a) 3-tier inspection bay - 8 lines, (b) Lifting Bay - 2 lines (with 30T EOT Cranes). One line in Lifting Bay has standard pit line (c) Office cum maintenance building.
- Shed 2 (200 m x 98 m covered shed): 12 m wide TXR/Rake commissioning bay, 25 m wide, Workshop, 17 m Pit

- Wheel Lathe: M/s Hegenscheidt make an underfloor wheel lathe for in-situ turning of wheel profiles.
- Automatic Coach Washing Plant: M/s Aquafrisch make
- Automatic Coach Washing Plant for drive-through
- Exterior washing of rakes.
- ETP: The Automatic Train wash plant integrated with
- ETP and RO system for the reuse of water.
- Bogie Shop, Traction Motor cum RMPU Bay & Wheel Shop. Also includes dedicated Office cum maintenance building.
- Electric Sub Station: Provided with rectifier-type transformer for DC traction supply for train operation
- Stabling Lines: 12 lines 200 m length with catwalk facility - for stabling of Rakes.

Fig. 7.58: Views of Joka Depot

Fig. 7.59: Internal View of a Workshop Building

Fig. 7.60: Internal and External Views of Work Shop Buildings

7.32 Ballastless Track Design – Structural Design, Codes & Standards

On Bridges/Viaducts, the track is installed on Plinth Beams to minimise the imposed Loads from the track supporting structure. With the third rail, a specific fixing arrangement has been provided in the form of pedestals.

UIC 774-3 – Track-Bridge interaction

- For the Track resistance of the loaded track for consideration of CWR forces.

UIC Recommendations for the design and calculation of ballastless track

RDSO - Procedure for Safety Certification and Technical Clearance of Metro Systems Euronorm EN 1991-2

- Types of Loads considered are from

IRS Concrete Bridge Code

- Types of Loads & Load Combinations
- Section design **BS 5400 - 2**
- Lurching Force

Bridge Rule –

- Centrifugal Force
- Raking Force

IS 875 - 3

- Wind Force

7.32.1 Structural Design

Dead load - the self-weight of the track structure, mainly reinforced concrete.

Superimposed Dead Load - Consists of the weight of all materials forming loads on the structure that are not part of the structural elements.

Rail UIC60- 60 kg per meter of rail with 18kg each per fastener assembly

Train Live load (LL)- Based on the axle load value at the maximum capacity, as per specification, the train live load is considered

Dynamic load (DY)- The train's live load is increased by an impact load factor, which accounts for dynamic train loading conditions

Lurching Force (LF)- Lurching forces are caused by the train rotating slightly about its axis, as per BS 5400-Part 2, clause No., 8.2.7, 56% and 44% of the train load is applied which causes a moment at rail level corresponding to an increase & decrease of vertical load by 6% of the maximum axle load per rail

Centrifugal Force (CF)- consists of the weight of all materials forming loads on the structure that are not part of the structural elements.

Wind on live load (WL)- wind forces are taken to act outwards in a horizontal direction at the level of the center of gravity of the train.

Wind Load is calculated as per IS 875 (Part-3), clause N

Racking Force (RF)- The racking force is a force acting horizontally, at the top of the rails, perpendicular to the centre line of the track. It is applied to both straight track and curved track. The racking force is always combined with a vertical traffic load.

As per Bridge Rules clause No. 2.9.1, the racking force value is: RF = 5.88kN/m.

Derailment (DR)-Derailment load is considered only for the design of the derailment guard.

• As per ACI 358.1R92, clause No. 3.5.2, the derailment load is a transverse load which corresponds to the application of 50% of one vehicle weight (live load) applied horizontally as 22a 5m long uniform impact load on the member used as a derailment wall

Earthquake (EQ)- Earthquake load in the transverse direction due to live load is to be considered. Minimum depth required from Proposed Rail Top Level to Viaduct Top Level is as follows (total sum has been rounded off to the nearest 5 mm):

Minimum clearance between the Kinematic Envelope and the top of the cable duct has been achieved by considering the infringement while designing and is a: At the Viaduct land acquired between Taratola and Majerhat, and relocation of various buildings done. • Calcutta Electricity Supply Company (CESC) • C-Comens-private party (G+2)

• MINT-Central Government (G+0) • Old Age Home-NGO (G+1)

Location: Rail Height 172, Rubber Pad 10, Base Plate with Rubber Pad 3, Minimum Depth of Plinth 185 –Total Rail Centre to Track centre 22,425 mm

Delay in acquisition/permission of land for Joka Depot, Irrigation department land along the Corridor.

7.33 Quality Control

Quality control (QC) in metro construction is a comprehensive system of inspections, testing, and verification processes to ensure that all materials, workmanship, and finished work comply with specified safety standards, regulatory requirements, and design specifications. The goal is to identify and correct defects proactively to ensure the long-term durability, safety, and operational efficiency of the metro system.

Key Components of Quality Control

The quality management system (QMS) for metro projects, is guided by standards like ISO 9001, encompasses both Quality Assurance (QA) and Quality Control (QC).

1. **Materials Inspection and Testing**

   Concrete, Steel/Reinforcement &Source of Supply

2. **On-Site Supervision and Workmanship Assessments**

   Regular Inspections, Dimensional Verification, Non-Conformance Reporting

3. **Samples and Testing**

   Taking of Samples, Testing and Laboratory Facilities at Site and Third-Party Verification

4. **Documentation and Planning**

   Quality Control Plans (QCPs), Inspection and Test Plans (ITPs) & Audits

5. **Advanced Technologies and Monitoring**

   Real-time Monitoring, Building Information Modeling (BIM)

6. **Independent Oversight**

Laboratory setup at Pailan has been included in Photos, Figs.7.

Fig. 7.61: Laboratory Facilities for Quality Assurance

7.34 Impediments

The prime decision was to hand over the Tram Lines. Complete removal of tram lines, OHE & even tram coaches from the site was decided. Others were Kolkata Port Trust (KoPT) – •Rehabilitation centre-NGO (G+3)

* Shifting of the Tram line

* Removal of unauthorised structures from the Irrigation land *Shifting of WBSEDCL cables

*Shifting of WBSETCL cables *Shifting of PHE pipelines

*Dismantling of C. Commence Building *Shifting of the Gas pipeline

*Shifting of the Boundary Wall of the CESC Building *Shifting of KMC pipelines from stations

* Dismantling of Old Age Home *Shifting of Rehabilitation India Building

Fig. 7.62: Removal of Kolkata Tram

Fig. 7.63: Transportation of Old Tram

Fig. 7.64: Relocation of Old Age Home (Original)

Fig. 7.65: Old Age Home (New)

Fig. 7.66: Overhead HT Lines (Before Removal)

Fig. 7.67: After removal of HT Lines

7.35 Passenger Safety

7.35.1 On Stations

i. Tactile Paving
ii. Signage
iii. CCTV Surveillance System
iv. Public Address System
v. Lifts & Escalators
vi. AFC Gates
vii. Fire Detection & Suppression
viii. Smoke Extraction
ix. Emergency Lighting

7.35.2 On Platform

i. Portable fire extinguishers
ii. Emergency Tripping System (ETS)

7.36 Permanent Way

Total Track length: 14.07 Km

Ballasted Track: 7.654 km with 300mm Ballast cushion

Ballast Less Track: 6.417 km

Sleeper density: 1540/km in yards and 1660/km in Test track

» Turnouts (1 in 8.5): 44 no's (RH T/O- 20 no's & LH T/O-24 no's & DS-1 no)

» Sleepers (60 Kg): 7913 nos'

Long Sleepers for 3rd rail: 2126 no's

Short Sleepers for 3rd rail: 518 no's

BLT for Rake unloading Line: 1 no (Length: 147T M)

7.37 Fire Safety

Length of Header pipe: 4.3km, IS-3589 & IS-1239, dia 200 mm and 150 mm

Total nos of hydrant points: 52 outside the Shed area

Total Hydrant points: 25 nos inside Shed-I

Total storage capacity of Fire water: 2,00,000 litres

Fire pumps: 4 no's (Jockey pump of 4 kW, 88 m, head,10.8 Cum/Hr: 2 no's, Diesel pump: 97Hp, 88 m head, 171 Cum/Hr flow, and Electric pump: 75 kw, 88 m head, 171Cum/Hr discharge capacity.

Fire Extinguishers (CO_2 Type 4.5 kg Fire Extinguishers, ABC Type 6 kg Fire Extinguishers, Foam Type 9 Litre Fire Extinguishers)

Sprinkler system in the administrative building: Total header length: 550 m,

7.38 Green Initiatives

Grid-connected Solar power Plant: 1400 kWp.

Effluent Treatment Plant (ETP): 35 KLD

Sewage Treatment Plant (STP): 11KLD

Energy efficient light fitting: LED light

Polycarbonate sheet: Sheet provided in the roof & panel walls of Sheds and other PEB structures

Variable Refrigerant Flow (VRF) type AC system

Wind-driven Turbo Vents

Horticulture: Tree plantation (200 Nos), Turfing 6500 Sqm

750V DC Aluminium 3rd rails system.

7.39 Ventilation System

A positive-pressure ventilation system was provided

3 Nos. 15000 CFM Fresh Air Intake Fans installed

2 Nos. 15000 CFM Exhaust Fans provided

Motorised Fire Damper provided in the Ventilation Duct

The ventilation system is integrated with the fire protection system.

Fig. 7.68: Permanent Way

Fig. 7.69: CRS Team During Trial Run

Short Briefs of Other Kolkata Metro Lines In Progress

a) Metro line 4 – Yellow Line (Ref. Wikipedia)

It currently consists of a short operational section between Noapara and Jai Hind (Kolkata Airport) that was inaugurated on 22 August 2025. The full line will be 22 km long with 14 stations, running from Noapara to Suripukur north of Barasat in North 24 Parganas. It is being built by RVNL, Senbo & ITD. It will run mostly on underground tunnels. The track is elevated until Jessore Road from Noapara and then goes underground towards Suripukur metro station. It has an interchange at the Kolkata Airport with the Line. The development work from New Barrackpore to Barasat has been stalled for the last eight years since it was sanctioned because of encroachments, and the Airport Authority of India objected to the elevated stretches from New Barrackpore to Barasat. Later, after objections from AAI, a decision was taken to convert this into an underground stretch. In June 2019 sanction for material modification for underground construction at an additional cost was sent to the Railway Board by Metro. For this underground tunnelling, AAI has agreed to shift its fuel tanks, buildings, and cables, and the project is expected to start in 2025. In February 2021, the Hon'ble Prime Minister gave his nod to construct this stretch underground, giving respite to thousands of families who were supposed to be evicted. Construction of an underground integrated station for the Orange Line and Yellow Line has begun at NSCBI Airport land, 150

In August 2025, before the inauguration of the Noapara–Biman Bandar section, it was understood that some changes in line alignment from Micheal Nagar to Barasat would be made, which would be constructed underground using a Tunnel boring machine that would lead to an increase in the total length of the line from 18 km to 22 km, adding three new stations beyond Barasat. These changes were made due to objections from AAI regarding the use of tall cranes and to avoid significant encroachment problems.

The Yellow Line has been divided into three phases.

- Phase 1: Noapara - Biman Bandar (Under trials, commercial services to start from August 2025)
- Phase 2: Biman Bandar - New Barrackpore (under construction till Michael Nagar; Michael Nagar - New Barrackpur metro stretch resumes as AAI permits 18m cranes and external construction)
- Phase 3: New Barrackpore - Barasat (Planned, construction to start)
- Phase 4: Barasat - Suripukur (Proposed)

b) Metro line 5– Pink Line (Ref. Wikipedia)

The Kolkata Metro Pink Line signifies an exciting expansion of the city's metro network. While the construction of this metro corridor is pending, it holds tr`emendous potential for improving connectivity between North Kolkata and the central city areas. The planned expansion, officially designated as Kolkata Metro Line 5 or the Pink Line, is set to cover a 12.5-kilometre route, smoothly linking Baranagar with Barrackpore. The blueprint includes provisions for 11 stations along this route, catering to the diverse needs of commuters.

The stations on the Kolkata Metro Pink Line route will connect Kamarhati, Agarpara, Panihati, Sodepur, Subhash Nagar, Khardaha, Tata Gate, Titagarh, Talpukur and Barrackpore. The Baranagar Metro Station is already open.

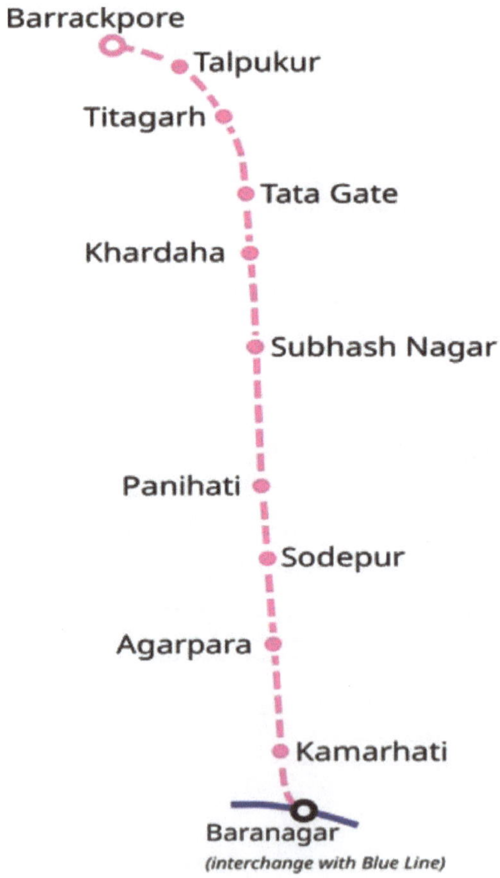

c) Metro line 6 – Orange Line (Ref. Wikipedia)

Orange Line is the sixth Line of Kolkata Metro. It will connect New Garia with Netaji Subhas Chandra Bose International Airport via two major satellite towns of the Kolkata metropolitan region, Salt Lake and New Town. It currently operates between Kavi Subhash and Beliaghata. This line is one o`f the most important stretches of the Kolkata Metro because it connects the southern and eastern parts of the city with Kolkata Airport. The total distance of this route is 29.87 km. It has an interchange with the Blue Line at Kavi Subhash and will eventually connect with the Green Line near IT Centre and the Yellow Line at Jai H`ind.

The Kavi Subhash–Hemanta Mukhopadhyay section of the Orange Line was inaugurated in March 2024 by the Hon'ble Prime Minister, along with the truncated sections of Taratala–Majerhat under the Purple Line and Esplanade–Howrah Maidan under the Green Line. A week after the inauguration and a day after the commercial run started on the Kavi Subhash-Hemanta Mukhopadhyay section, the completion of a 76 m viaduct gap of a single-span open-web girder steel bridge over the Tagore Park culvert. This completed the viaduct up to Beliaghata metro station. Trial runs have started over the Hemanta Mukhopadhyay-Beleghata metro station, and the plan was to open this stretch by the end of 2024. RVNL has faced enormous challenges while working. In one such incident, it incurred an enormous loss of over 5 crore rupees due to the lack of clearance of traffic from the Kolkata Police and work being stalled for over 60 days. The Hemanta Mukhopadhyay–Beleghata Orange Line extension was finally opened in August 2025, along with the truncated sections of Noapara–Jai Hind under the Yellow Line and Esplanade–Sealdah under the Green Line.

Diagram of Line 6: Orange line route map

The orange line includes 22 elevated stations, 1 underground station and 1 at-grade. The terminal Airport station is the underground one. This station includes a stabling yard and it will be the largest underground facility in India. It is 550 m long and 41.6 m wide and facilitates stabling and reversal of rakes. Green and Orange Lines under construction at Salt Lake Sector-V, November 2017. An urban garden developed by converting the unused space under the metro rail using the vertical garden technique.

Annex: 1 Excerpts of Tender by MTP-R for Kolkata Metro-Line 1

METROPOLITAN TRANSPORT PROJECT(RLYS)
12, MISSION ROW (4TH FLOOR)
CALCUTTA-1.

TENDER DOCUMENTS

Tender No.Civil/121

CONSTRUCTION OF SUBWAY STRUCTURES EXCLUDING DIAPHRAGM WALLS FROM KM. 12.340 TO KM. 12.691 - SECTION-14A.

Approximate contract value : Rs. 71 Lakhs

Earnest Money : Rs. 20,000/-

Date of opening : 27.7.78

Issued to :

Messrs Chatterjee Polk Pvt LTD, 59B Chowringhee Road
Congress Bhawan, 3rd Floor, Cal-20

Alongwith 3 extra copies of Annex-4 marked as duplicate, triplicate, tenderer's copy and necessary drawings.

(Not transferable)

V. K. D. 15/6/78

===

Chief Engineer,
Metropolitan Transport Project (Rlys.)
Calcutta-1.

Fig.A3.1: Part Plan Showing Category 'A'

Excerpts

(With Original Clause Numbers)

1.07(a) (1), the Tenderer is required to quote rated against the items appearing in Annex. 4 both in "Words" and "figures". The quantities are shown in the Annex. 4 are given only for guidance, are approximate, and are subject to variation according to the needs of the MTP(Rlys). However, MTP(Railways) accepts no responsibility for their accuracy. The railway does not guarantee ordering for work under each item furnished therein. If there be any variance in rates quoted by the Tenderer between "Words" and figures", that quoted in words will be treated as final

0182 All excavation shall be done by mechanical means, except for the top two meters, not only to increase the out-turn but also to discharge the concentration of manual labour force in huge numbers in the space-bound work sites, which may lead to unhygienic conditions. Keeping in view the location of struts and vertical restraints, small-size bulldozers or similar equipment capable of excavating within the restricted conditions of the cut shall be utilised. The contractor shall spell out, along with his tender, the type of machinery that is going to be used, indicating the size and other characteristics of such machine. No space will be available for dumping excavated spoils on the surface near work. The contractor shall remove such lifted muck directly and immediately to such disposal grounds as may be allotted by the Engineer.

1.07(a) (1), the Tenderer is required to quote rated against the items appearing in Annex. 4 both in "Words" and "figures". The quantities are shown in the Annex. 4 are given only for guidance, are approximate, and are subject to variation according to the needs of the MTP(Rlys). However, MTP(Railways) accepts no responsibility for their accuracy. The railway does not guarantee ordering for work under each item furnished therein. If there be any variance in rates quoted by the Tenderer between "Words" and figures", that quoted in words will be treated as final.

1.07(b) Alternative Designs

i. MTP(R) may consider the contractor's designs and schemes for construction. For this purpose, the Tendered must submit if he desires that his design and scheme be considered fully detailed drawings of his schemes and all calculations to justify his claim as to the adequacy of his proposals. The design criteria are furnished in the Annex. 7 may be taken into guidance for submission of the alternative designs.
ii. While submitting a quotation as per his designs, the tenderer must follow the itemised schedule as per Annexure-4, clearly understanding that the specifications and rates follow the special and Standard Specifications referred to.
iii. If any item of work in his design does not correspond with any of the items of schedule as per Annexure-4, he must give full specification of such new item. It is understood that

the rate quoted against such new items are complete rates of work as made ready for use, and no extra payment by way of carriage, handling, royalty, additional accommodation works necessary to bring these new items into use, etc., will be paid for. If these new items need any materials that are normally proposed to be supplied free of cost by the MTP(R) as in the case of departmental design, the quantities of requirement of such materials per unit measurement as well as for the whole work shall be denoted in the tender. Any extra requirement for the job beyond these declared quantities will have to be paid for by the contractor to the MTP ®. at rates as per para.

iv. In the event of the contractor's specification being inadequate and ambiguous, the Chief Engineer, MTP ® 's decision will be final.

v. The Tenderer must estimate the quantity of each item in his design carefully at the time of quoting, as these figures will be considered for evaluating his offer. If in actual execution, any lesser quantity is done, the quantity actually done will be paid for, while if the quantity actually executed is more than the quantity indicated in the tender, then the quantity to be paid for will be confined to the quantity satisfaction of the MTP(R) that need for such extra quantity arose out of any subsequent decision of the MTP (R) not included in the tender papers.

vi. If the tenderer's design is accepted, he shall have to submit detailed working drawings on reproduction paper together with two spare prints to the MTP (R), and work can be undertaken only after such drawings have been approved. The Contractor, on completion of the work, shall submit on reproduction paper all completion drawings certified as "made to the Engineer for record, together with three spare prints.

2.6 Fencing work sites

i. The contractor shall provide strong and elegant welling or fencing around the construction areas with corrugated iron sheeting or timber planks or any other appropriate material to a sufficient height at no extra cost of MTP (R) to keep away unauthorised persons entering into the working areas or to keep away inquisitive public gathering in numbers and obstructing the everyday life in the area or causing obstruction to the works. The height should be such that it provides complete physical and visual isolation from road users.

ii. The contractor shall ensure that all construction activities, except the movement of transport vehicles, are confined within such protected fenced areas and that at no time are plants, machinery and vehicles stable or debris or construction materials dumped outside such fenced areas in the course of the work, causing inconvenience to the public.

iii. The boundaries along which such fencing is to be put up shall be settled by the Engineer.

iv. If the contractor fails to put up such fencing, the Engineer shall have the option to set right the deficiencies at the contractor's cost.

2.7 Temporary decking for facilitating construction

Whenever it becomes necessary for the contractor to provide temporary decking over trenches for movement of his men and vehicles purely for the use of the contractor's men vehicles are to be provided and maintained by the contractor for which no separate payment will be made by the MTP (R). The MTP (R) shall have the right to use these dockings for the movement of its men and vehicles. It is under the annexe.4 will be made only for such decking which is necessary for the movement of the public.

2.8 Road diversions and tram line diversions

i. Such works that require traffic regulation/diversion shall be taken up by the contractor only after arrangements for such regulations/diversions are fixed up with the concerned authorities by MTP(R).
ii. The sequence and program me for the work shall be drawn up by the contractor in consultation and with the approval of the engineer in such a way that dislocation of the traffic, or traffic diversion through alternative routes, is the minimum and for the shortest duration possible. Once such work affecting road or tram traffic is taken up, it shall be the endeavor of the contractor to complete the same expeditiously so that the inconvenience to the road users is the minimum and for the shortest duration.
iii. wherever there is an obstruction to or regulation/diversion of the traffic the contractor shall provide all necessary traffic control signs and signals/that may be required by the road authorities or the police, and shall operate, man and maintain them efficiently and to arranging the police personnel where they have also to man such points shall, however, be matter between the MTP(R) and the police authorities.
iv. Tramway Lines: Diversions of Tramway lines or supporting the same on deckling when required, shall be done by the contractor as per the directions of the Engineer at respective rates payable under item 4.09 (c) of Annex.4.
v. A barricade of adequate strength shall have to be put up by the side of tramway lines as per Para 3.12(B)(xii) and no extras payment for the same shall be made. These barricades shall be clear of the moving dimensions of tramways with due regard to the possibility of over-hanging passengers of tram cars.

2.9 Use of public highways

The contractor shall take all reasonable precautions to prevent the deposit or accumulation of mud, filth, rubbish, or debris on the highways of public places and shall, from time to time, or as instructed by the Engineers or the authorities concerned, remove any material which he may have deposited at no extra cost from such places.

Quantity of Cement to be considered for payment

	Nature of work	Quantity of Cement	
1.	Controlled concrete	M-10	4.06 bags/M^3 of Concrete.
2.	-do	M-150	5.80 bags/M^3 of Concrete.
3.	-do-	M-200	7.32 bags/M^3 of Concrete.
4.	-do-	M-250	10.00 bags/M^3 of Concrete.
5.	-do-	M-200	8.00 bags/M^3 of Concrete.

For the controlled concrete works, if the contractor requires more cement than the above-set limits, he shall obtain specific sanction from the Engineer for such extra cement required for Bonafede to be charged to the contractor at rates as given in sub-para) below. Hence, it is for the contractor to design the mixes for the controlled concrete in the most economical manner to be within the limits of the quantities of cement laid down.

i. The contractor shall arrange and operate, all necessary plants, machineries and equipment necessary for successful and timely completion of the work, Apart from the various plants and equipment's to be arranged the contractor for successful completions of work, the contractor's attention is particularly drawn to the fact that he shall have to arrange for the necessary power road rollers and vibrating type rollers for trickling the back fill and road works without any delay and at no extra cost to MTP(R).
ii. If in the opinion of the Engineer, equipment/plants brought by the contractor are not suitable for the work concerned, the engineer shall have the right to order the contractor to replace them with suitable plants/equipment. In the interest of public convenience, the Engineer may insist on a specific way of execution of the work.

2.19 Sub-soil investigations:

Sub-soil data obtained from boreholes are given in the drawing listed out in Annexe 5. These or any other data obtained subsequently are made available to the contractor in good faith.

2.20 Advertising

i. The contractor shall treat the contract and everything within it as private and confidential. In particular, the contractor should not publish any information, drawings, or photographs relating to the work and should not use the site for advertising purposes, except with the written consent of the Engineer and subject to such conditions as he may prescribe.
ii. The contractor shall provide all facilities to the MTP(R) or its authorised representative to collect information, photographs or make movies of any stage of work, including the working of the contractor's plant, machine and equipment, etc., for record, publication or any other purpose.

2.21 Diversion of services/utilities

i. The following underground/overhead utilities generally come in the way of construction: -

a. telephones and other communication lines.
b. power cables and power lines.
c. water mains and service connections.
d. sewers, mains, and service connections.
e. surface drains.
f. Tram lines.
g. Gas mains and service connections.

2.22 Surface drains

The contractor shall ensure that the system of surface drains now prevailing at the site to serve the needs of the locality is affected to the minimum. Wherever such drains are affected, arrangements for proper drainage shall be carried out as per the direction of the Engineer and as per approved plans under the relevant items in the schedule of items.

2.23 Ventilating shafts and drainage sumps

i. The location and details of ventilating shafts and sumps have not been finalised. However, the contractor shall have to complete the work as per detail, which will be fixed by MTP (R) during the work progress. These works shall be paid for as relevant items of Anne.4. If, however, MTP (R) does not finalise the details of these shafts or sumps or does not ask the contractor to undertake the work before the works as per Annex.4 have been completed, then the contractor shall not be found to undertake the work concerning the ventilating shafts and sumps.
ii. Any phase working (in stages) found necessary for constructing the ventilating passages, shafts and drainage sumps shall be done by the contractor. As directed by the Engineer, without any claims for extra payment.

2.24 Damage to adjoining properties

i. An important requirement of the construction efforts shall be to complete the work on time without causing damage to adjoining properties. The method and sequence of construction, type of plant and machinery deployed, speed of execution, etc., shall all be based on specifications laid down and calculated to achieve this to the test of knowledge available with the MTP(Rlys). If, however, the contractor has any other solution to offer, the same will be duly considered.

2.25 work prepared off the site

The contractor shall give the engineer written notice of the preparation or manufacture at a place off the site of any material or component to be used on the works, stating the place and time of preparation or manufacture, so that the engineer may inspect at all stages of the production process. Failure to give such notice may result in the rejection of the material or component.

2.26 Suspension of works during bad weather

i. The contract period may extend over a few monsoon seasons. The period of completion allowed for the work assumes that during the monsoon season, progress on many items of work will not be possible. The contractor should, therefore, work during seasons; all work entrusted to him is completed during the monsoon, and only such work is safe to carry out during such periods.

ii. The piles shall be held in position by adequate temporary waling/struts. The sheet piles shall be driven to correct vertical alignment to lines and levels shown on the approved drawing. Any creep occurring shall be eliminated in a manner approved by the engineer. The sheet piles shall not be out of plumb by more than 1 in 60 in any direction. The waling/strutting is payable under items 4.07/4.08.

iii. The joints of piles shall be grouted or caulked as necessary during excavation to make them watertight, and this is included in the rate.

iv. No water jetting, etc., at the bottom of the sheet pile will be permitted to facilitate easy driving of the sheet piles.

v. Any piles which are driven out of alignment or twisted, broken, bent or otherwise damaged in driving shall be withdrawn and fresh piles driven by the rate includes all such contingencies.

3.03 temporary walling with 'H" piles and wooden lagging

i. (a) The 'H' piles shall be of either single 'H' sections or built-up 'H' sections. The contractor should plan his work so that the piles are reused as often as possible, and rates should be quoted accordingly. While it is in the contractor's interest to take over as small a quantity of piles as possible, he is responsible for ensuring that the work is completed according to the phasing and schedule required by MTP (R). The payment rate under item 4.02(a) shall be for the finished work done, irrespective of the number of uses of the piles and the quantity of piles extracted. It is in the contractor in his interest to ultimately extract the 'H' piles to derive the benefit of reimbursements, vide clause 2.11(vI) (a) to (e). The rate against item 4.02(a) under Annexe 4 is expected to be a net rate duly considering the credit likely obtained by the contractor for the released 'H' piles.

   (b) fabrication, required for built-up 'H' piles, is included in the items.

ii. The 'H' piles shall be installed either by driving with a pile-driving rig using a hammer or a Vibro sinker. A helmet of appropriate design should be used to protect the top of the pile.

iii. The depth to which the 'H' piles would be taken shall be indicated by the Engineer or as shown in the drawing. Depending upon the circumstances, the Engineer shall have the power to change such depth during the work progress.

iv. Timber lagging shall be with hardwood planks of approved thickness to be able to stand the expected solid pressure. The width of planks shall ordinarily be 150 mm to 225 mm and will depend upon the type of soil and method of insertion. The edges and ends of the planks shall be neatly finished to ensure tight joints. The length of the planks shall be enough to have a minimum bearing of 60 mm over the flanges of the 'H' piles.

v. The planks shall be inserted between the 'H' piles, starting from the top and working towards the bottom of the cut as excavation proceeds. It shall be ensured that the planks are inserted, and the side of the cutting is protected before the excavation for that area commences. For this purpose, the insertion of planks should precede at least 200 mm ahead of the excavation level. Such insertion shall be done by manual excavation of the soil in between the flanges of two adjoining 'H' piles, inserting the lagging plank from an angle, raising and jamming it to the plank inserted earlier, and tightly rear and packing so that there is no yielding of ground from behind. The excavation done behind the pile face shall be just sufficient wooden wedges driven between the flanges of the 'H' piles and timber lagging. It shall be ensured that wedges are released just before backfilling is done to facilitate the withdrawal of the 'H' piles after the work is finished.

3.04 Diaphragm walls

i. The diaphragm wall for supporting earth to allow excavation of the trench shall generally be 60 cm thick, made with cement concrete of M 200 strength, and reinforced as required to withstand the permanent and temporary states of stresses that will arise during the life of the structure (including live load where this is incidental). The engineer shall specify the live load where this is incidental. The Engineer shall specify the maximum panel length of the trench for the diaphragm wall, depending upon soil condition and surcharge load and locality, and this shall be binding on the contractor.

ii. Sequence of work: The sequence of work shall broadly be as follows:

   a. Construct guide walls in reintroducing ordered concrete to the required section and depth.
   b. Excavating trench between guide walls to the required depth in panel length as approved, keeping the excavated trench always filled with a bentonite slurry solution of approved specific gravity.
   c. Lowering and installing reinforcement cages.
   d. Laying concrete of the desired strength through Tremie pipes to the required height, including installation of the required formwork to form walls in panels.

iii. Guide Walls:

The guide walls shall be reinforced concrete with continuous reinforcement, as per MTP (R)'s design. As in the case of the diaphragm wall, the MTP (R) shall also supply the steel, cement, and ballast/chips for the guide walls free of cost.

iv. Trench excavation:

Trench excavation shall be done by any of the accepted methods, with suitable slurry wall equipment of Kelly Grab or rotary cutting head type. Before resorting to mechanical means, trench excavation shall be done manually up to the depth as may be desired by the engineer to expose the existence or otherwise of the utilities. The excavated spoil may be removed by the grab or reverse circulation method. The equipment used for the trench shall be compact, mobile and capable of excavating true to plumb and alignments. The excavated face of the trench wall should be smooth and without pockets. The maximum allowable tolerance for the completed diagram wall face from the designed plane shall be 200 mm, including all bulges, misalignment, defects in verticality, etc., all combined. The trench excavation equipment should be either crawler mounted or moving on rails, the primary consideration being it should be possible to keep precise alignment. The excavated spoils shall be disposed of as ordered up to a lead of 1 km. Extra loads, if any, beyond 1 km. Shall be paid as per item 4.05(b). The payment for trench excavation shall be on a square meter basis as per item 4.03(a), and the full depth of the trench from the road/ground level up to the designed depth shall be taken into consideration for arriving at the payable area. No payment will be made for extra depth beyond the designed depths.

v. The contractor shall arrange all equipment required for the work, including trench excavation equipment, cranes, trench pipes, concrete chutes, and all other incidental items.

vi. (a) Bentonite slurry: The bentonite earth used for preparing the solution shall be of approved quality and shall have the required thixotropic attributes to hold the excavation sides of the trench firmly without any collapse or erosion, the specific gravity of the solution wall be as approved and the contractor shall provide the necessary laboratory facilities at site for checking the quality of the bentonite earth and the specific gravity of the solution causing the course of the work.

Freshly mixed bentonite slurry.

The density of the freshly mixed bentonite slurry shall be measured daily as a check on the quality of the slurry being formed. The measuring device is to be calibrated to read within 0.005 g/ml.

Bentonite slurry was supplied to trench excavation.

In average soil conditions, the following tests shall be applied to the trench, and the results shall generally be within the ranges stated in the table below:

Item to be Measured	Range of results at 20 Deg. C.	Test method
Density balance	Less than 1.10 g/m1	Mud density balance
Viscosity method	33.75 seconds	Marsh cone
Shear strength (10 min gel strength) pH	1.4 to 1 N/M 9.25-12	Spherometer pH indicator paper strips.

Tests to determine density, viscosity, shear strength, and pH value shall be carried out initially until a consistent working pattern has been established, considering the mixing process, any blending of freshly mixed bentonite slurry and previously used bentonite slurry, and any process which may be employed to remove impurities from previously used bentonite slurry.

vii. **Concrete**

a. Concrete shall be controlled concrete and placed through tremie pipes of adequate diameter in such a way that at any time, the tip of the tremie pipe is at least 0.3 m below the top level of concrete around. The slump of Rev.2 concrete shall be 150mm minimum and range "150 mm to collapse" to allow the concrete to flow freely and fill up the trench but, at the same time, shall not get mixed up with bentonite slurry. The concrete mix and water-cement ratio shall be decided by the contractor, but the final concrete shall be dense and give the minimum guaranteed cube strength as per the prescribed or approved
b. Concrete shall be placed continuously by one or more Tremie pipes, and care shall be taken during placement to avoid contamination of the concrete.
c. The extraction of stop ends shall be carried out at such time and in such a manner as to avoid causing damage to concrete places against it.

3.19 Iron ore filling

i. Iron ore may be required to be filled up by the side and top of the box to provide the necessary dead weight to act against the structure's floatation due to hydrostatic pressure. The Iron ore shall be of approved quality with a minimum specific gravity of 4.5 when blended and rammed as detailed below and free from chlorides, sulphates, sulfides, nitrates, and other ingredients that can corrode or attack concrete.

Annex: 2

ORIGINAL

Annex.4
Civil/121

SCHEDULE OF ITEMS AND QUANTITIES.

Construction of Subway Structures excluding Diaphragm Walls from Km.12.340 to Km.12.691 - Sec.14/A.

Item No.	Brief description of work	Unit of payment.	Approximate quantity.	Rate of payment in 'Words' and 'figures' Rs. P.
1	2	3	4	5
.01	(a) (i) Providing walling with steel sheet piles with imported Lerssen Section (All Contractor's labour and materials except sheet piles which will be supplied by MTP(R) free of cost).	Tonne	Nil	Deleted
	(ii) - Do - with indegenous 'Z' Section piles - do -	Tonne	10 Tonnes	-
	(b) (i) Withdrawing Lerssen sheet piles after completion and delivering released piles to MTP(R) or stacking for reuse.	Tonne	Nil	Deleted
	(ii) - do - indegenous piles - do -	Tonne	Nil	Deleted
	(c) (i) Splicing of Larssen sheet piles to get the desired length (All Contractor's labour and materials except steel for cover plates which will be supplied by MTP(R) free of cost.)	Each joint	Nil	Deleted
	(ii) - Do - Indegenous piles - do -	Each Joint	If any	
.02	(a) Supply, driving, and installing 'H' piles (or fabricated 'H' piles) to support timber lagging (All Contractor's labour and materials).			
	(i) With new 'H' piles.	Tonne	5 Tonnes	
	(ii) With 'H' piles released from other contract sections of MTP(R).	Tonne	5 Tonnes	

Contd..

ORIGINAL

- 2 -

Annex.4
Civil/121

1	2	3	4	5
4.02	(b) Withdrawing 'H' piles after use and delivering released piles to MTP(R) or stacking for reuse.			
	(i) New 'H' Piles	Tonne	5 Tonnes	
	(ii) 'H' Piles released from other contract sections of MTP(R) .	Tonne	5 Tonnes	
	(c) Supplying and providing timber lagging to 'H' piles or at other similar location to form temporary walling (All Contractor's labour and materials).	Sq. Metre	150 Sq. Metres	
	(d) Extra over item 4.02(a) for fabrication in strengthening of 'H' Sections including all rivetting/bolting/welding etc. complete as necessary (All contractor's labour and materials except steel which will be supplied through item (a) above).	Tonne	If any	
	(e) Splicing 'H' piles to get the desired length (All Contractor's labour and materials including steel for cover plates).	Each Joint	5 Nos.	
4.03	(a) to (e): Item pertaining to Diaphragm wall.		Nil	Deleted.
4.04	(a) Driving and installing steel posts of 'H' sections (or fabricated 'H' section) to serve as vertical restraints for struts for wider excavations (All contractor's labour and materials except 'H' sections which will be supplied by MTP(R) free of cost).	Tonne	400 Tonnes	
	(b) Dismantling the above 'H' sections after use and delivering the same to MTP(R).	Tonne	200 Tonnes.	

Contd...

Annex. 2

Design of Diaphragm Wall in soft clay and sandy Soil/conditions

For Soft Clay Conditions

1.0 In standard Calcutta deposit t Soft/Medium clay soil conditions.

1.1 Representative Section considered: C S 9 between k/m 7/800 and 8/600
(Chandni Chowk and Esplanade Station)

1.2 Profile of cut and soil strata profile

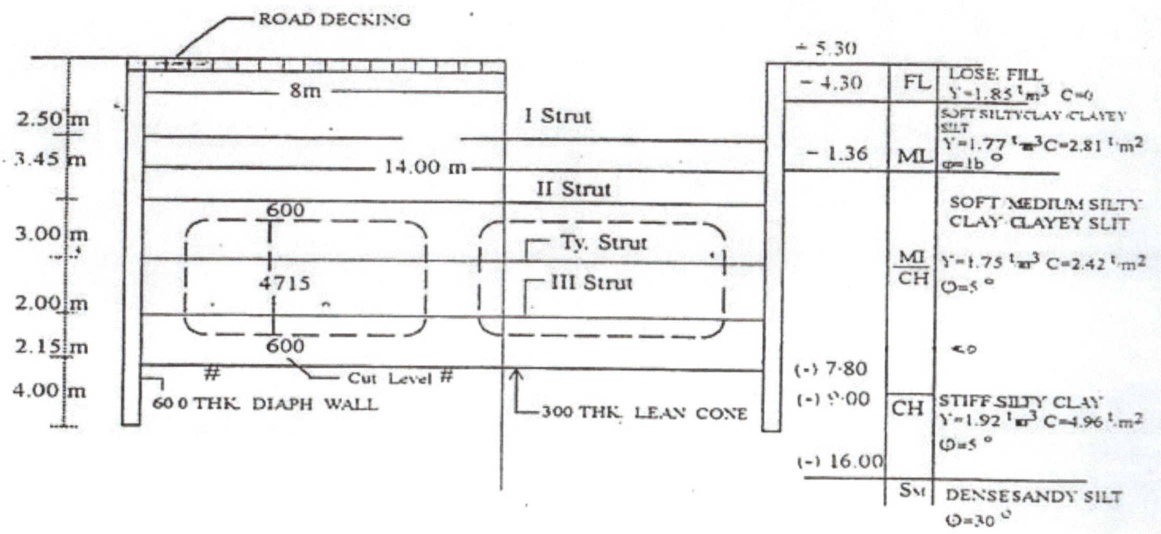

Fig. A2.1: Profile with Cut and Soil Data

2.0 Design of Diaphragm Wall (DW)

2.1 Design Criteria: Metro Rly's Design criteria for Diaphragm walls adopted

2.1.1 Reference Books/Codes followed

a. Theoretical soil mechanics – Terzaghi
b. Soil Mechanics in Engineering. Practice – Terzaghi & Peck
c. Foundation analysis and design – Bowles
d. State of the art report—Seventh international conference on soil mechanics and foundation engineering—by R.B. Peck.
e. Is 456 IS codes for Plain & R.C. Concrete.
f. IS 800 IS code for using structural steel in general building construction.
g. Design Aids for R.C. Concrete to IS 456-78

2. 1.2 Structural Design done by Limit State Method.

2.2 Factors considered in design

2.2.1 Pressure

The apparent Earth Pressure diagram suggested by Terzaghi & Peck for computing strut Toads Braced cuts is considered with a correction factor of 0.80 for DW designs and 0.90 for strut loads and wale designs.

Max Earth pressure (cut portion)

: KA rH

Where

KA: $l\text{-m}\ \frac{4c}{rH}$

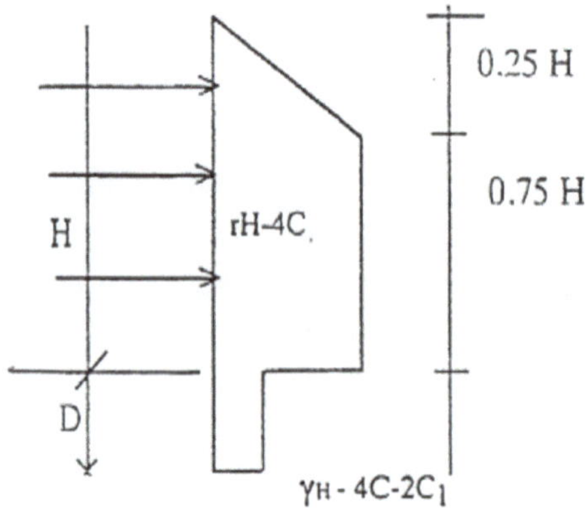

Fig.A2.2: Earth Pressure as per Terzechi & Peck

Considering m = 1.0

(as per pecks report)

$KA = l - \frac{4c}{rH}$

$kArH = (l - \frac{4c)rH}{rH})$

P = rH-4c

Below cut level: P = rH-4c-$2c_l$

Where c = undrained shear strength of soil above-cut level

And c1 = undrained shear strength of soil below-cut level.

The GWT is assumed to be at ground level, and as (Phi) is considered zero, no separate hydrostatic pressure is considered.

2.2.2 lateral pressure due to building surcharge of

g1=(Phi):5.50 $^{t}/m^{2}$ acting from a depth of I.50m

below G.L. and from a distance of 2.0 m from the outer edge of the wall. Considered.

2.2.3 Lateral pressure due to road surcharge of

$g_{2(Phi)}$ = 2.o t/m^{2} (live load surcharge due to vehicular traffic by the side of the cut) considered.

2.2.4 Superimposed Loads:

Liye and dead bad reactions on Diaphragm Wall due to road/Construction designs Considered for diaphragm wall penetration and structural designs.

2.3 The stability of DW was checked for the following

i. Stability of bottom of cut against heaving.
ii. To carry superimposed dead loads.
iii. Safety during various sequences of excavation and box constructions.

3.0 Check for depth of Penetration of DW to ensure the stability of the bottom of the cut against heaving. (Ref. Para I.07 Metro Design Criteria)

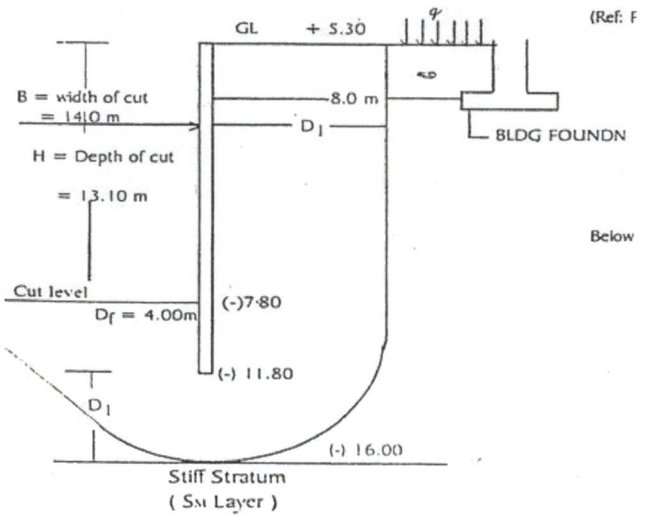

Average Soil data.

Cut Portion

r = 1.85 t/m3

C = 2.34 $^{t}/m^{2}$

Phi = 5°

Below cut

$$R = 1.85\ ^{t}/m^{3}$$

$$C = 4.32\ ^{t}/m^{2}$$

$$phi = 5°$$

To check for heave at (-) 11.80

RL of still stratum: (-) I6.00

D1 = radius of slip circle = 16 - 11.80 = 4.20 m.

[< B √2 = $\frac{14.00}{\sqrt{2}}$ = 9.90m]

Downward loads at (-) 11.80 due to slip wedge 4.20 m wide.

i. Due to Earth: 1.85x17.IO = 3I.64$^{t}/m^{2}$
ii. Surcharge loads:(Q)
    a. due to road/Crane: 2.00$^{t}/m^{2}$
    b. due to buildings: 3.00$^{t}/m^{2}$

    5.00$^{t}/m^{2}$

iii.
    a. Loads due to decking

    L.L+D.L: I 3.00$^{t}/m$

    b. Self wt. of/D/wall

    17. 10x0.60 x 2.40: 24.62$^{t}/m$

    c. Self wt. of Struts

    $\frac{4x.10.00}{3.0} \times \frac{0.50}{2}$: 3.33$^{t}/m$

    d. 2.3 h of strut load transferred to D/wall in transverse direction

    $\frac{2.50}{100}$ x 200: 5.00$^{t}/m$

    Total: 45.95$^{t}/m$

Less load carried by the bearing of D/wall:

(9c x b)

9x4.96x0.60= (-) 26.78t/m

Netload on soil wedge: 19.17^t/m

Load intensity at (-) 11.80 = $\frac{19.17}{4.20}$ = 4.56^t/m^2

Total Down ward load) 41.20^t/m

(i+ii+iii)

[r (11 + Df) + C $\frac{(H + Df)}{Df}$

iv. Relieving load:

Due to the shear strength of soil along the wedge:

2.34 x $\frac{(13.10.2.52) + 4.32 \times 4.00}{4.20}$ = (–) 10.00^t/m^2

[$\frac{C}{Df}$ (H + Dr – $\frac{2c}{r}$)

Net disturbing force

(i+ii+iii+iv) = 31.20^1/m^2

v. Stabilising load at (-) 11.80

ult bearing Capacity = Nc + rDf N_Q)

= 4.96X7.00 + 1.85X4.00X2.00

= 49.52t/m2

(Nc, N_Q Terzaghi Bearing Capacity factors with Skemptons corrections for B/D of cut)

vi. F.O.S against bottom heave:

$$\frac{CNc + rDf\ NQ}{r(H+Df) + \frac{C}{D1}(H + Df) - \frac{C}{D1}(H + Df - \frac{2C}{r})} = \frac{49.52}{31.20} = 1.59$$

>1.50

Adopt 4.0m penetration O.K.

4.0 Check for D/wall with 4.0 penetration to carry superimposed loads:

i. Loads on D/wall including self wt. of D/wall: 45.95t/m.
ii. Ult. Bearing Capacity due to skin friction and end bearing:

Skinfriction : 2.34x13.10

+ 2x4.32x4.00 = 65.21t/m.

End Bearing : (9cb)

: 9x4.9bx0.60 = 26.78t/m

Total = 91.99 t/m

E.O.S against sinking:

$\frac{91}{4} \times \frac{99}{95}$ = 2.00 >1.50 O.K.

5.0 To check the adequacy of the 600mm thick R.C. Diaphragm wall for various stages of excavation.

5.1. Stage – Excavation is done up to 3.00m below GL to erect the strut at 2.50m.

Adopting Rankine's formula for active and passive pressures

$$PA = Ka\ rH - 2c\ \sqrt{Ka} + Kaq$$

$$Pp = Kp\ rH + 2c\ \sqrt{Kp}.$$

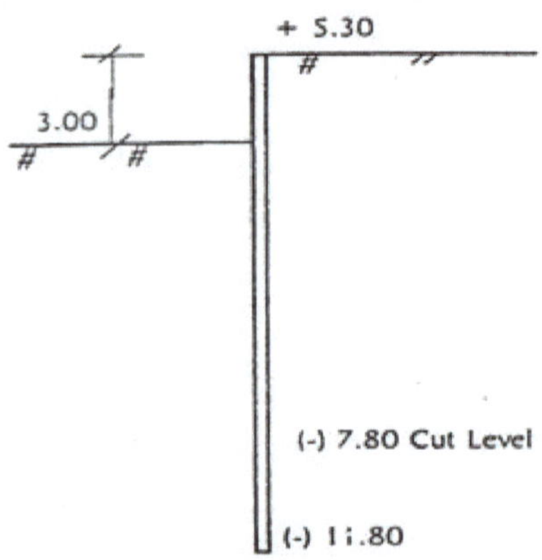

$$Ka = \frac{I\text{-sin } \phi}{I\text{+sin } \phi} = 0.84 \text{ for } \phi = 5°$$

$$Kp = \frac{I + \text{Sin } \phi}{I - \text{Sin } \phi} = 1.19 \text{ for } \phi = 5°$$

Active Pr:

PA at GL = 0.00

At (-) 7.80 = 0.84x1.85x13.10x2.34x0.92+084x5.00

= 20.25 t/m2

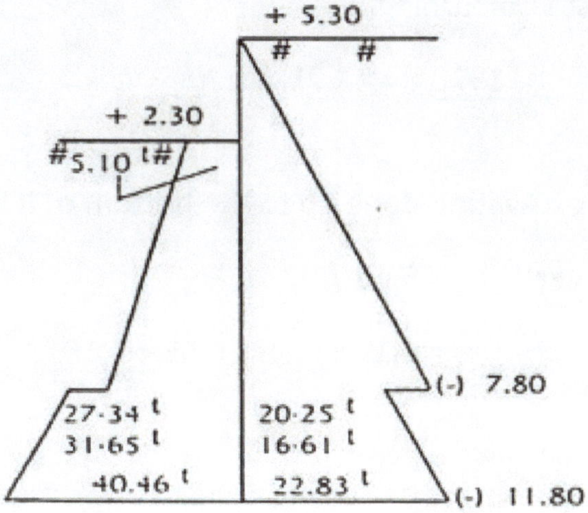

for layer below (-) 7.80

= 0.84 x1.85x13.10 2x4 32x0.92 + 0.84x5.00

= 16.61 =$^t/m^2$

at (-) 11.80 = 16.61 + 0.84 x 1.85 x 4.00 = 22.83 $^t/m^2$

passive Pr:

at cut level i.e. (+) 2.30 = 2x2 34x1.09

= 5.10 $^t/m^2$

for layer below (-) 7.80 = 1.19x1.85x10.10+2x4.32x1.09

= 31.65 t/m2

at (-) 11.80 = 31.65 + 1.19 x 1.85 x 4.00 = 40.46 t/m2

Equating moments about base

MA = $\frac{20.25}{2}$ x 13.10 x 8.37 ++ $\frac{16.61x4.002}{2}$ + $\frac{6.22x4.002}{2 \times 3}$

2

= 1260 tm.

Mp = 5.10x1.10x9.05

+ $\frac{22.24}{2}$ x 10.10 x 7.37

+ $\frac{31.65 \times 4.00\ 2}{2}$ + $\frac{8.81}{2}$ x $\frac{4.00\ 2}{3}$

= 1570 tm

F.O.S against Overturning

$$\frac{Mp}{MA} = \frac{1570}{1260} = 1.25 > 1.20 \quad \text{O.k.}$$

5.1.1 Stage 2 – Strut installed, excavation done up to the bottom of II Strut.

Pp at cut lever = 5.10t/m2

at (-) 7.80 = 1.19 x 1.85 x 6.65 + 5.10

= 19.74t/m 2

for layer below (-) 7.80

=1.19x1.85x6.65+2x4.32x1.09

= 24.06t/m 2

at (-) 11.80 = 24.06 +1.19 x1.85 x4.00 = 32.87 t/m2

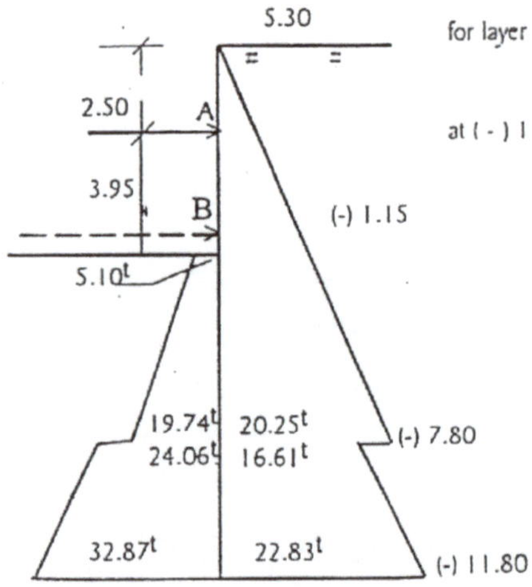

Taking Moments about A

$$MA = \frac{20.25}{2} \times 13.10 \times 6.23 + 16.61 \times 4.00 \times 12.60 + \frac{6.22}{2} \times 4.00 \times 13.27$$

= 1828.55 tm.

$$Mp = 5.10 \times 6.65 \times 7.28 + \frac{14.64}{2} \times 6.65 \times 8.38 + 24.06 \times 4.00 \times 12.60 + \frac{8.81}{2} \times 4.00 \times 13.27$$

= 2101.26 tm.

$$F.O.S = \frac{Mp\ 2101.26}{MA\ 1828.55} = 1.15 < 1.20$$

The Strut at B shall be installed by local trenching to minimise wall deflections.

Load Strut at A for this condition

$$RA = Pp - PA = \frac{19.74 + 5.10}{2} \times 6.65 + \frac{24.06 + 32.87}{2} \times 4.00 \left[\frac{20.25}{2} X 13.10 + \frac{16.61 + 2283}{2} X 4.00\right]$$

= 15.07t/m

5.1.2 Stage (3) – II Strut installed and excavation done up to the bottom of III Strut (Ty.)

With two struts in position, the active pressure Changes from Rankine's state to braced cut pressure, as explained in part 2.2.1.

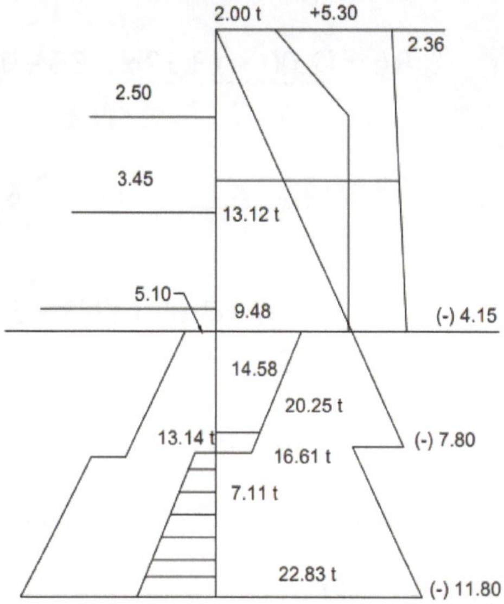

PA at cut level (-) 4.15

= 1.85x9.45-4x2.34+5.00

= 13.12 t/m2

PA as per Rankines Pressure at (-) 4.15

= 0.84x1.85x9.45-2x2.34x0.92+0.84+5.00

= 14.58 t/m

Pp at (-) 4.15

= 5.10 t/m2 at (-) 7.80

= 5.10+1.19x1.85x3.65 = 13.14 t/m2 for layer below (-) 7.80

= 1.19x1.85x3.65+2x4.32x1.09

= 17.45 t/m2

at (-) 11.80

= 17.45+1.19x1.85x4.00=26.26 t/m2

Net pressure below cut-

PA at (-) 4.15 = 14.58-5.10 = 9.48 t/m2

at (-) 7.80 = 20.25-13.14=7.11t/m2

Pp at (-) 7.80 = 17.45-16.61 = 0.84 t/m2

at (-) 11.80 = 26.26-22.83 = 3.43 t/m2

Taking Moments about B.

$$Ma = 13.12 \times \frac{3.502}{2} = 7.11 \times 3.65 \times (3.50 + 1.83) + \frac{2.37}{2} \times 9.65 \times (3.50 + 1.22) = 239.10tm$$

$$Mp = 0.84 \times 4.00 \times (3.50 + 3.65 + 2.00) + 2.59 \times 4.00 \times (3.50 + 3.65 + 2/3 \times 4.00)$$

= 132.44 tm.

Net moment: 239.10-132.44 = 106.66 tm.

MR Capacity of 60 cm D/wall = 60.00 tm. Therefore, III Strut (Ty.) shall be inserted by restricted excavation up to 3.00m below II strut and then by local trenching.

Strut loads for these conditions:

$$RA = \left[\frac{2.00 + 13.12}{2} \times 2.36 + 13.12 \times 0.14 + \frac{13.12 \times 3.45}{2}\right] 0.90 = 42.31\text{ t/m}$$

$$RB = \frac{13.12 \times 3.45}{2} + \left[13.12 \times 3.50 \times (1.75 + 3.65 + 2.00) + 7.11 \times 3.65 \times \left(\frac{3.65}{2} + 2.00\right) + \frac{2.37}{2} \times 3.65 \times (2/3 \times 3.65 + 2.00)\right] \frac{1}{9.15}$$

$$= 72.73\text{ t/m}$$

5.1.3. Stage (4) – 3 Struts installed and excavation done for insertion of 4th strut:

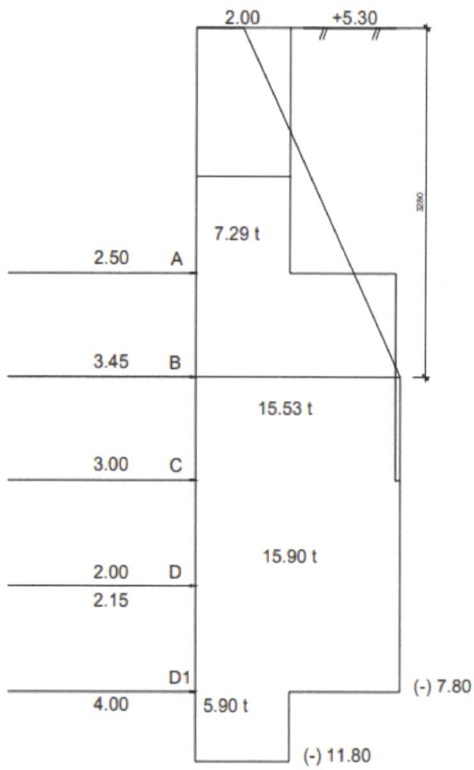

Braced cut Pr at (-) 5.65

= 1.85x10.95-4x2.34+5.00

= 15.90 t/m2

P_A at (-) 5.65

= 0.84x1.85x10.95-2x2.34x0.92+0.84x5.00

= 16.91 t/m2

Pp at (-) 5.65 = 5.10 t/m2

at (-) 7.80 = 5.10+1.91x1.85x2.15

= 9.83 t/m2

For layer below (-) 7.80

= 1.19x1.85x2.15+2x4.32x1.09 = 14.15 t/m2

at (-) 11.80 = 14.15+1.19x1.85x4.00 = 22.96 t/m2

Net Pr. Below cut

PA at (-) 5.65 = 16.91-5.10=11.81 t/m2

PA at (-) 7.80 = 20.25-9.83=10.42 t/m2

PA at below (-) 7.80 = 16.61-14.15 = 2.46 t/m2

Pp at (-) 11.80 = 22.96 – 22.83 = 0.13 t/m2

Taking moments about 'C'

$$M_A = \frac{15.90 \times 2.00\ 2}{2} + 10.42 \times 2.15 \times 5.08 + \frac{1.39}{2} \times 2.15 \times 2.72 + \frac{2.46}{2} \times 4.00 \times 5.48$$

$$= 176.63 \text{ tm} > 60.00 \text{ tm:}$$

Therefore, the Strut shall be installed by local trenching:

5.1.4. Stage (5) – All Struts installed and excavation done up to final cut level for Mud matting:

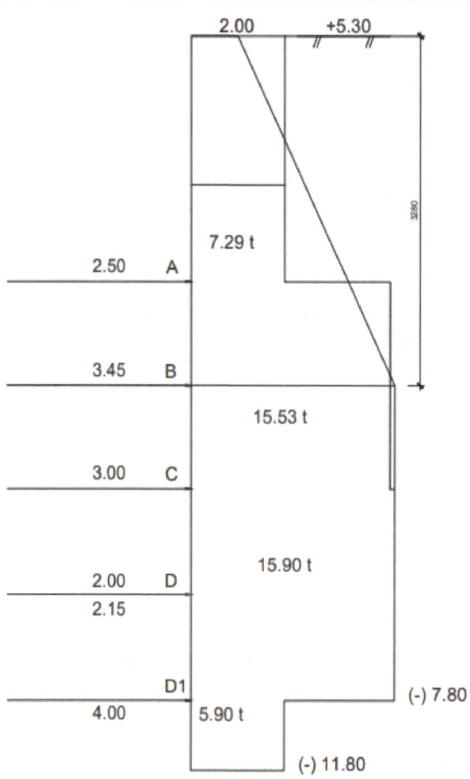

Modified Braced cut Pr upto cut level:

(Considering 80% Pr for forces on D/wall)

$$P = (rH - 4C + q)\ 0.80$$

$$= (1.85 \times 13.10 - 4 \times 2.34 + 5.00)\ 0.80$$

$$= 15.90 \text{ t/m2}$$

Below cut level

$$= 15.90 - 2 \times 4.32 \times 0.80$$

$$= 8.99 \text{ t/m2}$$

Simply fying Pr. Diagram and analysing by moment the distribution method for forces on D/wall

$$M_A = \frac{7.29 \times 2.50\ 2}{2} = 22.78\text{tm}$$

$$MAB = MBA = \frac{15.53 \times 3.45\ 2}{12} = 15.40\text{tm}$$

$$MBC = MCB = \frac{15.90 \times 3.00\ 2}{12} = 11.93 \text{ tm}$$

$$MCD = MDC = \frac{15.90 \times 2.00\ 2}{12} = 5.30 \text{ tm}$$

$$MD = \frac{15.90}{2} \times 2.15\ 2 = 8.99 \times 4.00 \times 4.15 = 185.98 \text{ tm}$$

(Excessive)

Regulating excavation up to final cut level in small stretches (say 3.0 m) and completing mud matting 300mm thick before further excavation and assuming a support at mud mat level.

$$MDD\ 1 = \frac{15.90 \times 2.15\ 2}{12} = 6.12\ tm$$

$$MD1 = \frac{15.90 \times 0.15\ 2}{2} + 8.99 \times 4.00 \times 2.15 = 74.49\ tm$$

Analysing for moments by moment distribution method

Stiffness ratio: AB=0.87, BC=1.33, CD=2.00, DD1=1.4

Distribution factor

B1 – 0.40

B2 = 0.60

C1 = 0.40

C2 = 0.60

D1 = 0.59

D2 = 0.4

Free span Bm Span AB: 15.40 x 1.50 = 23.10 tm

BC: 11.96 x 1.5 = 17.90 tm

CD: 5.30 X 1.50 = 7.95 tm

DD: 6.12 X 1.50 = 9.18 tm

Shear Force

Above A = 7.29 x 2.50 = 18.23

Below $A = \frac{15.53 \times 3.45}{2} + \left(\frac{22.78 - 11.48}{3.45}\right) = 30.06^t$

Above B = 26.78 – 3.25 = 23.50 ^{t}

Below $B = \frac{15.90 \times 3.00}{2} - \left(\frac{13.43 - 11.48}{3.00}\right) = 23.19^t$

Above C = 23.85 + 0.66 = 24.51^t

Below $C = \frac{15.90 \times 2.00}{2} - \left(\frac{14.78 - 13.43}{2.00}\right) = 15.24^t$

Above D = 15.90 + 0.66 = 16.56^t

Below $D = \frac{15.90 \times 2.15}{2} - \frac{(77.49 - 14.78)}{2.15} = 12.07^t$

Above D = 17.09 + 29.16 = 46.26 t

Below D = 15.90 c 0.15 + 8.99 x 4.00 = 38.35^t

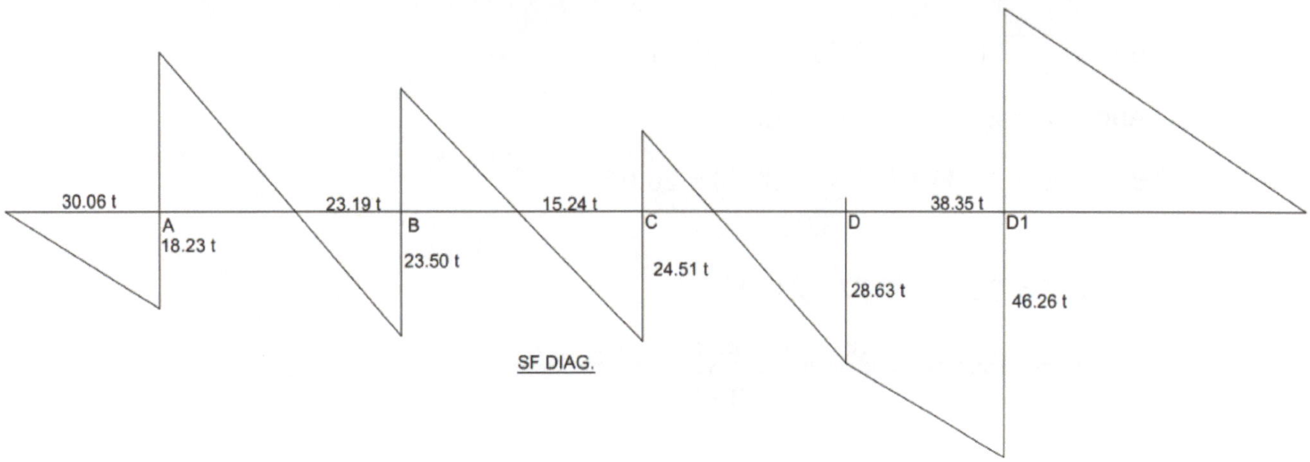

SF DIAG.

5.15 Stage (6) Mud matting done and base aft cast. Strut at D deloaded

Considering simplified Pr. Diagram as in 5.1.4 and analysing for moment

MA = 22.78 tm

MAB = MBA = 15.40 tm

MBC = MCB = 11.93 tm

$MCD = MDC = \frac{15.90 \times 3.55^2}{12} = 16.70$ tm

$MD = \frac{15.90 \times 0.60^2}{2} + 8.99 \times 4.00 \times 2.60 = 96.36$ tm

Stiffness Ratio	Dist factor
AB = 0.87	B1 – 0.40
BC = 1.33	B2 – 0.60
$CD = \frac{3.00}{3.55} = 0.85$	C_1 0.61
	C_2 0.39

Span Bms

AB = 23.10 tm

BC = 17.90 tm

CD = 16.70 X 1.50 = 25.05 tm

Shear Force:

Above A = 18.23 ^{T}

Below A = 26.78 + (22.78 – 16.48) = 28.61^{t}

Above B = 26.78 – 1.83 = 24.95 t

Below B = $23.85 + \frac{(16.48 - 8.18)}{3.00} = 26.62^{t}$

Above C = 23.85 – 2.77 = 21.08^{t}

Below C = $\frac{15.90 \times 3.55}{2} - \frac{(96.36 - 8.18)}{3.55} = 3.38^{t}$

Above D1 = 28.22 + 24.84 = 53.06 t

Below D 1 = 15.90 x 0.60 + 8.99x4.00 = 45.50t

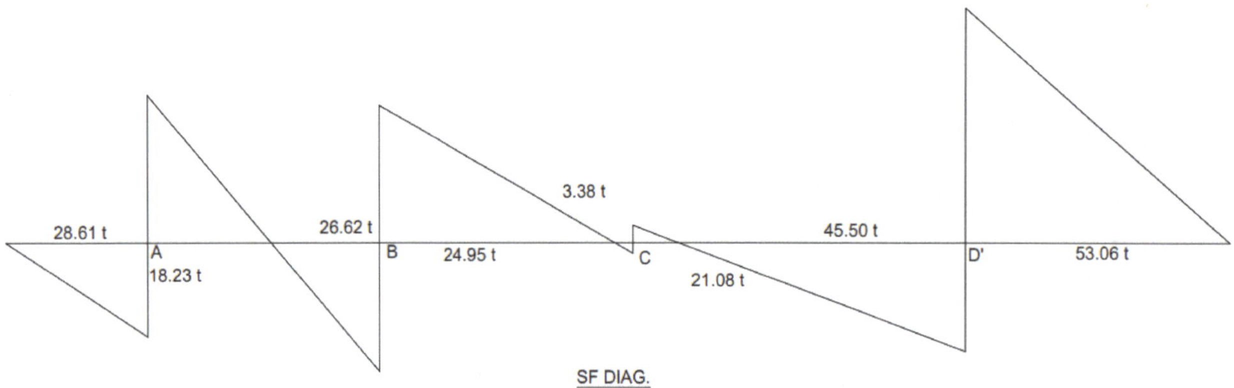

SF DIAG.

5.1.6 Stage (7) side wall continued & strut at 'C' deloaded

Adopting same simplified Pr. Diag and analysing for forces

MA = 22.78 tm

MAB = MBA = 15.40 tm

MBD1 = MC^{1} B = $\frac{15.90 \times 6.55^2}{12}$

= 56.85 tm

MDt = 96.36 tm

Stiffness Ratio: Dist Factor

AB = 0.87 $\quad \frac{0.87}{1.33} = 0.65$

BD 1 = $\frac{3.00}{6.55} = 0.46$ $\quad \frac{0.46}{1.33} = 0.35$

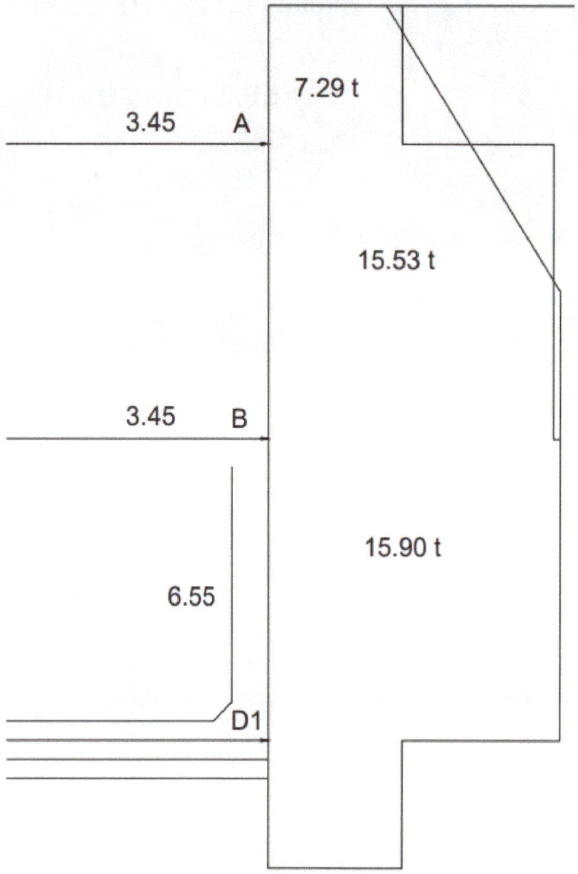

Shear Force

above A = 18.23 t

below A = $26.78 - \frac{(28.21 - 22.78)}{3.45} = 25.21t$

Above B = 26.78 + 1.57 = 28.35t

Below B = $\frac{15.90x6.55}{2} - \frac{(96.36-28.21)}{6.55} = 41.67$ t

Above D = 52.07 + 10.40 = 62.47 t

Below D = 45.50

5.1.7 Stage (8) Box completed and back fill done

Assuming Rane's Active Pr. Conditions for permanent condition (assuming water table at G L)

Pa at B' = Ka r_sH-2cVKa + kaq+y_wH

($^{\psi}$ = 5°)

= 0.84 x 0.85 x 7 185 – 2 x 2.34 x 0.92 + 0.84 x 5.00 + 1.00 7 7.185 = 12.21 t/m^2

Pa at D = 12.21 + 0.84 x 0.85 x 5.315 + 1.00 x 5.315 = 21.32 t/m^2

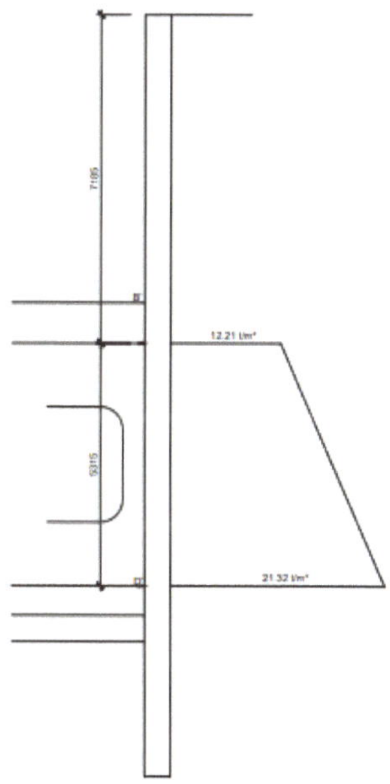

$$RB = \left[\frac{12.21x5.315^2}{2} + \frac{9.11}{2} x \frac{5.315^2}{3}\right] \frac{1}{5,315} = 40.2t$$

$$RD = \frac{(12.21+21.32)}{2} 5.315-40.52 = 48.59 \text{ t}$$

For zero Shear at X below Bl

$$40.52 - \left[12.21 \text{ x} + \frac{9.11}{5.315} \frac{x^2}{2}\right] = 0$$

$$0.86x2 + 12.21 \text{ x } 40.52 = 0$$

$$X = \frac{12.21\ 2 \pm 12.21^2 + 4 \text{ x } 0.86 \text{ x } 40.52}{2 \text{ x } 0.86} = 2.78 \text{ m}$$

$$M\ max = 40.52 \times 2.78 - \frac{12.21 \times 2.78^2}{2} - \frac{0.86 \times 2.78^3}{3}$$

$$= 59.30\ tm$$

SF max = 48.59 t

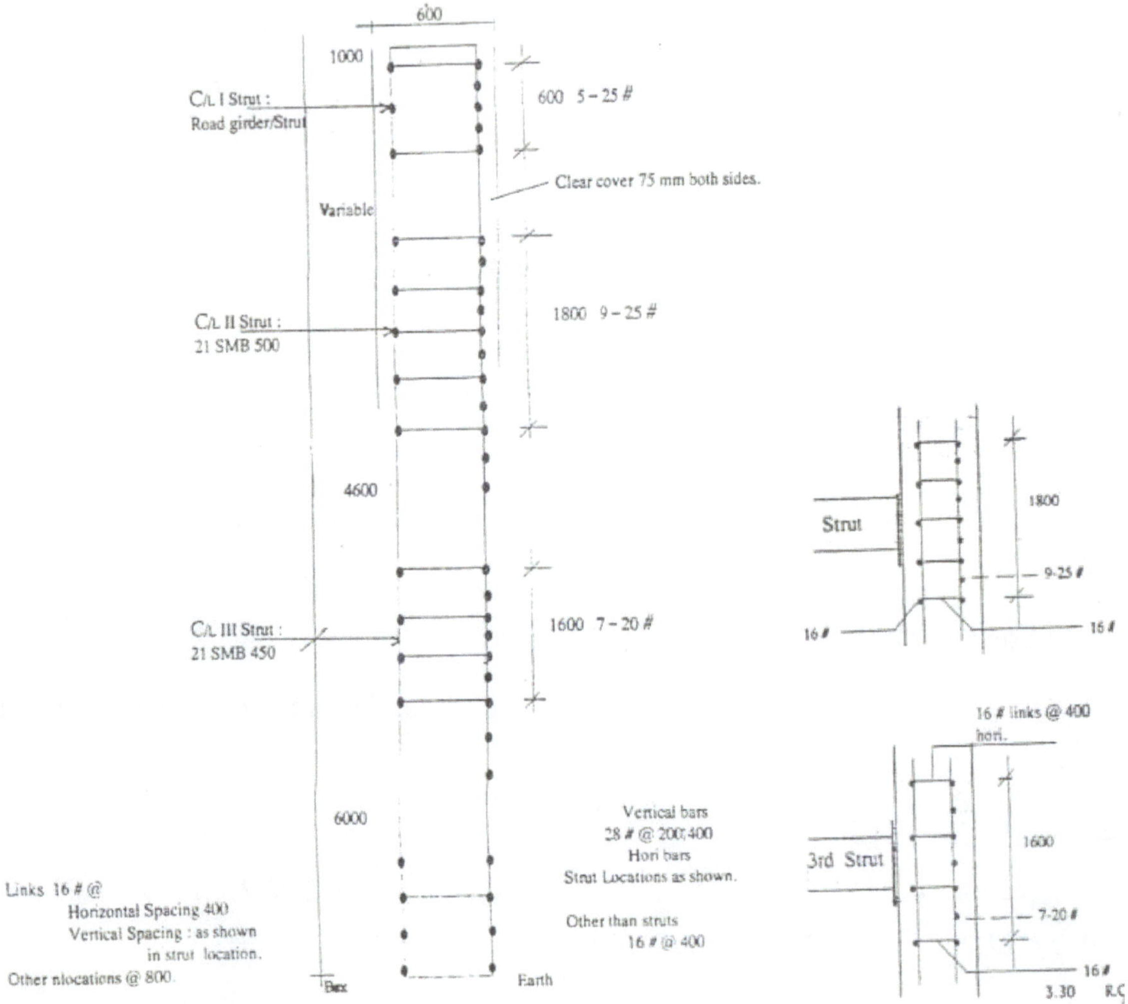

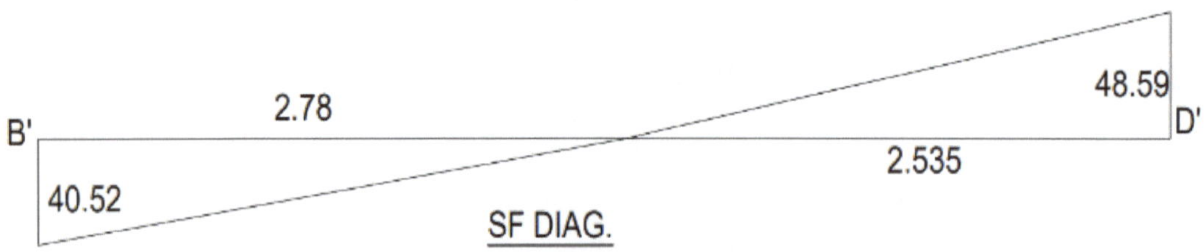

SF DIAG.

6.0 Diaphragm wall detailing:

6.1 Data.

Grade of Concrete M20

Reinforcement:Torsteel Fe415

Design: By Limit state Method (as per Is 456-1978 and SP-16)

: Rectangular sec-reinforced equally both sides.

$fck = 20N/mm^2$

Load Factor (partial safety factory) yf = 1.50

$fy = 415\ N/mm^2$

$$\frac{d}{D} = \frac{75}{600} = 0.13$$

a. For a max. moment = 96 tm:

Moment at critical sec-

(edge of mudmat support)} $\frac{8.99x4.00^2}{2} = 71.92$ tm

$$\frac{Mu}{fckbD^2} = \frac{1.50x71.92x9.807x10^6}{20x1000x6002} = 0.15$$

P = Axial Load on D/wall = 45.95 $^t/m$ (para 4.0)

$$\frac{Pu}{fckbD} = \frac{1.50x45.95x9.807x10^3}{20x1000x600} = 0.056$$

From chart 33 of SP-16 (Design Aids to IS 456-78)

P/fck = 0.10

$$As = \frac{0.10 \times 20 \times 1000 \times 600}{100x10^2} = 120cm^2$$

providing 32 # @ 130 c/c Area provided: 2x61-86 = 123.72 cm^2 O.K.

b. For M= 60.00 tm:

$$\frac{Mu}{fckbD^2} = \frac{1.50x60.00x9.807x10^6}{20x1000x600^2} = 0.12$$

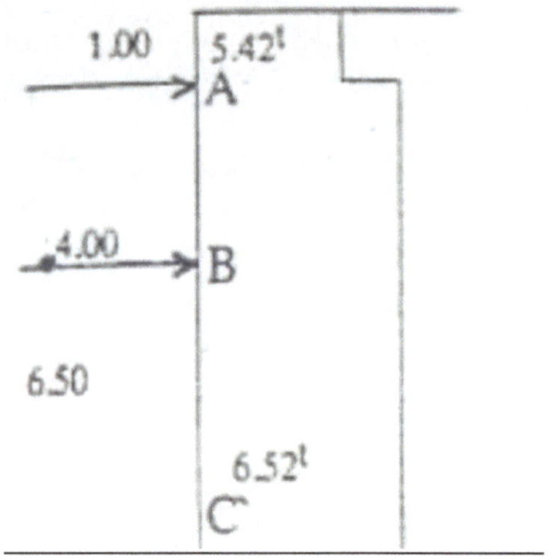

$$\frac{Pu}{fckbD} = \frac{1.50x45.95x9.807x10^3}{20x1000x600} = 0.056.$$

$$As = \frac{0.08x20x1000x600}{100x10^2} = 96\ cm^2$$

Providing 32 # @ 160 c/c

Area provided: 2x50.27 = 100.54 cm^2

c. For M = 30.00 tn.

$$\frac{Mu}{fckbD^2} = \frac{1.50x30.00x9.807x10^6}{20x1000x600^2} = 0.06$$

$$\frac{Pu}{fckbD} = 0.056$$

$$As = \frac{0.035x20x1000x600}{100x10^2} = 42\ cm^2$$

Providing 32 # @ 390 c/c Area provided 2x21.16 = 42.32 cm^2 O.K.

To matchcurtailments for various moments provide 32 # @ 130 C/c for 77tm & 60tm. And provide 32 # @ 390 C/c for 30 tm.

6.1.1.1 Shear reinforcements:

Shear Capacity of D/wall

Without shear reinf: KTc

[Para 39.2 – IS 456 – 78]

$100\ \frac{As}{Bd}$ for as = 120 cm^2 : $\frac{100 \times 120}{100 \times 50.9} = 2.29$

$KT_c = 0.81 \times 1.00$ N/mm^2

$100\ \frac{As}{Bd}$ for As = 96 cm^2 = $\frac{100 \times 96}{100 \times 50.9} = 1.83$

$KT_c = 0.78 \times 1.00$ N/mm^2

$100\ \frac{As}{Bd}$ for As = 42 cm^2 $= \frac{100 \times 42}{100 \times 50.9} = 0.8$

$KT_c = 0.57 \times 1.00$ N/mm^2

Shear Capacity in Tonnes:

: $0.81 \times 1000 \times 525$ = 43.36^t for 120 cm^2 As.

: 0.78×43.36^t = 41.75^t for 96 cm^2 As.

: $\frac{0.57}{0.81} \times 43.36$ = 30.51^t for 42cm^2 AS.

For max SF: 62.47^t at D^1 (centre of base slab)

Critical SF: at edge of slab: } 57. 70^t

Shear to be carried by shear reinf

$Vu - Tcbd = 57.70 - \frac{(0.81 \times 1000 \times 525)}{9.807 \times 10^3}$

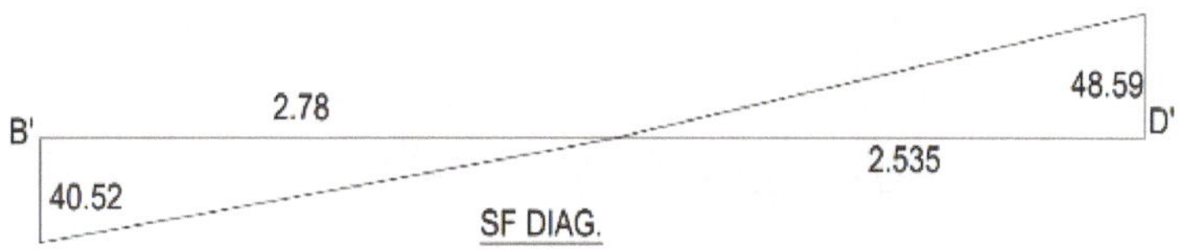

$= 57.70 - 43.36 = 14.34^t$

Providing 16 # stirrups @ -8L spacing reqd.

$Vus = \frac{0.87 \times fy \times Asv \times d}{Sv}$

$Sv = \frac{0.87 \times 415 \times 8 \times 201 \times 50.9}{43.36 \times 10^3 \times 9.807} = 717$ mm.

Provide 4L stirrups at 350mm c/c. vertically.

At other places provide minimum shear reinf.

$Sv = \frac{Asvfy}{0.4b}$

6.1.2 Design of wale reinforcements:

Adopting 3.0m panel widths

At strut location 'A'

$$\text{Max Load reaction/m} = (18.23 \times 30.06)\,\frac{0.90}{0.80} = 54.33^{t}$$
(from stage 5)

(assuming 90% forces for struts & wales)

Strut load: 3x54.33 = 163t (Adopt 2.1 SMB 500 strut).

M max (at edge of strut)

$$: \frac{54.33 \times 1.20^2}{2} = 39.11\text{tm}$$

$$\frac{Mu}{bd^2} = \frac{1.50 \times 39.11 \times 9.807}{1678 \times 539^2} = 1.18$$

Max SF = 54.33x1.20 = 65.20^{t}

$$Tv = \frac{Vu}{bd} = \frac{65.20 \times 10^3 \times 9.807}{1678 \times 539} = 0.71\ \text{N/mm}^2$$

$$Tc \text{ for } \frac{100 \times 34.36}{167.8 \times 53.9} = \frac{0.38}{(38.73t)} \text{ is } 0.42\ \text{N/mm}^2$$

Tc < Tu

Shear reinf is reqd.

Providing 16 #-4L stirrups spacing

$$Sv = \frac{0.87 \times 415 \times 4 \times 201 \times 539}{(0.71-0.42)1678 \times 539} = 597\ \text{mm}$$

Provide 390 mm c/c.

Provide up to $\frac{38.73}{54.33}$ = 0.71 m from end and nominal reint beyond.

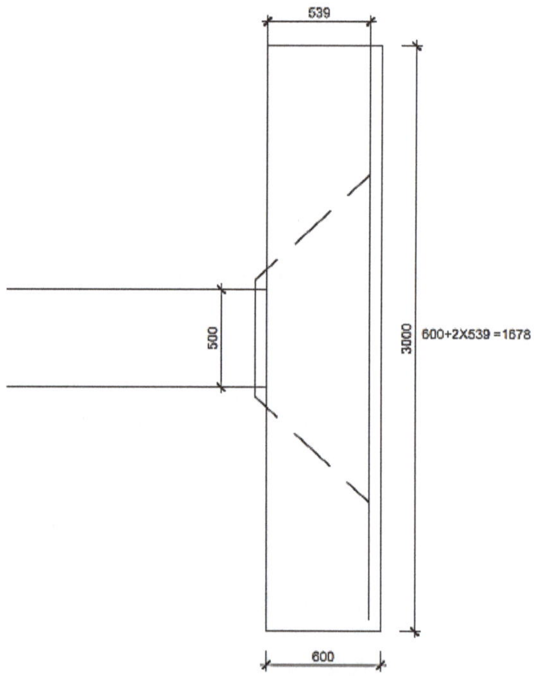

Sectional Elevation

b) for strut 'B'

$$\text{Max Load reaction: } (41.67 + 28.35)\,\frac{0.90}{0.80} = 79\text{t/m}$$
(from stage 7)

Strut load: 3x79 = 237t (adopt 2-15MB 600 strut)

$$\text{M max at edge} = \frac{79 \times 1.15^2}{2} = 52.24\ \text{tm}$$

$$\frac{Mu}{bd^2} = \frac{1.5 \times 52.24 \times 10^6 \times 9.807}{1778 \times 539^2} = 1.49\ N/mm^2$$

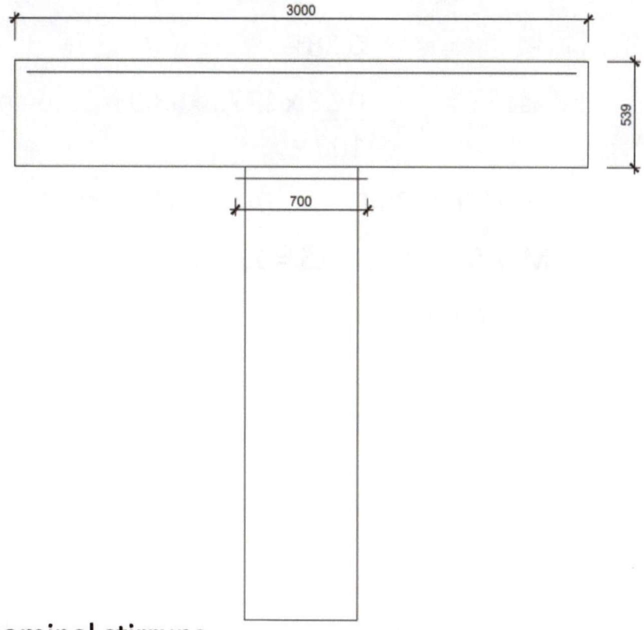

Plan

Pt: 0.457%

$$Ast = \frac{0.457}{100} \times 177.8 \times 53.9 = 43.80\ cm^2$$

Provide 8-28 # or 9.25 # ($A = 44.17cm^2$) O.K.

Max SF: 79x1.15 = 90.85t

$$Tc \text{ for } \frac{100 \times 44.17}{177.8 \times 53.9} = 0.46 \text{ is } 0.46\ N/mm^2 \quad (44.95^t)$$

providng 16 # - SL shear stirrup.

$$Sv = \frac{0.87 \times 415 \times 5 \times 201 \times 539}{(90.85-44.95)10^3 \times 9.807} = 434\ mm$$

Provide at 400 mm centres

Provide upto $\frac{44.95}{90.85}$ = 0.50m from end and beyond nominal stirrups.

c) For strut at 'C'

Max load reaction (from stage 4)

$$15.90 \times 1.50 + 15.90 \times 2.00 + \frac{11.81+10.42}{2} \times 2.15 + \frac{2.46}{2} \times 3.00$$

$= 67.33\ t/m^2$

Strut load: 67.33x3 = 202^t. (Adopt 2 – ISMB 600)

Adopt same reinf as for strut 'B' location.

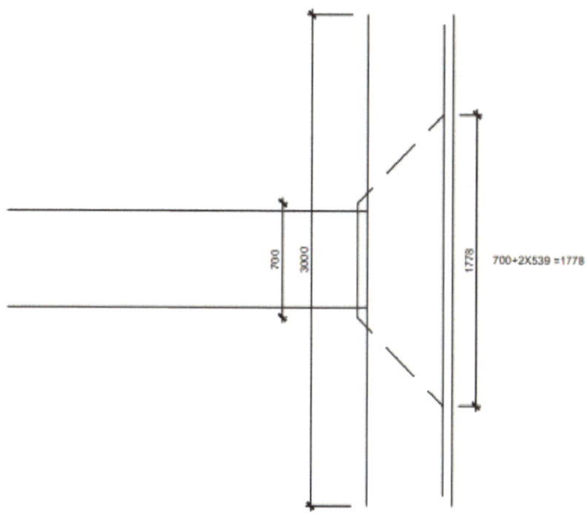

Sectional Elevation

d) for strut at D

Max Load reaction:

$$[15.90 \times 1.00 \times 15.90 \times 2.15 + 8.99 \times 4.00]\ \frac{0.90}{0.80}$$

(Considering load reaction from stage 5 before mudmatting is done in the final excavation stage) = 97 t/m

Strut load: 97x3 = 291 t (Provide 2 – ISMB 600 trut

$$M \text{ max at edge: } 97 \times \frac{1.15^2}{2} = 64.14\ tm$$

$$\frac{Mu}{bd^2} = \frac{1.50 \times 64.14 \times 10^6 \times 9.807}{1778 \times 539^2} = 1.83 \text{ N/mm}^2$$

$$Pt = 0.58\%$$

$$Ast = \frac{0.58}{100} \times 177.8 \times 53.9 = 55.6 \text{ cm}^2$$

provide 9-28# (Area: 55.41 cm^2)

Max SF = 97 x 1.15 = 111.55^t

Reinforcement Detail

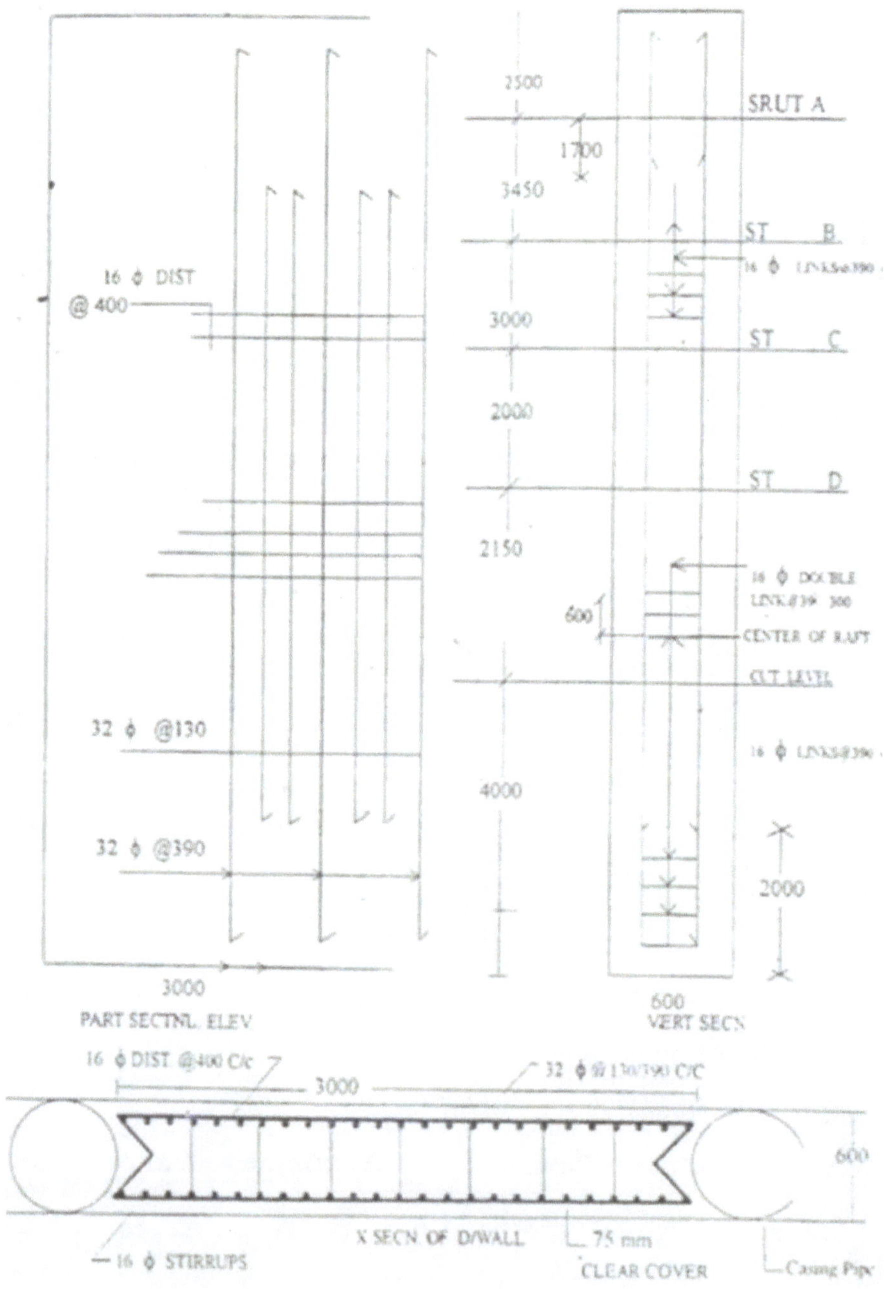

Tc for $\frac{100 \times 55.41}{177.8 \times 53.9}$ = 0.58 N/mm^2 (56.50^t)

providing 1b # 9L stirrups:

$$Sv = \frac{0.87 \times 415 \times 9 \times 201 \times 539}{(111.55-56.50)10^3 \times 9.807)} = 656 \text{ mm}$$

Provide 5-L stirrups at $656 \times \frac{5}{9}$ = 364 mm say (300 mm c/c)

6.1.3. Distributors:

Area reqd: $\frac{0.12}{100} \times 100 \times 60$ = 7.2 cm^2 (IS 456-78)

Provide 16 ø @ 400 Centres for each face.

Area provided: 2x5.03=10.06 cm^2

For wale reinf: provide 16# @ alternate main bar spacing:

6.2.1 Reinforcement drg:

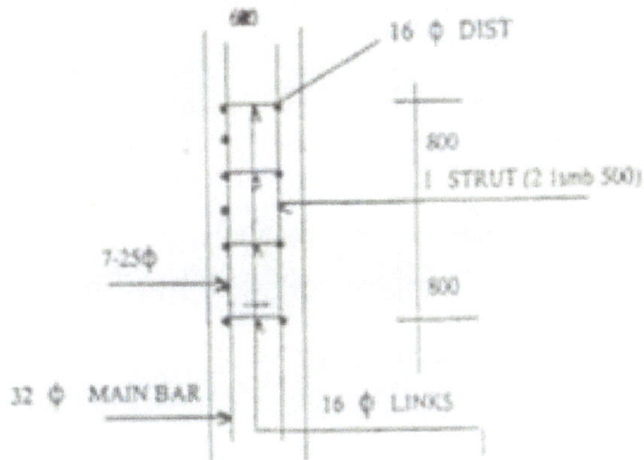

Links @ 533/260 for distribution of 780 mm from center of Panel on either side and 267/390 beyond

(b)

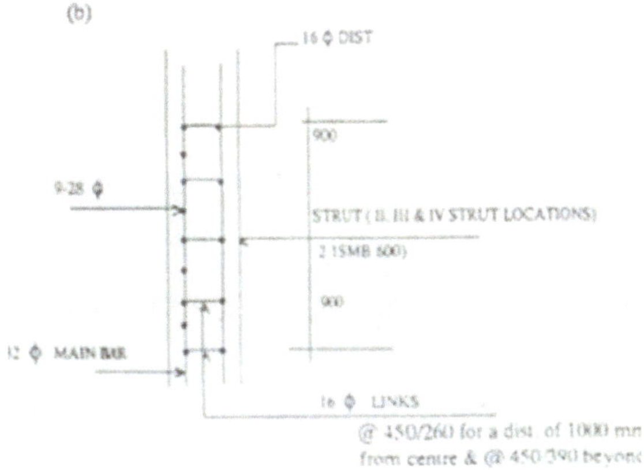

II Design of Diaph. Wall in Sandy soil.

1.0 Representative section considered: Cont. Sec. 17A.

Between Rabindra Sarobar And Tollyguage stations.

1.1 Profile of cut and soil strata.

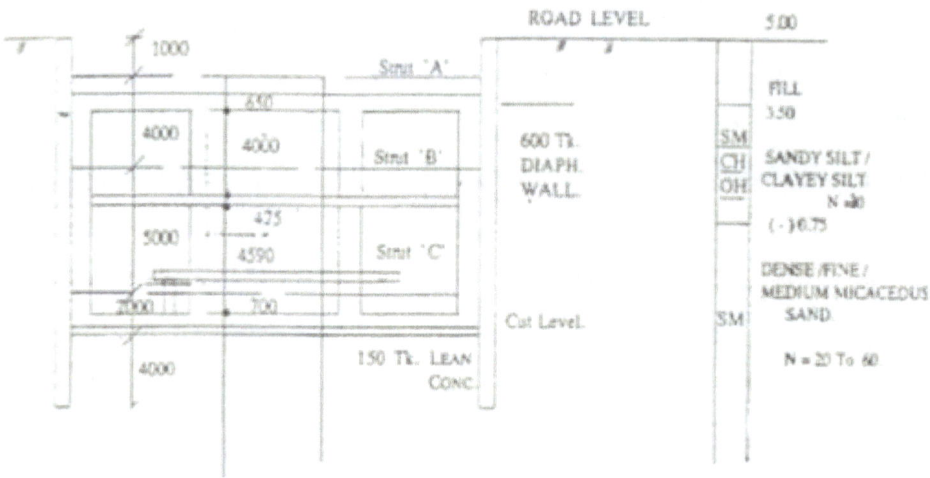

Fig.:

1.2 Data Considered

a) Soil: r = 1.85t/m^2

ϕ = 30° (Sandy silt Nave > 15)

b) Surcharge i) Road/crane: equivalent to 1.0 m ht of each

(Q) = 2.00 t/m^2

Building: 3.30 t/m^2

1.3 Lateral Pressures:

i. For braced cut: 0.65 karH (As per trezaghi Peck.)

$$(Ka = \frac{1 - \sin \phi}{1 + \sin \phi})$$

ii. For Retaining Wall: Active: Ka (rH + Q)

Passive: KprH.

1.4 Ground water table is assumed to be lowered below cut level by continuous dewatering during Construction.

1.5 A hinge is assumed @ 0.2L below cut level where L is the distance between last strut at cut level due to penectration into stiff stratum as per design criteria of M. Rly.

2.0 Design of Diaph wall (For max. depth)

2.1 Depth of Penetration:

Soil data:

R = 1.85 t/m^3

Ø – 30° (Sandy y silt)

(Lowering of water table below 'E' assumed)

Surcharge.

1. Road/Crane = 1.0 m ht of Earth = 2.00 ^{t}
2. Bldg = 3.30 t/m^2

Active Pr and C

: ka (r + 1 + q)

: 0.33 (1.85 x 10.00 + 2.00 + 3.30) = 7.85 t/m^2

Assuming 4.0 m Penetration

Active Pr. At the bottom of the DW: 7.85 + 0.33 x 1.85 x 6.00 = 11.52 t/m^2

Passive Pr.

At D = 0

at E: 3.00 x 1.85 x 4.00 = 22.20

Taking Moments about C

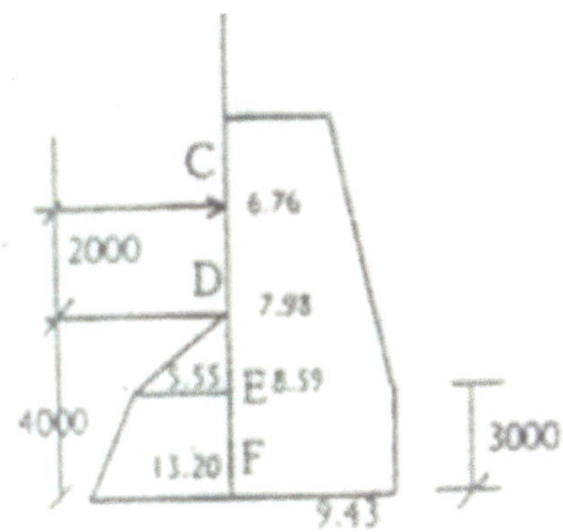

M act : $\frac{7.85 \times 6.00^2}{2} + \frac{3.67}{2} \times \frac{6.00^2 \times .2}{2}$ = 1.85.34 tm

M Pass: $\frac{22.20 \times 4.00}{2} \times (2.00 + \frac{2}{3} \times 4.00)$ = 207.20tm

F.O.S = $\frac{207.20}{185.34}$ = 1.117 < 1.20

Considering Capacity of Diaph. Wall (Say 40.00 tm).

F.O.S = $\frac{207.20 + 40.00}{185.34}$ = 1.33

2.2 to check for depth of penetration for carrying checking at minimum depth Load on Diaph wall:

a. Due to deck reaction: 16.00 t
b. Due to struts: $\frac{3x11.00x0.500}{3x2}$ = 21.52 t
c. Self wt.15.20x0.60x2.4 = 21.82 t, 40.64 ^{t}/m

Skin friction

Assuming lowering of WT below Diaph wall:

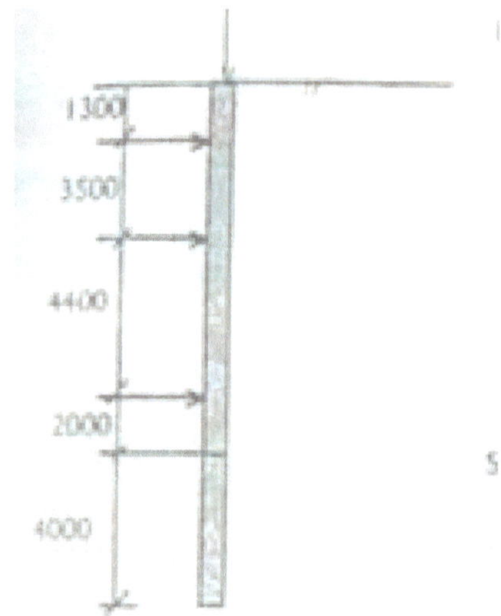

a. Out side: 15.20x1.85x tan 30° x 0.75 x 15.20 = 92.96
b. In side: 2.00 x 1.85 x tan 30° x 0.75 x 4.00 = 6.44

Total = 99.40^t

F.O.S = $\frac{99.40}{40.64}$ = 2.15 > 1.50 Safe

Skin friction:

Assuming WT. at cut level

Outside: $\frac{11.20}{2}$ x 1.85 x tan 30° x 0.75 x 11.20 = 50.47^t

Inside: 2.00 x 0.85 x tan 30° x 0.75 x 4.00 = 2.96^t

F.O.S = $\frac{72.95}{46.14}$ = 1.58 > $\frac{1.50}{\text{safe}}$

(without considering end bearing)

4.00m penetration is adequate

2.21 to check for depth of penetration with W.T. lowered up to the base of the cut only.

(1.0m below)

a. For 4.0m Penetration:

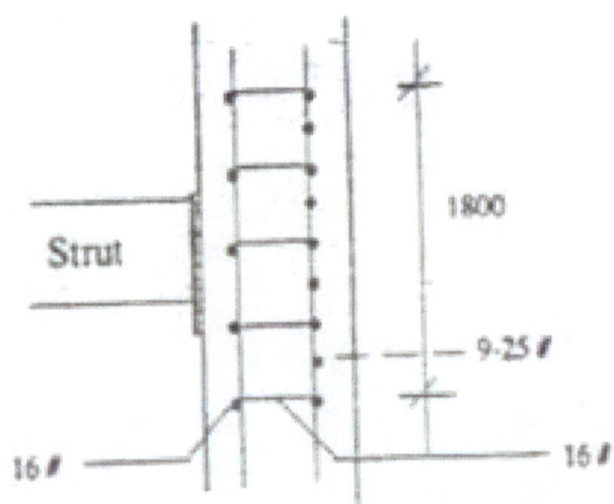

Active Pr.at D:6.76+0.33x1.85x2.00 = 7.98^t/m^2

at E:7.98+0.33x1.85x1.00 = 8.59^t/m^2

at F: 8.59+0.33x0.85x3.00 = 9.43^t/m^2

Passive Pr. At D=0

Pr. at E = 3.00x1.00 = 5.55^t/m^2

F=3.00x0.85x3.00+5.55 = 13.20^t/m^2

Taking moments of C 13.52

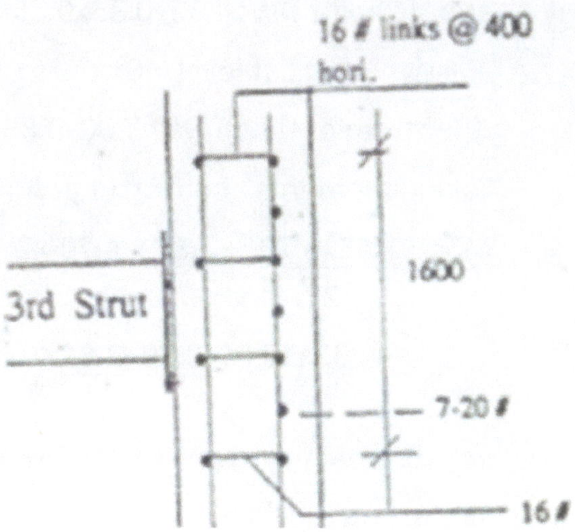

$M\ act: \frac{676x2.00^2}{2} + \frac{1.22}{2} x \frac{2.00^2}{3} x^2$

$+7.98 x 4.00 x 4.00 + \frac{0.61}{2} x 1.00 x 2.67$

+0.61x3.00x4.50+0.84x3.00x5.00 = 158.18 tm

2

$M\ pass = \frac{5.55x1.00x2.67}{2}+5.55x3.00x4.50$

$+\frac{7.65}{2}x3.00x5.00 = 139.72$ < M active Unsafe

For FOS = 1.20

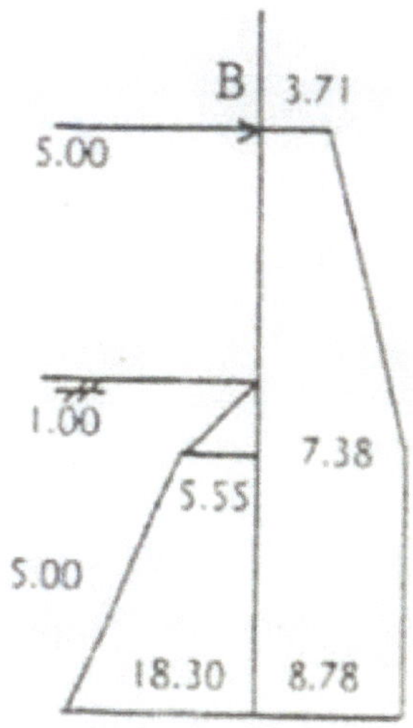

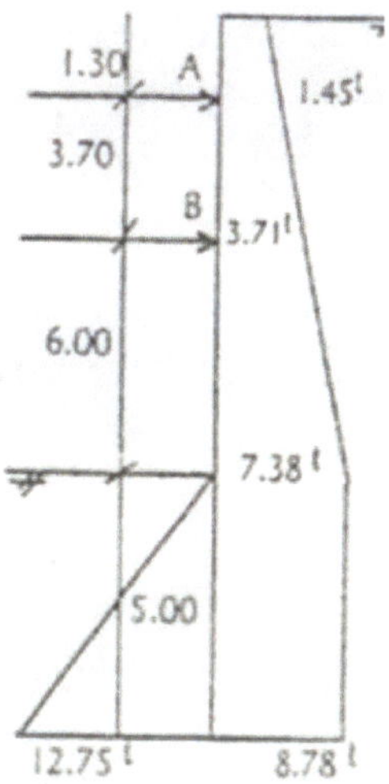

Moment on DW: 1.20x158.18 139.752 = 50tm

Take a moment as 40tm for design, allowing for wall continuity.

2.3.0 Design of DW (for max. depth) with 4.0m penetration

During – excavation

2.3.1 stage 1. After inserting one strut at 1.30m below GL to excavate up to 6.00m for insertion of 11 struts (W.T. assumed at cut level, i.e. (-) 1.30)

Active Pt.

at A = 0.33(1.85x1.30+2.00) = $1.45^t/m^2$

at cut level: 1.45+0.33x1.85x4.70 = $4.32^t/m^2$

at bottom of DW: 4.32+0.33x0.85x10.00 = 7.13^t/m^2

Passive pr at cut level = 0

At bottom of Diaph wall 3.00x0.85x10.00 = 25.50^t/m^2

Taking moments about strut at A

Mact: $\frac{1.45 \times 4.70^2}{2} + \frac{2.87}{2} \times \frac{4.70^2 \times 2}{3}$

$+4.32 \times 10.00 \times 9.70 + \frac{2.81}{2} \times 10.00 \times 11.37 = 615.89$tm

Mpass = 25.50x10.00x 11.37 = 1449.68tm

M act safe

BM in DW: negligible

2.3.2 stage 2. After inserting the II strut at 6.00m below GL, excavate 5.00m for further inserting the III strut. (W.T. lowered to cut level).

Active Pr.: at B: (1.85x5.0 +2.0)= 3.71 t/m^2

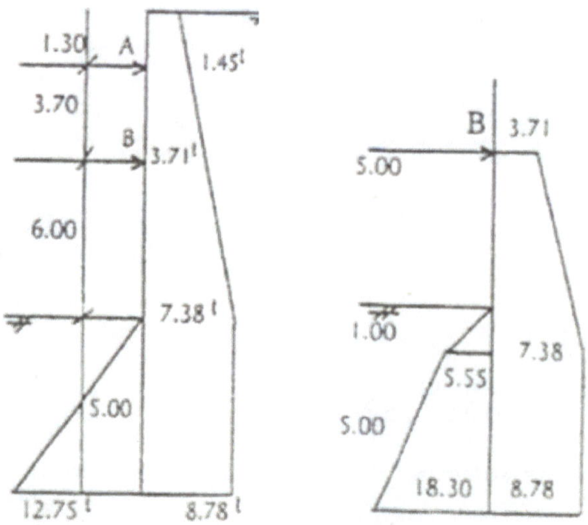

at cut level: 3.71+0.33x1.85x6.00 = 7.38 ^{t}/m^2

at bottom of DW = 7.38+0.33x0.85x6.00 = 8.78 ^{t}/m^2

Passive Pr.

at cut level: 0

at bottom of DW = 300x0.85x5.00 = 12.75 ^{t}/m^2

taking moments of B

M act = $\frac{3.71 \times 6.00^2}{2} + \frac{3.67 \times 6.00^2}{2} \times \frac{2}{3}$

$$+ 7.38 \times 5.00 \times 8.50 + \frac{1.40}{2} \times 5.00 \times 9.33$$

$$= 457.13 \text{ tm}$$

$$M \text{ pass} = \frac{12.75}{2} \times 5.00 \times 9.33 = 297.39 \text{tm}$$

Excavating up to 5.00 m below II strut (with W.T.lowered to 6.0m below II strut level)

Passive Pr. cut level = 0

1.0m below cut level = 3.00x1.85x1.00 = 5.55 $^t/m^2$

5.00 below cut level = 5.55 + 12.75 = 18.30$^t/m^2$

Taking moments of B

M act: 457.13 tm

$$M \text{ pass}: \frac{5.55}{2} \times 1.00 \times 5.67 + 5.55 \times 5.00 \times 8.50 + \frac{12.75}{2} \times 5.00 \times 9.33 = 549.00 \text{ tm}$$

Excavate up to 5.00m and insert strut by trenching.

2.3.3 for final excavation and all struts installed

Earth Pr.: (0.65 KarH ($^{\psi}$=30°)

=0.65x0.33x1.85x12.00=4.76 $^t/m^2$

Surcharge:

a. road/crane 0.33x2.00=0.66 $^t/m^2$ (1 m ht. of Earth) (from GL)
b. Building: 0.33x3.30 = 1.10 $^t/m^2$ (from, say 1.50m below GL)

Note: Nearest Building at 3.0m away from cut (stress in ten silty from Boussinessqs chart]

(0.3x11.00t/m^2 = 3.30$^t/m^2$)

Max m Bldg. Surcharge intensity at 1.5m below GL taken as 11.00 t/m^2 (Sandy soil)

All Struts in Position

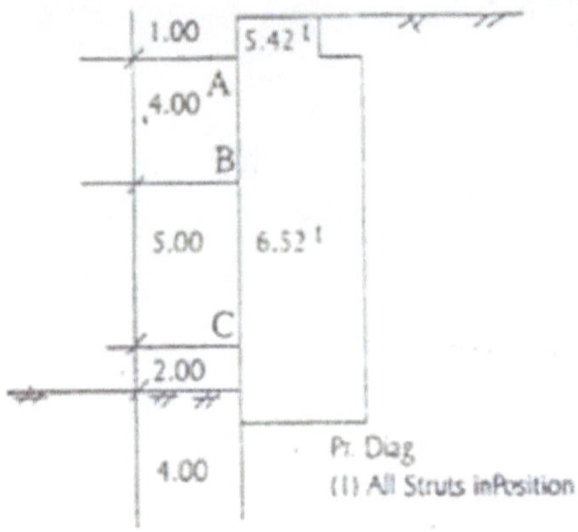

Sequence of Constr.

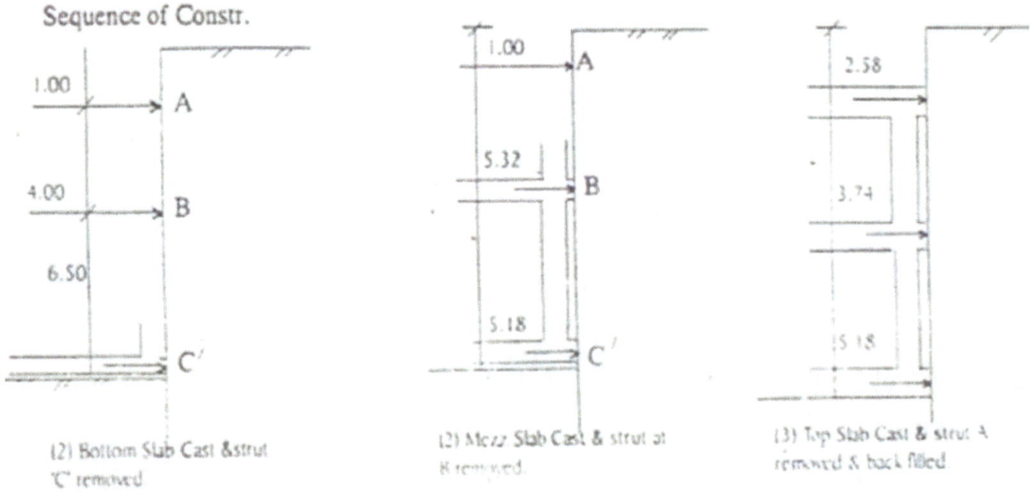

2.3.3.1 To analyse for case(i)

$$MA = \frac{5.42 \times 1.00^2}{2} = 2.71 \text{ tm}$$

$$MAB = \frac{6.52 \times 4.00^2}{12} = MBA = 8.69 \text{ tm}$$

$$MBC = \frac{6.52 \times 5.00^2}{12} = MCB = 13.58 \text{ tm}$$

$$MCD = \frac{6.52 \times 2.40^2}{12} = MDC = 3.13 \text{ tm}$$

Stiffness ratio

AB = 0.75, BC = 0.80 , CD= 1.25

Free BMS

AB: 8.69x1.50 = 13.04

BC: 13.58x1.50 = 20.37

CD: 3.13 x 1.50 = 4.70

Shear Force

Left of A = 5.42 t

Rt. of A = $6.52\times2.00-\frac{10.76}{4} = 10.35$ t

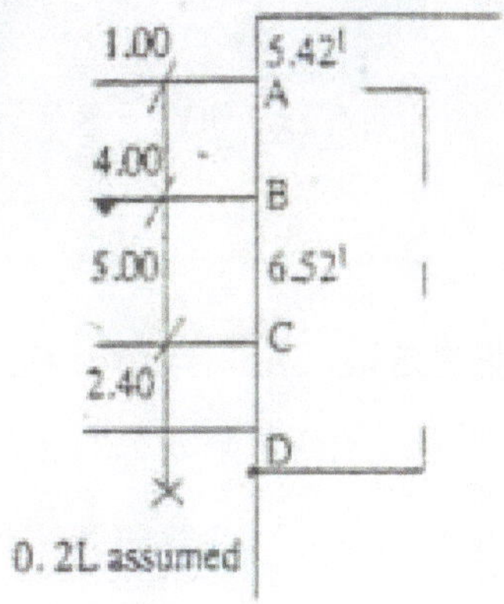

Left of B = 13.04+2.69 = 15.73 t

Rt. of B = $6.52\times2.50+\frac{2.77}{5.00} = 16.85\ t^{t}$

Left of C = 16.30-0.55 = 15.75 t

Rt. of C = $6.52\times1.20+\frac{10.70}{2.40} = 12.38^{t}$

Left of D = $7.82-4.46 = 3.36^{t}$

2.3.3.2 To analyse for case (ii)

MA = 2.71

MAB = MBA = 8.69

MBC' = MC'B = $\frac{6.52\times6.50}{12} = 22.96$ tm

SR AB = $\frac{3.00}{4.00} = 0.75$

BC' = $\frac{3.00}{6.50} = 0.46$

free BM: AB: 8.69x1.50 = 13.04 tm

BC' = 22.96x1.50 = 34.44 tm

Shear Force

Left of A = 5.42^{L}

rt of A = $6.52\times2.00-\frac{23.08}{4.00} = 7.27^{t}$

Left of B = $13.04+5.77 = 18.81^{t}$

rt of B = $6.52\times3.25+\frac{25.79}{6.50} = 25.16^{t}$

Left of C' = $21.19-3.97 = 17.22^{t}$

BM at 900 to left of B.

$2.71 + \frac{23.08}{4} \times 3.10$

$[\frac{6.52}{2} x 4.00 x 0.90$

$\frac{6.52}{2} x 0.90^2] = 8.75$ tm

2.3.3.3 To analyses for Case (iii) Perm condition.

Lat. Pr at A' = 0.33 (0.85x2.58+2.00+3.30++2.58 = 5.05 $^t/m^2$

at B' = 5.05+0.33x0.85x3.74+3.74=9.84 $^t/m^2$

at C' = 9.84+0.33x0.85x5.18+5.18 = 16.47 $^t/m^2$

MA'B = $\frac{5.05 x 3.74^2}{12} = \frac{4.79}{2} \times \frac{3.79^2}{15} = 8.12$ tm

MB'A' = 5.89+2.23 x 1.50 = 9.24 tm

MB'C' = $\frac{9.84 x 5.18}{12} + \frac{6.63}{2} \times \frac{5.18^2}{15} = 27.93$ tm

MC'B' = 22.00+5.93x1.50 = 30.89tm

SR: A'B' = $\frac{3.00}{3.74} = 0.80$	0.80 = 0.58	
	B	1.38
		0.42

B'C' = $\frac{3.00}{5.18} = 0.58$

Free Bm = A'B' = $\frac{5.05 x 3.74^2}{8} = \frac{4.79}{2} \times \frac{3.72^2}{7.8} = 13.12$ tm

(33.00) (11.40)

B'C' = $\frac{9.84 x 5.18^2}{8} = \frac{6.63}{2} \times \frac{5.18^2}{7.8} = 44.40$ tm

Shear Force

Sf rt of A = $\frac{5.05 x 3.74}{2} = \frac{4.79 \times 3.74}{2 \quad 3} - \frac{30.75}{3.74} = 4.21^t$

Left of B = 9.44+2.99x2+8.22=23.64^t

rt of B = $\frac{9.84 x 5.18}{2 2} + \frac{6.63 \times 5.18}{3} + \frac{30.75}{5.18} = 37.15$

left of B = 25.49+2x5.72-5.94=30.99^t

3.0 Design BM: 40.00tm

$$\frac{Mu}{fck\ bd^2} = \frac{1.50x40.0x10^5}{200x100x60^2} = 0.083$$

$$\frac{Pu}{fck\ bd} = \frac{1.50x20.00x10^3}{200x100x60} = 0.025$$

$Ast = Ast = 0.48x0.50x10^{-4}x200x100x60 = 28.00cm^2$

$$Eo = \frac{35.00x10^3}{18.2x0.8x53.9} = 45.00cm$$

provide 28#@200c/c both faces

A=30.78 sq.cm

E0 = 43.98 cm.

For 20.0 tm provide 28# @400 c/c

Provide 16 # distributors @ 400

Shear reinf.: 30.0 x 1000/(0.8x 53.9 x100) =6.95 sq.cm<7 sq.cm

Provide 16# links @ 400 horizontal and 800 vertical 3.20 Wale Reinf. 3.0 m panel

i) Top Strut – Max load/m = 20.0 t; 60.0t (Strut/Girder) Eff. Width: 600 mm

Mu/bd2= 14.54, Ast =19.94 sq.cm

Eo= 20x1.3x1000/(18.2x0.8x53.9) = 33.0

Provide 5-25 # A=24.54sq.cm Eo=39.2 cm

Max. load/m = 44.0 tm

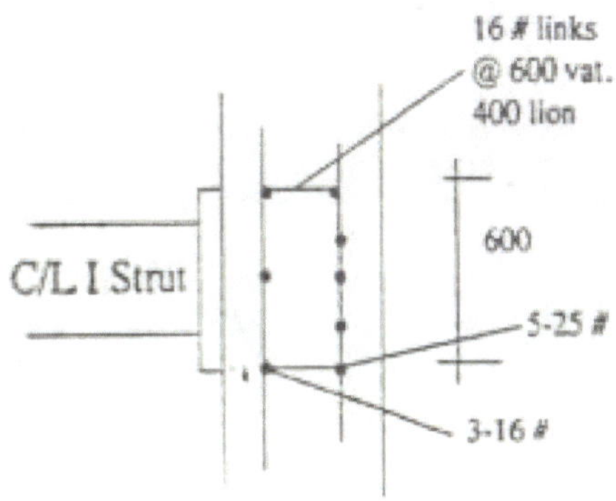

ii) Middle Strut: Strut Load = 132 t (Use 2- ISMB strut

Eff. Width: 1800

Mu/bd2 = 9.09, Ast=21,70 sq.cm, Eo=67.28 cm

Provide9-25 #, Ast=44.17 sq.cm, Eo=70.6 cm

Shear Stress=3.97 kg/sq.cm < 7 kg/sq.cm

iii) Third Strut: Strut Load/m = 28.00 (2-ISMB 450)

Strut Load= 84.00 t

Efft. Width == 1600 mm

Mu/bd2 =6.52, Ast=13.70 sq.cm

Eo= 43 cm, Provide 7-20# A=21.69 sq.cm

Shear Stress=2.81 kg/sq.cm

Provide nominal reinforcement.

Reinforcement Details

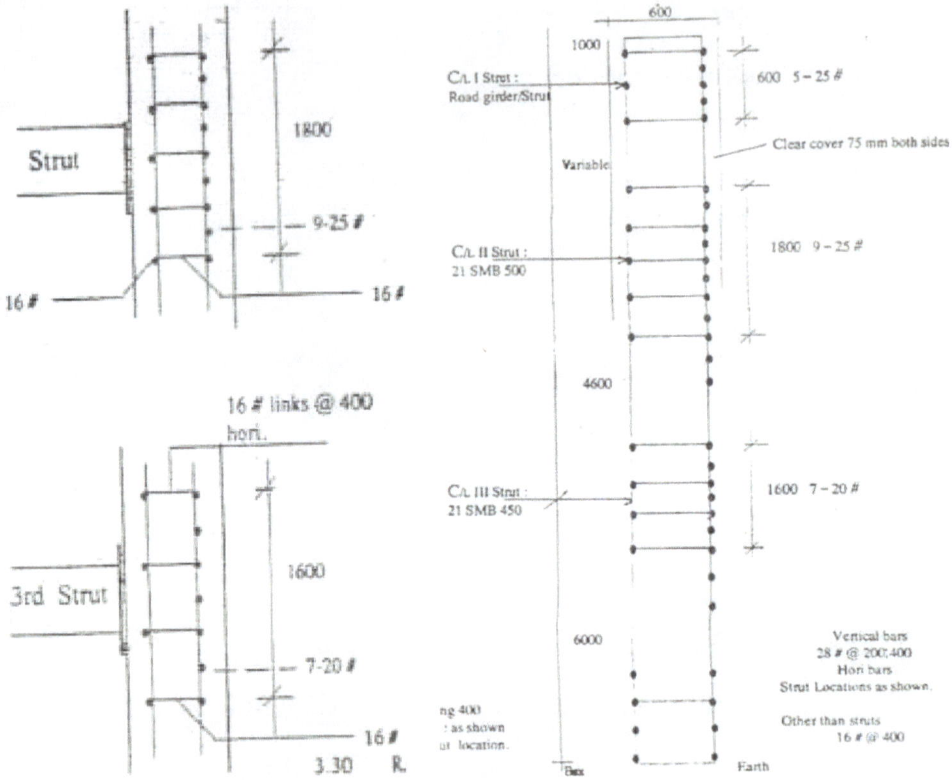

Annex: 3 Building Condition Survey

A3.1 Introduction

A building condition survey was carried out within the influence zone of the corridor construction. In the case of tunnels, this was reckoned as 25m to 40m from the tunnel's centerline on either side (tunnelling through TBM). In the station locations, the cut-and-cover method was being employed, and here, too, building condition surveys were undertaken.

The underground Stations and the ramps had been executed by cut & cover. One was almost forced to weave a transport corridor underground, and for elevated corridors through narrow existing roads. The built environment extended above and below the ground and did not allow unrestricted freedom for the engineers. Underground construction could always impact on the foundation units of buildings above or at least disturb the soil that supports the foundations with unpredictable consequences. This also helped reduce technical complexity and legal risks that may arise.

The 16.45 km east-west corridor between Salt Lake Sector-V and Swabhumi is elevated, and the remaining section between Swabhumi and Howrah Maidan is underground. The elevated section was designed with support from the existing median wherever it was possible, and the structural system consists of piers supported on pile foundations and a superstructure with a PSC-segmental box. The underground section passing through a densely built-up area was planned to be constructed using the Earth Pressure Balancing (EPB) Tunnel Boring Machine (TBM) to avoid acquisition and create minimum hindrances to existing traffic. However, for underground stations, the Cut-and-Cover method had to be used. The tunnel depth varied to about 20m below the Hooghly Riverbed. Underground stations were designed based on top-down construction. There are 12 stations, six elevated and six underground. The construction of tunnels by TBM and underground stations by cut and cover was prone to damage buildings adjacent to the alignment. The old, weak, cracked buildings and properties were more vulnerable to such damage.

Therefore, a building condition survey of all the buildings in the alignment's influence zone was necessary to take proper precautionary measures to prevent/minimise the extent of damage during the construction of tunnels and stations. Accordingly, a condition survey was undertaken to document the existing buildings within the influence zone and their conditions, depicting the damage before the tunnelling and deep excavation commences.

A3.2 Reconnaissance Survey

A survey team comprising a coordinator, four section leaders, and field supervisors was formed to conduct the building condition survey. Before the fieldwork began, a team of senior officials and section leaders conducted a reconnaissance survey along the proposed alignment of the Metro Data Forms to assess the methodology to be adopted during the actual field survey.

A3.3 Methodology

To inspect buildings within the influence zone of the proposed Metro tunnel to assess the existing health of the buildings for the exposed portions and foundations, where drawings are available. Reconnaissance of the building condition to make an initial assessment and to conduct a general study in different portions of the building by local enquiry and visual inspection, including measuring the width and length of the crack, area of damage viz. seepage, spalling of plaster, exposed reinforcement of the superstructure wherever possible as well as through desk study from available drawings. Identifying the location of the buildings from road kerb and adjacent structures for ease of future reference.

Digital photos of the building and the damage observed, such as cracks, spalling of concrete/plaster, exposed reinforcement, vegetation growth, seepage, etc.

Enquiring about the type of foundation from the owners and occupants of the buildings and residents. Collect available drawings from appropriate authorities/owners/occupants for desk study regarding the foundation of the building.

A3.4 Field Investigation

During the initial fieldwork stage, the survey teams faced objections from house owners/occupants and were reluctant to share information regarding the existing status. However, with much persuasion, after understanding the usefulness of such an investigation of age-old buildings, including buildings more than 200 years old.

A few buildings were observed to exist in the influence zone of the proposed tunnel but were missing from the available Topographical Survey Map. A few buildings were found to exist on the Topography Map, but their addresses did not match the list. Temporary structures with tile/tin/asbestos roofs were generally not considered for the survey influence zone. The condition of a few buildings was found to be very alarming. Nineteen such buildings were identified and included in the survey. The field investigations were compiled in Data Forms for each building, including photographs of front and side views, different types of damage, and overloading, if any.

A3.5 Building Condition

i) Under Construction (UC):

Buildings that were noticed to be still under construction are at different stages of completion

ii) Good:

Buildings that were under proper routine maintenance as evident from visual inspection and based on the information by its owner/occupants and having sound health with minimum damages in the form of dampness, minor cracks in plaster and in auxiliary components, chajjas, tiny plants, which could be uprooted during routine maintenance work. It was noted that the coefficients for seismic force had changed when the code was revised in 2002, and are much higher according to the IS:1984 code, demanding a retrofit.

iii) Average

These buildings had already started showing distress and had medium types of damage, such as cracks, spalling of plaster from walls and slabs exposing reinforcements, damage to wall or concrete chajjas, roof slabs, and vegetation growth. These buildings need immediate attention to be r

Iv) Bad

Major type of damage in the form of large cracks in walls and concrete restored to good condition. portions, exposure of reinforcements with traces of corrosion, vegetation growth with complex roots penetrating the walls, and spalling of plasters, thus allowing rainwater to seep into the mortars appearing in the buildings but the primary structural load-bearing members were not deformed and damaged excessively, have been categorised as bad.

v) Vulnerable:

Where the members of concrete, steel, timber, and brick masonry become very thin and lose their sizes appreciably, or slabs and beams sag excessively due to non-repair over a large span. Cantilevered roof slabs were thoroughly cracked, and road-facing walls had huge openings. The thin walls of the top floor were severely damaged. There was no cement mortar in many brick masonry joints that were almost open, and a vertical crack in the wall from top to bottom was noted. Arch lintels were cracked. The lintel beam in the main entrance was thoroughly cracked. Vertical cracks were observed in columns and pillars on the side facing the road. Steel beams and columns were rusted. Lintels cracked. Diagonal shear cracks were found in walls running perpendicular. Points where steel joists support, most components were damaged and condemned by the railway authority. Structures that show unattended thick and significant vegetation, these structures were included in the category of vulnerable structures. In many parts of the structures, extensive cracks had already developed in load-bearing brick masonry walls, leading to part or total loss of cement mortar between the brick elements. Buildings with severe damage, such as large cracks, spalling of plaster or concrete, exposed corroded reinforcement, steel beams, rotten wooden beams and runners, immense vegetation growth, collapse of chajjas, balconies, portico columns, etc., posing a threat to the stability of the structure, had been categorised as "Vulnerable." Photographs of such damage were taken wherever possible. During the survey, no drawings could be obtained from the owners or Civic Authorities, and thus, the assessment of the buildings was purely based on field investigation observation only.

As an outcome of the Building Condition Survey, some buildings were considered vulnerable. Structures where diagonal shear cracks have already developed in load-bearing brick masonry walls due to diagonal tension and buildings that were in average or poor condition, also included under the vulnerable category.

A3.6 Survey-by-Afcons-Infrastructure (Design- Build Contractor for Howrah Maidan-to-Sealdah)

Briefs-of-Survey-Outcome

- It is a proven fact that no underground work, including tunnelling, can be done without some impact on the structure falling in the influence zone.
- Generally, the influence zone is prominent up to 12 meters from the center of the tunnel. Beyond that, the settlements are less than 10mm. However, this depends upon soil strata and the depth of the tunnel.
- While tunnelling beyond the vent shaft, one has to encounter buildings on both sides of road.
- 27 buildings need to be taken care of during tunnelling.
- 9 out of 27 buildings are identified as critical/vulnerable are need extensive care during tunnelling.

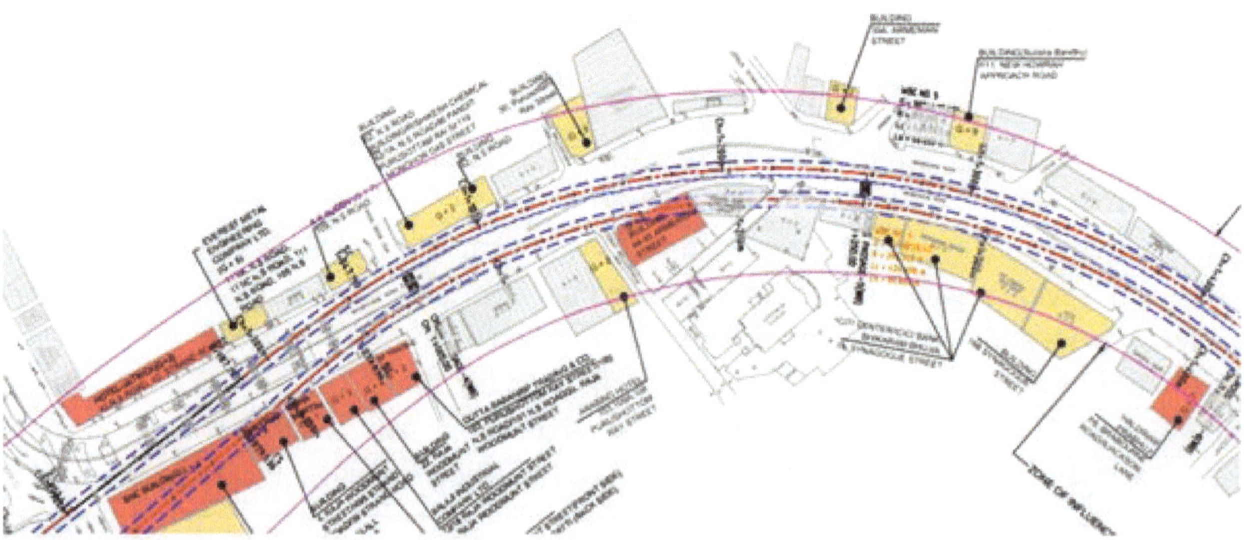

Fig.A3.1: Part Plan Showing Category 'A'

The survey is aimed at capturing Category "A"/Critical Buildings on both sides of the metro corridor. Red-marked buildings have been identified as Category "A"

S.No	Building Name	Address
1	Hotel Jatrik	42,N.S.Road, 42, Strand Road
2	Building 2 Raja wood mound St	2 Raja wood mound St
3	Building, 4 Raja wood mound St	4 Raja wood mound St
4	Usha Martin	6/8 Raja wood mound St
5	Balaji Industries	12/18 Raja wood mound St
6	Building 22 Raja wound mound street	22 Raja wound mound street
7	Dutta Bhaban/ BP trading	24 Raja wound mound street
8	Building 44/45	44/45 Armanian Street
9	Haldhiram Prabhu Ji	P8, Brabourne Road / 8 Jackson Lane
10	Building Besides David Synagogue	109A/1A Canning Street
11	Hollywood Institute of Management & Clusters of Shops besides Bethel Synagogue	20/A Brabourn road

These buildings are in deplorable condition. There is evidence of dislodged materials falling from the buildings. Red-marked buildings have been identified as Category "A" Buildings and need extensive care during tunnelling. These buildings are in deplorable condition.

The Reasons for Building Condition Resurvey

- The building condition survey was conducted in the year 2010-2011 for the first time.
- After restarting work, it was felt that the building condition had deteriorated substantially after the passage of 7 years since the first survey.
- Basic reasons for the change in building conditions can be summed up to the following reasons.
  - Buildings have crossed their design life/useful life.
  - Recent earthquakes (2 times in 2015 and 2016)
  - Fire event (especially in 4-Raja Woodmount Street)
  - Self-settlement of buildings due to change in effective stress owing to water table variations (seasonal and depletion)
  - Fast deterioration of buildings due to the absence of maintenance
- Visual inspection to assess the condition of the building
- Detailed high-quality digital photographic documentation showing all major distress of the building
- Distress mapping of existing buildings, including documentation of distress like cracks, delamination, geometric deformation of structural components and out of plumb, etc.
- The buildings that need care are divided in two categories

Category "A"/Critical Buildings:

These buildings are in **very poor** condition and need to be

- Demolished, or
- Extensively repaired and propped, and evacuated if need be based on the building response to the tunnelling, or
- Reasonably repaired, evacuated, and extensively propped during tunnelling
- Restricted pedestrian/vehicle movement adjacent to the buildings under administrative control

Category "B" Buildings:

These buildings are in **poor** condition and need to be

- Reasonably repaired, or
- Reasonably propped during tunnelling
- The residents will need to be evacuated if there is distress in the building beyond the tolerable limit.
- Restricted pedestrian/vehicle movement under administrative control

It would be recommended to seek fitness certificates from building owners or from civic authorities for "A" and "B" category buildings. The fitness certificates are mandatory for buildings older than 40 years and are mandatory per building by-laws. Sample of the Layout of Critical buildings along the alignment

Fig.A3.2: Building Surveyed

Salient features of building

- Length of building – **58m**
- Width of building – **21m (Avg.)**
- Number of stories – **G+3**

- Height of building – **15.7m height (4.2+4.0+4.0+3.5)**
- Type of foundation – **Brick Masonry with steel Beam**
- Age of Building – **105 years (Approx)**
- Type of Usage – **Commercial Complex**
- Property edge distance from c/l of Tunnel - **Min.-3.63 m Max.-12.98 m**

Building at Raja Woodmont Street

Damages identified

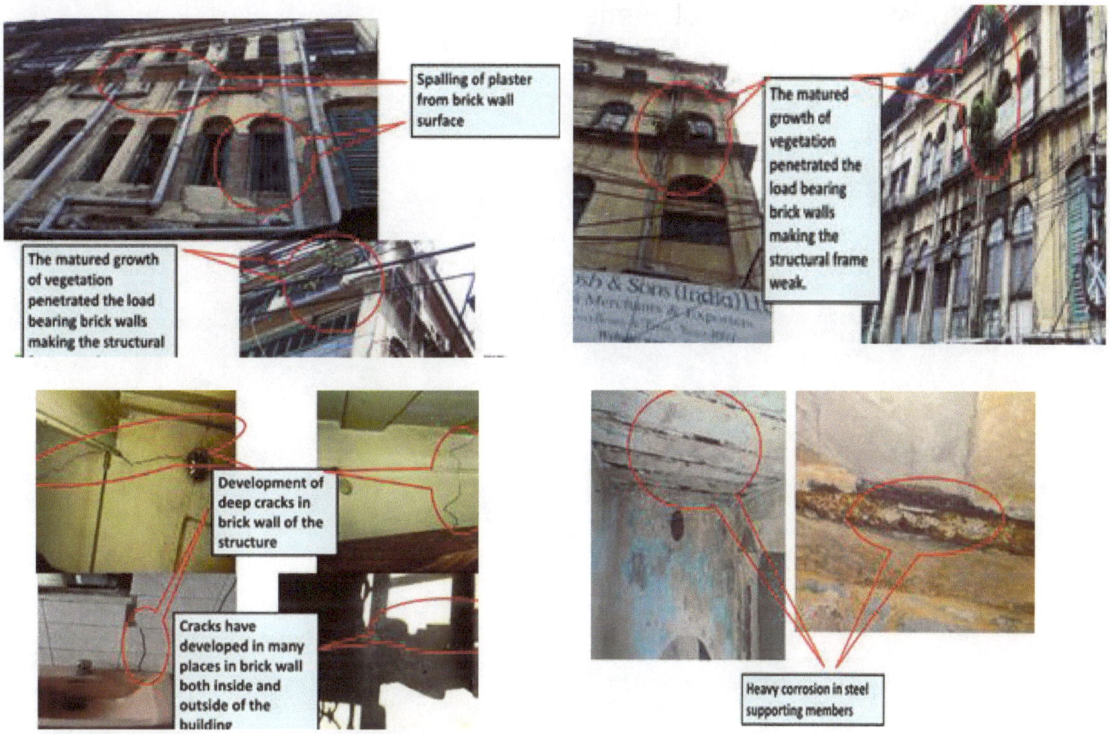

Figs. A3.3: Building at Raja Woodmunt Street: Damages as marked

Fig.A3.4: View of a Deteriorated Building

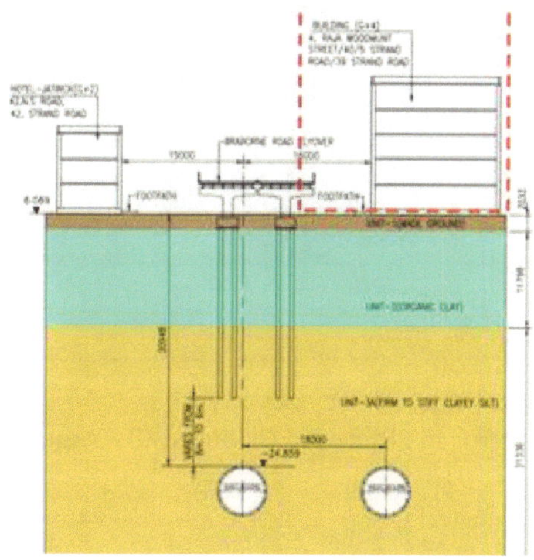

Fig.A3.5: Engineering Section – Building Above the Proposed Twin Tunnel

Salient Features

- Length of building – **19m**
- Width of building – **16m**
- Number of stories – **G+3**
- Height of building – **15.6m height (4.3+4.1+3.8+3.4)**
- Type of foundation **– Brick Masonry with steel Beam**
- Age of Building – **109 years (Approx)**
- Type of Usage – **Commercial Complex**
- Property edge distance from c/l of Tunnel -

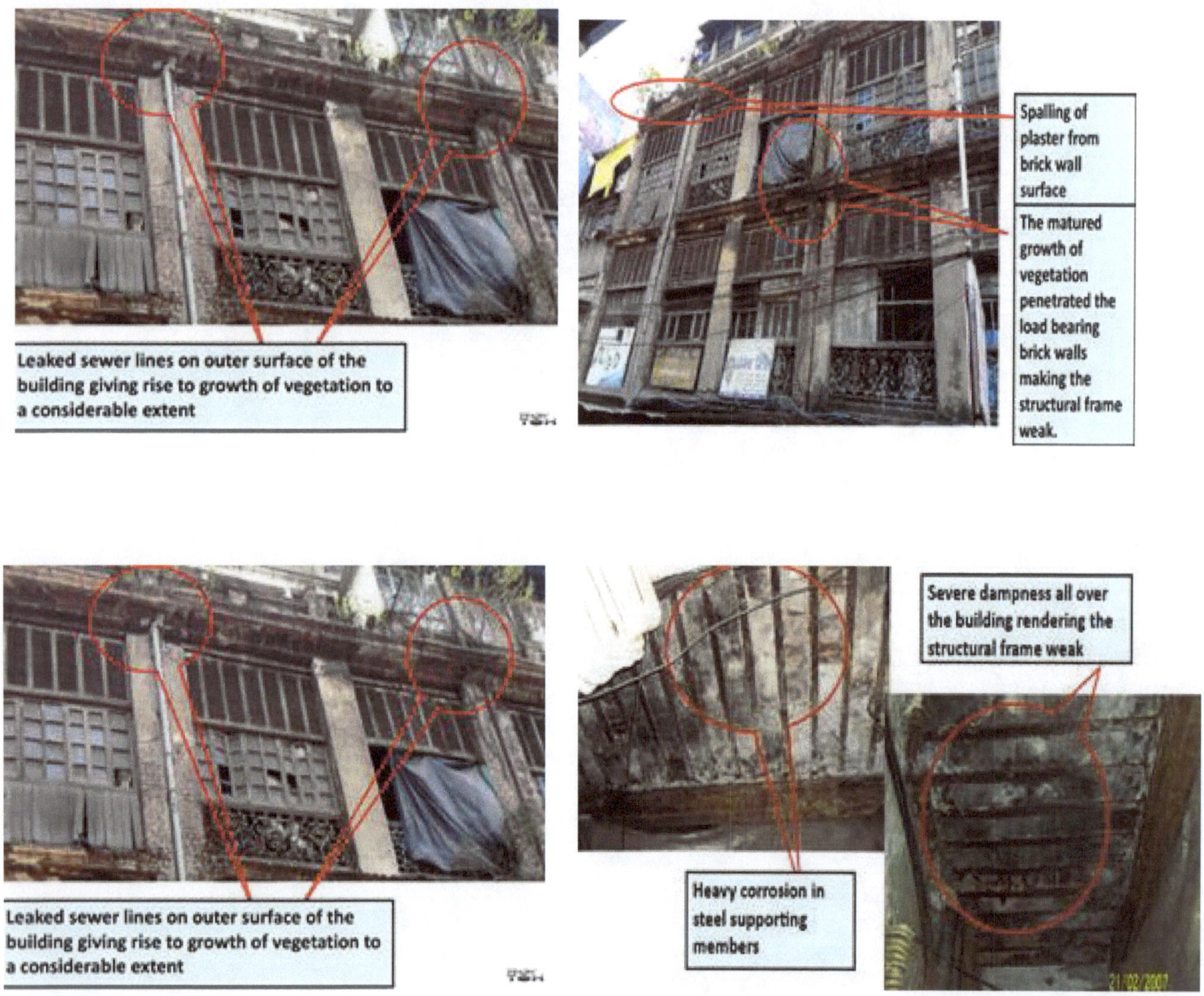

Fig. A3.6: Photos Showing Extensive Damage

Table: A3.1

dress	Type of structure (Name of the building)	Age of the building	Usage	Reference (Chainage) as per survey map	Remarks
66, N. M. Road Kolkata-700054	G+2	>60 years	Residential	A251R	The cantilevered roof slabs are thoroughly cracked, and road-facing walls have huge openings. Perhaps the additional top floor has caused overloading.
47, N. M. Road Kolkata-700054	G + 1	>50 years	Residential cum commercial	A246R A247R, A248R	A cluster of buildings showing wide diagonal cracks in walls running perpendicular to the road. The thin walls of the top floor are very severely damaged. There are no cement mortars in many brick masonry joints, which are almost open.
24A, N.M.Road Kolkata-700054	G+1	101 years	Residential	In between A204R & A206R	Vertical crack in the wall facing the road from top to bottom. Arch lintels cracked.
23A, 23B&23C N.M.Road Kolkata-700054	G+1	174 years	Residential	In between A199R & A201R	Vertical crack in the wall facing the road. Wooden projected beam in lousy condition, fibres exposed. The lintel beam in the main entrance is thoroughly cracked.
19 N.M.Road Kolkata-700054	G+1	54 years	Residential	A197R	Vertical crack in the wall facing the road. Vertical cracks in columns and pillars. Steel beams and columns are rusted. Lintels cracked.

dress	Type of structure (Name of the building)	Age of the building	Usage	Reference (Chainage) as per survey map	Remarks
18/2, N.M. Road Kolkata-700011	G +1	59 years	Residential	In between A195R & A197R	The balcony is sagging, and the supporting steel angles are badly rusted. The wall facing the road is badly cracked.
16, N.M. Road Kolkata-700011	G +2	150 years	Residential cum commercial	A194R	The mortar between bricks is absent. Walls running perpendicular to the road have diagonal shear cracks. Points where steel joists support walls are also badly cracked.
164, Rly. Qtr. N.M.Road Kolkata-700054	G +3 Rly Qtr	>100 years	Residential	Ch-5+400	All components damaged have already been declared condemned by the railway authority.
170A, Baithak Khana Road Kolkata –700012	G+0	>100 years	Commercial	Front part of A139L	Lintels cracked. Brick joints are mortarless. Open cracks due to thick vegetation growth. Vertical cracks in the walls facing the road.
134, B.B Ganguly St. Kolkata-700012	G+1	109 years	G.F.-commercial Others-residential	A134R	Columns cracked at the end of the canopy. Canopy beams are also badly rusted—mortarless brick joints.
162, B B Ganguly Street Kolkata-700012	G+1 (Series of shops, both GF & 1st F.)	>75 years	Commercial	A130L	Walls perpendicular to the road were severely cracked—mortarless brick joints. Bricks dislocated in front and side walls. Thick old vegetation walls. Thick old vegetation growth. RSJ beams rusted.
129/1 B.B Ganguly Street Kolkata-700012	G+1	200 years	G.F.-commercial Others-residential	A134R	There is no roof slab. Big banyan tree on walls. Mortarless brick joints. Vertical cracks in walls.

dress	Type of structure (Name of the building)	Age of the building	Usage	Reference (Chainage) as per survey map	Remarks
126 ABC BB Ganguly Street Kolkata -700012	G+3	>150 years	Residential cum commercial	A131R	Concrete beams cracked. Steel joists rusted. Thick vegetation growth.
169B, B. B. Ganguly Street Kolkata -700012	Basement + G+2	>200 years	Residential cum commercial	At the back of 169A	The sagging roof, a portion of the slab, and the wall fell off. Severe crack on walls—no inhabitants on the 1st and 2nd floors, which are completely deserted.
117, B. B Ganguly Street Kolkata -700012	G+1	>100 years	Commercial	A120R	Mortarless brick joint. Thick vegetation growth. Walls perpendicular to the road have thin, eroded bricks.
101 B.B Ganguly Street Kolkata -700012	G+1	>100 years	Commercial Others-residential		Steel angles are badly rusted. Thick vegetation growth. Mortarless brick joints. Wooden beams are badly damaged.
93/1A, B. B Ganguly Street Kolkata -700012	G+2	>100 years	Commercial		There are diagonal cracks in the wall facing the road, thick vegetation growth, badly damaged cantilever slabs, and a very bad condition of the top floor.
284A/284B, B. B Ganguly Street Kolkata -700012	G+2, G+1 (Arambagh Opticals)	>100 years	G.F.-commercial Others-Residential	A74L	Sagged roof and wooden runners. Extra steel rods provided between runners also sagged. Mortar is missing between brick joints.
300, 301C, D, F B.B. Ganguly Street Kolkata-700012	G+2 (Sulekha works, Kol. Police Resv. Force and others.)	>80 years	G.F.- Shops Rest- Police barrack	A67L	Deep open vertical cracks in the walls. Load-bearing columns cracked. Brick chunk dislocated.

Sl No.	Address	Type of structure (Name of the building)	Age of the building	Usage	Reference (Chainage) as per survey map	Remarks	
V1	66, N. M. Road Kolkata-700054	G+2	>60 years	Residential	A251R	The cantilevered roof slabs are thoroughly cracked, and road-facing walls have huge openings. Perhaps the additional top floor has caused overloading.	
V2	47, N. M. Road Kolkata-700054	G + 1	>50 years	Residential cum commercial	A246R A247R, A248R	A cluster of buildings showing wide diagonal cracks in walls running perpendicular to the road. The thin walls of the top floor are very severely damaged. There are no cement mortars in many brick masonry joints, which are almost open.	
V3	24A, N.M.Road Kolkata-700054	G+1	101 years	Residential	In between A204R & A206R	Vertical crack in the wall facing the road from top to bottom. Arch lintels cracked.	
V4	23A, 23B&23C N.M.Road Kolkata-700054	G+1	174 years	Residential	In between A199R & A201R	Vertical crack in the wall facing the road. Wooden projected beam in lousy condition, fibres exposed. The lintel beam in the main entrance is thoroughly cracked.	
V5	19 N.M.Road Kolkata-700054	G+1	54 years	Residential	A197R	Vertical crack in the wall facing the road. Vertical cracks in columns and pillars. Steel beams and columns are rusted. Lintels cracked.	
V6	18/2, N.M. Road Kolkata-700011	G +1	59 years	Residential	In between A195R & A197R	The balcony is sagging, and the supporting steel angles are badly rusted. The wall facing the road is badly cracked.	

Sl No.	Address	Type of structure (Name of the building)	Age of the building	Usage	Reference (Chainage) as per survey map	Remarks	
V7	16, N.M. Road Kolkata-700011	G +2	150 years	Residential cum commercial	A194R	The mortar between bricks is absent. Walls running perpendicular to the road have diagonal shear cracks. Points where steel joists support walls are also badly cracked.	Sheet No.3
V8	164, Rly. Qtr. N.M.Road Kolkata-700054	G +3 Rly Qtr	>100 years	Residential	Ch-5+400	All components damaged have already been declared condemned by the railway authority.	
V9	170A, Baithak Khana Road Kolkata –700012	G+0	>100 years	Commercial	Front part of A139L	Lintels cracked. Brick joints are mortarless. Open cracks due to thick vegetation growth. Vertical cracks in the walls facing the road.	
V10	134, B.B Ganguly St. Kolkata-700012	G+1	109 years	G.F.-commercial Others-residential	A134R	Columns cracked at the end of the canopy. Canopy beams are also badly rusted—mortarless brick joints.	
V11	162, B B Ganguly Street Kolkata-700012	G+1 (Series of shops, both GF & 1st F.)	>75 years	Commercial	A130L	Walls perpendicular to the road were severely cracked—mortarless brick joints. Bricks dislocated in front and side walls. Thick old vegetation growth. RSJ beams rusted.	
V12	129/1 B.B Ganguly Street Kolkata-700012	G+1	200 years	G.F.-commercial Others-residential	A134R	There is no roof slab. Big banyan tree on walls. Mortarless brick joints. Vertical cracks in walls.	

Sl No.	Address	Type of structure (Name of the building)	Age of the building	Usage	Reference (Chainage) as per survey map	Remarks	
V13	126 ABC BB Ganguly Street Kolkata -700012	G+3	>150 years	Residential cum commercial	A131R	Concrete beams cracked. Steel joists rusted. Thick vegetation growth.	
V14	169B, B. B. Ganguly Street Kolkata -700 012	Basement + G+2	>200 years	Residential cum commercial	At the back of 169A	The sagging roof, a portion of the slab, and the wall fell off. Severe crack on walls—no inhabitants on the 1st and 2nd floors, which are completely deserted.	
V15	117, B. B Ganguly Street Kolkata -700 012	G+1	>100 years	Commercial	A120R	Mortarless brick joint. Thick vegetation growth. Walls perpendicular to the road have thin, eroded bricks.	
V16	101 B.B Ganguly Street Kolkata -700 012	G+1	>100 years	Commercial Others-residential		Steel angles are badly rusted. Thick vegetation growth. Mortarless brick joints. Wooden beams are badly damaged.	Sheet No.:4
V17	93/1A, B. B Ganguly Street Kolkata -700 012	G+2	>100 years	Commercial		There are diagonal cracks in the wall facing the road, thick vegetation growth, badly damaged cantilever slabs, and a very bad condition of the top floor.	
V18	284A/284B, B. B Ganguly Street Kolkata -700 012	G+2, G+1 (Arambagh Opticals)	>100 years	G.F.-commercial Others-Residential	A74L	Sagged roof and wooden runners. Extra steel rods provided between runners also sagged. Mortar is missing between brick joints.	
V19	300,301C, D, F B.B. Ganguly Street Kolkata-700 012	G+2 (Sulekha works, Kol. Police Resv. Force and others.)	>80 years	G.F.- Shops Rest- Police barrack	A67L	Deep open vertical cracks in the walls. Load-bearing columns cracked. Brick chunk dislocated.	

Acknowledgement

The Author would like to express his sincere gratitude and thanks to Professor SS Chakraborty, Former CMD, CES, for encouraging and giving free hand to get involved in Transportation and the Mass Transit System globally, and Dr AK Chatterjee, a well-known Expert from Kolkata, for suggesting a book on Kolkata Metro concerning the first Metro in the Country, North-South Metro, particularly when very few persons involved in the project are alive. My sincere thanks to Mr Rajesh Prasad, former Director Operations & Advisor, IRCON and Akshay Sharma, Director-Projects, PITCMRL, for their contribution.

I convey my sincere thanks to Mr GR Madan, former GM, Kolkata Metro & Director-MRTS for MMR DA, Neeraj Jain, former CE, Biswanath Dewanjee, Former ED, KMRCL. I am grateful to my colleagues and former MTP experts, namely, BK Ghosh, MK Bhattacharjee, Chandurkar, and BG Kale; system experts, Satyen Bhowmik & V Narayan; and civil/Structural experts. My sincere thanks to my colleagues and fellow professionals who helped me from time to time, by providing data to the best of their knowledge, namely G Ranganath Babu, C Sankarlingam, KC Thapliwal, M Pramanik, S Chatterjee, Soumya Ray, Makarand Khare and Amal Bhattacharya. Thanks to Jitendra Rastogi, Mukul K Lodhi, and Partha P Banejee. My sincere thanks to my colleague Tanmoy Chakraborty, who provided continuous assistance during my tenure at E-W Metro and throughout the writing of this book. Also, thanks go to Auyon Roy for his coordination with the Publisher for publication.

The author drew upon the Detailed Project Reports by DMRC and CES-led consortia, Reports by GC & IC on Mumbai Metro-Line 3, along with reports by RITES, BMRCL and Wikipedia. And the aurhor used the work experience of CES as part of GC, DDC and of RUPL (Author's team) as LDC.

References

1. Dr. BC Roy, 2004. Planning of Transportation System in an Indian Metropolis
2. Dr. BC Roy, Kolkata Metro, including East-West Metro
3. Dr. BC Roy & Dr N Som, 2000, Behaviour of Braced Cut in Soft Soil Condition, Calcutta Metro
4. Dr. BC Roy & A. Basu, Evolution of Metro Rail Projects in India: Design and Construction Practices
5. Biswanath Dewanjee, First Tunnel across Major River Hooghly for East-West Metro Kolkata
6. PRK Murthy, Metro Rail System in MMR: Planning, Implementation, and Challenges
7. Dr. BC Roy & S Wagh; Underground Transportation Tunnel: Cut–and–Cover Approach
8. Dr. BC Roy & S Wagh; Globalisation of Structural Design and Construction in Developing Countries
9. Dr. BC Roy, December 2003, Thane MRTS-Station Area and Way Structure
10. Dr. BC Roy, September 2004, Planning of Transportation System in Indian Metropolis
11. BK Ghosh & Dr. BC Roy, August 2006, Kolkata Metro – Planning and Design (North-South Metro)
12. V Narayan, S Bhowmik and Dr. BC Roy, August 2006, Kolkata Metro-Design and Construction (North-South Metro)
13. Dr. BC Roy, 2006, Top-down Construction of Four-Level Integrated Intersection Planning, Design & Construction
14. Dr. BC Roy, 2008, Transport Infrastructure in the Indian Urban Environment
15. Dr. BC Roy, 2009, Mass Rapid Transport in Indian Urban Areas
16. Dr. BC Roy, 2011, Planning Aspects of East-West Metro and Depot
17. Dr. BC Roy, 2013, Kolkata Transit System including East-West Metro
18. Dr. BC Roy, 2013, Keynote Address on East-West Metro
19. Dr. BC Roy, 2014, Mass Transit System & Evolution of Kolkata Transport System
20. Dr. BC Roy, 2015, Urbanisation and Growth in the Indian Transportation Scenario
21. Dr. BC Roy, 2016, Urbanisation in India: Mass Transit System
22. Dr. BC Roy +1, 2023, Nairobi MRTS Conceptualisation: Transit-Oriented Development for Sustainable Transport
23. Dr. BC Roy, Dr R Ekambaram, 2021, Performance Assessment on Influence Zone of Existing Structures for Metro Construction
24. Dr. BC Roy & Biswanath Dewanjee, Maiden Endeavour for creating a Transportation Tunnel Below Major River in India
25. Hemant Sharma, BC Roy & B Dewanjee, Kolkata Metro, including the East-West Metro
26. Amal Bhattacharya, 2024, Influence of Process Performance During the Construction Stage- A Comparative Case Study of Kolkata East-West Metro and Pune Metro Line 3.
27. Amal Bhattacharya & VN Heggade, 2004, Fast Track Construction of Kolkata Metro-EWE 1.

28. Harsimran Singh & Soumya Roy, A Unique Metro Station in Kolkata- (Line 3) AM Shete, Dr SP Patil, Prof. SS Sathe, 2020, Seismic Analysis of Pune Metro Tunnel
29. Sandeep Bharadwaj, 2021, Analysis of Chawri Bazar Section of Delhi Metro Tunnel
30. B&S Engineering Consultants, Elevated Viaduct Yamuna River Near Kalindi Kunj-Botanical Garden, Ph 3, Delhi Metro
31. B&S Engineering Consultants, Elevated Viaduct & Six Stations-Karkarduma – Gokulpuri Corridor, Line 7, Delhi Metro
32. B&S Engineering Consultants, Elevated Viaduct & Ramp on H Nizamuddin - Mayur Vihar Corridor, Line 7, Delhi Metro
33. B&S Engineering Consultants, Viaduct for Automatic Walkway/Travellators at Dhaula Kuan, Punjabi Bag and Rajouri Garden, Delhi Metro, Phase 3.
34. Standardisation and Indigenisation of Metro Railways, Systems and Subsystems, 2013, Ministry of Urban Development, GOI
35. Design Basis Report for Elevated Station, 2019, CMRL
36. Design Basis Report for Viaduct,2016, CMRL
37. Detailed Project Report for Pune Metro,2015, DMRC
38. Construction Specifications for Phase 4, Civil Works, DMRC
39. Madan Magdum and Bilavari Karkure: Finite Element Modelling and Design of Underground Metro System
40. Rajesh Prasad - Soil Engineering Challenges in Implementation of Metrowerks in the City of Kolkata
41. Rajesh Prasad – Geotechnical and GI Data for Metro Line 3, Kolkata in the City of Kolkata
42. Rajesh Prasad – Design Standard and Planning for Metro Line 3, Kolkata
43. Mumbai Metro Planning Group (MMPG) Mumbai: Metro Study- Seventh Rail Corridor, December 1997
44. CES-led Consortium Detailed Project Report: Thane MRTS, 2001
45. CES-led Consortium: Detailed Project Report of Andheri- Ghakopar Metro Corridor
46. Wikipedia: Metros in Kolkata, Delhi, Mumbai, Bengaluru and others
47. DMRC: Detailed Project Reports for Metros in Delhi, Bengaluru, Pune, and Others
48. R.S. Grover- Former Member, Electrical – Presentation on Kolkata Metro
49. DMRC Detailed Project Report Ph 4-Delhi Metro
50. DMRC Signalling and Telecommunication Report by Abhisek Garg
51. First Section of Delhi Metro Phase III Operational with Ticketing System from Thales (thalesgroup.com)
52. Ministry of Urban Development, GOI - Traction Systems, General Power Supply Arrangements and Energy Efficient Systems for Metro Railways
53. Ministry of Urban Development, GOI - Standardisation of Signalling and Train Control Systems for Metro Railways.
54. Interim Consultancy Report for Mumbai Metro Line 3.

55. Presentation on Kolkata Metro UG 1 – AFCONS; Atkins
56. O. Vora, A. More & E. Thinakaran – Mumbai Metro Line 3: Initial Drive of Tunnelling Activities
57. Amalendu Gope, Arnab Sur and Ravi Ranjan Kumar – Challenges for NATM Design of Grant Road Station in Mumbai Metro Line 3
58. Vinay K Pandey and SC Mishra
59. SK Gupta, MG Khare & GVR Raju – Geotechnical Challenges in Design and Construction of Cut-and-Cover Tunnel- Mumbai Metro Line 3
60. SK Gupta, MG Khare & JD Calentano – Design and Construction of Bored Tunnels for Mumbai Metro Line 3
61. Dr MG Khare – Instrumentation, Monitoring & Reporting

About the Author

Dr. BC Roy

Dr. BC Roy (Bidhan Chandra Roy), FNAE, Ph.D. (Engg), MTech (Structures). He is a fellow of Major institutions worldwide, such as the Institution of Structural Engineers (UK), Indian National Academy of Engineers (INAE), International Association for Bridge & Structural Engineering (IABSE), Institution of Civil Engineers (UK), American Institution of Civil Engineers, Institution of India and many others. Dr over 50 years of extensive experience in providing engineering design consultancy services in India & abroad, spanning the transport infrastructure sector, including bridges, Roads, highways, flyovers, railways, metro rail, Industrial, residential, commercial & sports facilities. Dr Roy was the Team Leader for an International Award-Winning Nivedita Setu. He took a lead role for First (N-S) Metro in India, the first metro across a major river in India. DPRs for Thane MRTS & Andheri-Ghatkopar Metro, and DDC & LDC of metros in Delhi, Mumbai, Chennai, Bangalore, Gurgaon and Ahmedabad. He led a team for the integrated transport system in Nairobi, Bangalore, Thiruvananthapuram, Indore, Rajkot, and other. During his long career, he had been involved in developing several innovative structures and several high-profile projects. This is his fourth book. He is a former Board Member of CES, Dy CE of CPPL, Transport Advisor of AECOM, and on the Board of Directors of RUPL and RUICPL.

www.ingramcontent.com/pod-product-compliance
Lightning Source LLC
LaVergne TN
LVHW070943210926
848482LV00002B/78